LEGUMES IN DRY AREAS

Editors

D. Kumar
A. Henry
K.P.R. Vittal

Central Arid Zone Research Institute,
Jodhpur 342 003 (India)

Indian Arid Legumes Society
Jodhpur 342 003

&

Scientific Publishers (India)
Jodhpur 342 001

Correct Citation
Legumes in Dry Areas (*Eds. D. Kumar, A. Henry & K.P.R. Vittal*).
Pub. Indian Arid Legumes Society and Scientific Publishers (India), Jodhpur, India.

Print : 2022

ISBN 978-81-7233-604-2 (Scientific Publishers (India))

Published by
Indian Arid Legumes Society,
CAZRI Campus,
Jodhpur 342 003, India
&
Scientific Publishers (India)
5-A, New Pali Road, P.O. Box 91
Jodhpur - 342 001
E-mail: info@scientificpub.com
www.scientificpub.com

PREFACE

Arid legumes comprising moth bean, mung bean, cowpea, cluster bean and chickpea occupy more than 4.0 million hectare area in the agro-eco sub region of arid zone. Mothbean, even through decreased in area to about 0.85 m hectare but increased in production to 0.18 million tons. Similarly, the productivity jumped to 210 kg ha^{-1}. Even though the coefficient of variation decreased with area but increased with production. There it signifies that the crop is relegated to more marginal areas. The increase in yield was only due to adoption of improved technology like varieties and cultural practices. Most of the fertile area under moth bean was replaced by mungbean. Thus the coefficient of variation of production decreased consequently with increase in area. The present production is 266 kg per hectare. Cowpea, even though, was a neglected crop with less area, it recorded an area increase but fall in productivity. Production could be increased with expansion in area. However, this was not followed with varietal replacement. Chickpea, an irrigated crop, recorded a significant increase in area of production and productivity but the point of concern is increasing coefficient of variability. This requires the attention of the scientists and policy makers. Clusterbean commercial crop with a large coefficient of variation associated with area and production earning worth Rs. 1126 crores by exporting 1.89 lac tons of guar gum. Increase in area but decrease in production and productivity could be attributed to mismatch between available length for growing period and selection of varieties. Thus the scenario of legumes production represents area of less stability and high variability. Not much information is available on horse gram in this tract. With these crops a response to micro-nutrients like zinc was also noticed in the farmers field.

Legumes are not only essential for supplement to cereal based diet as proteins but also important for its fodder values. These crops are sown as mixed or inter crop. The rotations of legumes and millets followed almost on 2/3rd of area but not on regular basis. These crops are popular because of their indigenous post harvest value addition. The moth bean, being a crop of the region inherently housing wide adaptability to environment in future it would have a greater role in scenario of climate change.

The proceeding arranged in various chapters emanates from the 3rd symposium on Arid Legumes, held at Central Arid Zone Research Institute, Jodhpur from 28-30 June, 2008. This symposium had wide participation of scientists from Rajasthan, Gujarat, Haryana, Punjab, Delhi, Uttar Pradesh,. Chhatisgarh, Uttara Khand, Madhya Pradesh to Maharashtra, Tamil Nadu and Karnataka. This documents the available genetic resources of arid legumes and its need of conservation and success stories on improvement aspects. Based on physiological responses recorded on biotic and abiotic stresses, future thrusts arrived at. Avenues on biotechnological approach were examined. Socio-economic aspects including parameter of value addition, export potential etc. were discussed. In this symposium it was worth to notice that the personnel from industries joined in discussions. Strategies are identified for enhancing the productivity of the crop and disseveration of information for overcoming the constraints. It is hoped that the papers contained this volume will be found useful by scientists, extension workers, farmers and policy makers equally in the field of arid legumes in drylands.

D. Kumar
A. Henry
K.P.R. Vittal

Dt. 6th May, 2009
Jodhpur

CONTENTS

Strategies to enhance agricultural production in hot arid western plain region of Rajasthan

N.L. Joshi and Amal Kar

Central Arid Zone Research Institute, Jodhpur 342 003

ABSTRACT

The dominantly sand-covered and drought-stricken hot arid western plain region of India (also called the Western Dry Region) occurs in western part of Rajasthan state, and is a part of the Thar Desert where agricultural production system is severely constrained by environmental, biotic and socio-economic factors. High human and animal pressures on its fragile land resources have aggravated the desertification problems in all its three zones. Traditional rainfed kharif crops are still grown in the region, but irrigation has made vast changes in land use and agricultural production, with mixed consequences for rural economy and environment. While net sown area has increased by ~45% during the last five decades, culturable waste has declined by 27% and fallow lands by 30%. Market-driven economy has given impetus to growing of cash crops, including crops that are high water demanding, which has led to depletion of aquifers, and use of poor quality water. A shift in the region's large livestock base is also being witnessed. The grazing lands are now highly degraded, resulting in a gradual decline in sheep population and increase in goat population. Crop residues provide more support to the bovines, encouraging farmers to replace cattle with buffalo. Studies suggest that the region is more suited for a livestock-based economy, but development in the sector is not satisfactory due to inadequate management, health care, market linkages, etc. There is an urgent need to improve the quality of livestock breeds, and assure the supply of feed and fodder, especially through reclamation of the vast wastelands, improvement of the status of pastures, and establishing facilities for the processing and marketing of perishable produces. The current over-emphasis on cropland productivity needs to be replaced with more sustainable practice of agro-forestry and alternate land uses that would not only maintain the land productivity, but will also help in sustaining the rural economy.

Key words: Agro-climatic zones, arid, constraints, crops, desertification, drought mitigation, livestock, options, production.

INTRODUCTION

The hot arid western plain region of India (19.084 million ha), also called the Western Dry Region, occurs in the western part of Rajasthan state that contains the dominantly sandy Thar Desert. The region is flanked in the north by the Irrigated north-western plain zone (in Rajasthan and Haryana) of the Trans-Gangetic Plain Region, in the south by the north and north-west Gujarat zones of the Gujarat Plains and Hills Region, and in the east by the semi-arid eastern plain zone and the sub-humid southern plain zone of the Central Plateau and Hills Region. In the west the region borders Pakistan, where the Thar Desert continues till the fertile Indus Plains are reached further west. Based on the climatic, edaphic and terrain characteristics the region has been further subdivided into three zones: (a) Arid western plain (zone I), (b) Transitional plain of inland drainage (zone II), and (c) Transitional plain of Luni basin (zone III; Fig. 1). Districts included in each zone and the total area of each zone is given in Table 1.

Table 1. Agro-climatic sub-zones in Western Dry Region

Sl. No.	Zone	Districts	Area (million ha)
I.	Arid Western Plain	Barmer, Jaisalmer, Bikaner, Jodhpur (excluding Bhopalgarh and Bilara tehsils), Churu (excluding Churu, Taranagar and Rajgarh tehsils)	12.416
II.	Transitional Plain of Inland Drainage	Jhunjhunun, Sikar, Nagaur, Churu (only Churu, Taranagar and Rajgarh tehsils)	3.699
III.	Transitional Plain of Luni Basin	Pali, Jalor, Jodhpur (only Bhopalgarh and Bilara tehsils), Sirohi (only Sheoganj and Revdhar tehsils)	2.969

The climate of the region being arid, erratic behaviour of the meagre rainfall (100-450. mm; ~90% during July-September), extreme temperatures (often >45°C in the peak of summer and sub-zero in winter) and high summer winds (>30 km h^{-1} during sandstorms in summer) are the perpetual climatic problems to reckon with, especially for agriculture. There is a distinct rainfall gradient from east to west that is best reflected in the arid western plain where the mean annual rainfall varies from 100 mm in the westernmost part of Jaisalmer district to 370 mm to the east of Jodhpur, most of it coming during July-September. Some winter rain (10-20% of total annual) is more common in the northern districts of Bikaner, Churu, Jhunjhunun and than in the southern part. Drought is and will remain a major determinant of agriculture in the region. Rainfall analysis of the region for last 103 years shows that drought occurred in 57 years (24 moderate, 21 large, 10 severe and 2 very severe), and 46 years had no drought. Zone I experienced 55 drought years (32 large to very severe), while zone II experienced 58 (32 large to very severe), and zone III 58 (36 large to very severe). However, during the last four decades, zone I has experienced more number of droughts, especially in the drier Barmer and Jaisalmer districts [1].

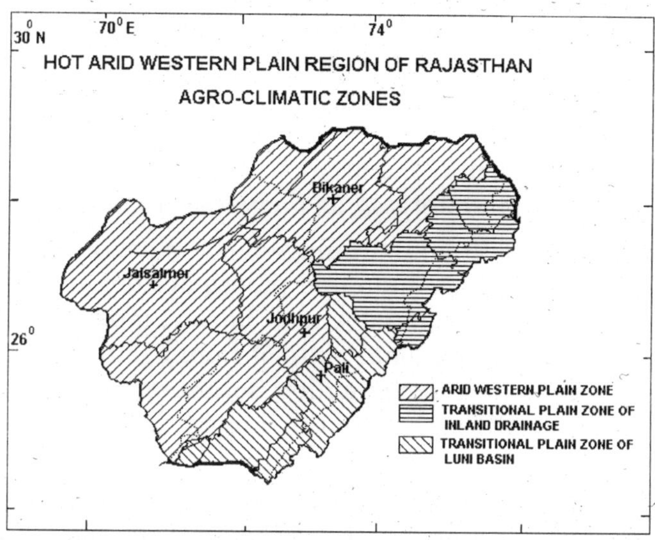

Fig. 1. Agro-climatic zones in Western Rajasthan

The terrain is dominantly sandy, with sand dunes of 10-30 m average height, interspersed with interdune plains of different sizes covering more than 60% area. A vast rocky-gravelly upland with isolated hills cuts through the dune landscape with a NE-SW orientation in Barmer-Pokaran Jaisalmer-Ramgarh tract, while another occurs in Bikaner-Kolayat tract (zone I). The ephemeral Luni River and its tributaries from the Aravalli Ranges along the eastern margin of the region drain the southeastern part (zone III) where the landscape is relatively flat with moderately deep alluvium. By contrast the streams from the hill ranges in the northeastern part are few, less integrated and mostly lost in the dune-covered areas (zone II).

The soils are dominantly sandy, with 60-90% fine sand and 2-10% of silt-clay in the topsoil. These are generally low in organic carbon, low to medium in available phosphorus and medium to high in available potassium. Most of the micronutrients are adequate, but deficiency of Zn, Mn and Fe are reported from some of the intensively cropped irrigated fields, especially in the eastern half of the region.

Due to very high human and animal pressures on the fragile landscape land degradation problems are also high. About 10.5 million ha area is affected moderate to severely by wind erosion, ~2 million ha by water erosion and 0.1 million ha by salinization [2]. Added to this is the problem of fast depleting groundwater reserve that has created acute problem of water availability in many parts of zones II and III.

Pearl millet, sesamum, pulses and castor are the major kharif crops, and irrigated wheat, mustard and rapeseed are the rabi crops. Commissioning of the Indira Gandhi Canal network (IGNP) along the western part of Bikaner and Jaisalmer districts in zone I has vastly improved the agricultural scenario in the canal command area, introducing high-value and water-intensive irrigated rabi crops like wheat and mustard in the areas where even mono-cropping of kharif crops was difficult due to harsh terrain and adverse climate.

The transitional plain of inland drainage has a mean annual rainfall of 350-500 mm, where pearl millet, sesamum and pulses are dominantly grown during the kharif, and groundwater-irrigated wheat, barley, mustard, and un-irrigated gram during the rabi. The transitional plain of Luni basin also has mean annual rainfall of 350-500 mm, where pearl millet, sorghum, maize, castor, sesamum and pulses are usually grown during the kharif, and groundwater-irrigated wheat, barley and mustard during the rabi.

Land use in the Western Dry Region is dominated by croplands (~64% of the total reporting area during 2001-02, a normal monsoon year) that include net sown area (50%), current fallow (6%) and other fallow (8%). Only 6% area is sown more than once. There is also large inter-annual variation due to uncertainty of rainfall. During drought years the net sown area dwindles, depending upon the severity of drought. For example, during the drought of 2002-03, net sown area dwindled to ~30%. Despite the above, there has been a major shift in land use during the last five decades. Net sown area has increased by ~45%, while culturable waste has declined by 27% and current other fallow lands have declined by ~30%. Area under non-agricultural uses has increased by 65%, while barren and unculturable land has partly been re-classified as permanent pastures, causing the former to decline by 60% and increasing the area under the latter by ~80%. The maximum change has taken place in double cropped area that has increased by >600%, especially due to expansion of canal irrigation in zone I, and due to groundwater irrigation in all the three zones.

There are large inter-zonal variations in the land uses. Thus, the maximum cropland is in zone II (85.6%), followed by zone III (72.2%), and zone I (54.7%). Net sown area is also distributed in a similar fashion (74.0% in zone II, 53.5% in zone III and 40.8% in zone I), but current fallow is maximum in zone III (8.2%), followed by zone II (7.0%) and zone I (5.2%), while other fallow lands are maximum in zone III (10.5%), followed by zone I (8.7%) and zone II (4.6%). The dominance of shallow and gravely soils in zone III is the major reason for higher fallow lands in it. The better-endowed zone II has 13.5% area under double cropping, especially through groundwater irrigation, followed by 10.3% in zone II (mostly groundwater-irrigated) and a meagre 2.9% in zone I (both canal-irrigated and groundwater-irrigated). Among the irrigated croplands, zone III has only 9.4% of net irrigated area (15.7%) under double cropping, while zone I has 61.5% area of net irrigated area (2.9%) under double cropping. In zone II only 29.2% of the net irrigated area (15.3%) is under double cropping. These disparities highlight the inherent resource constraints in the region, despite the expansion of irrigation and other infrastructural facilities [3].

Present and Future Production

Crop Sector

The major crops in the region are pearl millet, sorghum and wheat as cereals; moth bean, moong bean as kharif pulses; groundnut, sesame, castor, rape and mustard, and Taramira as oilseed crops. Other important crops are chilli, clusterbean, cumin and cotton. The region also produces some fruits and vegetables, and also livestock-based products (including milk, wool and meat). Because of the climatic and other constraints, the production rarely matched the demand, although presently growing awareness of farmers to market-driven economy and increasing access to facilities are creating gluts in some products in some years (e.g., mustard in 2006).

Broadly, the current production of cereals in the region is 2.52 million t, against a demand of 2.84 million t [3]. Zone I is estimated to have a demand of 1.12 million t, while the production in good rainfall year (e.g., 2001-02) is 0.82 million t, thus leaving a shortfall of 27% from the demand in the zone. In zone II there is presently an estimated surplus of 3% over the zonal demand, while in zone III there is an estimated shortfall of 14%.

Considering the strength of the region in having large wastelands (~36% of the total area), which include about 20% culturable wastes, there is vast scope for promoting horticultural crops, as well as fodder trees, shrubs and grasses for developing the livestock-based farming system, especially in zone I where ~49% area has been classified as wasteland. Compared to this, zone II has ~13% area under wastelands, and zone III ~20%.

Strength of the region also lies in production of high-value crops like isabgol, cumin, henna, etc. Value addition of the raw products will help to enlarge the market of these products. There is good scope for promoting medicinal and aromatic plants on wastelands, including plants like *Citrullus colocynthes* (Tumba) and *Aloe vera* that are good sand binders, as well as can use poor quality groundwater (e.g., isabgol in lower part of Barmer district). In other words, diversification of agriculture from the current food-crop-centric and high-water-use approach, to a judicious crop-livestock-based farming system that gains from the zone's strength using improved agro-technologies, will increase the production base and add value to the products to meet market demands. This also calls for strengthening of the infrastructural and credit systems to strengthen the livelihood of the stakeholders.

Livestock Sector

According to the livestock census of 2003, the Western Dry Region has a total livestock of 21.12 million. Zone I has 10.21 million heads of livestock (48.3% of the total in the region), while zone II has 7.14 million (33.8%), and zone III 3.77 million (17.9%). More than 60% of crossbred cattle (87,903) and 57% of buffaloes (1.50 million) inhabit zone II, while 57% of the total cattle reside in zone I (1.90 million). Among the small ruminants, 56% of sheep (3.8 million) and 48% of goats (3.8 million) reside in zone I. Among other animals, 68% of the donkey and 60% of the camels inhibit zone I, whereas 66% of mules reside in zone II.

It is apparent that livestock number is strength of zone I, thriving mostly on the extensive rangeland, including natural pasture of *Lasiurus sindicus* in the northern part of Jaisalmer that is undergoing fast transformation due to canal irrigation. Zone I also contributes the highest milk and wool products, as well as mutton and goat meat. Unfortunately, despite the potentials for livestock-based economy in the zone, development in the sector is not satisfactory due to lack of adequate management, health care, market linkages, etc. There is an urgent need to improve the quality of the livestock breeds, assure the supply of feed and fodder despite the high frequency of droughts, especially through reclamation of the vast wastelands through plantation of trees and shrubs that can fulfill the requirements of fodder, fuel, etc., improving the status of permanent pastures and other grazing lands, and establishing processing facilities and markets for the perishable produces. Such efforts will not only open up the job opportunities for the rural youth in the zone, but will also assure better income to the livestock-raisers [3].

Zone II with 60% of the crossbred cattle in the region and 55% of buffaloes, produces higher quantity of milk and milk products, as compared to Zone I. Increased availability of crop residues from the irrigated croplands despite the degradation of grazing lands, can further enhance the productivity of livestock.

Adequate numbers of animal product processing units for value-added products is urgently required in this area.

Zone III has about 4% of cross-bred animals and 25% of indigenous cattle Due to better availability of water and fodder resources, this zone supports 29% of the buffaloes in the region, as well as 25% of the sheep population. Upgradation of indigenous livestock breeds and other technologies interventions like supply of feed-processing units, have immense scope.

Constraints in Achieving Enhanced Productivity

The general problem of this region with large populations is, essentially, one of human ecology. In the absence of adequate scientific knowledge, and the need to feed large human and animal populations, the natural resources are being over-exploited, resulting in an ecological imbalance. Irrigated agriculture using surface water brought from outside the region showed the possibilities of maximizing water use for cropping and for profit. In the areas not covered by canal network, the inherently meager and fluctuating groundwater began to be exploited, especially as rural electrification progressed.

The constraints of the region can be grouped into three categories, viz., environmental, biotic and socio-economic. The environmental constraints are related to low and erratic precipitation, frequent droughts, intense radiation, occurrence of frost, high wind velocity, poor soil physico-chemical conditions, limited water for human and animal use, and for irrigation. Irrigation has become the most problematic environmental issue. In the canal-command areas where irrigation is practiced for 60 to 80 years, the groundwater table has risen by 17-25 m. In recent decades the problem has aggravated more in the Indira Gandhi Canal Stage-I command and in the Ghaggar plains, where the average rise is of groundwater table is 0.8 m per year. This has brought groundwater near to the root zone of crops in about 0.1 m ha productive land. The situation in IGNP Stage-II is expected to be more precarious as 34% of the area has a hardpan barrier at less than 2.0 m depth. Ultimately water logging and salinity are likely to spread in these areas.

The biotic problems are mainly because of pressure of human and animal population on marginal and sub-marginal lands, low plant productivity, and damages due to insects and pests. The genetic potential of landraces is very low here, and their response to improved management practices is also poor. Spread of improved varieties is very slow. Low and erratic crop stand is a major cause of low yields. Poor germination from low-quality seeds, low soil moisture and soil crusting lead to seedling death, reduced plant stand and low grain yields. Weeds compete with crops for water, nutrient, air and space, and cause considerable losses in the poorly managed fields. Parasitic weed *Striga* attaches itself to the roots of sorghum/millets from where it draws its moisture and nutrient requirements, inhibiting plant growth and reducing yields. Low level of mechanisation and increased cost of labour are hindrances to effective weed control. The production is also constrained by insects, diseases and birds. The birds attack the crops mainly during the grain filling stage.

The socio-economic constraints include inadequate exposure to adoption of modern farming technologies and to the merits of conservation of natural resources/environment. Most farmers living in arid are poor and their risk-bearing capacity is very low. They have small and fragmented production units. Their poor economic condition coupled with unreliable precipitation tends to keep the use of inputs such as improved seed, chemical fertiliser, pesticides and hired labour to a minimum. Lack of alternate vocation other than cropping and animal husbandry, fragmented holdings, semi-nomadic occupations, etc., are the other constraints.

Overall, the problem is very complex as the crops are subjected to wide range of soil-moisture conditions and to diverse interplay of environmental factors. The immense variability in soil moisture due to rainfall variations can be grouped into four broad categories [4]: (1) Crops are subjected to soil moisture deficit at any stage of growth due to erratic distribution of rain. (2) Early rains, followed by prolonged dry period, results in progressive decline in soil moisture reserve, thus restricting crop growth and yield. (3) Soil moisture stress at critical crop growth stages reduces the yield. (4) Delay in onset of monsoon adversely affects the rainfed crops. Apart from the above four situations, rains may practically fail completely, leaving no chance for crop production.

The major challenge in rainfed agriculture in the region is, therefore, to minimise the yearly variations in crop yield due to aberrant weather conditions and to stabilise the production at a reasonably acceptable level. Due to highly uncertain temporal and spatial distribution of the monsoon rains, crops usually face moisture stress of different intensities at different growth stages. Therefore, long-term strategic approaches to efficiently conserve and utilise the rain water on the one hand and in-season tactical approaches to mitigate the adverse effects of weather aberrations on the other are needed. Worthwhile outcome from the specific in-season corrections could be expected only when a sound management strategy for soil, water and crop is adopted.

Options to Enhance Productivity

Major agronomic manipulations to increase input use efficiency

Efficient management of rainwater: This essentially involves improvement of moisture regime in root zone of crop/tree through in-situ/inter-plot water harvesting techniques, which can be achieved through proper tillage, timely weeding and hoeing, placement of subsurface barriers (bentonite, clay, etc.), mulching, etc.

Selection of efficient crops and varieties matching with rainfall: Western part of zone I, not put under canal irrigation, is more suitable for livestock farming, while the rest of zone I and the zones II and III are better suited for arable farming, coupled with livestock husbandry. This natural variability can be utilized and improved upon for efficient input use and natural resource management. For ensuring input use efficiency, improved crop varieties that are drought-resistant, high-yielder, and have other important traits, as well as genetically improved indigenous breeds of livestock need to be promoted.

Efficient and judicious use of irrigation water: Surface water is scarce in the region, and potable groundwater is almost exhausted in many parts due to overuse for irrigation. Yet, faulty water management for irrigation continues, which will further worsen the situation. Adoption of advanced methods of irrigation, including drips and sprinklers, and judicious use of water for orchards of Ber, pomegranate, Aonla, Gunda, etc., and in the less water-demanding, high-value commercial crops like cumin, isabgol, mustard, etc., are urgently needed [5].

Plant protection and post-harvest care: For improving the input use efficiency it is also necessary to protect the crops from pest attack. High application of chemical pesticides has proved dangerous not only for the plants, but also for consumers through the food chain. Integrated pest management, using plant residues (e.g. mustard residues for soil solarization and neem products for bio-control), is now advocated to ensure environment-friendly products.

Intercropping systems: Growing of several crops with pearl millet is a traditional practice in the region to reduce the risk of monsoon failure, but no systematic planting is carried out in this system. Although several crops are sown together, planting one or two legumes with pearl millet is the most prevalent and successful. In this intercropping system, paired planting of pearl millet is done keeping two rows of pearl millet at 30 cm apart and two pairs at 70 cm apart. One or two rows of grain legume intercrops are planted in the inter-pair space of pearl millet. Studies show that double-rows of moth bean, clusterbean and mung bean planted in the inter-spaces of paired rows of pearl millet (30/70 cm) yielded 381, 381 and 458 kg ha^{-1} additional grain without adversely affecting the yields of pearl millet. Assessment of land equivalent ratio (LER) indicated that maximum advantage of 54% could be achieved by growing double-rows of moth bean in the interspaces followed by the double-rows of sesame (31%), single-rows of moth bean (30%), and double-rows of clusterbean (21%) and mung bean (19%). Growing two rows of mung bean though resulted in comparatively lower LER (1.19) or only 19% better land utilization but gave better economic returns. Intercropping of a short duration crops with long duration one (horticulture/silviculture) is expected to make the best use of resources [5].

Fertilizer for intercropping: Determination of the optimum fertilizer dose in an intercropping system is still debatable. CAZRI's studies show that legumes contribute very little (5 kg N ha^{-1}) to the base crop nitrogen requirement, and that the yield of principal crop increases with increased levels of nitrogen (60- 90 kg depending on rainfall) but the yield of intercrop (i.e., mung bean) gets adversely affected, possibly because the increased vegetative growth of pearl millet at higher nitrogen levels suppresses the growth of intercrop [6]. So, nitrogen fertilizers should be applied basal at the time of sowing and should be top-dressed as per need of the cereal crop only (viz., pearl millet).

Intercropping *visa-a-vis* strip cropping: In sub-normal rainfall years legumes as intercrops of pearl millet perform inferior than under strip cropping, due to the higher overall population in the system. In strip cropping four rows of pearl millet could be alternated with four rows of legumes [7]. In next year the rows could be shifted in such a way that legume rows occupy the space of preceding pearl millet rows. Such rotation of rows could be followed year after year to derive additional yield benefits in pearl millet through addition of organic matter due to leaf fall of legume (i.e., clusterbean).

Triplet planting system: The average productivity of pearl millet in arid region is low than in other areas due to high moisture stress at maturity stage. Reduction of population and change of planting system has been found to help in achieving sustained production. A triplet planting system has been developed by skipping every fourth row in the normal planting. Triplet planting system without any population adjustment (125 x 10^3 plants ha^{-1}) provides as good yields as uniform planting [8] in normal rainfall (409 mm) and during moderate drought years (260 and 193 mm), and significantly higher yield in a severe drought year (152 mm). The yield losses due to 25% less sown area in the triplet system, is compensated by higher yields in border rows. The triplet system with adjusted population (keeping recommended population of 1.6 x 10^5 plants ha^{-1}) gave 7-36% higher yields beside 25% saving in fertilizer input over uniform rows in different years.

Plant population: Plant population in the intercropping system also influences sustainability. Moderate density of pearl millet (133 x 10^3 plants ha^{-1}) with low density of mung bean resulted in the highest sustainable yield index

(SYI) of 0.37 for pearl millet which indicated that at least 37% of the maximum observed yield over years was assured with high probability in intercropping system as against 23% in sole cropping of mung bean [9].

Supplemental irrigation: Giving an irrigation to stabilize and improve yields of a crop, which is otherwise dependent on rainwater, is termed supplemental irrigation. Giving a supplemental irrigation during prolonged drought periods by using underground water, even if of poor quality, or rainwater harvested in tanks, can make all the difference between a total crop failure and good yields. If the crops face drought at sensitive growth stages giving supplemental irrigation during this period is the best potential use of limited water supply [10].

Alternate land use systems : CAZRI's studies show that a farming system approach with alternative land use systems, appropriately synergized with other farm enterprises like arable cropping, dairy, poultry, etc., has better prospect in the arid region. These farming systems so evolved should be suitably dovetailed with the management of village common/degraded lands in watershed/index catchments, to rebuild and conserve the natural resource base in a holistic fashion to meet the requirements of food, fiber, fodder and fuel besides providing income for livelihood. Agroforestry, agri-horticulture, agri-pasture, farm forestry, silvi-pasture, etc., are the most suitable options. The proportions of different alternate land use systems developed at CAZRI [11] for an area of 7 ha are agroforestry (30%), farm forestry (20%), agri-horticulture (25%), agri-pasture (15%) and silvi-pasture (10%). Such apportioning of the farmland can provide sustainable crop yield, maintain soil fertility and increase the availability of high quality fodder for livestock. Selection of multipurpose trees and shrubs, catering to the needs of food, fodder, timber, firewood, etc., and for erosion control is central to the concept.

Agroforestry systems: *P. Cineraria*, H. *integrifolia* and *H. binata* are suitable for agroforestry systems in arid region. Agroforestry systems with *P. cineraria* and *Tecomella undulata* with various dryland crops are also promising. In general, tree density of about 100-200 plants ha^{-1} is optimum for minimum interference with the yield of various dryland crops like clusterbean and cowpea.

Agri-Horticulture systems: In agri-horticulture systems with jujube (*Z. mauritiana*) the density of 200 plants ha^{-1} is optimum for growing dryland crops in the interspaces of the fruit trees. At this density level both the components have not shown any yield reductions. Among the dryland crops clusterbean showed better performance in drought years, whereas mung bean was the superior crop in good rainfall years. However, cowpea showed yield stability in most of the years. The system besides giving good yield of cowpea and fruit tree sustained 700-1000 animal (goat and sheep) days ha^{-1} yr^{-1}. Under such system, tall growing crops like clusterbean and pearl millet were more adversely affected due to the crop-tree competition than cowpea, mung bean and moth bean due to shade effect of jujube. Soil fertility improved under both agro-forestry and agri-horticulture systems.

Agri-pasture and silvi-pasture systems: The grass component in agri-pasture and silvi-pasture systems is more profitable than arable farming. Intercropping of grasses like *L. sindicus*, *C. ciliaris*, *C. setigerus* and *D. annulatum* with legumes like moth bean and mung bean in the ratio 2:1 have been found compatible. The average grain (130-280 kg ha^{-1}) and fodder (350-600 kg ha^{-1}) yields from the system are satisfactory. However, intercropping of grasses and dryland crops is sometimes not feasible and under such conditions strip cropping of grasses and kharif legumes in 1:2 ratio is recommended with a strip width of 5-10 m. The grass component also improves the soil organic matter and start giving production from the establishment year onwards.

Keeping 10-15% of the total land holding as fallow for 2 to 3 years is a normal practice among the farmers of arid areas. Such fallow lands may be developed for silvi-pasture system. This practice of lay farming will improve soil fertility and productivity of succeeding crops. *A. lebbeck, T. undulata, C. mopane, A. senegal, Z. nummularia* and *Z. rotundifolia* are some important tree species that are compatible with the grasses. Among pasture legumes, *C. ternatea* and *L. purpureus* are compatible with *L. sindicus* and *C. ciliaris* [12]. Integration of shrub like *Z. nummularia* with buffale grass (*C. ciliaris*) strips in 1:2 ratio gives higher live weight (33 kg ha^{-1} yr^{-1}) and wool production (5.65 kg ha^{-1} yr^{-1}) from ovine than sole pasture.

Crop-enterprise substitution option: In the rainfed areas of the western part of zone I, receiving annual rainfall of 200 mm and less, it is better to substitute crop farming with livestock farming with sheep and goats. In the eastern part, receiving 200-300 mm rainfall, livestock farming, supported with fodder-based farming system (agri-pasture, agro-forestry, silvi-pasture), is a better option. In areas receiving 300-400 mm rainfall, agro-forestry and agri-horticulture need to be advocated. More emphasis is to be given to rainwater harvesting and its utilization for growing high-value crops.

Multi-enterprise systems: Livelihood security of the inhabitants in the region is dependent on a multi-enterprise system because of the high risk of failure for any single enterprise. Therefore, traditionally the farmers of the region are tuned to multi-enterprise system. To make the system sustainable, and at the same time more productive and remunerative, incorporation of various farming system components as per the needs of farmers and market trends is essential. It should encompass components like crops, trees, grasses, shrubs and animals in suitable proportion. Also, medicinal and industrial plants like Senna, Henna, *Aloe vera*, etc., can be included as per the zonal suitability and farmers' perception, competence and market demand.

Improving livestock-based economy: Since livestock rearing is a major occupation in the region, following two major issues need to be addressed: ensuring feed and fodder security and correcting the infertility problem in animals. Establishment of village fodder banks, improving the fodder quality through various treatments and feed processing will assure the availability of nutritious feed, even during drought and other scarcity periods, and assured supply of animal products irrespective of drought. Nutritional security of the animals will not only ensure better milk yield, but also quality meat production. Establishment of a network of co-operative dairy industry, backed up by an efficient transport network and energy supply, will help to generate enough income from milk and milk-based products in the region.

The future thrust for farming systems research and development for Indian arid zone should aim to develop sustainable farming system modules for major micro-farming situations of the region. On-farm research in farming system perspective should be given greater priority for various rainfall situations. For sustainable development of arid tract the adoption strategies of farming systems should be, to focus on effective public-private-partnership in an integrated area development approaches like watershed, index catchment, cluster village management etc.

REFERENCES

1. Narain, P. and Kar, A. (eds.). 2005. *Drought in Western Rajasthan: Impact, Coping Mechanism and Management Strategies*. Central Arid Zone Research Institute, Jodhpur. 104 p.
2. Kar, A., Moharana, P.C. and Singh, S.K. 2007. Desertification in arid western India: Current scenario. In, *Dryland Ecosystem: Indian Perspective* pp. 1-22. (K.P.R. Vittal, R.L. Srivastava, N.L. Joshi, Amal Kar, V.P Tewari and S. Kathju (Eds.). Central Arid Zone Research Institute, Jodhpur and Arid Forest Research Institute, Jodhpur.
3. CAZRI. 2007. *CAZRI Perspective Plan: Vision 2025*. Central Arid Zone Research Institute, Jodhpur, 94p.
4. Lahiri, A.N. (1978) Soil water - plant relations and production optimisation from arid areas. In *arid zone research and development* pp. 223-228. H.S. Mann (Ed.) Scientific Publishers, Jodhpur:
5. Bhati, T.K. and Joshi, N.L. 2007. Farming systems for sustainable agriculture in Indian arid zone. In, *Dryland Ecosystem: Indian Perspective* pp. 35-52 (K.P.R. Vittal, R.L. Srivastava, N.L. Joshi, Amal Kar, V.P Tewari and S. Kathju (Eds.). Central Arid Zone Research Institute, Jodhpur and Arid Forest Research Institute, Jodhpur.

6. Joshi, N.L. 2008. Performance of mung bean as component crop in pearl millet mungbean intercroping as influenced by nitrogen application. J. Arid Legumes 4: 48-54.

7. Mahender Singh and Joshi, N.L. 1994. Performance of pearl millet based intercropping systems under drought conditions. Arid Soil Research and Rehabilitation. 8: 277-283.

8. Joshi, N.L. 1990. Triplet planting system in pearl millet for sustained production and better energy use in the Indian arid zone. In *Abstracts - International Symposium on Natural Resources Management for a Sustainable Agriculture*, Vol. 1&2, pp. 140-141. Indian Society of Agronomy, New Delhi.

9. Joshi, N.L. 1999. Effect of component densities on the performance of Pearl millet – Mung bean intercropping system under different rainfall situations in arid environment. Annals Arid Zone. 38: 15-25.

10. Kathju, S., Garg, B.K., Vyas, S.P. and Lahiri, A.N. 1993. Effect of supplemental irrigation and soil fertilization onpearl millet under drought. Arid Soil Research and Rehabilitation. 7: 317-326.

11. Bhati, T.K. 1997. Integrated farming systems for sustainable agriculture on drylands. In *Sustainable Dryland Agriculture* pp. 102-105. CAZRI, Jodhpur.

12. Bhati, T.K., Bhati, G.N. and Shankarnaryan, K.A. 1986. Legume-shrub grasslands in western Rajasthan. *Annual Progress Report*, CAZRI, Jodhpur.

Achieving self-sufficiency in cereals, pulses and oilseeds in India: the way forward

B.B. Singh

Visiting Professor, G.B. Pant University of Agriculture and Technology, Pantnagar

Indian agriculture is in a great paradoxical situation. From a near famine situation and import of over 10 million tons food grains in 1965-66, India reached a net exporter status of food grains from 1990 to 2002 with over 40-60 million tons buffer stock but India is now back to food importing status with over 250 million Indians suffering from under-nutrition and malnutrition And this is happening in spite of the fact that India has 93 ICAR supported agricultural institutes, 43 State Agricultural Universities, 74 All India Coordinated Research Projects, 500 Krishi Vigyan Kendras (KVKs), and hundreds of seed companies, agro-industries and NGOs spread through out the country employing over 7000 agricultural scientists and over 40,000 extension workers. Numerous research papers, conferences, symposia, national planning sessions and farmers' commission report notwithstanding, the food grain yields are static for the last several years. With the static food production and continuously growing population, the per capita food availability has gone down from 469 g in 1961 to less than 422 g now. Consequently, India is importing millions of tons wheat, pulses and oilseeds!

Why has the food production stagnated in India? The answer is simple. Under pressure from a severe food deficit situation from 1950 to 1966, Indian government invested heavily on improved seeds, irrigation, fertilizers, pesticides, improved machineries, infrastructures and market and extension support and achieved green revolution within a few years. This increased the food production from 82 million tons in 1961 to over 212 million tons in 2000 which led to a large buffer stock of food grains. This probably caused a general complacency and the Indian government drastically reduced investments on fertilizers production, pesticides, irrigation, market support and extension services which arrested growth in food production. Consequently, the average yields of cereals, pulses and oil seeds in India are now less than half compared to China and other countries which have continued investing in agriculture. With only 130 million ha cultivated land, China is using about 34 million tons of fertilizer nutrients whereas India, with 160 million ha of cultivated land, uses only about 17 million tons of fertilizer nutrients. This amounts to about 260 kg fertilizers/ha/year in China and about 100 kg/ha/year in India. Also, bulk of the 100 kg/ha fertilizer currently used in India, is based mainly on nitrogenous compounds and little of phosphorus and potash causing imbalanced fertility and soil deterioration. Moreover, the total amount of pesticides used in India is about 500g/ha/year compared to over 3 to 8kg/ha/year in many countries. This is well reflected by the average cereals yields of about 2.5 tons/ha and giving about 210 million tons total food grains from 160 million ha in India compared to about 5.1 tons/ha cereals yield and 413 million tons total food grains in China from only 130 million ha!

There is a direct relationship between nutrients applied and grain yield obtained. For every ton of grain yield, the major cereals remove about 50 to 60 kg of nutrients (N+P+K+Ca+S+Zn+Fe+Mg+Mn etc.) per ha. If only 100 kg/ha fertilizer is used, as it is done in India, the expected yield would not be more than two tons/ha and soil nutrients would be mined beyond that yield level causing poor soil fertility. This is what is happening in India. To obtain an average yield of 5t/ha at least 250 kg/ha nutrients must be applied. Also, the fertilizers have to be applied in desired proportion of nitrogen, phosphorus, potash and other micronutrients like sulphur, iron, zinc etc. In the long term fertility trials at Pantnagar, use of about 120 kg N + 40 kg P205 + 40 kg K20/ha supplemented with some FYM, has sustained the rice yield over 6 ton/ha level for 25 years whereas the yields have declined below 3 t/ha level in the treatments receiving only 120 kg N /ha. Phosphorus and potash fertilizers are more expensive in India and therefore, farmers use less P&K causing imbalance and poor yields.

What should be done to restore the positive growth in food production? Achieving self-sufficiency and even surplus production in cereals, pulses and oilseeds is possible if the Indian government resumes substantive investments in agriculture and well focused and targeted research projects on crop and pest management. The current Indian population of 1.1 billion is expected to stabilize at about 1.5 billion in 2050. To meet the additional food requirement, the current total food production in India of about 220 million tons can easily be raised to 350 million tons in next 15-20 years just by adequate fertilization and good crop and pest management with the same amount of water and land. The current cereals, pulses and oilseeds yields in India are about 2.5 t/ha, 0.70 t/ha and 1.0 t/ha compared to 5.1 t/ha, 1.7 t/ha and 3.0 t/ha in China respectively. The crop yields in India are also much lower than the world averages. For example, the average yield of soybean in India is about I t/ha compared to 2.2 t/ha world average and 2.8 t/ha in Brazil and USA. The difference is mainly in the use of sufficient and balanced fertilizers and better pest and crop management. This is evident within India also where the wheat yields in Punjab and Haryana are much higher than those in because of relatively more fertilizers and better crop management. The deterioration of soil fertility and more due to imbalanced fertilization (more urea and less phosphorus and potash) and burning of crop residues rather than returning them to fields as a source of organic matter. The front line demonstrations in India using improved varieties and recommended crop management practices have shown over 5 t/ha yields of cereals and over 2t/ha of oilseeds and pulses. Thus, the technology is available. It just needs to be popularized by making adequate inputs available and providing extension and market support. The specific suggestions to increase food production in India are:

i). Maximize cereals yields in irrigated areas by ensuring adequate fertility (at least 120:40:40 NPK + micronutrients + 1 t/ha FYM) and good crop and pest management. Taking two crops of cereals/year with even 3.5 tons/ha from each crop in irrigated lands would amount to 350 million tons from the current 50 million ha irrigated area in the country. The actual potential is 500 million tons.

ii). Introduce short duration pulses in the existing niches in the wheat-rice, and rice-rice systems. Currently over 10 million ha land in wheat-rice system can be available to plant short duration pulses from April to June with a potential of producing over 10 million tons of additional pulses. Some prototype varieties of 60-day mung bean and cowpeas are available but more funds need to be invested for focused research and development of these crops for these areas.

iii). Raise the current soybean yields from 1 t/ha to 2 t/ha in M.P, Maharatsra, Gujrat and other soybean states by training and maximization demonstrations using adequate fertility (mainly P, K, micronutrients and some FYM) and crop management inputs. This would raise the current soybean production from about 8 million tons to 16 million tons without bringing extra land under soybean.

iv). Maximize production in drier and marginal areas by focusing on only those crops which perform well there e.g. selected arid-zone legumes and millets. Also, the initial growth of cotton is slow and therefore, intercropping cotton with early cowpea bean has been very successful in Cameroon and Zambia. This yields about 1-1.5 ton /ha of cowpea within 65 days with little or no reduction in cotton yield. Similar studies need to be conducted under Indian conditions.

v). Maximize production of groundnut, mustard, pigeon pea, chickpea and lentils by promoting improved disease resistant varieties, balanced fertility and good crop management in areas where these crops do well.

vi). In view of the rapid spread and widespread popularity of 'Bt-crops' for insect resistance and 'Roundup-ready crops' for herbicide tolerance around the world, Indian government should obtain these technologies and these genes should be deployed in all the major crops in India. The herbicide tolerant soybean crops have become very popular in USA, Brazil and Argentina because the weed management has become very easy. Weeds are a major problem in soybean and other crops in India. Some have argued that India has a large labor force but in practice, they are either not available on time or manual weeding is not possible due to continuous rains and bad field conditions. Therefore, it would be highly desirable to develop herbicide tolerant crops in India. The controversies about GM crops and related trade barriers would disappear within the next 10-15 years as the world leaders and general public gets better educated on the pros and cons of biotechnology .

vii). railways, assured electricity and water supply, availability of fertilizers and other inputs and remunerative price for farm produce.

viii). Develop a 50 years master plan for improving agriculture in India. Strengthen and raise the standards of agricultural teaching and research beyond the routine activities in the universities and coordinated projects. Research projects should focus on adequate inputs and well managed agriculture in irrigated areas and rapid agro-industrial growth for value addition and post harvest management of the food grains and crop residues.

ix). Some of the State Agricultural Universities should be classified as National Agricultural Universities so that best students and teachers in the country could be brought together for achieving international excellence in agricultural teaching and research..

x). Introduce a system of achievement oriented recognition and reward policy in place of time bound promotions in universities and research institutions.

Needless to emphasize, a strong agricultural base with complete food self-sufficiency and adequate buffer stock is the bedrock of the progress in any nation. The rapid industrial growth and IT revolution came to India only after the green revolution was achieved and India became free from massive food imports. In the rapidly changing geo-political and global economic situations including the use of food crops for fuel, depending upon food imports may be full of uncertainty and highly risky. Therefore, the Government of India should restart investing on more fertilizer production, irrigation schemes and market and extension support to farmers and revive the green revolution that was suspended several years ago.

Phytosanitary issues for trade promotion of arid crops

R.K. Khetarpal and Kavita Gupta

Division of Plant Quarantine, National Bureau of Plant Genetic Resources, New Delhi-110 012

Ever since the WTO came into existence in 1995, international trade of plants/ plant products has acquired a new dimension which involves compliance to legal standards in the form of Agreement on Application of Sanitary and Phytosanitary (SPS) Measures for effective regulation. This implies that countries need to update their quarantine or plant health services to facilitate import! export. The SPS Agreement concerns the application of food safety and animal and plant health regulations which should be based on science, applied only to the extent necessary and not discriminate between countries with similar conditions. The Agreement refers to the three international organizations for standard-setting whose activities are relevant to its objectives: the Codex Alimentarius Commission (CAC) on food safety, the World Organization of Animal Health on human and animal health, and the International Plant Protection Convention on prevention, detection and eradication of plant pests. Till date, twenty nine International Standards on Phytosanitary Measures (ISPMs) have been developed and several others are at different stages of development.

- export, all countries need to make suitable arrangements for inspection, treatment and issuance of Phytosanitary Certificate after thorough examinatioll of the plant! planting material to meet the quarantine requirements of the importing country which is issued by the regulatory authority of the exporting country. Also, most of the countries have import regulations, requiring Import Permits issued by the plant protection service of the exporting country which spells out the conditions under which the importer may receive a desired plant material. This procedure is followed during trade of all agricultural commodities.

The arid crops grown in Western Rajasthan, Gujarat, and Southern parts of Haryana include millets like sorghum and pearl millet; grain legumes like guar, moth, cowpea, moong; spices like cumin, coriander, fenugreek, chillies, oilseeds like rape/ mustard and other crops like onion, garlic, etc. All these crops have a great potential for export directly or as processed commodities and export promotion has also been receiving due attention. Export performance of the crops grown in the region, especially of spices and millets has been on the increase leading to the establishment of a Sanitary and Phytosanitary Laboratory in Mandhor (Jodhpur) region of Rajasthan. Seed spices are mainly cultivated in Rajasthan and Gujarat and a sizable quantity is exported every year to countries like Japan, USA, Malaysia, Singapore, UAE, Saudi Arabia, South Africa, etc. earning valuable foreign exchange for the country.

With the advent of the SPS Agreement, India also geared up its regulatory framework to face the challenges of free trade and the previously existing regulation on plant quarantine was superseded by a more comprehensive Plant Quarantine (Regulation of Import into India) Order 2003. Under this Order, the' Additional Declarations/ Special Conditions' of import commodities for freedom from regulated pests is based on a mandatory pest risk analysis (PRA). So far, i.e., till April, 2008 ten amendments of the PQ Order 2003 have been notified to the WTO revising definitions, clarifications regarding specific queries raised by quarantine authorities of various countries, with revised lists of crops under the Schedules IV, V VI and VII. Besides, 21 national standards have been developed on various phytosanitary issues involved.

The major phytosanitary issues for trade in arid crops/ products include:

Pest risk analysis: PRA is a systematic way of gathering, evaluating, recording and disseminating information leading to recommendations of an action. The PRA process consists of three stages (a) Stage 1- Initiation identifies pathways and pests to be analyzed; (b) Stage 2- Pest Risk Assessment- gives estimates on risk based on the probability of entry and establishment of pest and its potential impact and (c) Stage 3- Pest Risk Management-identifies management options to reduce the risk to an acceptable level. The guidelines for PRA are covered under the ISPM-2, ISPM-11 and ISPM-21 which ensure that all restrictions in trade are based on the actual assessment of risks and are not arbitrary or discriminate against any exporting countries with the same pest status.

Pest free areas for growing crops: The easiest way of management of pest risk is by importing produce from an area that has been declared as pest-free. The ISPM 4 and 10 provide guidelines on the establishment and maintenance of pest free areas (PFA) which may be the entire country or just an uninfested part of a country situated within a generally infested area. Efforts need to be made for identifying PF As for exportable crops of the region to facilitate trade.

Maximum permissible limits of pesticide residues: Since agricultural produce is also traded in fresh form, pesticide residues are undesirable and need to be minimized and kept within permissible limits i.e. below the prescribed maximum residue limits (MRL) of the importing country. The CAC has fixed MRL or tolerance limits of various pesticides on agricultural produce. MRLs are very important in international trade as excessive residues are not accepted by importing countries and act as non-tariff barriers. Effective plant protection schedules of exportable crops that do not leave high residues need to be developed. Application of pesticides in the field is a phytosanitary requirement to manage the pest, but the residue left is a sanitary issue. Thus, the SPS requirements are closely interlinked and cannot be mutually exclusive, despite the different agencies dealing with them.

Disinfestation treatments: Post harvest treatments need to be carried out to mitigate the risk where inspection alone is not adequate or the produce is not from a PF A. Disinfestation treatments either as a single or a combination of treatments can also be used to reduce the post-harvest phytosanitary risk. The ISPM 14 provides the guidelines for use of combination of treatments/ system approach for treatments to eradicate pests on the exportable commodities. Typical treatments given to the agricultural commodities are fumigation, cold & high temperature, hot air/ water, irradiation, and lor combinations thereof.

Post-harvest deterioration of crops: Post-harvest rotting caused by fungal infection such as carcinogenic aflatoxins and other secondary pests carried over in incipient form from the field gets flared up under improper storage conditions. Many countries strictly adhere to the minimum permissible limits prescribed by CAC for aflatoxin contamination. Hence, Codex standards or the standards prescribed by the importing country need to be strictly followed to boost our exports particularly of the products that can harbour such toxins.

Trade promotional research: In order to reduce post-harvest losses at production sites, standardization of packing line operations and proper packaging of different commodities need urgent attention. Also, there is a need to reprioritize research based on critical gaps to meet the requirement for export of organic produce and for residue free integrated pest management technology .

Need for Upgradation of National Phytosanitary Regulatory System: Efforts are needed to streamline our regulations as per international norms to enhance our agricultural trade. Some of the areas that need immediate attention are strengthening of PRA unit, establishment of a nationally coordinated plant pest surveillance system, establishment of diagnostic facilities for

Idetection and identification of exotic pests, strengthening domestic quarantine/ emergency services to check spread of introduced pests into newer areas and develop quality assurance machinery to ensure compliance with SPS requirements of the importing country.

To boost exports of arid crops, India needs to study the phytosanitary requirements of target countries on exportable items and develop disinfestation technology to meet the exact specifications prescribed by the importing countries. Standards need to be developed on priority to facilitate the export of several exportable arid crops. Besides, arid horticulture also has an immense potential and efforts need to be made for optimum utilization of the rainfed and marginal lands through horticultur

Currently, India needs to improve the quality of its exports an y conform to the international standards as of late, there have been complaints from the EU countries over the high residues of pesticides and chemicals in Indian products. Therefore, special attention needs to be paid to acceptable additives/ preservatives/colours and to tolerance levels of mycotoxins in the material to be exported. EU had rejected some grape and shrimp containers from India during the last financial year due to high residues of pesticides and antibiotics in them. There is a need to check the quality standards that the Indian producers follow and improve the infrastructural facilities existing here.

llie liberalized trade policies of the government under WTO have increased the responsibility on quarantine personnel for properly achieving their objectives of excluding exotic pests or to carry out eradicative measures. The risk of pest introduction and the means to stop the establishment of these pests into new areas continues to be a major concern of phytosanitary/ quarantine officials. Unless proper phytosanitary measures are taken, pests could get transported allover the globe, get establisIled in new areas and devastate crop production. On the other hand there are tremendous opportunities to export agricultural products if they meet the international quality standards and overcome phytosanitary constraints. This is possible only if the government agencies involved in SPS issues work in complete coordination with each other, to meet the specifications of quality/codex standards for export of indigenous crops and their products to give agricultural export the necessary push.

Indian natural resins and gums - An potential source of livelihood in forests, arid regions and mouuntains

Bangali Baboo

Director, Indian Institute of Natural Resins and Gums, Ranchi -834'010

1. Resins and Gums

All the natural resins are of vegetable origin with the exception of lac and similar insect exudations which harden on exposure to air. They are insoluble in water but usually dissolve readily in alcohol, ether, carbon bi-sulphite and certain other solvents. Natural gums come out as exudates from the trunk of a gum producing tree as a response to injury. Gums also extracted from seeds, seed coats, micro organisms etc. They are generally insoluble in oils or organic solvents like alcohols, ether, hydrocarbons etc. However, they are soluble in water or atleast become soft and swollen when mixed with water forming sticky substances. Natural gum-resins, as the name implies, consist of natural mixtures of gums and resins in variable proportions and therefore, partly soluble in water and have a penetrating and characteristics odour and taste. Certain gum-resins contain small amount of essential oil. They are called oleo gum-resins e.g. asafoetida.

2. Application Areas

Resins are largely used in surface coating formulations for several applications like wood furniture polishes/varnishes, paints, lacquers, food and pharmaceuticals, adhesives, insulations, cosmetics, handicrafts, jewelry etc. Some of the resins like lac are very good source of several bio-active compounds of various uses. The gums and gum-resins are mostly used in food (thickening/gelling agents, stabilizers, emulsifiers), pharmaceuticals, cosmetics, textiles, chemical industries. In several application areas there are no substitutes for these natural products while in some synthetic alternatives are available and used.

3. Production and Demand

Indonesia, India and China are among the World's major producers of gums and resins. India produces annually about 2,81,000 t of these valuable natural products. Of these, about 55,500 t are resins, 2,24,000 t of gums and about 1500 t of gum-resins. India is traditionally largest producer of lac, guar gum and karaya gum. In recent years, due to back to the nature trend there has been a revival of interest in natural resins and gums extracted from forests by rural and tribal people who depend on these resources to sustain their livelihood. Two-fold increase in demand is expected due to realization for eco- friendly and safe natural materials for human contact and consuption.

4. Export and Import- India

Annual average export during 2001-02' to 2005-06 was Rs 7.848 million. This included Rs 1,371 million of resins and Rs 6,363 million of gums. The average imports during the same period was Rs. 1,497 million. This included Rs. 866 million of resins and Rs 289 million of gums. Major export items are lac, guar, karaya, asafoetida and major items are lac, rosins, gum-arabic, asafoetida and olibaennum.

5. Role in National Economy and Livelihood Support

The gross value of goods and services provided by the forestry sector is estimated at an average of Rs. 26,330 crores i.e. about 2.37% of GDP. Of this recorded forest revenue about 60% comes from minor forest products including gums and resins. The sector supports about 50 million population inhabiting forests 'and sub-forest areas and 70% of employment in the sector is in minor forest produce (about 1.6 million mandays). It is an admitted fact that neither the forests nor the tribals and poor inhabiting these should be removed for environmental protection. The only approach appears to be developing minor forest products like natural resins and gums based economic activities in these areas of uplift the poor and maintain required forest cover or vegetation.

6. Ecological Approach for Economic Development

Forests are not only important source of subsistence, employment, revenue, earnings, raw material to a number of industries but also play pivotal role in ecological balance, environmental stability, bio-diversity conservation, food security and sustainable development. Extensive tree wealth exists outside continuous forested areas. Vegetation plays the most critical role in maintaining and regulating the living as also the non-living components of any eco-system. Besides this, there is revival of interest in natural resins and gums due to increasing recognition of their immense social, environmental and industrial development possibilities. For sustained economic development in and outside forest areas as also deserts degraded lands, human intervention for plantations of economic value is essential. The interventions so far, largely focused on exploitation of these resources and plantations, if any, of less-economic value trees such as fast growing tress. Plantation of economic value is expected to ensure sustained devel0pment with stable environment.

7. Processing, Value Addition and Marketing - An opportunity for Enhanced Income and Livelihood Support

Gums and resins are low volume, high value produce. These can be processed to add value in quality for higher returns. In some products value addition through primary processing alone results in 3 times higher returns. Developing products of commercial use would further augment returns, employment and export earning. To achieve this an organized marketing support would be crucial. In India lac and guar gum are processed but for most of other resins and gums processing and value addition is meager.

8. Goal

Harnessing science, engineering and marketing through research and extension initiatives to support ecological for economic development of forests, sub-forests, mountains and deserts. Natural resins, gums and gum-resins are non-toxic, bio-degradable and eco-friendly for various applications. India being a mega bio-diversity centre, stands to gain and establish its leadership in this sub-sector.

Status of breeder seed production of a arid legumes in Rajasthan Agricultural University

P.R. Kothari, Gupta, P.C. and Deepak Gupta

Rajasthan Agricultural University, Bikaner

Introduction

Pulse crbps play an important role in Indian agriculture. Besides being rich in protein, they sustain the productivity of the cropping system. Their ability to use atmospheric nitrogen through biological nitrogen fixation (BNF) is economically more sound and environmentally acceptable.

Twentieth century has been as era of green revolution in India. Wheat and Rice were considered as importallt crops and major emphasis was given on research and developmental aspects of these crops, while pulses were ignored and marginalized to poor soils. In consequence to this, there has not been much improvement in production of these crops. However, releasing the importance of pulses in sustainable agriculture and human nutrition, Govt. of India constituted an Expert Committee to draw action plan for increasing pulse production.

Pulses occupy 68.32 million hectare area and contribute 57.51 million tonnes to the world's food basket. India shares 35.2% area and 27.65% of the global production. Per cent contribution of pulses in total food grain production in India has declined during the last three decades whereas it has increased for rice and wheat.

In India, more than a dozen of pulses are grown, common pulses are chickpea, lentil, field pea and rajmash during *rabi,* and pigeonpea, mungbean, mothbean, cowpea, horse gram and ricebean during kharif season. These crops are grown on 22-24 million hectare area, producing 13-15 million tonnes of the grains in different years. The production of pulses reached to its highest (14.91 million tonnes) with the area 23.5 million hectares during 1998-99. Sixty per cent of pulse area is in *rabi* and 40 per cent in kharif season. There was considerable decline in pulses production during 1999-2002. This decline was mainly due to reduction in area under *rabi* pulses because of early cessation of monsoon and failure of winter rains.

The most important states for pulses are Madhya 'Pradesh, Rajasthan, Maharastra, Uttar Pradesh and Andhra Pradesh, which together accounts for 82 per cent of the production from an area of about 74 per cent. In Rajasthan, pulses are grown on 3.44 million hectares area, producing 1.62 million tonnes of grain on an average of last 5 years upto 2001-02, out of which 1.86 million hectares area comes under *kharif* pulses with production of 0.39 million tonnes. The area production of *kharif* pulses raised upto, 2.35 million hectares and 0.65 million tonnes, respectively during 2001-02 due to good monsoon season. Among the *kharif* pulses mothbean, moongbean, urad and cowpea are major pulses.

Mothbean

Mothbean does not claim to be a national pulse because its contribution in terms of area and production is to the tune of 5.9 and 1.6 per cent, respectively. It is important crop of arid and semi-arid regions, especially Northern-Western desertic regions of Indo-Pak subcontinent. Ecologically, it is an annual legume of dry and warm habitats and is characterized as the most drought hardy annual legume in arid region. Mothbean with deep and fast penetrating root system, can survive upto 30-40 days in open fields, experiencing fast depletion of soil moisture, in concomitant with atmospheric temperature peaking to more than 35°C. These adaptive features emboded in mothbean against harsher and unhospital growing situatinos for unspecified intervals, have led this crop to be recognized as the arid legume. Multi-adaptive and adjusting features have scaled mothbean as the only alternative annual crop of the sand dunes, requiring no inputs and negligible agronomic aftercare. This crop is an essential component sub-segment of cropping systems, prevalent and common, in arid zone like, agri-horticulture, silvi-pastoral, agro-forestry, mix-cropping, intercropping, sole cropping etc.; Therefore, this crop is a part of all systems of texturally common poor lands representing the holding of common men, characterized with limited physical and financial resources. Mothbean is a source of food, fodder, feed, green manuring and used as pasture hence, seres as a multipurpose crop. Being pulse, it is a cheap source of vegetable protiein for balancing the nutritional deficiency, most commonly occurring on less productive soils, on which financially less equipped, great chunk of people have been depending for their livelihood green manUflng an u.sed as pasture, it is a cheap source of vegetable prof.

At the country level, area and production of mothbean have been fluctuating due rainfall intensity, its distribution and shifting and substitution of mothbean to more assured and remunerative crops. Area, production and productivity of mothbean at the country level appear to have been settled at 13.52 lakh hectares, 2.41 lakh tonnes and 215.26 kg ha^{-1}, respectively. Among the states, Rajasthan occupying 11.55 lakh hectare, contributes 85 per cent of the total area of mothbean in India, whereas, production contribution of this state being almost 78.4 per cent. Thus, fluctuations in area and production of mothbean in Rajasthan would be a deciding factor ror the fate at the country level.

Table: Area, Production and Productivity of mothbean in Rajasthan

S.No.	Year	Area (lakh hactares)	Production (lakh tonnes)	Productivity (kg ha^{-1})
1	1993-94	13.36	1.51	112
2	1994-95	12.92	3.31	256
3	1995-96	12.12	1.55	119
4	1996-97	12.20	3.10	254
5	1997-98	11.23	2.99	266
6	1998-99	8.76	1.04	118
7	1999-2000	7.71	0.29	39
8	2000-01	9.61	1.41	144
9	2001-02	13.04	3.09	237
10	2002-03	4.31	0.19	.44
11	2005-06	12.28	1.49	122
	Average	10.69	1.82	155.55

Breeder seed indent and production of mothbean varieties by RAU, Bikaner

Variety	2002-03		2003-04		2004-05		2005-06		2006-07	
	Indent	Prodn	Indent	Prodn	Indent	Prodn	Indent	Prodn	Indent	Prodn
RMO-40	14.00	9.50	13.30	38.00	16.48	26.96	20.96	26.00	26.62	30.00
RMO-225	6.50	2.50	9.55	24.00	4.45	5.25	14.85	19.00	12.95	15.00
RMO-257	5.00	1.30	12.75	24.00	2.80	13.90	5.00	13.90	4.5	8.00
RMO-435	-	-	7.20	5.25	3.90	3.90	12.40	21.00	15.85	70.00
RMO-423	-	-	-	5.35	0.07	1.50	1.80	1.00	3.8	5.00
Total	25.5	13.3	42.8	96.6	27.70	51.51	55.01	80.9	63.72	128.0

Mungbean

Another important *kharif* pulse is moongbean, which contain 25 per cent protein of high digestibility and superior quality. Around the world, it is cultivated on 5.0 million hectares /with production of 2.5 million tonnes, while in India it ois grown on 30.41 lakh hectares with production and productivity of 11.73 lakh tonnes and 380 kg ha^{-1}, respectively. Crop is popularly grown in Maharastra, Andhra Pradesh, Rajasthan. In Rajasthan, it is cultivated on 5.54 production 1.10 lakh tonnes and productivity 199 kg ha^{-1}.

Area and Production of mungbean in India

State	Area ('000 ha)	Production ('000 m.t.)
Maharastra	768.8	29.5
Andhra Pradesh	519.0	173.2
Rajasthan	554.0	110.0
Orissa	276.5	134.2
Bihar	195.8	106.2
All India	3041.7	1172.7

Breeder seed indent and production of mungbean varieties by RAU, Bikaner

Variety	2002-03		2003-04		2004-05		2005-06		2006-07	
	Indent	Prodn	Indent	Prodn	Indent	Prodn	Indent	Prodn	Indent	Prodn
RMG-62	1.00	2.78	2.30	5.75	2.58	2.58	5.00	5.50	5.50	110.00
RMG-268	0.35	1.63	0.90	1.45	2.80	2.00	5.00	8.75	8.50	50.00
RMG-344	0.50	1.56	0.95	1.45	3.40	2.50	5.00	12.70	10.50	20.00
K-851	-	2.20	-	6.00	-	-	-	-	-	
Gangotri (G1)	0.35	0.05	-	4.00	2.0	2.76	-	3.50	-	2.00
Jamnotri (G8)	1.75	0.85	2.95	8.00	4.00	17.82	8.00	39.00	8.50	9.00
MUM-2	4.80	8.00	3.90	8.50	8.45	9.27	11.00	33.00	-	-
RMG-492	-	-	0.50	7.03	1.40	1.41	3.00	7.00	8.00	6.00
GM-3	-	-	-	0.63	-	-	-	-	-	-
Total	8.75	17.07	11.5	39.28	24.63	38.34	37.00	109.45	41.00	197.00

Guar

Second. important crop under arid legumes is Guar. The crop is known for exceptionally high adaptation towards poor and erratic rains, low inputs and less care, soil enrichment properties ropping systems, in industrial use in many ways and other These qualities have made it the most favoured crop of marginal farmer in the arid areas. Guar is grown in almost 23.30 lakh hectares area in India. The maximum contribution of states in respect of area is shared by Rajasthan (18.18 lakh hectare), followed by Gujarat (2.27 lakh hectares), Haryana (1.27 lakh hectare) and Punjab (0.14 lakh hectare). In other words, Rajasthan alone comprises almost 78.04 per cent of area. Whereas, other states show quite negligible contribution. However, 78.0 per cent cent area of guar in Rajasthan contributes 61.19 per cent production to the national basket of guar. The Productivity of Gguar in Rajasthan is lowest (233.7 kg ha^{-1}) against the national productivity (322.7 kg ha^{-1}). Thus, guar yield at the national level could be increased by increasing the productivity in Rajasthan and increasing the area in other states like Haryana, Gujarat and Punjab and other non-traditional areas, where productivity is already higher.

India and States	Area (lakh ha)	Production (lakh ton)	Productivity (kg ha^{-1})
India	23.30	11.98	428.0
Rajasthan	19.85	7.33	370.0
Gujarat	2.27	1.15	668.8
Haryana	1.20	1.08	740.0
Punjab	0.14	0.0065	922.0

Table : Breeder seed indent and production of guar varieties by RAU, Bikaner

Variety	2002-03		2003-04		2004-05		2005-06		2006-07	
	Indent	Prodn	Indent	Prodn	Indent	Prodn	Indent	Prodn	Indent	Prodn
RGC-936	15.30	19.80	15.66	36.75	19.46	27.41	10.86	82.00	18.00	110.00
RGC-197	-	13.00	0.80	4.00	0.70	30.51	0.50	27.50	0.40	60.00
RGC-471	1.00	1.50	3.80	3.50	-	-	-	-	-	
RGC-1002	6.60	6.35	8.50	6.10	1.80	12.0	4.60	28.00	8.93	60.00
RGC-1003	4.20	7.95	7.70	14.00	4.30	27.50	4.25	30.50	9.90	60.00
RGC-1017	0	2.30	-	17.00	6.30	11.50	4.10	52.00	7.75	40.00
RGC-986	5.30	4.90	7.00	18.00	5.40	20.90	6.05	22.00	5.50	20.00
M-83	0.03	0.76	0.09	2.00	0.40	0.68	0.15	5.00	0.25	3.00
RGC-1031	-	-	-	-	-	-	-	-	7.50	25.00
RGM-112	-	-	-	-	-	-	-	-	-	20.00
Total	8.75	17.07	11.5	39.28	24.63	38.34	37.00	109.45	41.00	197.00

Cowpea

Fourth crop under arid legumes is Cowpea. Cowpea is a versatile kharif pulse crop because of its smothering, drought tolerance and soil restoring properties short duration fast growth and low water requirement characteristics of cowpea make it suitable for growing and low water requirement characteristics of cowpea make it suitable for growing in dry land area. Cowpoea is called "vegetable meat' because of its high amount of protein of better biological value. The crop forms quality forage with good feeding vale coparable to Lucerne In Rajasthan on an average cowpea is grown on 1.10 lakh hectares with production 0.28 lakh tonnes and productivity 251 kg ha^{-1}.

Breeder seed indent and production of cowpea varieties by RAU, Binkaner

Variety	2002-03		2003-04		2004-05		2005-06		2006-07	
	Indent	Prodn	Indent	Prodn	Indent	Prodn	Indent	Prodn	Indent	Prodn
RC-19	4.20	3.75	3.00	1.55	1.50	2.20	0.80	2.40	1.50	15.00
RCV-7	0.03	0.09	-	0.06	1.20	0.68	1.60	1.10	1.00	4.402
RC-101	2.50	1.70	1.60	0.80	2.10	1.88	2.75	5.50	7.50	20.00
Total	6.73	5.54	4.60	2.41	4.8	4.76	5.15	9.00	10.00	39.4

Source: Seed plant Govt. of Rajasthan, 2008

Seed replacement rate (SRR%): Existing, Desirable and Plan upto 11-12

Crop	SRR%				Desirable SRR (%)	08-09	09-10	10-11	11-12
	04-05	05-06	06-07	07-08					
Mothbean	1.5	1.5	1.19	3.36	25	8	15	20	25
Mungbean	5.06	8.11	9.44	14.94	50	28	30	35	40
Guar	2.27	2.49	3.75	7.38	50	11	18	20	25
Cowpea	2.17	6.81	9.75	9.88	25	19	25	30	35

Source: Seed plan, Govt. of Rajasthan, 2008

Quality seed requirements

It is needless to emphasize the importance of quality seeds in production and productivity of any agricultural crop and about 10-15 per cent replacement rate of seed is desirable. The seed replacement rate of pulses in general being hardly 2-5 per cent is miserable. With the available production technology and varieties released, the production can be enhanced by 15-20%, if quality seeds are made available to the farmer. For the supply of good quality certified seed to the growers, a seed production chain is necessary. The chain comprises of breeder seed-foundation seed-certified seed. The basic requirement for production of quality seed, an end product to be available with the farmers, is the production of breeder seed. Unless and until the provision for producing the required quantity of pure breeder seed is made, the targeted objective of producing quality seed becomes hollowo Department of Agriculture and cooperation (DAC) and Indian Council of Agricultural Research (ICAR) regulating seed chain.

Table : Quality seed present status and requiement in Rajasthan.

Crop	Total Area under cultivation (lakh ha)	Certified seed distributed during 2002-03 (qtls.)	If 50% of area under certified seed then seed required (qtls)	Breeder seed required to produce targeted certified seed (qtls) (1:625)
Moth bean	10.53	4155	78975	126.36
Mung bean	5.54	9857	41550	66.48
Guar	19.85	1740	148875	238.2
Cowpea	1.10	432	8250	13.2

The low seed replacement rate might be improved if the conversion of beeder seed to foundation seed and certified seed is enhanced by proper monitoring the seed multiplication chain.

Genetic resources and its role in improvement of arid legumes

S.K. Sharma[1], S.K. Mishra[1], N.K. Dwivedi[2], N.K. Gautam[1] and Gopala Krishnan, S.[2]

[1] National Bureau of Plant Genetic Resources, Pusa Campus, New Delhi - 110 012;
[2]National Bureau of Plant Genetic Resources, Regional Station, Jodhpur - 342003

ABSTRACT

Genetic resources are the reservoirs of useful genes, which could be of immense use for the genetic improvement of crop plants. The National Bureau of Plant Genetic Resources, New Delhi is involved in collection, characterization, evaluation, conservation, documentation and distribution of different crops including the four arid legumes (cowpea, guar, mothbean and horse gram) germplasm in India. Altogether, 11,207 germplasm accessions of the four arid legumes have been conserved in the National Gene Bank at NBPGR, New Delhi. This paper describes in brief the various activities related to management of genetic resources of arid legumes namely germplasm collections, evaluation, conservation, distribution and the contribution of the indigenous germplasm towards improvement of arid legumes.

Introduction

Plant genetic resources form the unique and irreplaceable source for further genetic improvement and for incorporation of higher diversity in crops and cultivars in agriculture. The genetic diversity of agricultural crops is represented by landraces, cultivars, genetic stocks and elite germplasm (domesticated genepool) as well as by wild and weedy relatives of cultivated plants (wild genepool). All these materials form a gene pool of agricultural crops, which is used to improve important traits, to broaden the genetic base of cultivars and also as a source of new diversity for agriculture.

Key words: Area legumes, genetic resources, improvement.

Origin of arid legumes

According to Vavilov [1], about 160 species of domesticated plants have originated in India; besides this, over 320 species of wild relatives are also known to occur in India [2]. Among the arid legumes, mothbean [*Vigna aconitifolia* (Jacq.) Marechal] and horse gram [*Marcotyloma uniflorum* (Lamb.) Verdc.] are of Indian origin [3, 4]; guar [*Cyamopsis tetragonoloba* (L.) Taub.], probably domesticated in north-western India [5]; and cowpea [*Vigna unguiculata* (L.) Walp.] is of African origin with India being a secondary centre of diversity [6]. The north-eastern region of India and Western Ghats are the secondary centre of diversity for vegetable type cowpea and the central India for grain type cowpea. Since long, they have been an important part of the cropping system in the semi-arid and arid regions of India, subjected to continuous natural and artificial selection, resulting in creation of considerable diversity in the germplasm.

Since, 1970's awareness has been generated to identify, collect and conserve the fast depleting and irreplaceable crop genetic resources for use by present and future generations. The systematic exploration and germplasm collection programme of National Bureau of Plant Genetic Resources, New Delhi has resulted in establishment of sizeable collection of these arid legumes. Besides collection and conservation, enrichment of the germplasm base is also done through augmentation of germplasm through the import of germplasm from abroad; genetic stocks and released varieties from within the country. Another important activity of NBPGR is the need based distribution of germplasm from time to time for different institutes/state agricultural universities for utilization in crop improvement programmes. The present paper deals with various genetic resources related activities in relation to arid legumes namely, exploration and collection as well as augmentation, characterisation and evaluation, registration of unique germplasm and distribution of the germplasm in India. Genetic resources have played a vital role in improvement of arid legumes and part of the discussion is also devoted to the contribution of genetic resources in crop improvement.

Genetic resources of arid legumes

Germplasm exploration and collection has resulted in collection of 5125 accessions of guar, 2019 accessions of mothbean, 4338 accessions of cowpea and 2514 accessions of horse gram from different parts of India. Altogether, 11,207 germplasm accessions of the four arid legumes including guar (3714), mothbean (1540), cowpea (3398) and horse gram (2554) are conserved at the National Gene Bank, NBPGR, New Delhi (Table 1).

The germplasm collections represents the wide spectrum of variability displayed in these crop species and has been collected from various regions of the country as well as from abroad. Efforts are also in progress to identify the gaps in the germplasm collections and need based targeted collection trips have been carried out to collect germplasm of specific importance. As a result, germplasm of adak guar (37 acc.), adak moth (8 acc.) and *Vigna trilobata* (6 acc.) have been collected from north western Rajasthan. Adak guar, a primitive weedy form of guar with typical shattering habit, high tolerance to drought and profuse branching habit and adak moth, with shattering habit, black seeds and dense hairs is very hardy in nature.

Table 1. Germplasm of arid legumes conserved at the National Gene Bank, NBPGR, New Delhi (as on April 2008).

Crop	Cultivated				Wild spp.	Total
	Indigenous collections	Exotic collections	Genetic Stocks	Released varieties		
Guar	3675	39	2	42	32	3714
Mothbean	1502	38	2	14	11	1540
Cowpea	2350	1048	1	37	3	3398
Horse gram	2545	9	1	14	1	2554

Besides germplasm collection from within the country, NBPGR is also involved in augmenting the germplasm with introduction of germplasm from abroad. A total of 1134 germplasm accessions of the arid legumes have been introduced from various countries which includes guar (39), mothbean (38), cowpea (1048) and horsegram (9).

Cyamopsis senegalensis (EC533475) - wild relative of guar imported from USA.

Cyamopsis serrata (EC533476) - wild relative of guar imported from USA.

Some of the important introductions with respect to arid legumes include; the wild relatives of guar namely *C. senegalensis* and *C. serrata* from USA, EC587822 (cowpea) an early maturing accessions with resistance to Cowpea Aphid Borne Mosaic Virus and bacterial blight.

Germplasm evaluation

Germplasm characterization and evaluation is one of the important activities needed in order to mine the utilizable variability in the germplasm collection. Germplasm evaluation helps in identifying the extent of variability for different traits and also in finding out trait specific germplasm accessions in the collections. The range of variability observed in the germplasm collections of the arid legumes for different agro-morphological traits are presented in the Table 2.

Besides the agronomic traits, specific traits are important with respect to each of the arid legume for their improvement. One such trait is gum content in guar and IC116601 (34.38%) collected from Bikaner, Rajasthan possessed the highest gum content among all the germplasm collections characterized [7].

Table 2. Variability for different agro-morphological characters in the arid legumes.

S. No.	Character	Range			
		Guar	Mothbean	Cowpea	Horse gram
1.	Days to 50 % flowering	25 - 76	32 – 84	25 - 76	54 - 122
2.	Days to 50% maturity	66 - 128	57 – 105	50 - 163	79 - 183
3.	Plant height (cm)	55.0 - 238.0	11.7 - 49.3	30.0 - 390.0	17.0 - 137.0
4.	Branches/ plant	0 - 29	2 – 12	1 - 13	4 - 16
5.	Clusters/ plant	2 - 86	9 – 73	1 - 107	5 - 59
6.	Pod length (cm)	1.6 - 17.0	2.2 - 5.3	10.0 -100.0	2.8 - 8.3
7.	No. of pods/ plant	4 - 412	25 – 128	1 - 231	10 - 175
8.	No. of seeds/ pod	2 - 15	2 – 10	4 - 24	3 - 9
9.	Seed yield/ plant (g)	1.2 - 71.0	1.6 - 19.1	3 - 115	3.1 - 15.5
10.	100 seed weight (g)	1.5 – 5.3	1.5 – 4.6	3.5 - 19.7	2.2 - 16.2

The identification and characterization of trait specific germplasm in guar is being carried out at NBPGR Regional Station, Jodhpur. Several accessions for determinate character, photo-insensitiveness and branched all node cluster bearing habit have been identified and are being characterization for further utilization.

Documentation of genetic resources

Thirteen catalogues of germplasm evaluation of the arid legumes guar (5), mothbean (2), cowpea (4) and horse gram (2) have been published by NBPGR. These catalogues form a wealth of information on various descriptor traits for these crops and serves as a valuable guide for selection of germplasm for utilization in the arid legumes improvement programme.

Registration of germplasm

Registration of genetic stocks and elite germplasm serves as a useful tool for effective utilization through germplasm exchange. Till date, six germplasm accessions of arid legumes including guar (2), mothbean (2), cowpea (1) and horse gram (1) have been registered for novel/ unique traits (Table 3).

Table 3. Germplasm of arid legumes registered for unique/ novel traits

Crop	INGR No.	Donor/other identity	Unique/ novel feature
Guar (2)	04037	GDM-1	Determinate plant type
	08027	CH/55-2	Foliaceous bracts in the inflorescence
Mothbean (2)	01024	CZM-32	Drought tolerance
	04095	RMM-12	Single stem, early maturity, high influx of sodium ions in root from soil
Cowpea (1)	04038	CAZC-B	Black seed coat colour
Horse gram (1)	02007	GN-1882	High protein content (> 35 %)

Genetic resources in improvement of arid legumes

The ultimate objective of all the activities in management of plant genetic resources is to provide for its use in the crop improvement programmes. With increasing constrain in crop production, breeders continually need new and diverse germplasm to find specific traits to maintain or improve yields [8]. Among different types of crop genetic resources, released varieties and breeding lines contribute more to crop breeding. Wild relatives and genetic stocks, however, are expected to play a greater role in the future [9].

The wide range of variability in the germplasm collections have been utilized for varietal development. Particularly, the indigenous germplasm collections have made immense contribution in the improvement of arid legumes. Table 4 shows the various breeding methods employed in breeding arid legumes and also the contribution made by indigenous germplasm, since 79 varieties released till date are direct selections from the indigenous germplasm collections. A perusal of the various methods used in arid legumes improvement reveals that direct selection from indigenous germplasm is the predominant methodology in varietal improvement. Although, hybridization and selection has led to success in crops such as guar and cowpea, there is a need to strengthen the hybridization programme to recombine desirable traits in these crops. Mutation breeding has also helped in improvement of arid legumes and 12 varieties have been released through induced mutation. However, in horsegram, all the varieties released till date have been developed through direct selection from indigenous germplasm. Therefore, there is an urgent need to initiate hybridization in horsegram to develop superior recombinants.

Table 4. Breeding methods employed for improvement of arid legumes.

Breeding methodology (Varieties developed)	Crop and varieties			
	Guar (33)	Mothbean (13)	Cowpea (45)	Horsegram (17)
Selection from indigenous germplasm	25 varieties: Grain purpose (18): FS 277, HG 75, HG 128, HG 258, Durgajay, Durgapura safed, Maru Guar, RGC 197, RGC 471, Kutch 8, Malosan, GG 1, PLG -85, Suvidha, Naveen, Sona, AG 111, AG 112 Dual purpose (2): Bundel 1, Bundel 2 Fodder purpose (3): HFG 119, HFG 156, Guara 80 Vegetable purpose (2): Pusa Navbahar, Durgabahar	7 varieties: Type-3, Baleshwar-12, Jwala, Jadia, Gujarat Moth-1, IPCMO-912, IPCMO-880	31 varieties: Assem, Bundhel Lobia-1, C 152, Chharodi, Co1, Co 3, Co 4, Co 5, Cowpea 78, Cowpea 88, FS 68, Gomti, GC1, GC2, GC3, JC 2, JC 5, JC 10, K 11, KBC 1, Lalitha, No.2-1, No. 5-19-4-1, No.10, NP 3, Paiyur 1, Pusa Phalguni, PTB 1, RS 9, Type 2, Type 5269	17 varieties: Co 1, Paiyur-1, Paiyur-2, Hebbel Hurali 1, Hebbel Hurali 2, HPK-2, HPK-4, Madhu, PDM-1, VZM1, Birsa Kulthi-1, VLG-1, Maru Kulthi, KBH-1, PHG-9, AK-21, AK-42
Hybridization and Selection	7 varieties: Grain purpose: HG 365, HG 563, RGC 936, RGC 986, RGC 1002, RGC 1003, RCG 1017	1 variety: CAZRI Moth – 2	8 varieties: Co 2, Cowpea 74, KM1, PTB 2, RC19, S 288, S 248, Gujarat cowpea	Nil
Induced mutation	1 variety: Vegetable purpose: M-83	5 varieties: Maru Moth – 1, RMO-40, RMO 257, FMM-96, CAZRI Moth – 1	6 varieties: V-16 (Amba), V37 (Shrestha), V38 (Swarna), V130 (Pusa Safed), V240 (Rambha), V585 (Pusa Sampada)	Nil

The germplasm evaluation programme has focused on identification of germplasm for specific traits which can be utilized in improvement of the legume. For example, in guar, the unbranched determinate types with lesser ground spread have the potential advantage over the branched type in that they can accommodate higher per hectare plant populations and are more synchronous in maturity. The unbranched types may also find its use in mixed cropping with pearlmillet under rainfed farming. The photoinsensitive accessions can be utilized in breeding cultivars suited for spring planting under north Indian conditions where assured irrigation can be provided [10]. Therefore, the utilization of the genetic variability in the arid legumes germplasm will not only help in increasing the productivity of these crops in the dry areas of the country but also help in sustaining the production.

Distribution of germplasm

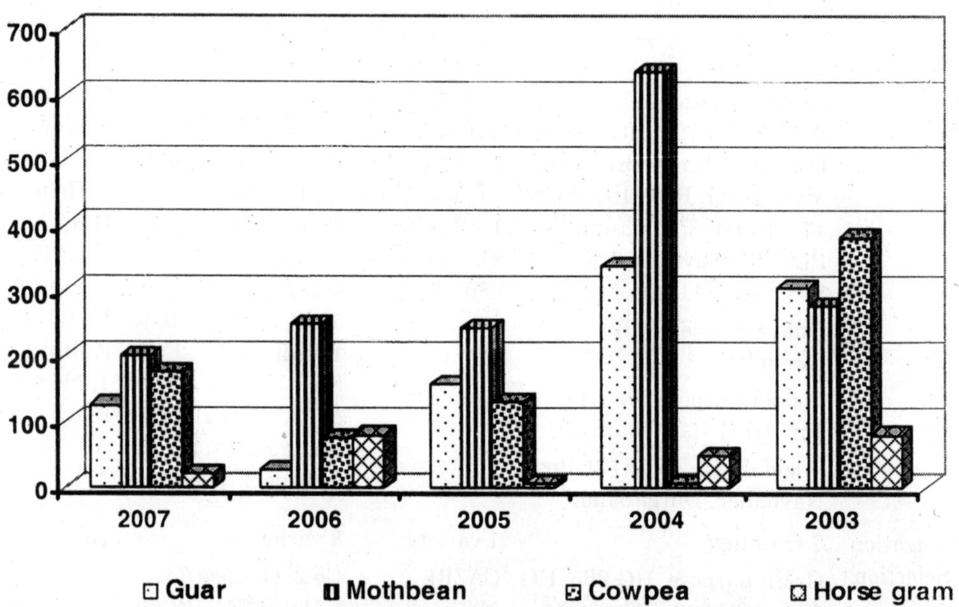

Figure 1. Distribution of germplasm of arid legumes during the last five years.

During the last five years, a total of 3561 germplasm accessions of arid legumes namely guar (951 acc.), mothbean (1607 acc.), cowpea (769 acc.) and horse gram (234 acc.) have been distributed to research workers based on their indents (Fig. 1). Besides this, germplasm for multilocational evalu-ation have been distributed through national network project on arid legumes, which includes guar (100 acc. in 4 locations), mothbean (100 acc. in 3 locations) and horse gram (100 acc. in 4 locations) during 2006.

Future Perspectives

A large number of germplasm of arid legumes has been collected and conserved; therefore further collection efforts will be focussed on trait specific germplasm to fill the gaps in the existing collection.

Even though a large number of accessions of arid legume crops are conserved in the genebanks, evaluation to identify better genetic stocks in all the accessions is important. This can be addressed to some extent by identifying the core collections which will ease the work on evaluation and utilization of *ex situ* conserved germplasm effectively. Documentation of valuable indigenous knowledge, registration of unique germplasm and DNA fingerprinting are also important in order to effectively utilize the diversity available in the genetic resources of crop plants.

REFERENCES

1. Vavilov, N. I. 1926. Studies on the origin of cultivated plants. Bull. Appl. Bot. Genet. Pl. Breed. 16: 1-248.
2. Arora, R. K. 2000. Wild relatives of cultivated plants. In: Singh, N. P., Singh, D. K., Hajra, P. K. and Sharma, B. D. (eds) Flora of India. Introductory volume part II. Botanical Survey of India, Calcutta, pp 218-234.
3. de Candolle, A. P. 1825. Leguminosae. In: *Prodromus Systematis Naturalis Regni Vegetabilis 2*, (*Phaseolus* 390-396), Paris. pp. 93-524.
4. Vavilov, N. I. 1951. The origin, variation, immunity and breeding of cultivated plants. Chron. Bot. 13: 1-366.
5. Hymowitz, T. 1972. The trans-domestication concept as applied to guar. Econ. Bot. 26: 49-80.
6. Simmonds, N. W. 1970. Evolution of crop plants. Longman and Co., London, U.K. 339p.
7. Dwivedi, N. K., Bhandari, D. C., Neelam Bhatnagar and Dabas, B. S. 1999. Characterising clusterbean (*Cyamopsis tetragonoloba* (L.) Taub.) germplasm for yield and quality traits. Annals of Arid Zone **38**(2): 151 – 156.
8. Duvick, D. N. 1986. Plant breeding: Past achievements and expectations for the future. Economic Botany, **40**(3): 289-297.
9. Li, Y., Wang, J., Cao, Y., Gao, W., Fang, J. and Lou, X. 1998. The use of genetic resources in crop improvement: Lessons from China. Genetic Resources and Crop Evolution **45** (2): 181 – 186.
10. Gopala Krishnan, S. and Dwivedi, N. K. 2007. Genetic stocks of clusterbean [*Cyamopsis tetragonoloba* (L.) Taub.] – assessment and utilization. *Journal of Arid Legumes* 4(2): 121 – 126.

Characterization of mungbean [*Vigna radiata* (L.) Wilczek] germplasm in arid and semi-arid region of India

N. K. Dwivedi, Gopala Krishnan, S. and Raju Ram Meghwal

National Bureau of Plant Genetic Resources, Regional Station, CAZRI Campus, Jodhpur-342 003, Rajasthan (India)

ABSTRACT

Mungbean [*Vigna radiata* (L.) Wilczek] is a major pulse crop in the Indian sub-continent. The productivity of this crop can be improved through exploitation of variability and evaluation of 90 indigenous accessions collected from different sites of Gujarat (22), Madhya Pradesh (07) and Rajasthan (61), and 132 exotic accessions introduced from Taiwan and China, was carried out at this regional station (26° 18' N latitude and 73° E longitude, altitude 224m) in augmented incomplete block design with 3 checks (M1319B, Pusa 105 and Pusa Vishal) during Kharif 2003, 2004, 2005 and 2006 respectively. The trials were unprotected for diseases. The data on 9 quantitative and 18 qualitative characters were recorded. A wide range of variability was observed for all the morphological characters except in terminal leaf length, leaflet shape, flower colour and flowering tendency. About 90.00% and 95.00% accessions of indigenous collections, and 86.3% and 91.6% accessions of exotic collection exhibited resistance to yellow mosaic virus and leaf crinkle virus, respectively. High variability was observed for traits such as 50% flowering, plant height, clusters per plant, pods per cluster, seeds per pod and seed weight per plant. Twenty six exotic accessions showed pot shattering habit. The accessions with superior desirable traits can be utilized in different breeding programmes for improvement of particular character or for the development of a new verity.

Key words: Mungbean, germplasm, arid, semiarid region.

Introduction

In the Indian sub continent mungbean [*Vigna radiata* (L.) Wilczek] is a major pulse crop. The diversity of this crop is found maximum in India which is the primary center of origin of this crop (1,2,3). The productivity of this crop requires genetic improvement through exploitation of germplasm available. A thrust for germplasm collection, augmentation and evaluation is of utmost significance to have a dynamic breeding programme. Production through improvement of pulse crops is necessary (4). The superior donor parents are urgently required for incorporating desirable characters in the pulse crops (4, 5). National Bureau of Plant Genetic Resources, Regional Station, Jodhpur has been working towards the build up genetic resources in mungbean from India as well as from abroad since last twenty five years. The present paper deals with characterization, preliminary evaluation and identification of promising accessions with desirable traits.

Material and Methods

The preliminary characterization and evaluation of 220 accessions comprising of 90 indigenous accessions collected from different sites of Gujarat (22), Madhya Pradesh (07) and Rajasthan (61), and 132 exotic accessions introduced from Taiwan and China, was carried out at NBPGR regional station (26°18' N and 73°E, altitude 224) with temperature 10°C to a maximum 46°C during Kharif season of 2003-2004, 2004-2005, 2005-2006 and 2006-2007. The soil type is sandy. The gemplasm accessions were raised in an augmented incomplete block design with 3 checks (M 1319B, Pusa 105 and Pusa Vishal) in each block. The accessions were grown in 2 m rows with a row spacing of 60 cm and plant to plant spacing of 15 cm. The trials were unprotected for diseases. The data on 9 quantitative and 18 qualitative characters were recorded using minimal descriptors developed by NBPGR(6). All the quantitative data were analyzed statistically for mean and range. The qualitative characters have been reported as such and their frequencies were only calculated.

Results and discussion

Indigenous mungbean: The 90 indigenous accessions of mungbean germplasm were characterized for 9 quantitative and 18 qualitative characters. Wide range of variability was recorded in both qualitative and quantitative traits. All the accessions possessed greenish yellow flowers, entire leaflet shape, small terminal leaf length and showed asynchronous flowering tendency (Table 1). No twinning tendency was reported (7). Sen and Ghosh (8) reported that twining growth habit is dominant in this crop. However, variation was observed in plant vigour and growth habit, leaf colour, branching pattern, leafiness, twining tendency, raceme position, pod colour and shape and seed colour. The range of variability and frequency observed for qualitative traits are presented in Table 1. Five accessions showed pod shattering habit while nine and five accessions were infected with YMV and leaf crinkling disease respectively. Similar observations were also reported earlier (7, 9, 10, 11). The quantitative characters also showed wide range of variation. The result such as descriptive statistical analysis is presented in Table 2. The characters of plant height, clusters/ plant, pods/ cluster, seeds/ pod, pods and seed yield/ plant were highly variable based on the variance. Days to 50% flowering ranged from 30 -59 days. It is reported in early studies that the earliness is due to additive action (12, 13).

Table 1. Frequency distribution of mungbean accessions for various qualitative characters.

Characters	Frequency	Frequency	Frequency
		Indigenous (90)	Exotic (132)
Early plant vigour	Poor	17	09
	Good	62	79
	Very good	11	44
Plant growth habit	Erect	79	101
	Spreading	11	31
Leaf colour	Light green	21	13
	Intermediate green	39	80
	Dark green	30	39
Terminal leaf length	Small	90	130
	Medium	00	02
Leaflet shape	Entire	90	132
Plant surface	Glabrous	83	89
	Pubescent	07	43
Branching pattern	Central	57	97
	Basal	26	26
	Top	07	09
Flower colour	Green Yellow	90	132
Leafiness	Sparse	24	42
	Intermediate	42	78
	Abundant	24	12
Flowering tendency	Asynchronouns	90	132
Twining tendency	None	53	132
	Slight	23	00
	Intermediate	11	00
	Pronounced	03	00
Raceme position	Above canopy	76	93
	Intermediate	14	39

	Glabrous	03	17
	Peburulent	86	115
Pod colour (mature)	Straw	17	27
	Tan	03	06
	Brown	15	30
	Brownish black	53	53
	Black	02	16
Pod shape	Straight	69	109
	Curved	21	23
Shattering habit	Absent	85	106
	Present	05	26
Seed colour	Light green	39	59
	Dark green	11	15
	Greenigh yellow	08	08
	yellow	01	00
	Dark green mosaic	03	16
	Bluish green	05	15
	Brown mosaic	03	09
	Light green mosaic	20	10
Diseases	Yellow mosaic virus	09	18
	Leaf crinkle virus	05	11
	None	76	103

Table 2. Range and mean for various traits in mungbean germplasm and checks.

S. No.	Characters	Range		Mean	Checks		
		minimum	maximum		M1319B	Pusa 105	Pusa Vishal
1.	50% flowering (days)	30.00	59.00	41.64	43.50	41.00	39.50
2.	Plant height (cm)	21.99	88.69	51.64	35.57	45.89	40.70
3.	Primary branches	01.70	05.15	03.58	03.62	03.59	03.47
4.	Pods/ cluster	02.00	05.40	03.67	03.81	03.76	03.95
5.	Clusters/ Plant	03.90	17.60	07.37	07.10	07.55	06.80
6.	Seeds/ pod	06.30	12.65	10.05	09.72	09.79	10.28
7.	Pod weight plant (g)	05.39	37.15	18.49	20.78	18.49	16.05
8.	Seed weight (g)	03.04	21.34	10.23	12.33	10.93	10.44
9.	100 seed weight (g)	03.22	08.87	04.88	04.80	03.87	04.18

Exotic mungbean: The 132 exotic accessions were also characterized for 9 qualitative and 18 quantitative characters. Wide range of variability was noticed in both qualitative and quantitative characters. All the accessions possessed greenish yellow flowers, entire leaflet shape and showed asynchronous flowering tendency (Table 1). No twining tendency was observed. Variation was observed in plant vigour and growth habit, leaf colour, branching pattern, leafiness, twining tendency, raceme position, pod colour and shape and seed colour. The range of variability and frequency observed for qualitative traits are presented in Table 1. Twenty six accessions showed pod shattering habit while eighteen and eleven accessions were infected with YMV and leaf crinkling disease respectively. The quantitative characters showed wide range of variation. The characters such as plant height, clusters/ plant, pods/ cluster, seeds/ pod, pods and seed yield/ plant were highly variable as in indigenous green gram germplasm.

Table 3. Mungbean accessions showing high seed yield (> 15 g) in comparision to that of checks.

S. No.	Accession Number	Source	Seed yield/ Plant (g)	Days to 50% flowering	Branches/ plant	Clusters/ Plant	Pods / cluster	Seeds/ pod	100 seed wt
1.	IC311445	Banaskantha (G, I)	15.65	30	5.4	11.6	4.2	10.0	5.84
2.	IC324021	Kachchh (G, I)	20.46	43	4.8	21.6	4.4	11.6	3.82
3.	IC325782	Pali (R, I)	20.52	60	4.8	16.0	3.4	10.6	3.79
4.	IC325788	Pali (R, I)	23.35	39	5.4	22.8	4.4	13.6	3.36
5.	IC325791	Pali (R, I)	15.88	58	4.4	11.4	3.4	9.0	3.61
6.	IC325828	Pali (R, I)	17.14	38	4.2	14.6	2.6	12.2	4.07
7.	IC333213	Barwani (MP, I)	19.65	37	3.9	14.6	3.8	11.8	4.63
8.	EC511388	China	15.09	39	2.6	7.5	4.2	9.9	6.15
9.	EC520037	Taiwan	19.33	33	2.5	7.9	5.3	8.7	6.51
10.	EC538111	Taiwan	15.20	38	2.6	8.3	4.0	9.6	6.75
Checks									
a.	M1319B		12.33	44	3.62	7.10	3.81	9.72	4.80
b.	Pusa 105		10.93	41	3.59	7.55	3.76	9.79	3.87
c.	Pusa vishal		10.44	40	3.47	6.8	3.95	10.28	4.18

G - Gujarat; MP - Madhya Pradesh; R - Rajasthan; I - India

Promising accessions

Accession IC325788 collected from Pali district of Rajasthan had highest seed yield (23.35 g) with 50% flowering (33 days), primary branches (5.4), clusters/ plant (22.8), pods/ cluster (4.4), seeds/ pod (13.6) and 100 seed weight (3.36 g) (Table 3). While among exotic germplasm accessions EC520037 had highest seed yield (19.33 g) with 50% flowering (33 days), primary branches (2.5), clusters/ plant (7.9), pods/cluster (5.3), seeds/pod (8.71) and 100 seed weight (6.51 g) (Table 4). Maximum per plant pod yield (139.16 g) and seed yield (23.35 g) was recorded in the indigenous accession IC325788 collected from Rajkiabaskhurd, Pali (Rajasthan). Accessions EC538111 had highest 100 seed weight (8.87 g) with days to 50% flowering (38.5 days), primary branches (2.4), clusters/ plant (5.1), pods/ cluster (2.5), seeds/pod (9.8) and seed yield/ plant (7.25 g), while EC520029 and EC520042 ranked second and third respectively (Table 4). Based on the screening for different qualitative and quantitative traits of 90 accessions of indigenous and 220 accessions of exotic germplasm, promising lines identified for different characters are listed in table 5 and 6 respectively.

Table 4. Exotic Mungbean accessions showing high 100 seed weight (>8 g) in comparision to that of checks

S. No.	Accessions	Source	100 seed wt (g)	Days to 50% flowering	Branches/ Plant	Cluster/ Plant	Pods/ Cluster	Seeds/ pod	Seed yield/ plant
1.	EC512793	China	8.87	38.50	2.4	5.1	2.5	9.8	7.25
2.	EC520029	Taiwan	8.35	33.00	2.6	6.8	2.9	8.7	8.18
3.	EC520042	Taiwan	8.18	33.50	2.5	6.9	4.3	9.7	12.58
4.	EC528094	Taiwan	8.10	37.50	3.1	7.2	3.3	10.7	12.94
5.	EC512790	Taiwan	8.03	43.00	3.9	3.5	3.4	9.0	5.27
Checks									
a.	M1319B		4.80	44	3.62	7.10	3.81	9.72	4.80
b.	Pusa 105		3.87	41	3.59	7.55	3.76	9.79	3.87
c.	Pusa vishal		4.18	40	3.47	6.80	3.95	10.28	4.18

Table 5. Promising accessions of indigenous mungbean identified during evaluation.

Trait	Promising accessions
Days to 50% flowering (< 35 days)	IC311445 (34), IC311446 (34), IC333153 (33)

Plant height (>100 cm)	IC325782 (120.3), IC325828(109.0), IC325833 (103.6), IC415598 (108.3), IC324005 (122.9), IC324012 (108.9), IC324021(116.0)
Primary branches (> 5)	IC415598 (5.3), IC311437 (5.2), IC311445 (5.4)
Clusters/ plant (> 20)	IC370515 (33.2), IC324021 (21.6)
Seeds/pod (> 12)	IC325770 (12.8), IC325799 (12.8), IC325823 (15.0), IC325828 (13.4), IC325833 (12.2), IC333153 (12.8), IC329039 (14.0), IC323998 (12.6), IC324036 (12.8)
Pods weight / plant (>25 g)	IC325782 (37.5), IC325787 (34.3), IC325788 (39.2), IC325833 (30.6)
Seed yield/ plant (>15 g)	IC311445 (24.19), IC324021 (20.46), IC325788 (23.4), IC325828 (20.8), IC333213 (19.7), IC325733(27.1)
100 seed weight (>4.5 g)	IC418452 (9.90), IC418454 (7.30)

Table 6. Promising accessions of exotic mungbean identified during evaluation.

Trait	Promising accessions
Days to 50% flowering (<33 days)	EC511352 (32.0), EC520036 (31.5), EC520038 (31.0), EC520039 (31.5), EC520041 (31.0), EC528607 (32.0)
Plant height (>50 cm)	EC528606 (55.8)
Primary branches (>4)	EC528606 (4.9), EC512792 (4.8), EC520020 (4.7), EC520039 (4.6), EC538133 (4.6)
Clusters/ plant (>10)	EC511375 (10.2), EC520014 (10.6), EC520023 (12.4), EC520032 (10.2), EC528606 (10.5)
Seeds/ pod (>11)	EC512792(11.3), EC528091(11.7)
Pod weight/ plant (>25 g)	EC511352 (27.33), EC528092 (30.33), EC512792 (35.15)
Seeds yield/ plant (>15 g)	EC511388 (15.09), EC520021 (15.05), EC520037 (19.33), EC538111 (15.20)
100 seed weight (>8 g)	EC512793 (8.87), EC520029 (8.35), EC520042 (8.18), EC528094 (8.10), EC512790 (8.03)

These accessions with superior desirable traits can be utilized in different breeding programmes for improvement of a particular character or to develop a new variety with multiple desirable traits for the benefit of the farmers of the mungbean growing areas of Gujarat and Rajasthan. The seeds of all these accessions have been deposited in the National Gene Bank at NBPGR, New Delhi for long term conservation and kept in Medium Term Storage facility at this Regional Station of NBPGR as active collections and for further research and utilization.

EC520036 – A good bearing early maturing mungbean accession.

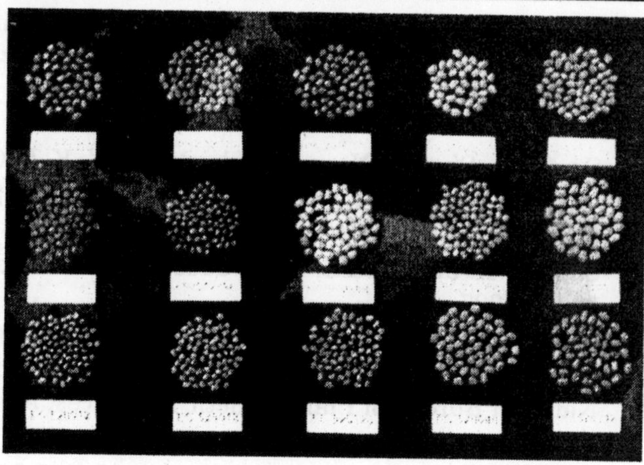

Variability in seed colour, shape and size among the exotic accessions of mungbean.

FUTURE THRUST

1. Germplasm collection from unexplored areas.
2. Utilization of superior accessions for improvement of specific trait.
3. Hybridization programme for obtaining promising recombinants.
4. Incorporation of disease and pest resistance.
5. Identification of accessions with tolerance to abiotic stresses.
6. Further studies for plant type and quality.

REFERENCES

1. De candolle A., 1884. *Origin of cultivated plants*. Heffner publishing company, New York.
2. Vavilov, N. I. 1926. Origin, variation, immunity and breeding of cultivated plants (translated by K. Starr Chestet). *Chronica Botanica* 12: 1-364
3. Zhukovsky, P. M. 1950. *Cultivated Plants and their wild relatives* (Translated by P.S. Hudson, 1962). Commonwealth Agricultural Bureau, Farnham Royal, England, P.1-107.
4. Paroda, R. S. 1990. Production and improvement of pulse crops In: Genetic improvement of pulse crops. Ed. Jafar Nizam and Irfan A. Khan and S.A. Farook.
5. Dwivedi, N. K., Neelam Bhatnagar and S. Gopala Krishnan. 2005. Performance of exotic pulses in arid and semi-arid region of India. *Indian J. Pl. Gent. Res.* 18(1): 54-55.

6. Mahajan, R. K., Sapra, R. L., Srivastava, U., Singh, Mahendra and Sharma, G. D. 2000. Minimal descriptors (for. Characterization and Evaluation) of Agri-Horticultural crops (Part-I). National Bureau of Plant Genetic Resources, New Delhi. P 1-230.
7. Sunil, N., Sivaraj, N. and Chakrabarty, S. K. 2003. Characterization and evaluation of mungbean [*Vigna radiata* (L.) Wilczek] germplasm from Andhra Pradesh, India. *Indian J. Pl. Gent. Res.* 16(1): 18-20.
8. Sen, N. K. and Ghosh, A. K. 1959. Genetic studies in greengram. *Indian J. Genet.* 19: 210-227.
9. Kawalkar, T. G., Bisht, I. S., Mahajan, R. K., Patel, D. P., Gupta, P. N. and Chandel, K. P. S. 1996. Catalogue on greengram [*Vigna radiata* (L.) Wilczek] germplasm. National Bureau of Plant Genetic Resources, New Delhi. P 1-361
10. Arora, R.K., Chandel, K.P.S. and Joshi, B.S. 1973. Morphological diversity in *Phaseolus sublobatus* Roxb. *Curr. Sci.* 42: 359-361.
11. Chandel, K.P.S. 1980. Evolutionary studies in *Vigna radiata*, *V. mungo* and other species. M.Sc. Thesis, Department of Plant Biology, University of Birmingham, England. P 1-100.
12. Chowdhury, J.B., Choudhury, R.B. and Kakar, S.N. Studies on the genetic variability in moong (*Phasoelus aureus*). *J. Res. Punjab Agric. Univ.* 8: 169-172.
13. Ramanujam, S., Tiwari, A. S. and Mehra, R.B., 1974. Genetic divergence and hybrid performance in mungbean. *Theor. Appl. Genet.* 45: 211-214.

Heterosis in mungbean [*Vigna radiata* (L.) Wilczek]

C.G. Intwala, P.S. Vashi, V.S. Patel and V. Khandelwal

Department of Agricultural Botany, N. M. College of Agriculture, Navsari Agricultural Universtiy, Navsari 396 450 (Gujarat).

ABSTRACT

Heterosis studies for seed yield and its related components in 40 F_1 hybrids derived from line x tester analysis involving four females and a set of ten males, indicated an appreciable amount of heterosis for almost all the characters. Pods per plant was found to be the most heterotic component followed by seed yield per plant. A fairly high degree of heterosis was also observed for clusters per plant, pods per cluster, seeds per pod and branches per plant. Among all the crosses, VRMGg-1 x SM-52 manifested highest heterosis with high seed yield per plant followed by crosses GM-4 x Pusa bold-1, GM-4 x Pusa 9632 and GM-3 x Pusa 9632. Heterotic effects for seed yield per plant was found to be associated with heterosis for number of pods per plant, number of clusters per plant and number of pods per plant.

Key words: Mungbean, heterosis, line x tester.

INTRODUCTION

Exploitation of heterosis on commercial scale in major field crops has increased both production and productivity per unit area of land. Green gram being a highly self pollinated crop, utility of heterosis *per se* may not be of much use but cross combinations showing excellent heterosis can be used in developing high yielding pure lines. The study of heterosis will provide the basic information regarding the breeding methodology to be employed for varietal improvement. Therefore, the present investigation was carried out to know the direction and magnitude of heterosis in mungbean.

MATERIALS AND METHODS

The experimental materials for the present investigation consisted of 54 entries including 4 females, 10 males and their 40 crosses. These were evaluated during summer-2001 at three locations *viz.*, Navsari, Achhalia and Vanarasi. The trials were conducted in Randomized Block Design (RBD), with three replications. The parents and F_1s were represented by a three row plot of 30 plants spaced at 45 x 15 cm. All the agronomical practices and plant protection measures were followed as and when required to raise a good crop. Observations were recorded on five competitive plants for nine characters in each entry. Heterosis was estimated for all the characters over better parent (BP) as per standard procedure.

RESULTS AND DISCUSSION

The range of heterosis for seed yield and its components is given in Table-1. The degree of heterosis varied considerably for yield components. Heterosis for seed yield ranged from -10.12 to 58.12 per cent. However, pods per plant showed the highest range of heterosis (-10.75 to 84.06 per cent). On the basis of mean performance, SM-56 was found to be the best parent for seed yield per plant and clusters per plant. For major yield contributor i.e., pods per plant, SM-52 was the superior among all the parents which was also good for branches per plant.

Based on the mean performance, the F_1 VRMGg-1 x SM-56 was found to be the best for seed yield per plant, which was also the best for all the major yield contributors *viz.*, pods per plant, clusters per plant and branches per plant.

The magnitude of heteosis over better parent for seed yield and its components indicated that all the characters distinctly differed for upper limits in both positive and negative directions. The maximum amount of heterosis was observed for pods per plant (84.06%) followed by seed yield per plant (58.12 %) and pods per cluster (47.29%). Seed yield per plant decides the economic worth of the hybrids, hence ten most promising combinations on the pooled basis for different characters were analysed (Table 2). Interestingly all the ten promising combinations showed highly significant positive heterosis for seed yield [1,2,3].

Table 1: Range, heterosis and most heterotic crosses for nine characters in mungbean (data pooled over locations)

Characters	Range		Heterosis (%)	Best hybrids	
	Parents	Crosses		Based on mean performance	Based on BP heterosis
Seed yield per plant	5.52 to 10.42	5.17 to 14.24	-10.12 to 58.12	VRMGg-1 x SM-56	VRMGg-1 x SM-52
Branches per plant	1.80 to 3.07	1.84 to 4.10	-26.22 to 38.87	VRMGg-1 x SM-56	VRMGg-1 x Pusa bold-2
Pods per cluster	3.27 to 4.68	2.87 to 5.75	-28.71 to 47.29	GM-3 x Pusa 9632	GM-3 x Pusa bold-2

Clusters per plant	3.46 to 6.28	3.96 to 8.44	-18.24 to 44.33	VRMGg-1 x SM-56	GM-3 x Pusa 9632
Pods per plant	11.52 to 23.41	13.21 to 41.09	-10.75 to 84.06	VRMGg-1 x SM-56	GM-3 x Pusa 9632
Pod length	6.07 to 7.84	6.13 to 7.17	-15.16 to 5.80	Co-4 x SM-52	VRMGg-1 x SM-7
Seeds per pod	8.47 to 11.07	8.44 to 12.32	-15.46 to 14.24	GM-3 x SM-49	GM-3 x Pusa bold-1
Harvest index	31.63 to 55.43	21.60 to 54.83	-56.95 to 12.45	GM-3 x SM-56	GM-3 x SM-52
100-seed weight	4.23t o 6.13	3.61 to 6.61	-26.75 to 9.06	GM-4 x Pusa bold-2	GM-3 x Pusa bold-2

Table 2 : Ten promising crosses for seed yield per plant, their mean performance and heterotic effects for different characters in Mungbean (data pooled over locations)

Hybrids		Seed yield per plant	Branches per plant	Pods per cluster	Clusters per plant	Pods per plant	Pod length	Seeds per pod	Harvest index	100-seed weight
VRMGg-1 x SM-56	A	14.24	4.10	5.39	8.44	41.09	6.64	8.60	54.66	5.13
	B	36.75**	38.72**	28.31**	34.51**	76.68**	-9.12**	-5.03*	7.04**	-7.79**
VRMGg-1 x SM-52	A	14.18	3.96	3.54	8.43	39.42	6.50	8.97	48.71	4.96
	B	58.12**	28.99**	-12.26**	41.34**	68.39**	5.22**	1.77	0.67**	-8.60**
GM-3 x SM-56	A	13.61	3.31	5.17	7.97	36.91	6.34	8.34	54.83	5.01
	B	30.61**	12.03**	23.15**	26.90**	58.70**	-13.22**	-11.65**	7.39**	-9.99**
GM-4 x SM-56	A	13.28	3.36	5.52	8.87	35.54	6.48	8.62	50.02	5.60
	B	27.52**	13.53**	31.48**	41.24**	52.84**	-5.36**	-9.98**	-2.04**	0.70
GM-4 x Pusa-9632	A	12.94	2.00	5.13	7.63	32.38	6.67	8.74	47.87	5.56
	B	39.80**	-20.00**	18.16**	41.65**	63.05**	-1.32**	-8.70**	-8.43**	1.32
GM-3 x Pusa-9632	A	12.85	2.89	5.75	7.78	36.56	6.78	8.90	47.41	5.78
	B	38.84**	23.81**	32.35**	44.33**	84.06**	0.33	-5.76**	-9.32**	5.36**
VRMGg-1 x Pusa-9632	A	12.82	3.47	4.54	7.73	36.23	6.83	8.67	54.72	4.95
	B	38.54**	23.81**	4.48	43.51**	82.41**	1.15	-1.64	4.67**	-9.82**
GM-4 x Pusa Bold-1	A	12.81	3.56	5.16	7.91	31.90	6.90	10.83	52.58	6.22
	B	40.66**	20.75**	28.89**	36.40**	51.34**	-8.27**	12.85**	-5.15**	3.32*
VRMGg-1 x	A	12.53	3.83	4.44	8.26	33.32	6.42	10.93	52.02	5.43

Pusa Bold-1	B	37.48**	30.19**	10.19**	42.34**	64.36**	-14.62**	13.89**	-6.16**	-9.78**
GM-3 x Pusa Bold-1	A	12.30	3.46	5.34	7.40	34.34	6.60	10.97	51.87	6.05
	B	34.92**	17.36**	44.88**	27.59**	69.40**	-12.26**	14.24**	-6.43**	0.46

A = Mean performance, B = Heterosis (%)*, ** Significant at 5 and 1 per cent probability level, respectively

Component wise examination of the crosses revealed that all the ten crosses depicted significant positive heterosis for clusters per plant, nine crosses for branches per plant and eight crosses for pods per cluster over their respective better parents. Thus, clusters per plant, pods per plant, branches per plant and pods per cluster were the main contributors to seed yield per plant. This clearly revealed that heterosis for seed yield per plant was due to heterosis for individual yield components or additive or synergistic effect of the component traits or alternatively due to the multiplicative effect of partial dominance of component traits [4,5,2].

There can not be any gene system for yield *per se* and the yield as an end product is due to the interaction between the different yield components [6]. In general, the top ten high yielding crosses which also manifested highly significant, positive heterosis of higher magnitude were combinations of either high x high or low x high gca combiners. Thus, it was observed that parents VRMGg-1, GM-3, GM-4, SM-52, SM-56, Pusa bold-1 and Pusa 9632 were promising parents which resulted high heterosis for seed yield and its related traits. Further, exploitation of crosses involving these parents could be more rewarding for breeding programme of mungbean.

REFERENCES

1. Aher, R. P., Sonawane, V. P. and Dahat, D. V. 2000. Heterosis in mungbean (*Vigna radiata* (L.) Wilczek). *Indian J. Agric. Res.*, **34**(2): 134-137.
2. Reddy, L. N., Reddysekhar, M, Raja Reddy, K. and Hariprasad Reddy, K. 2003. Heterosis for yield and its components in mungbean (*Vigna radiata* (L.) Wilczek). *Legume Res.*, **26**(4): 248-253.
3. Reddy S.M., Reddy, K.R. and Reddy C.R. 1993. Heterosis for yield and yield components in mungbean (*Vigna radiata* (L.) Wilczek). *Indian J. Genet.*, **54**(1): 1-5.
4. Hegde, V. S., Parmeshwarappa, R. and Goud, J. V. 1996. Heterosis and inbreeding depression in mungbean (*Vigna radiata* (L.) Wilczek). *Mysore J. Agric. Sci.*, **30**(2): 101-105.
5. Reddy, K. H. P. 1998. Heterosis in green gram (*Vigna radiata* (L.) Wilczek). *Ann. Agric. Res.*, **19**(1): 26-29.
6. Grafius, J. E. 1959. Heterosis in Barley. *Agron J.*, **51** : 515-554.

Development of new plant types in guar [*Cyamopsis tetragonoloba* (L.) Taub.] for enhancing productivity and grain quality

Gopala Krishnan, S. and N.K. Dwivedi

National Bureau of Plant Genetic Resources, Regional Station, CAZRI Campus, Jodhpur – 342 003, Rajasthan, INDIA
e-mail: gopal_icar@yahoo.co.in

ABSTRACT

Guar (*Cyamopsis tetragonoloba* L. Taub.), is a multipurpose drought tolerant summer-annual legume cultivated primarily in India and Pakistan. With increasing constrain in the production of guar, breeders continually look for new and diverse germplasm with specific traits to enhance crop productivity. New plant types with unique characters and altered growth habit will play an important role in improving the crop yield. All the cultivated guar varieties possess indeterminate growth habit, and are branched with alternate cluster bearing habit. In the present paper, we present the identification and characterization of new plant types in guar with specific traits. The new plant type guar include determinate all node bearing plant habit, and branched all node bearing habit. The determinate plant type includes three different types namely (a) leaf determinate, (b) stem determinate and (c) inflorescence determinate types, which differ for various morpho-agronomic traits. Accessions with branched as well as all node bearing habit have also been identified which can be utilized in improving guar yields. These new plant types of guar will play a major role in enhancing the productivity as well as quality of the grain produce.

Key words: Clusterbean, new plant type, quality, productivity.

INTRODUCTION

Guar [*Cyamopsis tetragonoloba* L. Taub.] is an annual drought hardy leguminous crop cultivated mainly in semi-arid and arid tracts of India and Pakistan for the galactomannan gum extracted from its endosperm. The guar gum is widely used in textile, paper, petroleum, mining, cosmetics and food industries [1]. The success of any crop breeding program primarily depends upon the availability of genetic variability. A wide array of variability for different agro-morphological characters has been reported in guar [2, 3, 4]. Genetic stocks for several important agro-morphological traits have been identified in clusterbean which can be utilized in the breeding programmes to suit different needs [4]. In the present study, we report the identification, characterization and development of new plant types, which would play an important role in guar improvement.

Conventional plant types in guar

The conventional cultivars in guar are typically indeterminate with respect to growth habit and are either branched or unbranched. The unbranched plant types produce a single, erect primary axis and are devoid of any secondary axis. The cluster bearing habit of the unbranched types vary from continuous/ all node to alternate bearing. On the other hand, the branched types is classified into erect branching (plant produces one or more branches), basal branching (plant produces branches at the base of the primary axis), and branching (branching throughout the primary axis) [5]. With respect to cluster bearing habit, the branched type produces clusters in alternate/ irregular manner.

The indeterminate growth habit of guar causes continual flowering and results in heterogeneous pod maturation, thus making it difficult to decide when to harvest the seeds. Flowering can continue upto 7 months if the environmental conditions are favourable for plant growth, thereby affecting the quality of the grain. An estimate under experimental field conditions shows that immature and black seeds in the grain produce from the indeterminate cultivars can be as high as 16%. Table 1 shows the number of green and immature pods, immature and black seeds produced from a typical indeterminate plant.

Table 1. Pod maturity in a plant type with indeterminate growth habit.

Portion of the plant	No. of clusters	No. of pods		No. of grains		Proportion of ill filled grains (%)
		Mature	Immature	Filled	Ill filled	
Top 1/3rd	7	18	1	0	41	100.0
Middle 1/3rd	6	53	3	219	70	24.2
Bottom 1/3rd	9	98	0	497	27	5.2

It is clear from the table that majority of the immature and black seeds are contributed by the pods produced the top 1/3rd of the indeterminate plant. These pods are green with half filled/ immature seeds resulting in 100% immature/ black seeds from the uppermost portion of the plant. This proportion gets reduced in the middle 1/3rd and is negligible in the lower most 1/3rd of the plant. This is due to sequential flowering nature of the guar plant.

Restructuring the plant type

Restructuring of plant type has played a key role in enhancing the productivity of wheat and rice. Jain [6] suggested that major gains in yield could be achieved by developing a plant type with high harvest index, response to increased plant population per unit area, and early maturity. He further suggested that improvement in harvest index would likely to be associated with determinate, erect and compact growth habit. This plant type would intercept more sunlight and permit large population per unit area. Sedgley et al. [7] also emphasized that plant type suited for for high input environments should have erect growth habit and limited branching.

MATERIALS AND METHODS

During the germplasm evaluation in kharif 2006, determinate plants produced by spontaneous mutations were identified in the guar germplasm. A total of 91 single plant selections were made from 27 germp-lasm accessions. These single plant selections were grown during summer 2007 to study the stability for determinate growth habit and also for studying reaction to photosensitivity. During this characterization trial, some spontan-eous mutants with branched as well as all node cluster bearing habit were identified and were characterized during kharif 2007. The stable determinate lines were then characterized during kharif 2007.

The characterization of the 91 germplasm accessions were carried out in augmented randomized block design with 3 checks namely GDM-1, an unbranched, alternate cluster bearing, determinate genetic stock from CAZRI, Jodhpur; PLG-85, an unbranched, alternate cluster bearing indeterminate cultivar; RGC-197, an unbranched, alternate cluster bearing, indeterminate cultivar. The characterization trial was carried out during summer 2007 with supplementary irrigation and repeated under rainfed condition during kharif 2007. Five rows per accession were raised in 2 m long rows at a spacing of 45 x 15 cm. The crop was raised following recommended agronomic practices. The accessions were characterized for different traits as per the standard descriptor [8]. The statistical analysis of the characterization data was carried out as per standard procedure for computing mean, range, and co-efficient of variation (both phenotypic and genotypic).

RESULTS AND DISCUSSION

With respect to flowering habit, guar shows two distinct types namely determinate/indeterminate types. In determinate plant type, the stem terminates into inflorescence whereas the indeterminate plant type exhibits continuous apical growth. Most of the germplasm collections possess indeterminate growth habit and all the varieties released till date in India are indeterminate [4]. Determinate and spreading variants derived from mutagenic treatment with Kitazin and Saturn has been reported by Rao *et al.* [9]. They found that determinate guar was single stemmed, dwarf with broad and smooth leaves while the spreading type was very short and greatly reduced in plant height (9.6 cm). The dominance of indeterminate growth habit over determinate growth is also reported.

In our characterization study involving the 91 determinate germplasm lines, we found that the determinate character showed variations in termination of the apical growth. The determinate guar was further differentiated into three types on the basis of the stem termination namely (i) Guar Leaf Determinate Mutant (GLDM) – in which the stem terminates into leaf, (ii) Guar Stem Determinate Mutant (GSDM) – in which the stem terminates into an racemous inflorescence and (iii) Guar Inflorescence Determinate Mutant (GIDM) – in which the stem terminates into a cluster of inflorescence (Figure 1).

Figure 1. Different types of determinate mutants identified in guar.

Out of the 91 single plant selections identified, only 89 selections showed true to type determinate character while two selections were indeterminate in nature. Among the stable determinate mutants characterized, three germplasm selections were leaf determinate and the remaining 86 selections were stem determinate in nature. The inflorescence determinate mutants were identified during summer 2007 and eighteen single plant selections were identified as inflorescence determinate in nature. But only 10 plant progenies, among these were found to be stable for inflorescence determinate character. The leaf determinate mutant in guar was more vigorous and produced broader leaves, stouter stem and possessed alternate cluster bearing tendency in comparison with that of stem determinate and inflorescence determinate mutants. The inflorescence determinate plant types had good cluster bearing ability.

GLDM-2 (IC11380/P_1-3) was found to be very superior for cluster length (12.8 cm) and seed yield among the leaf determinate accessions and it produced pink seeds in comparision with GDM-1 which produces gray seeds. A wide range of variability was observed for various agro-morphological traits among the determinate plant types during kharif 2007 (Table 2).

Table 2. Variability for agro-morphological traits in determinate guar germplasm characterized during kharif 2007.

S. No.	Character	Mean	Range		C.V. (%) - Phenotypic	C. V. (%) - Error
			Min.	Max.		
1.	Plant height (cm)	72.4	48.7	99.4	10.8	12.3
2.	Clusters/ plant	17.4	6.0	30.2	3.8	5.0
3.	Cluster length (cm)	8.7	5.3	15.7	1.9	18.3
4.	Pods/ cluster	8.9	3.7	15.4	2.2	18.3
5.	Pods/ plant	105.6	36.2	204.8	34.6	11.6
6.	Pod length (cm)	5.6	4.4	6.9	0.4	4.3
7.	Seeds/ pod	7.8	5.7	9.5	0.7	7.4
8.	Seed yield/ plant (g)	9.3	1.5	21.1	4.1	20.4
9.	100 seed weight (g)	1.95	1.54	2.74	0.20	8.49

Among the stem determinate germplasm accessions characterized GSDM-46 (IC421811-5) was found to be superior for yield. The stem determinate guar accessions were all unbranched, erect, all node cluster bearing, early maturing and uniform at maturity. However variability was observed in plant height, length of the terminal inflorescence and other morpho-agronomic traits in the determinate accessions (Table 2). All the determinate accessions tested during summer 2007 exhibited photo-insensitiveness.

GSDM 46 (IC421811-5) – A unique determinate genotype

GSDM 46 (IC421811-5) is an early maturing high yielding determinate guar accession with the following desirable attributes

Important features

Photoinsensitive; early termination of stem; early and uniform maturity

All node cluster bearing habit; average no. of clusters / plant – 18.8; mean cluster length – 9.11 cm

Average no. of pods/ cluster – 8.7; no. of pods/ plant – 144.4

Mean seed yield/ plant – 17.74 g

GSDM 46 (IC421811-5) – An early maturing high yielding determinate guar accession.

Advantages of determinate growth habit

These new plant types of guar with unbranched, erect, all node cluster bearing ability, photo-insensitiveness and determinate growth habit has several advantages over the indeterminate cultivars. The important advantages of these new plant types are:

- The determinate growth habit curbs excessive vegetative growth and reduce the risk of lodging.
- The identification of new germplasm with determinate growth habit will allow guar to be developed as a crop that is well suited to intensive agronomic-management conditions.
- The single stemmed plant type combined with uniform maturity of the determinate guar genotype will suit the for modern agriculture practices such as mechanized harvesting using combiners.
- The early and uniform maturity of determinate guar accessions will be a highly valuable in crop rotation, thereby expanding the crop niche.
- The photo-insensitiveness of the determinate accessions can be utilized in breeding cultivars suited for raising summer crop of guar with supplementary irrigation under north Indian conditions.
- The single stemmed determinate types are more uniform in maturity and thereby help in reducing the problem of black seeds, which reduces the quality of the seeds.
- The superior performing single stemmed all node bearing determinate guar lines can be utilized in breeding genotypes suited for mixed/ intercropping.

Branched and All Node Cluster bearing Guar

A wide range of variability has been reported for several morphological characters in guar [3, 4]. But there have not been a single report on any type of branching behaviour [5] producing clusters in every node (branched as well as all node cluster bearing habit in a single genotype).

The crop geometry in guar combined with the genetic nature of these traits probably did not permit the co-existence of these two traits. In the branched plant types namely erect branching (EB) and branching (B) described by Hymowitz [5], the production of primary branches from the nodes on the main stem does not allow all node cluster bearing habit as only either the branch or the clusters can be produced from a single node. In case of basal branching (BB), the traits in question namely basal branching [10] and all node cluster bearing are recessive characters [11]. Hence, recombination breeding for combining these two characters is not only difficult but has never been attempted previously. For the first time, we report the identification and characterization of branched all node cluster bearing guar genotype.

BANG 3 – A unique guar genotype

BANG 3 (IC421844-4/P_1-3) is a superior high yielding branched as well as all node cluster bearing guar genotype.

Important features

Basal branching habit; photoinsensitive

Three primary branches and three secondary branches

All node cluster bearing habit

Average no. of clusters in main stem – 11

Total no. of clusters/ pl. – 23.8

No. of pods/ pl. – 83.3

Average seed yield/ pl. – 8.1 g

BANG 3 (IC421844-4/P_1-3) - a unique good performing branched all node bearing guar accession.

Plant progenies from the single plant identified from the germplasm IC421844-4/P_1 (with basal branching as well as all node bearing habit) were grown and further selection was exercised in the subsequent generation. This has led to the identification of four stable branched genotypes named as BANG (Branched All Node bearing Guar).

BANG 4 – A unique guar genotype

BANG 4 (IC421844-4/P_1-4) is a superior high yielding branched as well as all node cluster bearing guar genotype.

Important features

Basal branching habit; photoinsensitive

Four primary branches and four secondary branches

All node cluster bearing habit

Average no. of clusters in main stem – 15

Total no. of clusters/ pl. – 63

No. of pods/ pl. – 188.2

Average seed yield/ pl. – 12.2 g

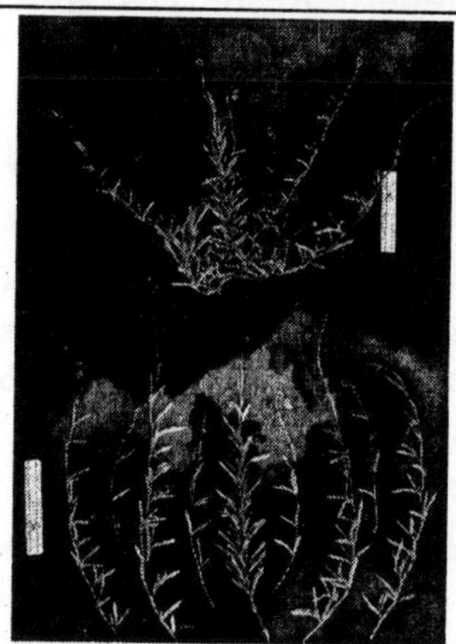

BANG 4 (IC421844-4/P_1-4) – a superior Branched All Node bearing Guar.

For highest seed yield, Paroda and Saini [12] recommended that the guar plant should have early maturity, determinate growth type, optimum plant height and branching number, and high numbers of clusters per plant with each cluster having the maximum number of pods. A new plant type in guar with optimum basal branching, all node bearing habit improves the number of clusters produced per plant and the early maturity, uniform plant maturity complemented by determinate habit will not only improve the per hectare productivity but also help in maintaining better grain quality.

Conclusion

The determinate growth habit in guar provides several advantages over its indeterminate counterpart. A number of crops such as cowpea, faba bean, lupins, pigeon pea and soybean have gained significant advantage with determinate growth habit. The identification of new germplasm possessing unbranched, erect, all node cluster bearing ability, photo-insensitiveness and determinate growth habit will allow guar to be developed as a modern crop that is well suited to intensive agronomic management conditions. The branched, determinate, all node bearing types has higher yielding capacity which has the potential for improving productivity in guar. In the near future, the new plant types of guar described here is expected to play a major role in enhancing the productivity as well as quality of the grain produce.

Acknowledgement

The authors are thankful to Dr. S. K. Sharma, Director, National Bureau of Plant Genetic Resources, New Delhi for providing all the necessary facilities and his constant encouragement. The technical assistance of Sh. Kheta Ram and Sh. Bhatta Ram are also thankfully acknowledged.

REFERENCES

1. Whistler, R. L. and T. Hymowitz. 1979. Breeding, genetics and physiology. In Guar: Agronomy, production, industrial use and nutrition. Purdue University Press, W. Lafayette, IN, USA. pp. 35-56.
2. Dabas, B. S. 1975. Studies on inheritance of quantitative characters and gum content in guar [*Cyamopsis tetragonoloba* (L.) Taub.]. Unpublished Ph. D. Thesis, I.A.R.I., New Delhi, India.
3. Mittal, S. P., Dabas, B. S., Thomas, T. A. and Chopra, D. P. 1977. Assessment of genetic stocks of guar [*Cyamopsis tetragonoloba* (L.) Taub.]. Proceedings of the First ICAR Guar research workshop, CAZRI, Jodhpur. Pp. 58 – 68.

4. Gopala Krishnan, S. and N. K. Dwivedi. 2007. Genetic stocks of clusterbean [*Cyamopsis tetragonoloba* (L.) Taub.] – assessment and utilization. Journal of Arid Legumes 4(2): 121 – 126.

5. Hymowitz, T. 1963. Studies on guar-*Cyamopsis tetragonoloba* (L.) Taub. Ph.D. diss. Oklahoma State Univ., Stillwater, OK 74078, USA, pp. 87-91.

6. Jain, H. K. 1975. Breeding for yield and other attributes in grain legumes. Indian Journal of Genetics and Plant Breeding 35:169–187

7. Sedgley, R. H., Siddique, K. H. M. and Walton, G. H. 1990. Chickpea ideotypes for Mediterranean environments. In: van Rheenen HA, Saxena MC, Walby BJ, Hall SD (eds) Chickpea in the nineties: proceedings of the second international workshop on chickpea improvement. ICRISAT, Patancheru, AP, India, pp 87-91.

8. Mahajan, R. K., Sapra, R. L., Srivastava, U., Singh, Mahendra and Sharma, G. D. 2000. Minimal descriptors (for. Characterization and Evaluation) of Agri-Horticultural crops (Part-I). National Bureau of Plant Genetic Resources, New Delhi. P 1-230.

9. Rao, R. M., Murthy, P. K. and Rao, D. 1982. Note on determinate and spreading variants in cluster bean. Current Science 51(19): 945-946.

10. Liu, W., Hou, A., Peffley, E. B., Auld, D. L. and Powell, R. J. 2006. The inheritance of a basal branching type in guar. Euphytica 151: 303-309.

11. Sohoo, M.S. and Gill, K. S. 1976. Inheritance of seven qualitative characters in clusterbean [*Cyamopsis tetragonoloba* (L.) Taub.]. Genetica Agraria 30: 185-190.

12. Paroda, R. S. and M. L. Saini. 1978. Guar breeding. Forage Research, 4: 9-39.

Accelerating production of horsegram through adoption of HYVs in NW himalayan hills: *opportunities and constraints*

S.R.K. Singh[1], Gyanendra Singh[2], Vinay Mahajan[3], K.S. Hooda[4], Narender Kumar[5] and H.S. Gupta[6]

1. Present Address: Sr. Scientist (Ag Extension), Zonal Co-ordinating Unit for TOT, Zone-VII, Jabalpur
2. Sr. Scientist (Plant Breeding), Vivekananda Parvatiya Krishi Anusandhan Sansthan, Almora -263 601 (UK)
3. Head & Pri. Sci. (CID), VPKAS, Almora
4. Sr. Scientist (Plant Pathology), V.P.K.A.S., Almora
5. Present Address: Sr Scientist, IIPR, Kanpur
6. Director, VPKAS (ICAR), Almora- 263 601

ABSTRACT

The present study was conducted during *kharif* 2007 to find out the reasons for gaps between potential and realised yield in horsgram. Nine improved horsegram production technologies (IHPT) were tested in six villages for studying the extent of adoption and constraints. Results showed majority of the farmers (68.33%) were medium level adopters. Among technologies, adoption of organic manure was highest (65.0%) followed by plant poulation (46.67%), appropriate seed rate (40.0%) and line sowing (37.50%), etc. Major constraints in adoption were non-availability of quality seeds, lack of motivation for new technology, and leaning towards subsistence farming. Suitable strategies need to be worked out to make farmers high adopters of the technologies. Besides, factors responsible for low adoption need to be addressed for sustaining adoption of horsegram production technologies. This will help in enhancing the horsegram production in the region, which would, in turn, impart nutritional security and improve health of the hill farmers.

INTRODUCTION

Horsegram (*Macrotyloma uniflorum* [Lam.] Verde), commonly known by different names in different languages, viz., Kulaththa (Sanskrit), Kulthi, Gahat (Hindi), Kurtikalai (Bengali), Kulthi (Marathi), Kalathi, Kulit (Gujrati), Muthira (Malyalam), Ulavalu (Telgu), Lollu (Tamil) (1). It is branched, trailing or sub-erect annual pulse and is an important pulse crop of the Uttarakhand hills. During 2006-07 in Uttarakhand hills, about 7689 metric tones of horsegram was produced from 10890 hectares of land with the average productivity of 706 kg/ha (2). There exists a substantial yield gaps (55-62%) between yield at research farm and farmers' fields.

Horsegram as traditional crop occupies a distinct position and well defined role in the food and nutritional security of tribal, backward and hilly areas because it provides access to sufficient food for livelihood at household level and medicinal use. The primary limitation in pulses for low yields are indeterminate growth habit in local cultivars, limited source supply to economic sink leading to flower dropping to poor pod setting (3, 4). The institute had developed 4 varieties in horsegram (VL *Gahat* 1, VL *Gahat* 8, VL *Gahat* 10 and VL *Gahat* 15) are quite popular among farmers of hilly areas.

Several studies indicated that the adoption of improved technologies gives high yields and income to the farmers. Higher horsegram production can be achieved by adoption of improved technologies by the large number of farmers (5). It has been observed that improved technologies are not accepted by the farmers at a time and also to full extent. In this context, the study was conducted with the objectives to enumerate inquisitiveness of the farmers, to measure the extent of adoption of IHPT and investigate the relationship between socio-economic characteristics with the adoption and delineate the constraints perceived by the farmers.

METHODOLOGY

Locale of the study: This study was conducted during *Kharif* 2007 in the institute-adopted villages' viz., Dhaspad, Palauli, Kafli (Almora district), Satkhol, Darim and Pukhrad (Nainital district) of Uttarakhand. These villages were adopted under institute extension programmes in which the improved technologies developed by the institute were provided to the adopted farmers for their acceptance and adoption. The study area represents rainfed growing conditions.

Data collection and analysis: An extensive survey was conducted in the adopted villages. A sample of 120 farmers was drawn from the adopted villages using proportionate random sampling technique. Based on the expert's opinion, nine IHPT were selected for studying the extent of adoption and perceived constraints. All the selected farmers were interviewed personally using a well-structured interview schedule. An adoption index (Obtained score*100/ Max. attainable score) was developed on the basis of which the respondents were categorized into low, medium and high level adopters. For analysis of collected data, descriptive statistics (frequency and percentage) was used.

FINDINGS AND DISCUSSION

1. Socio-economic profile of the sampled farmers

In order to know the background and socio-economic status of the respondents, it is important to analyze these characteristics. In all 10 socio-economic variables were studied. The analyzed data showed that majority (68.00%) of the respondents were from middle-age group (31-50 years). The maximum proportions of the respondents (48.59%) were educated upto Xth class, whereas 35.11 per cent were Undergraduate and above and only 16.30 per cent were illiterate. Majority (62.00%) of the horsegram growers were operating in smaller holding (1.10 to 2.4 acres) and 28.00 percent of them in marginal holdings (upto 1.0 acres). Majority of the respondents (69.00%) had medium level of farming experiences (3-5 years). The maximum proportions of the respondents (49.30%) were of middle-income group (Rs. 9100 to 20000) followed by 35.42 percent low-income group (< Rs. 9100) and 15.28 percent high-income group (> Rs. 20000). About three-fourth (76.51%) of the respondents had medium level of economic motivation. A large majority (94.35%) of the respondents was participated in one organization and 59.00 percent were members in more than two organization. Regarding extension contacts, majority (65.89%) of the respondents had medium level of extension contacts. Majority (59.62%) of the respondents had medium exposure of mass media. Majority of the respondents (58.75%) possessed two to four farm implements, followed by 30.00 percent (above four farm implements).

2. Correlates of adoption

With the assumption that adoption of the IHPT may have relationship with various socio-economic characteristics of the respondents, Pearson product moment correlation coefficient 'r' was computed and compared. The results are presented in Table 1.

Data depicted in Table 1 indicates that correlation coefficients of seven variables viz., age, education, land holding, farming experiences, annual income, extension contacts and mass media exposure were significant at 0.01 level of probability with adoption of IHPT. Among these, only age was found negatively correlated which infers that young farmers had higher adoption score than old-aged growers. It is clear from the above findings that there were significant relationship between majority of the socio-economic characteristics of respondents and the extent of adoption of IRPT. The variables showing positive and significant relationship need greater attention on the part of extension agency to enhance the adoption of IHPT in the area.

Table 1: Correlates of adoption of IHPT (N = 120).

Variables	Correlates	Values			Coefficient of correlation (r)
		Mean	S.D.	Range	
X_1	Age (yrs)	39.12	9.24	24-56	-0.604**
X_2	Education (years of schooling)	2.73	0.76	2-4	0.550**
X_3	Land holding (acres)	33.67	14.82	14-78	0.435**
X_4	Farming experiences (yrs.)	4.68	1.78	2-9	0.628**
X_5	Annual income (Rs. in 000)	14.67	6.18	5-36	0.432**
X_6	Economic motivation	2.07	0.76	1-3	0.007^{NS}
X_7	Social participation	0.87	0.90	0-3	0.040^{NS}
X_8	Extension contacts	1.95	0.75	1-3	0.608**
X_9	Mass media exposure	2.15	0.67	1-3	0.525**
X_{10}	Farm implements	3.27	1.37	1-6	0.078^{NS}

3. Overall adoption of IHPT

The extent of adoption in respect of IHPT was studied by adding individual score received on different practices. On the basis of the total score, they were categorized into three categories and the results are presented in Table 2 & Fig 1. The data shows that majority of the respondents (68.33%) were had medium level of adoption of IHPT. Twelve and twenty percent of the respondents had high and low adoption level. Results suggest a need for greater extension effort to provide know-how of the IHPT to the respondents so that their adoption level would be enhanced. Moreover, it was also observed that farmers with more resources and extension contacts were able to adopt more technologies than other farmers.

Table 2: Distribution of the respondents on the basis of adoption of IHPT (N = 120).

Adoption Level	Frequency	Percentage
Low (< 4.0)	24	20.0
Medium (4.0 – 7.0)	82	68.33
High (> 8.0)	14	11.67

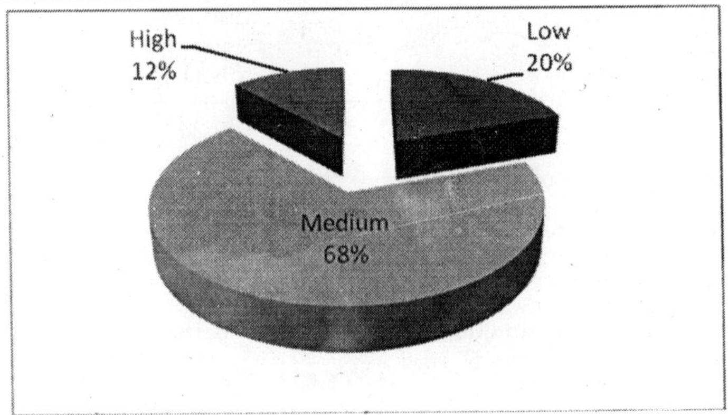

Fig 1: Adoption of IHPT in the sampled farmers.

4. Technology-wise adoption of IHPT

In order to ascertain extent of adoption of the IHPT, the responses of selected farmers were collected on nine selected technologies (Table 3).

Table 3: Distribution of respondents on the basis of technology-wise adoption (N = 120).

S No.	Recommended technology	Frequency	Percent	Rank
1.	High yielding varieties	32	26.67	VII
2.	Seed rate	48	40.0	III
3.	Line sowing	45	37.50	IV
4.	Seed treatment with fungicides	24	20.0	X
5.	Plant population	56	46.67	II
6.	Weedicide application	37	30.83	VI
7.	Organic manure application	78	65.0	I
8.	Fertilizer application	41	34.16	V
9.	Plant protection measures	35	29.16	VII

1. **Organic manure application:** Majority of the farmers (65.0%) adopted organic manure application for their horsegram cultivation. Limited supply of chemical fertilizers and easy availability of organic manure to the farmers were the major reasons for high rate of adoption.
2. **Plant population:** Empirical evidences showed that plant population plays important role in the high yields of the horsegram. Results showed that around 47.0 percent of the farmers have adopted the practice. This attracts greater attention of the scientists, as there is ample scope for enhancing the extent of adoption of this technology/practice.
3. **Appropriate seed rate:** It was adopted by 40.0 percent of the farmers in the study area. Besides, some considerable proportions of the farmers were also convinced with the technology but to minimize risk of re-sowing and fear of poor germination due to low soil moisture, they use 1.5-2 times higher quantity seeds than recommended rate i.e 20-25 kg/ha.
4. **Line sowing:** Thirty seven percent of the farmers adopted the line sowing (line spacing - 45 cm) practice despite they perceived it more labour intensive and time consuming. But they are convinced with the benefits of line sowing specially in weed and pest management. Rest proportions of the farmers were either partial or non-adopters of the technology due to their poor economic conditions and scanty labour and farm machinery availability to them.
5. **Fertilizer application**: During survey it was observed that pulse crops were being grown either as a border crop or marginal lands. Due to this, farmers are not much concerned with the fertilizer application to the crop. Results showed that only 34.16 percent of the farmers adopted recommended dose (20:40:20 N:P:K kg/ha).

Besides, farmers have adopted the other recommended technologies also, but their proportion is still low ($\leq$ 30.0%), and is a matter of great concern for the scientists. Suitable strategies could be worked out for further motivating the farmers for adopting the other IHPTs for enhancing horsegram production and productivity in the region. This in turn will lead to more protein availability to the hill farmers.

5. Constraints in the adoption of IHPT

Constraints are the hindrances in the adoption of the recommended improved technologies at the farmers' fields (Table 4).

Table 4: Distribution of the respondents on the basis of the constraints in the adoption of IHPT (N=120).

S No	Constraint	Frequency	Percent	Rank
1.	Non-availability of high yielding variety seeds	82.0	68.33	I
2.	Lack of motivation for new technologies	75.0	62.25	II
3.	Leaning towards subsistence farming	69.0	57.50	III
4.	Poor investment capacity	51.0	42.50	V
5.	Lack of awareness of new technologies	58.0	48.33	IV

1. **Non-availability of high yielding variety seeds:** This was the most important constraint reported by 68.33 percent of the respondents and ranked first. High yielding variety seeds were not available timely and adequately. Though VPKAS has made arrangement for seed availability through seed sale counter and farmer fairs, but still our reach is limited.
2. **Lack of motivation for new technologies:** Majority (62.25%) expressed that they are dubious in using new technology as many times they could not harvest the expected yield and benefits which discouraged them. Due to being complex, diversified and risk-prone agriculture, hill farmers are much concerned with their livelihood than profits.
3. **Leaning towards subsistence farming:** It has been observed that majority of the farmers are engaged in the subsistence farming rather commercial farming. This is due to the rainfed growing situation coupled with small and scattered land. Such situation compels hill farmers for subsistence farming which works barrier to the adoption of most of the improved technologies along with horsegram.
4. **Lack of awareness of the new technology:** This works as constraints in the adoption of most of the improved technologies. We know awareness happened to be the initial steps in the adoption process. Unless hill farmers will be aware with the IHPT, there will be no desired rate of adoption of the same.
5. **Poor investment capacity:** Due to existence of the above situation, hill farmers having poor investment capacity. This in turn affects the rate of adoption of the improved technologies in general and IHPT in particular.

CONCLUSION AND IMPLICATIONS

On the basis of afore-said discussions, though farmers are known with the importance of the improved technologies of the horsegram but due to rainfed farming, they are unable to realize it properly. Majority of the farmers are medium adopters which could be converted into high adopters. The constraints faced by the hill farmers should be taken care by the extension workers functional in hills so that farmers could easily adopt the IHPT and can enhance their income as well as health also.

The findings of this study are useful for the pulse breeders working in hills. Besides, extension worker and change agents can take care of the factors affecting adoption and can try for reducing the constraints faced by the farmers. Moreover, higher adoption of the horsegram technology will enhance the production, which in turn will lead to the sustainable livelihood and improved quality of life of the rural hill farmers.

REFERENCES

1. Shukla S K, V Mahajan and H S Gupta 2006. Horsegram in Uttaranchal – Status, Constraints and Prospects Chapter 7, pp 99-114. In "Horsegram in India Scientific Publishers (India), PO Box 91, Jodhpur, India.
2. Anonymous (2007) Directorate of Agriculture, Uttarakahand.
3. Mahajan, Vinay and Gupta H S 2006a Crop Improvement in Hill Agriculture: Approaches, Status and Prospects, Chapter 5 pp 83-99. In "Sustainable Production from Agricultural Watersheds in North Western Himalaya" (Editors: HS Gupta, AK Srivastava and JC Bhat) VPKAS Almora, Uttaranchal, Published by Vivekanada Parvatiya Krishi Anusandhan Sansthan, Almora Pp 342.
4. Mahajan, Vinay and Shukla S K 2006b Approaches to Improvement in Pulses and Oilseed. Chapter 10 pp 152-162. In "Sustainable Production from Agricultural Watersheds in North Western Himalaya" (Editors: HS Gupta, AK Srivastava and JC Bhat) VPKAS Almora, Uttaranchal, Published by Vivekanada Parvatiya Krishi Anusandhan Sansthan, Almora Pp342.
5. Thyagarajan, S. (2004). Horsegram production technology-adoption and constraints. Indian Journal of Extension Education, 40 (3&4): 44-47.

Leaf shape in mungbean [*Vigna radiata* (L.) Wilczek] – variation in leaflet lobation pattern

Gopala Krishnan, S.* and N. K. Dwivedi

National Bureau of Plant Genetic Resources, Regional Station, CAZRI Campus, Jodhpur - 342 003, Rajasthan, India

*For correspondence. e-mail: gopal_icar@yahoo.co.in

ABSTRACT

Flowering plants display a striking variation in leaf size and shape. A large number of mutations affect leaf size and shape without altering pattern. Mungbean [*Vigna radiata* (L.) Wilczek] produces leaves that are compound, imparipinnate and trifoliate with long petioles. The leaflets are generally ovate with entire margin. Both spontaneous and induced mutations affecting leaflet shape in mungbean has been reported. However, the variation in leaflet lobation pattern of mungbean has not been described in these studies. In the present paper, we report stable genotypes of mungbean varying in degree of lobation and their advantage for crop plants.

INTRODUCTION

Flowering plants display a striking variation in leaf size and shape. Leaves vary widely in linear dimensions, from millimeters to nearly 1 m; they also vary in shape, from simple leaf with entire, serrated or deeply lobed margins to compound leaves with pinnately or palmately arranged leaflets. Besides, the inherent control by genes, leaf shape in plants is also affected by factors such as plant ontogeny [1], gibberellins [2] and changes in environmental conditions namely light quality and quantity [3], air temperature and CO_2 levels [4], a phenomenon known as environmentally induced heterophylly [5]. A large number of mutations affect leaf size and shape without altering pattern. Normally, mungbean [*Vigna radiata* (L.) Wilczek] produces leaves that are compound, imparipinnate and trifoliate with long petioles. The leaflets are generally ovate with entire margin. Both spontaneous and induced mutations affecting leaflet shape in mungbean have been reported [6, 7, 8]. Based on leaflet lobation pattern, plants of *Vigna radiata* var. *sublobata* were categorized as entire to very sharply lobed [8]. However, such a variation in leaflet lobation pattern of mungbean has not been described. Here, we report the identification of spontaneous mutations producing trilobed leaflets in mungbean and development of stable genotypes varying in degree of lobation. A method for categorizing the lobation pattern in leaflets based on the degree of lobation is also proposed in order to classify the leaflets into deeply lobed, shallow lobed and unlobed categories.

Table 1. Characteristic features of three categories varying for leaflet shape in mungbean.

Accession Number	Shape of the leaflet	Leaflet length (cm)	Leaflet width (cm)	Terminal lobe length (cm)	Terminal lobe width (cm)	Degree of lobation	Terminal lobe width/ leaflet width
IC507341-1	Deeply lobed	13.04	10.24	9.01	2.91	0.69	0.28
IC507341-3	Shallow lobed	12.88	10.77	7.15	4.90	0.56	0.46
PDM-54	Unlobed	9.84	8.94	0.00	0.00	0.00	0.00

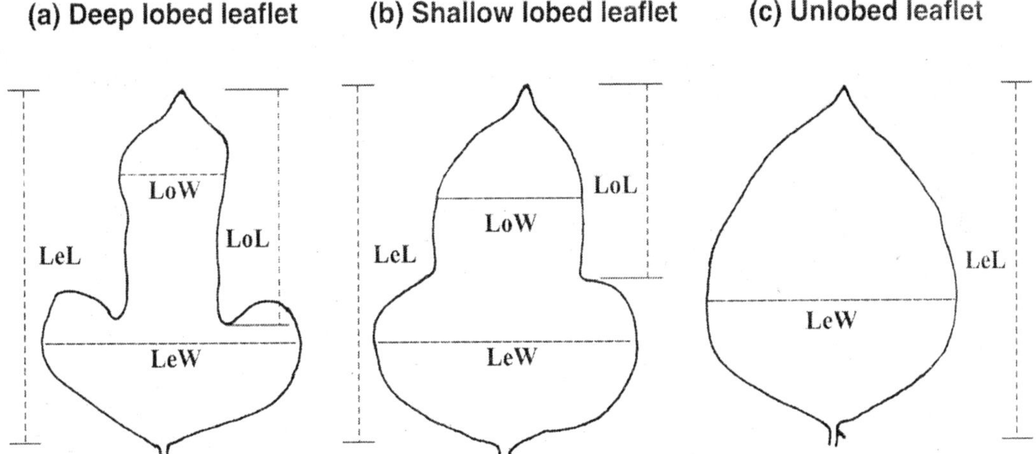

Where LeL - Leaf length; LeW - Leaf width; LoL - Lobe length; LoW - Lobe width

Figure 1. Measurement of leaflet parameters in mungbean leaflets

MATERIALS AND METHODS

A mungbean accession IC507341 (Collector number - PLM 233) was observed to produce trilobed leaflets during kharif 2006. Significant variation was observed for leaflet lobation pattern among the plants within this accession. Twenty three plants of this accession were harvested on single plant basis in order to ascertain stable inheritance and purify the lines based on lobation pattern. The single plant progenies were grown during kharif 2007 in two replications with PDM-54, a mungbean variety with entire leaflets as a check. Measurements on various leaflet characters such as leaflet length, leaflet width, terminal lobe length and terminal lobe width (Fig. 1) were made on the terminal leaflets of the first and fourth compound leaf from the apex of a plant. The measurement from the fourth compound leaf from apex was found to be consistent and observations were recorded on 10 plants per line which was subjected to further analysis. The degree of lobation was calculated as the ratio of lobe length to leaflet length.

RESULTS AND DISCUSSIONS

Based on the measurements made on the leaflet characters, the lines with trilobed leaflets were classified into two categories namely deep lobed and shallow lobed. The mean data for various leaflet characters of the accessions namely leaflet length, width; terminal lobe length, width; degree of lobation and the ratio of terminal lobe width to the leaflet width are presented in Table 1. Most of the lines produced deep lobed leaves and the lines with deep lobed leaflets were assigned number, IC507341-1. They produced deep lobed leaflets with small lateral lobes and a long terminal lobe with a mean lobe length of 9.01 cm compared to the leaflet length of 13.04 cm (Figure 2a). The degree of lobation was 0.69, while lobe width/ leaflet width ratio was only 0.28, reflecting the long and narrow nature of the terminal lobes in the leaflets of this accession. Only two lines produced shallow lobed leaflets with broad terminal lobes causing obscure lobation pattern and were assigned number, IC507341-3 (Figure 2b). The degree of lobation was only 0.56 compared to 0.69 in accession with deeply lobed leaflets. The plants belonging to this type was highly vigorous and the terminal lobe was very broad and therefore, has a higher lobe width/ leaflet width ratio (0.46). Some plants with shallow lobed leaflets also showed a delay in the expression of trilobed leaflet character in the leaves till the fourth trifoliate leaf. These plants with delayed expression of trilobed leaflets has been purified and used in further studies. The check, PDM-54 produced entire leaflets which were unlobed in nature (Figure 2c).

The inheritance of lobed leaflets in mungbean has been reported to be governed by a single dominant gene, *E* [6], which has been confirmed by various other workers [7, 10, 11]. However, inheritance of this trait by single gene with incomplete dominance [12] and two dominant genes [13] have also been reported. Observation on the tight linkage between purple hypocotyl and trilobate leaflets has been reported [8]. But in our studies, it was observed that the accessions with lobed leaflets produced green hypocotyl. The trilobed leaflet is heritable, with stable expression and complete penetrance, easy to score visually and therefore, can be used as a morphological marker in mungbean as well as in DUS testing.

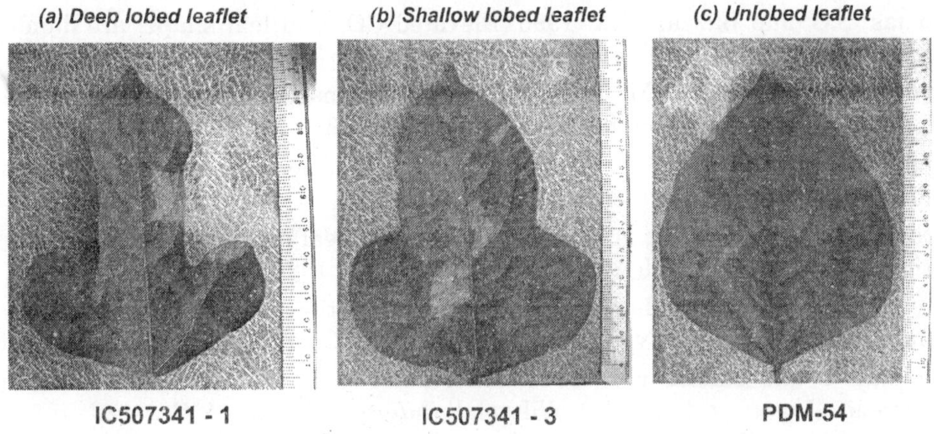

Figure 2. Variation in trilobed leaflet recorded in mungbean germplasm

Leaflet shape could be used to genetically manipulate canopy structure in order to alter light environments and increase plant productivity [14]. A highly dissected leaf margin decreases the effective size of the leaf. Small size or dissection thus increases the boundary-layer conductance. As a benefit, leaves suffer less extremes of temperatures, neither high temperature in high sun and low transpiration, nor low temperature at night under radiative cooling [15]. It has been reported that the leaf lobes is probably more associated to the adaptation to heat and water loss [16] and leaves with lobes have been found to convectively dissipate heat most easily as compared to other leaf shapes [17], which are advantageous in hot and dry environments. Traditionally, grain legumes such as mungbean have been grown on marginal lands of poor fertility and are still grown in edaphic conditions which are not very different from those in their native habits. Therefore, the trilobed leaflet character described above can be utilized in genetic enhancement of mungbean crop bred to suit dry regions.

Acknowledgement

We thank the Director, NBPGR, New Delhi for providing the necessary facilities for the study.

REFERENCES

1. Kerstetter, R.A. and Poethig R.S. 1998. The specification of leaf identity during shoot development. *Annual Review of Cell and Developmental Biology* 14: 373–398.
2. Gray, R.A. 1957. Alteration of leaf size and shape and other changes caused by gibberellins in plants. *American Journal of Botany* 44: 674–682.
3. Jones, C.S. 1995. Does shade prolong juvenile development? A morphological analysis of leaf shape changes in *Cucurbita argyrosperma* subsp. *sororia* (Cucurbitaceae). *American Journal of Botany* 82: 346–359.
4. Thomas, S.C. and Bazzaz, F.A. 1996. Elevated CO_2 and leaf shape: are dandelions getting toothier? *American Journal of Botany* 83: 106–111.
5. Byrne, M., Timmermans, M., Kidner, C. and Martienssen, R. 2001. Develop-ment of leaf shape. *Current Opinion in Plant Biology* 4: 38–43.
6. Singh, D. and Mehta, T.R. 1953. Inheritance of lobed leaf margin in mung (*Phaseolus aureus* L.). *Current Science* 22: 348.
7. Sareen, P.K. 1982. Genetic analysis of the trilobate leaf mutant in Mungbean (*Vigna radiata* var. *aureus* (L) Wilczek). *Current Science* 51: 1028–1029.
8. Sareen, P.K. 1985. Further genetic analysis of the trilobate leaf mutants in. mungbean (*Vigna radiata* var. *aureus* (L.) Wilczek). *Current Science* 54: 930–931.
9. Talukdar, T. and Talukdar, D. 2003. Inheritance of growth habit and leaf shape in mungbean [*Vigna radiata* (L.) Wilczek.]. *Indian Journal of Genetics and Plant Breeding* 63(2): 165–166.
10. Pokhle, Y.S. 1972. Spontaneous mutation in mung (*Phaseolus aureus* Roxb.). *Science and Culture* 38: 142.
11. Singh, K.B. and Sethi, S.C. 1970. A leaf mutant in mungbean (*Phaseolus aureus* Roxb.). *Science and Culture* 36: 348.
12. Sen, N. K. and Ghosh, A.K. 1959. Genetic studies in green gram. *Indian Journal of Genetics and Plant Breeding* 19: 210–227.
13. Singh, V.P., Yadav, R.C. and Saharan, R.P. 1984. Inheritance of lobed leaf in green gram. *Indian Journal of Agricultural Sciences* 54: 632–635.
14. Wells, R., Burton, J.W. and Kilen, T.C. 1993. Soybean growth and light interception: response to differing leaf and stem morphology. *Crop Science* 33 (3): 520–524.
15. Gutschick, V.P. 1999. Biotic and abiotic consequences of differences in leaf structure. *New Phytologist* 143: 3–18.
16. Niklas, K.J. 1989. The effect of leaf-lobing on the interception of direct solar radiation. *Oecologia* 80: 59–64.
17. Vogel, S. 1970. Convective cooling at low airspeeds and the shapes of broad leaves. *Journal of Experimental Botany* 21: 91-101.

Assessment of soluble protein content in some promising Moth bean germplasm accessions

Raju R. Meghwal, N.K. Dwivedi and H.S. Gehlot*

National Bureau of Plant Genetic Resource, Regional Station,
CAZRI Campus, Jodhpur – 342003:
*Stress Physiology Lab, Department of Botany, J.N.V. University, Jodhpur, Rajasthan.

ABSTRACT

Moth bean is one of the drought tolerant legumes, grown in the arid and semi-arid regions of the India. Variability for morphological characters of 108 germplasm accessions along with three checks (Jadia, Marumoth and RMO-40) was assessed during kharif 2006-07. Eight promising accessions, IC39688, IC402292, IC34249, IC311400 and IC311396 were promising in cluster per plant. IC39718, IC329044 and IC402285, IC311396, IC329044 were promising in pods per plant and pod yield per plant. Among these accessions, the highest cluster per plant was recorded in IC311396 (76.40) in comparison with Jadia (43.71). Pods per plant was highest in IC329044 (193.2) compared to Jadia (105.98). IC329044 had high pod (29.06g) and seed (21.5g) yield per plant. The soluble protein contents among the nine promising accessions of moth bean dry seeds was estimated through Bradford Protein Assay Method. The soluble protein content among these accessions ranged from 25.13 mg/g (IC34249) to 12.22 mg/g (IC311400). The highest soluble protein was observed in IC34249 (25.13 mg/g) in comparison with Jadia (16.96 mg/g). The superiority of morphological characters and soluble protein content in the promising accessions assumes great significance in crop improvement programmes.

INTRODUCTION

Moth bean, *Vinga aconitifolia* (Jacq.) Marechal, is the most drought hardy and temperature tolerant legume crop, grown in the arid and semi-arid regions of India and abroad (1). Being the most drought hardy and temperature tolerant among pulses, it is one of the most popular crop of arid and semi-arid zones of Rajasthan, which account for 80% of the total moth bean area of country. It is also known as moth, kidney bean, aconite bean, mat, mout, bhirni and dew gram. It has been reported to have originated in India (2). A wide range of variation was observed for qualitative and quantitative characters such as plant height, days to 50% flowering, no. of clusters per plant, pod yield per plant, seed yield per plant and 100 seed weight were recorded (3). The plant resembles a small mat, for it is a ground-hugging plant and only about a foot high. It serves as a multipurpose crop. Besides being rich in protein (24-26%) grains are also rich source of carbohydrates along with substantial amount of fibre, minerals and amino acids. In India, green pods are eaten as a vegetable, and the tiny seeds are eaten whole or spilt (4). The agro-morphological characters and soluble protein content in promising varieties of moth bean assumes great priority in crop improvement programmes. In the present study, we present the soluble protein content estimates in some of the promising accessions identified during the characterization programmes.

MATERIAL AND METHODS

A set of 108 germplasm collections collected from Gujarat (30), Haryana (3) and Rajasthan (75), were evaluated for their performance against the check variety Jadia, Marumoth and RMO-40, under rain fed situations during Kharif season of year 2006-2007. Variability in morphological characters of these accessions was recorded on five competitive plants at random from each of accessions. Eight promising accessions viz., IC34249, IC39688, IC39718, IC329044, IC402285, IC402292 (collected from Rajasthan), IC311396 and IC311400 (collected from Gujarat) were taken for study. IC39688, IC402292, IC34249, IC311400 and IC311396 were promising in cluster per plant. IC39718 and IC329044 promising in pods per plant and IC402285, IC311396 and IC329044 promising in pod yield per plant and IC329044 was superior in seed yield per plant also. The promising accessions identified for various agro-morphological traits were analysed for soluble protein content.

Estimation of Soluble Protein

Soluble protein contents in dry seeds of the eight promising accessions and one check of moth bean was estimated through Bradford Protein Assays Method (5, 6).

50 mg of dry seeds were homogenized in 10ml of phosphate buffer (pH-7.0) and centrifuged at 8000-10000 rpm for 15-20 minutes. 0.2 ml of supernatant was eluted and the volume was made up to 1.0 ml using phosphate buffer. 3.0 ml of CBBG-250 dye was added and mixed well. 2 ml of distilled water was added to the mixture and shaken well. The solution mixture was left at room temperature for 10 minutes and O.D. measurement was taken at 595 nm. A

blank solution made up of 1.0 ml phosphate buffer and 3.0 ml of dye was used as standard.

RESULTS AND DISCUSSIONS

Variability for various morphological characters was observed in the selected accessions viz., IC34249, IC39688, IC39718, IC329044, IC402285, IC402292, Jadia, IC311396 and IC311400 (Table1). Among these accessions the highest (76.4) and the shortest (40.6) cluster per plant were recorded in IC311396 and IC402285. Accession IC329044 had the highest pods (193.20) and seed yield (21.5) per plant. Variability was also noticed in 100 seed weight. Accession IC 402285 (3.01g) had more seed weight than others promising accessions. Soluble protein content among these promising accessions has shown distinct variation (Table 1). Highest soluble protein was observed in IC34249 (25.13 mg/g) followed by IC329044 (24.65 mg/g) and lowest soluble protein was observed in IC311400 (12.22 mg/g). The variability among morphology characters and soluble protein content in promising accessions of moth bean germplasm will help in improvement of seed quality and also assumes great significance in crop improvement programmes.

Table 1. Variability for agro-morphological characters and soluble protein content in some superior accessions of Moth bean

S.No.	Promising Accession	Plant height (cm)	Days to 50% flowering	Cluster/ plant	Pods/ plant	Pod yield/ Plant (g)	Seed yield/ Plant (g)	100-seed weight (g)	Soluble protein content (mg/g)
1.	IC34249	30.91	39	69.60	142.20	16.80	8.9	2.95	25.13
2.	IC39688	30.92	44	67.20	139.60	18.42	10.1	2.70	12.84
3.	IC39718	33.30	47	62.60	168.80	20.38	15.1	2.45	24.52
4.	IC311396	31.20	42	76.40	137.40	23.73	17.6	2.84	14.90
5.	IC311400	33.33	43	74.80	145.40	22.11	18.2	2.62	12.22
6.	IC329044	25.84	35	52.60	193.20	29.06	21.5	2.89	24.66
7.	IC402285	28.62	35	40.60	155.80	23.63	18.3	3.01	20.33
8.	IC402292	27.88	35	67.40	152.00	18.92	11.5	2.76	23.96
9.	Jadia	30.20	42	47.80	105.98	15.02	17.2	2.82	16.96

Standard Curve

The reference curve was prepared by using defatted Bovine Serum Albumin (BSA) (0.02 mg/0.1 ml) for standard proteins.

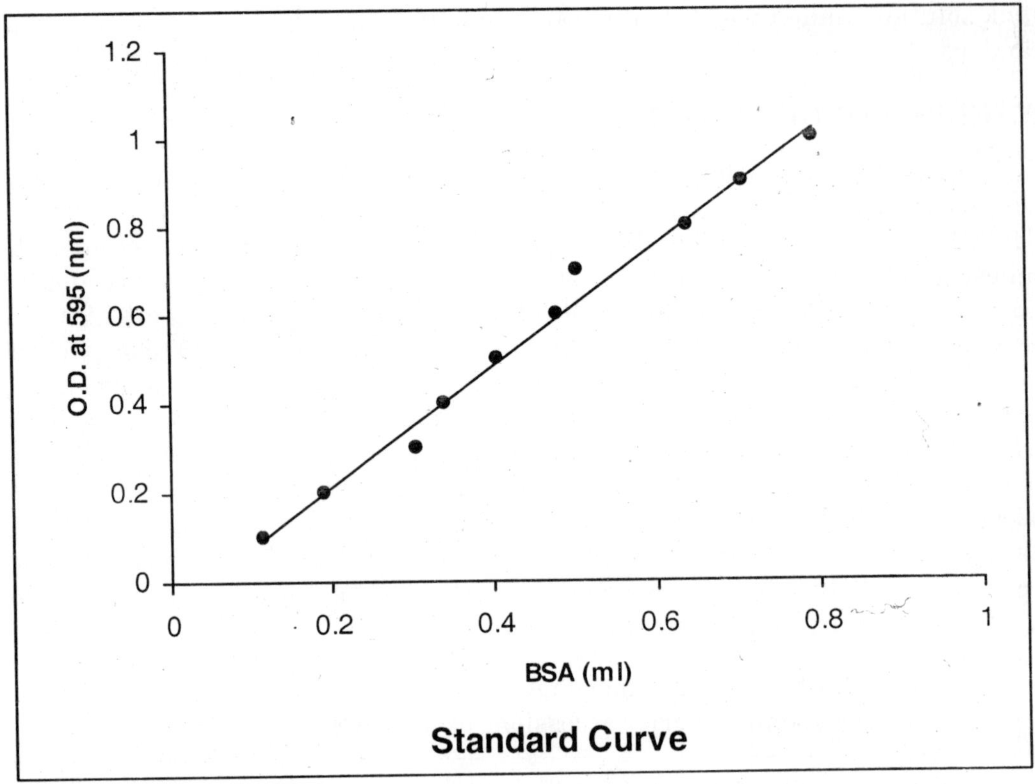

Standard Curve

APPENDIX

Solutions

CBBG-250 dye(Commassie Brilliant Blue G-250 dye) – 100 mg in 50 ml 95% ethanol add 100 ml 85% (w/v) phosphoric acid and makes the volume up to 1000 ml by adding double distilled water. Dark bottle was used for dye making. The dye can be stored at 4.0°C, protected from light.

Phosphate buffer (0.1M) - salts used.

1. NaH_2PO_4 (sodium dihydrogen phosphate) monobasic- 1.56g dissolved in 100ml of distilled water.
2. Na_2HPO_4 (Di sodium hydrogen phosphate) dibasic- 1.78g dissolved in 100ml of distilled water.

Standard BSA-(Bovine Serum Albumin) 100 ppm.

10 mg of defatted Bovine Serum Albumin dissolved in 100 ml of distilled water.

Acknowledgements

Authors express their sincere thanks to the Director, NBPGR, New Delhi, for providing facilities, and to Head, Department of Botany, JNV University, Jodhpur, providing facilities for soluble protein analysis, Dr. Gopala Krishnan S. for timely help and suggestions and to Shri Kheta Ram, T-5, for field assistance.

REFERENCES

1. DeCandole, A.: Origin of cultivated plants. Hafur Publication Company, New York (1884).
2. Vavilov, N.I. Origin, variation, immunity and breeding of cultivated plants (translated by K. Starr Chestet). *Chronica Botanica*. 13: 1-364 (1926).
3. Chopra, D.P., Thomas, T.A., Mehra, K.L. and Singh, Ranbir, 1981. Catalogue on Moth bean germplasm (series-2), NBPGR Regional Station, Jodhpur, 29p.
4. Kumar, D. and N. B., Singh, 2002. Moth bean in India. Scientific Publishers (India), Jodhpur.
5. Bradford, M.M. A rapid and sensitive for the quantitation of microgram quantities of protein utilizing the principle of protein-dye binding. *Analytical Biochemistry*, 72: 248-254 (1976).
6. Stoscheck, C.M. Quantitation of Protein. *Methods in enzymology*. 182: 50-69 (1990).

Breeding arid legumes for drought tolerance

A. Henry

Central Arid Zone Research Institute, Jodhpur, Rajasthan - 342003

ABSTRACT

A desired conceptual Model of crop variety that has been developed for irrigated agriculture, have to be made in case of rainfed crops, more particularly for the pulses grown in arid areas. The plant modifications for rainfed agriculture have to be designed so as to achieve a desired and balanced combination of resistance to soil and atmospheric drought on one hand and the plant productivity parameters on the other. It is necessary that a suitable maturity pattern that would enable the crop to overcome the long spells of drought at critical growth and developmental stages, should be incorporated alongwith the other factors conducive to a high harvest index and consistent yielding ability. In order to offer better insurance against crop failure it is necessary to develop suitable genotypes for mixed cropping/inter-cropping, taking into consideration various components of competitive ability and complimentarily. Plant breeding and genetical tools such as disruptive selection, mutation breeding, random mating and heterosis breeding would be useful in over coming undesirable character association in this regard and thus paving the way towards desirable gene combinations. Suitable procedures, to avoid genetic slippage and exhaustion of variability, have to be practiced.

Key words: Arid legumes, breeding, drought tolerance.

Introduction

An analysis of the conditions of crop growth on arid lands suggests that an acute ecological imbalance of the components of productivity is the sole complex factor limiting the stable agricultural production in these areas. Assuming that the environmental factors are beyond amelioration at this stage, the emphasis has to be laid on the proper management of soil water and plant factors. Crop varieties to be grown on arid regions should possess morphological and developmental characters combinations conducive to a consistent performance under moisture stress conditions. The term drought resistant is applied to describe those varieties which are able to grow and yield satisfactorily in areas liable to periodic drought.

According to May and Milthrope (1) the following three main types of drought resistance may be distinguished, viz.,

1. Drought escape or ability to complete the life cycle before being subjected to serious water stress.
2. Drought endurance (tile capacity to survive a drought) with high internal water content maintained during the period of drought by virtue of a deep root system or by reducing transpiration.
3. Drought endurance with low internal water content during the drought but with the ability to recover and grow rapidly when the soil water is replenished. This response is often called resistance to desiccation.

In drought escape or drought avoidance includes those plants which normally escape subjections to a long period without rain. Plant breeders have long been aware that the ability to mature before the onset of a long period of drought is possibly the most important factor in the drought resistance of crop varieties in seasonally arid areas.

Plant Productivity Vis A Vis Drought Resistance

A critical examination of most of the currently cultivated land races/ varieties of arid legumes reveals that they have been selected more for their ability to survive atmospheric and soil drought rather than for productivity. The attributes necessary for survival are not necessarily the same as those essential for productivity. It is a balanced linked up of the components of plant productivity with a considerable degree of drought resistance that would be desirable. The moisture availability period in arid areas varies from 70-90 days in different regions. Our approach is to isolate such a genotypes which successfully complete their life cycle with a higher harvest index within this period. A high degree of variability exist for this trait in arid legume crops. It has been observed that reduction in plant height lowered the dry weight of vegetative parts which resulted in an increased harvest index. Harvest index was positively correlated with grain yield but negatively correlated vegetative parts. In arid legumes a higher number of pods and grain weight number give the higher yields. In other words, apart from a good measures of drought tolerance and early maturity, a variety should possess a balanced combinations of sink capacity and grain weight in order to give consistent yields under moisture stress Conditions.

It has also been observed that drought resistance character like deep root system are often associated with prolonged vegetative growth resulting in late maturity. In order to combine deep root system with earliness of maturity and profuse bearing, a faster rate of root growth could be made use of

Stability of Performance

According to Eberhart and Russell (2) stability of performance of a crop variety is described by three parameters, viz., mean yield, regression of the mean yield on an environmental index and the deviation from the regression. A variety with a consistently high performance shows a high mean, regression nearing unity and measures a low degree of deviation from regression. These criteria can profitably be applied to screen out suitable crop varieties for rainfed arid areas as stability of performance should be an important character to be associated with such crop varieties. One of the way of achieving this is to take advantage of heterosis breeding as accomplished in case of pigeonpea. Another approach towards attainment of stable performance would be to make use of the additive genetic variance.

Genotypes for Mixed Cropping

Breeding suitable plant types for crop mixture is essential as the crop mixtures offer insurance against crop failure which is very common in years of acute drought in arid regions. Arid legumes are generally grown as an inter/mixed cropped with pearl millet-a staple cereal in the region. Mixed populations may yield significantly more than the mean of the two components. This could be brought about in more that one ways, viz., (i) through active competition when the superior competitor is also the more productive component of mixture (ii) through facilitation, this occurs when one or both components provide the other with some positive stimulation so that the latter grows better than it would have done in the absence of the former. (iii) If neither component provides the other with any positive stimulation but grows independently of it our in such a way the its reacts own to its own density than to that of other component.

Since varieties or sub-species have more common requirements than species or genera, we should expect the competitions to be more severe within species and less severe between species or genera. The genotypes suited for pure stand need not necessarily in a mixture may have to be generally non competitive and complimentary. Now it is known that various 9omponents are associated with competitive ability and whose best combination can be achieved through genetic manipulation, to evolve suitable genotypes for mixed cropping. The studies have shown that mixtures were generally more stable than the pure lines and the degree of stability apparently dependent upon the type of competitive interactions involved.

Breeding Procedures to Achieve Suitable Character Combinations

It has been revealed that selection for local adaptations in stress environment was found to be inferior as compared to selections for developmental traits in a favourable environment. The conventional methods such as selection for yield components, testing of advanced generations for drought resistance and emphasis on single plant performance in contrast to population performance have resulted in fixing of correlated gene combinations, including unfavourable ones, rapid exhaustion of useful genetic variation and prevention of new recombination's which automatically restricted further improvement. Whenever a desirable character combinations essential to over come the yield barriers could not be achieved by conventional breeding methods, induced mutants for new genes can be made use of .Negative character associations could also be over come by following disruptive selection procedures recurrent selection and biparental breeding methods.

Major thrust areas for breeding aspects

The major recommendation that are emerged for breeding programmes from crop improvement programms by various workers are listed below:

1. To define and understand the various levels of environmental stress to know the extent of stress injury or resistance in crops.
2. It is a pre-requisite to characterize the environment and extent of resistance to a given stress operative in a given environment. The proper understanding of environmental factor such as extent of precipitation and its distributions in relation to extent of drought, the probability levels with regards to predictability or rainfall, environmental variations for temperature fluctuations, nutrient availability in relation to moisture levels in the soil, duration for crop growth etc. is necessary to define research objectives in the specific area/region.
3. At present a number of morphological, physiological and biochemical attributes are involved in better performance under a given drought conditions. It is necessary to screen to available germplasm for these attributes so that available genetic variability could be exploited for enhancing the productivity.
4. The reliable screening techniques is essential for the following, viz.
   a. assessing the plant resistance at the critical developmental stages,
   b. these are quick to measure the parameters

c. these should be repeatable even in small sample size.
d. and are capable of screening large populations efficiently. This requires an interdisciplinary team approach.

5. Selection criteria are required for practicable breeding programme for achieving genetic advance after identification of relevant attributes for stable performance under drought conditions. This genetic stock could be intermated with the high yielding varieties. The products of stabilizing selection could be shuttled between 'stress' and 'non-stress" environments while effecting selection for desirable characters. Genotypes with recovery resistance, better rooting pattern, and more productivity as measured in terms of per day productivity could be used further for advancement by using appropriate breeding methodology.
6. There is an increasing gap between the yield obtained under optimal and stress environmental conditions. Dryland areas represent a major proration of the cultivated land in the country. The improvement of the yield dryland conditions/problematic areas requires larger research investments in spite of expected slow progress.

The important emphasis on evaluation of the new plant types in arid legumes are as follows:

Moth bean or Dewgram

1. To identify/evolve a variety having a semi-erect plant habit, dark green leaves, early maturity and high yield potential and resistance to yellow mosaic virus.
2. To identify suitable variety for inter/mixed cropping in pearl millet.
3. Breeding programmes for improvement of moth bean started at CAZRI, Jodhpur in .1975 [3]. In identification of suitable varieties for region indicated variety like T-2 and T-18 [4], IPCMO-880, IPCMO-186[5], IPCMO-371, PLMO-34, IPCMO-433 [6], holds promise for the region. In hybridization programmes breeding lines JMS-l, JMS-2, JMS-3 and JMS-4 indicated superior performance for different growing situation over other varieties [7]. In mutation breeding programme, using chemical mutaen EMS on seeds of locally adapted variety 'Jadia' led to the development of promising mutants like JMM-259, JMM211, JMM-60, which had higher number of pods, early in maturity and higher yield potential than its parent variety 'Jadia' and other checks varieties [8,9], out of these mutants, Mutant JMM-259 gave better performance over the check varieties in All India Coordinated trials and based on its performance it was released and notified for general cultivation by the Central Seed Committee in Dec. 1988 under the name of Maru Moth-l [10]. JMM-DBS-l was another promising mutant evolved [11]. The

estimation of genetic diversity in the land races indicated that there is exist enough genetic variability which can' be utilized development of high yielding varieties (Henry and Krishna, 1986). The development of high yielding varieties is basic requirement in Moth bean [12]. In evaluation of moth bean varieties for inter/mixed cropping with pearl millet, it was revealed that variety JMM-259 gave better performance in inter cropping while var. T-18 gave better performance in mixed cropping [14, 14b].

Clusterbean or Guar

1. To identify/evolve a variety, which can be adapted to the agro climatic conditions of the region at the same time their growth rhythms matching the rainfall pattern of the region.
2. To evolve varieties resistant to *Alternaria* blight, bacterial blight and dry root rot diseases.
3. To evolve variety having high galactomanan content in the endosperms of the seeds.
4. To evolve determine plant type.

Breeding programmes to bring about genetic improvement in clusterbean was started in erstwhile dry farming center at CAZRI [15]. More than 5 thousand germplasm of this crop were evaluated and characterized for different characteristics. Studies on genetic diversity in the indigenous material indicated that enough genetic variability and diversity is present in the available germplasm for utilization in the hybridization programme [16,17,18a,18b]. High yielding genotypes evolved through crossing programmes are being tested at various location through All India Coordinated Varietal trials. Clusterbean crop has a indeterminate growth habit, where flowering, vegetative growth and pod formation takes place simultaneously with the result at maturity upper portion of the plant remains green and upper portion pods also remains immature and result in dry matter loss at harvesting. To avoid such loss, it was envisaged to evolve a determinate plant type, which is not available in the existing germplasm, through mutation breeding. For the purpose seeds of variety FS-277 were treated with chemical Mutagen, EMS. Variants having determinate growth habit were selected from the segregating generations. These mutants gave better performance over parent FS-277 [19]. Its further testing with other promising varieties like HG75, G-1 indicated at par performance while with vars. Like RGC-936 and Suvidha, it gave better performance [20]. In the identification of the suitable varieties it was indicated the varieties like HG-75, HG-182, D, Safed, 2470/12, were suitable ones [21,22] among medium late maturity group. Among the single stemmed varieties, varieties like HFG-314, PLG-119, PLG-85 proved superior. Among early maturing group, varieties like

HG-7-2 (Naveen) and Suvidha proved superior [23]. Varieties were evaluated against different rhizobial strains [24].

Variant selected from the germplasm like 2470/12 gave better performance over check varieties in All India Coordinated varietal trials. Its further testing in adaptive trails and minikit trials indicated its superior performance over the other check varieties tested. Based on its superior performance it was released and notified for general cultivation by Central Seed Committee in Nov. 1986 under the name Maru-Guar [25]. In evaluation of clusterbean varieties for inter/mixed cropping with pearl millet, it was revealed that var. FS-277 gave better performance in inter cropping system over others [26].

Green gram (Mung bean)

The major emphasis is on the development of green gram varieties resistant to YMV and other diseases with stable performance and good cooking quality, maturing in 65 to 70 days, so that it can be grown in multiple cropping and in rotation with other crops.

In identification of suitable varieties for the regions, it was revealed that variety S-8 gave the most promising performance both under normal and late sown conditions followed by K-851 [27]. Of late other varieties which are giving promising performance are RMG-l1, RMO-131, RMG-146, P-7093, RMG-268, PGM-54 [28]. The insect pests scenario on different green gram varieties was worked out in detail by [29]. The 4 promising genotypes of Mungbean evaluated for different cropping systems with pearlmillet for four years. The results revealed that variety S-8 gave higher monetary returns both in inter/mixed cropping system over other varieties. The higher total productivity in inter/ mixed cropping was obtained with variety G-l which also indicated higher LER value [30].

Cowpea (Lobia)

1. To identify/evolve varieties which are more compact in their growth and which are relatively erect with a non trailing habit. This will result in higher seed yield of the crop.
2. To evolve varieties resistant to diseases and pests.

Large number of varieties tried in evaluation trials indicated that variety Charodi-1, having brownish seed colour with small medium seed size, gave the most promising performance over the other varieties both under normal and late sown conditions followed by FS-68 [31]. Variety K-11 was another variety which gave better performance under late sown conditions only. Efforts were made to improve the most suited variety Charodi-1 through Mutation breeding to evolve variants having plant type of Charodi-l with seeds having white seed coat colour and medium bold size. This led of the development of mutants with above characteristics, viz., CAZCW, CAZC-9, CAZC-10, CAZC-11 and are being tested in All India Coordinated Varietal trials to see its suitability over the various locations [32], CAZC-B was another Mutant having black seed coat colour and was promising over parent variety Charodi-1 [33]. Evaluation of different varieties in inter/mixed cropping with peal millet revealed that variety Charodi-1 also gave better performance in both the inter mixed [34]. V -265 and RC-80 were better for favourable season, while varieties VCM-8, VPC-124, GC-1 were suitable for adverse season [35]. Early to medium maturing varieties were found to give better performance in arid areas [36].

Horsegram (Kulthi)

This minor pulse crops is being grown in tribal areas of Pali and Jalore district and its adjoining area of Sirohi. Varietal improvement in this crop is in incipient stage. Efforts were made to collect the land races from its growing areas. High yielding variant plant selected were subsequently evaluated for various characteristics. Results revealed that genotypes PL-Selection, (brown seed coat colour), SN-selection (dark brown seed coat colour) and SU-Selection (light dark seed coat colour), were high yielding and stable genotypes for various situation [37]. Genotype PI-Selection was tested under diverse agroclimatic zones of the country over the years in All India Coordinated Varietal trials, and gave the highest mean performance, over the other varieties tested. Based on its superior performance genotype PL-selection was released and notified for general cultivation by central seed committee in year 1989 under the name Maru-Kulthi-1.

Chickpea (Gram)

It is grown to a limited extent in Ganganagar, Nagaur district under limited moisture supply conditions. The objective in varietal improvement of gram in order of priorities are high grain yield with good consumer acceptance parameters (Cooking quality, palatability, seed colour, size) and stability in performance by incorporating resistance to diseases like wilt, blight and rust and insect pests.

The testing of promising varieties of chickpea under limited moisture supply conditions in arid regions revealed that variety ICCC-42 and Manak (14-16 q ha-1) holds promise for the area.

Pigeon pea (Arhar)

With the availability of the extra early maturity strains in pigeon pea, it was envisaged to evaluate its performance under both rainfed conditions during monsoon season and under limited moisture supply conditions during winter season in the arid regions to see its suitability. The objectives for the improvement are that it should be of short duration and should carry resistance to the three major diseases, viz., wilt, sterility mosaic and *phytophthora* blight as well as to pests particularly pod borer and pod fly.

The evaluation of extra early germplasm during monsoon revealed that there exists genetic variability for all the important characters as well as genetic diversity among the genotypes so that same could be utilized with advantage in crossing programmes [38,39] The performance of different varieties of pigeon pea during winter season revealed that varieties 'Basant' and 'Bahar' are suitable or the post rainy season cultivation under limited moisture supply conditions [40].

Table I. The performance of improved var. of grain legumes at CAZRI, Jodhpur

Grain legumes	Varieties	Maturity (Days)	Mean productivity (Kg ha-l) (Long tenn)	State average (kg ha-l) (Long tenn)
Clusterbean	Branched medium late varieties			
	Maru GuarHG-75, G-l	97-105	750-850	216
Branched early varieties				
	Suvidha, Naveen, RGC-936	80-85	800-900	
Unbranched varieties				
	FS277, PLG-119, PLG-85, HFG-314	90-85	600-800	
Moth bean	Medium maturity varieties			
	CAZRI Moth-I, Maru Moth-I, Jadia, Jwala	80-85	400-600	150
Medium late varieties				
	IPCMO-880, T18	90-95	300-500	
Green gram	S-8, K-851	60-65	700-900	194
Cowpea	Charodi-l, FS-68	60-65	700-900	257
Horsegram	Maru Kulthi-l	100-105	600-700	

REFERENCES

1 May, L.H. and Milthr/ope, P.L. 1962. Drought resistance of crop plants. *Field crop Abst.* 15(3): 171-179.

2 Eberharat, S. A. and Russell. W.A. 1966. Stability parameters for comparing varieties. Crop Science. 36-40.

3 Singh R.P., Singh, K.C. and Henry, A. 1981. Mothbean Research at CAZRI (1975-1980). Div. Bull. No. 7. p. 18

4 Henry, A. and Daulay, H.S. 1983b. Genotype x environment interactions for seed yield in Moth bean, *Indian Journal of Agricultural Sciences* 53: 469-470.

5 Henry, A. and Daulay, H.S. 1987. Phenotypic stability for seed yield in moth bean and clusterbean. *Proceeding First Symp. on Crop Imp.* Vol. II pp. 137-145 (Eds.) K.S. Gill *et al. Crop Imp. Soc. of Indian* P AU, Ludhiana.

6 Henry, A. 1995b. Performance of moth bean genotypes on drylands of western Rajasthan. *Forage Res.* 21: 149-151.

7 Henry, A. 1995a. Phenotypic stability in moth bean for seed yield on drylands of western Rajasthan. *Forage Res.* 21: 58-61.

8 Henry, A. and Daulay, H.S. 1983a. Performance of induced mutant in moth bean. *Indian Journal of Genetic and Plant Breed* 43: 343-344.

9 Henry, A. and Daulay, H.S. 1983c. Genotype x environment interactions of induced mutants in moth bean. *Madras Agri. Journal* 70: 705-708.

10 Henry, A., Daulay, H.S. and Bhati, T.K. 1993. Maru Moth-I, A promising moth bean variety for arid and semi arid regions. *Indian Farming* 43: 21-24.

11 Henry, A. and Daulay, H.S. 1 990b. Induced dark brown mutants in moth bean. *Madras Agril. J.* 77: 501-504.

12 Henry, A and Krishna G. V.S.R. 186. Genetic divergence in mothbean, Indian Journal of Agricultural Sciences 56:298-299.

13 Henry, A. and Singh, R.P. 1985. Mothbean In Efficient management of dryland crops (Eds. V. Venkateswarlu and V. Balasubramaniam) CRIDA, Hyderabad pp-199-207.

14 Daulay, H.S., Henry, A. and Bhati., T.K. 2000. Screening of suitable genotypes of clusterbean, Moth bean and green gram in legume based cropping with pearlmillet. Current Agriculture: 24: 85-88.

15 Singh, R.P. and Henry, A. 1981. Guar Research at CAZRI (1971-1980) Div. Bull No. 6P. 20

16 Henry, A. and Mathur, B.K. and Daulay, H.S. 1984. Genetic divergence in clusterbean *Forage Res.* 10: 89-94.

17 Henry, A. and Daulay, H.S. and Krishna, G.V.S.R. 1986. Correlation, path coefficient analysis and genetic diversity in clusterbean. *Madras Agril. J.* 73: 11-16.

18 Henry, A. and Mathur, B.K. 1987. Correlation and path coefficient analysis in clusterbean. *Annals of Arid Zone* 26: 159-162.

19 Henry, A. and Daulay, H.S. 1987b. Induced mutants in moth bean and clusterbean. *Abst First Symp. on Crop Imp.* 23-27 Feb. 1987. *Crop Imp. Soc. of India* PAU, Ludhiana, pp. 109-110.

20 Henry, A. 1999. Performance of different maturity groups of cowpea varieties in dryland of western Rajasthan. Abst. In National Seminar on plant physiology of Interface of Agri-Horticulture and Industry org. by Indian Soc. of Plant physiology, New Delhi, RAU, Bikaner and Agril. Univ. Udaipur, Dec. 30-Jan. 1, 2000 at RCA, Udaipur, pp. 74.

20 Henry, A. 1999. Performance of determinate plant type mutant vis-a-vis indeterminate varieties of clusterbean on dryland of western Rajasthan. Abst. In National Seminar on "Plant physiology at interface of Agri Horticulture and Industry" organized by Indian Soc. of Plant Physiology, New Delhi. RAU, Bikaner and Agril. Univ: Udaipur - Dec. 30-Jan. 1,2000 at RCA, Udaipur, pp. 23.

21 Henry, A. and Daulay, H.S. 1983c. Genotype x environment interactions for seed yield in clusterbean. *Indian Journal of Agricultural Sciences53:* 467-468.

22 Kackar, N.L. and Henry, A. 1984. Genotype x environment interaction for seed yield in clusterbean. *Indian J. of Genetic and Pl. Breeding* 44: 465-468.

23 Henry, A. 1993. Performance of clusterbean genotypes on drylands of western Rajasthan *Forage Res.* 19: 338-340.

24 Daulay, H.S. and Henry, A. 1997. Response of clusterbean varieties to different rhizobial strains in drylands of west em Rajasthan. Forage Research: 23: 105-107.

25 Henry, A., Daulay, H.S. and Bhati, T.K. 1992. Maru Guar a promising clusterbean for arid regions. *Indian Farming* 42: 24-25.

26a Daulay, H.S., Henry, A. and Bhati, T.K. 1991. Studies on suitable genotypes of grain legumes for inter/mixed cropping with pearlmillet. In *Abst. Vol. II Golden Jubilee Symp. on Genetic Research and Education. Current Trends and the next fifty year* Feb. 12-15, 1991. *Indian Soc. of Genet & Pl. Breed* New Delhi pp. 414-415.

26b Daylay, H.S., Henry, A. and Bhati, T.K. 1998. Compatibility of promising varieties of clusterbean for inter/mixed cropping in drylands of western Rajasthan. Forage Research. 24: 13-16.

27 Henry, A. and Daulay, H.S. 1989. Technology for mung bean cultivation in arid regions. *Indian Farming* 39: 12-13.

28 Henry, A. 1997. Role of improved varieties of arid legume crops for increasing the productivity. Pp: 71-72. Abst. In National Symposium on "Recent Advances in Management of Arid Ecosystems", organized by Arid Zone Research Assoc. of India, at CAZRI, Jodhpur, 3-5 March, 1997.

29 Verma, S.K. and Henry, A. 1988. Intensity of insect pests on mung bean cultivars in arid Rajasthan. *Annals of Arid Zone27:* 71-74.

30 Daulay, H.S., Henry, A. and Bhati, T.K. 1999. Studies on suitable genotypes of green gram inter/mixed sowing in pearlmillet on drylands of western Rajasthan Current Agriculture 23: 93-96.

31 Henry, A. and Daulay, H.S. 1988e. Grow cowpea variety charodi-l in arid regions. *Indian Farming* 38: 9-10.

32 Henry, A. 2001. Performance of Induced Mutants in Cowpea. Pp. -225. Abst. In Hundred years of post Mendelian Genetics and Plant Breeding - Actrospects and Prospects organized by ISGPB, Nov. 6-9, 2001. New Delhi.

33 Henry, A. 2001. Performance of Induced black seed coat Mutant in Cowpea – Indian Journal of Genetics and Plant Breeding, 61: 381-382.

34 Daulay, H.S., Henry, A., Bhati, T.K. and Mathur, B.K. 2006 Compatibility of promising varieties of Moth bean and Cowpea for inter/mixed cropping in pearl millet in arid regions of western Rajasthan. Journal of Arid legumes 3: 44-48.

35 Henry, A. 1998. Relative Performance of Mung bean varieties on drylands of Western Rajasthan. Pp. 142-143. Abst. In National Symposium on Management of Biotic and Abiotic Stresses in Pulse Crops" organized by Indian Society of Pulses Research and Development, HPR, Kanpur (D.P.) June 26-28, 1998.

36 Henry, A. 1998. Relative performance of cowpea varieties on dryland of western Rajasthan. Pp. 151 Abst. In National Symposium on Management of Biotic and Abiotic Stresses in Pulse Crops" organized by Indian Society of Pulses Research and Development, HPR, Kanpur (U.P.) June 26-28, 1998.

37 Henry, A. and Daulay, H.S. 1988a. Genotype x environment interactions for seed yield in horsegram. *Indian Journal of Agricultural Sciences* 58:684-686.

38 Henry, A. and Krishna, G.V.S.R. 1990a. Correlation and path coefficient analysis in pigeonpea. *Madras Agril. J.* 77: 440-442.

39 Henry, A. and Krishna, G.V.S.R. 1992. Genetic divergence in pigeonpea. *Madras Agri. J.* 79: 41-44.

40 Henry, A. and Daulay, H.S. 1988b. Relative performance of pigeonpea genotypes in arid regions of western Rajasthan, India. *International pigeon pea Newsletter No.* 7,8-9.

Drought physiology of arid legumes

B.K. Garg and Uday Burman

Central Arid Zone Research Institute, Jodhpur – 342 003

ABSTRACT

Clusterbean, moth bean, cowpea and horse gram grouped together as arid legumes are known to possess different mechanisms of drought resistance to survive and produce under drought-prone environments. The development of short duration and early maturing varieties that enable the crop to escape severe soil moisture deficits is of paramount importance in these legumes. Most of these legumes display several mechanisms of dehydration avoidance to maintain turgor potential for normal metabolic functioning under water deficits. These comprise maximization of water uptake through deep and extensive root systems and reduction of water loss by appropriate shoot attributes and osmotic adjustment for continued photosynthesis and growth at low water potentials. Significant genotypic differences exist for different physiological attributes and drought resistance traits which need to be better understood for selection and development of drought tolerant genotypes with higher yield potential under water limited conditions. These legumes are generally most sensitive to water stress at the flowering stage and genotypic differences at this critical stage are most important. Maintenance of higher leaf turgidity, net photosynthetic rate and stomatal conductance at similar stress levels, along with least alterations in leaf metabolites in moth bean was also reflected in its higher yield stability under water stress as compared to clusterbean and other legumes.

In addition to drought tolerance characteristics, various improved cultural practices, optimum plant nutrition and adoption of drought ameliorative measures such as thiourea application must be considered to reduce drought impact and to obtain sustainable crop yields under arid environments.

Key words: Drought tolerance, genotype variations, photosynthesis yield improvement.

INTRODUCTION

Arid legumes comprising clusterbean, moth bean, cowpea and horse gram are the integrated components of dryland farming, being the main stay of pulse and forage production in arid and semi-arid regions of the country. Their average productivity is, however, very low as they are confined and concentrated in arid areas having poor fertility soils and scanty and erratic rainfall leading to frequent droughts. Therefore, it is of paramount importance to identify and develop drought tolerant genotypes based on sound physiological mechanisms that influence the performance of these legumes under water deficits (1,2,3). Furthermore, appropriate crop management and improved cultural practices are equally important to reduce the hazards of drought for ensuring stable yields and increased water use efficiency under drought-prone environments of arid regions (4,5,6). The available information on physiological aspects of drought tolerance in these crops and some of the approaches for yield improvement under drought are discussed here.

Drought Resistance

Drought resistance refers to the ability of a crop to produce its economic product with minimum loss in water-deficit environment relative to the water-constraint free management. According to Turner (7) the ability of a crop to produce satisfactory in areas subjected to water deficits has been termed as drought resistance. Basically three mechanisms namely escape, avoidance and tolerance are involved in drought resistance (8) while Turner (7) suggested that they could be classified into three major categories:

(1) *Drought escape:* It is the ability of a plant to complete its life cycle before acute soil and plant water deficit develops. This is an important strategy of matching phenological development with the period of soil moisture availability to minimise the impact of drought where the growing season is short and terminal drought predominates. This mechanism involves rapid phenological development (early flowering and maturity), developmental plasticity (variation in duration of growth period depending on the extent of water stress) and mobilization of preanthesis assimilates to grain.

(2) *Drought avoidance:* It is the ability of plants to maintain relatively higher tissue water potentials inspite a shortage of soil moisture. Drought avoidance is achieved by maintenance of turgor through increased rooting depth, efficient root systems and internal hydraulic conductance and by reduction of water loss through reduced stomatal conductance, reduced absorption of solar radiation by leaf rolling and reduced evaporative surface i.e. leaf area reduction. Clusterbean and moth bean survive under drought conditions by doing a balancing act between maintenance of turgor and reduction of water loss.

(3) Drought tolerance: It is the capacity of plants to withstand water deficit with low tissue water potential. Tolerance to dehydration is considered to arrive at the molecular level and depends upon membrane structure and enzyme activity. Basically it is due to the ability of the cells to maintain integrity of the cell membranes and prevent denaturation of proteins. However, crop plants including arid legumes use more than one mechanism at a time to resist drought.

Drought Tolerance Assessment

Drought is by for the most important environmental stress in agricultural and many efforts have been made to improve crop productivity under water limiting conditions. While natural selection has favoured mechanisms for adaptation and survival, breeding activity has directed selection towards increasing the economic yield of cultivated species. Meanwhile, fundamental research has provided significant gains in the understanding of the physiological and molecular responses of plants to water deficits but there is still a large gap between yield in optimal and stress conditions.

A number of physiological studies have identified some traits for which presence or expression is associated with plant adaptability to drought-prone environments. However, a critical aspect in all studies dedicated to drought tolerance is the assessment of the degree of drought tolerance of different genotypes. In many studies the identification of tolerant and susceptible cultivars is based on few physiological measures related to drought response. The difficulty in identifying physiological parameters as reliable indicator of yield in dry condition has suggested that yield performance over a range of environments should be used as the main indicator for drought tolerance (9). Several indices were proposed to describe yield performance of a given genotype under stress and non-stress conditions or in comparison with the average yield or the yield of a superior genotype. Clearly there are only few studies on this line in the arid legumes (10). Some authors have expressed the yielding capacity with regard to an environment-related physiological trait, such as canopy temperature (11) or water potential (12). Rizza et al. (13) proposed on integrated "Water stress index" (WSI) based on a simple soil water balance and the integrated reduction of plant transpiration relative to potential transpiration. They related yield potential and adaptability of cultivars under water stress to the intercept and slope of the linear regression of yield versus WSI.

Physiological Bases for Yield under Drought

The water content and the water potential of plant tissues are the most important physiologically relevant integrators of drought effects on plants (14). Once a drop in water potential develops, responses of a wide range of physiological processes are induced, either directly or by plant hormones signalling changes in water status. Physiological traits relevant for the responses to water deficits and/or modified by water stress span a wide range of vital processes (Table 1). As a consequence, it can be expected that there is no single response pattern that is highly correlated with yield under all drought environments. Furthermore, different crop developmental stages show different sensitivity to water stress that has also been found to be the case in arid legumes (15,16).

Table 1. Physiological traits relevant for response to drought conditions

Plant traits	Modulation under stress	Effect relevant to yield
Stomatal conductance	Decreases under stress	Less water consumption
Photosynthetic capacity	Reduction under stress	Calvin cycle enzymes, light reactions, biomass accumulation
Timing of phonological phases	Advanced flowering in wheat, barley, delayed in rice	Early/late flowering, maturity and growth duration, reduced grain number.
Starch availability	Inhibition of photosynthetic activity	Abortion, reduced grain number
Stem reserve utilization	Remobilisation and compensation for reduced current photosynthesis	Higher remobilisations for grain filling, high yield.
Stay green	Resistance to premature senescence	Delayed senescence, post-flowering drought resistance.
Single plant leaf area	Reduced under stress	Plant size and related productivity.
Rooting depth	Reduced total mass but increased root/shoot ratio	Higher/lower tapping of soil moisture.
Osmotic adjustment	Slow response to plant water potential	Accumulation of solutes/ions, increased membrane stability.
Antioxidative stress	Acclimation of defence system	Protection against active oxygen species
Accumulation of stress-related proteins	Accumulated under stress.	Protection of cellular structure and protein activities.

Several studies have addressed yield under drought stress as a function of single physiological traits, in attempts to understand which metabolic processes or morpho- physiological traits are crucial in ensuring high yield performance under a wide range of environments. But most of these studies relate to cereals crops like wheat, rice, maize and sorghum. Some attempts have been made in our laboratory to short list few of physiological traits contributing to high yield of arid legumes under stressful conditions of Indian arid zone (1,10). However, it will be pertinent to mention here some of the single physiological parameters associated to yield in cereals which include –

(1) Stomatal conductance in wheat cultivars (17).

(2) Stay green in maize, rice and sorghum (18).

(3) Osmotic adjustment (19,20).

(4) Carbon isotope discrimination (Δ) in C_3 species. (21).

(5) Selection for stem reserve contribution to grain filling (22).

However, these parameters have not received adequate attention in assessment of drought tolerance in arid legumes.

Molecular Markers for Drought Tolerance Traits

Many efforts have been dedicated in recent years to understand the genetic basis of physiological traits conferring advantages in dry environments, while less attention has been given for understanding the high yield stability in dry and wet environments. Single genes such as those controlling flowering time, ear type, plant height and osmotic adjustment may have important roles in adaptation to drought-prone environments.

During the last 10 years, the application of QTL analysis provided very good opportunities to identify chromosome regions controlling variations in almost all the physiological, morphological and developmental changes observed during plant growth in water-limiting conditions. Attention has particularly been paid to –

Genetic variation of the osmotic adjustment.

Genetic basis of stay green phenotype.

The ability of roots to exploit deep soil moisture to meet ET demands.

Reduction of leaf area to limit water use.

Isotope discrimination.

Non-stomatal water loss from the leaves e.g. through the cuticle.

Rate of leaf elongation in response to soil moisture and ET.

Stress induced LEA proteins.

Stomatal sensitivity to ABA.

Overproduction of free proline.

However, most of there efforts have been made in crops like wheat, rice, maize, sorghum and some temperate legumes and at present there is no specific information on genetic basis of drought tolerance in arid legumes. As drought is a very complex phenomenon, use of above biotechnological tools for increasing efficiency of selection for drought tolerance in arid legumes is also advocated.

Drought Tolerance in Arid Legumes

In arid and semi-arid regions because of low and erratic rainfall, sowing is generally delayed until there is sufficient soil moisture for stand establishment. Thus, the most important characteristics that enable crop plants to yield satisfactorily are the matching of phenological development with soil moisture supply. In regions with a characteristic terminal drought, extension of crop production in drier regions has been accompanied by the development of short duration varieties that enable the crop to escape severe soil water deficits. In clusterbean and moth bean seed yields were significantly more in early flowering/maturing genotypes under low rainfall conditions (23,24,25). Significant genotypic differences were observed in crop performance and various physiological attributes among early and late genotypes of moth bean under extreme dry conditions in the field (25). Early flowering (30-31 days), genotypes such as RMO-40 and RMO-257 consumed soil moisture at a slower rate during vegetative stage and conserved it for the critical flowering stage. In contrast late flowering (45-50 days) genotypes consumed water at a faster rate during the vegetative stage. Consequently early flowering genotypes produced higher seed yield and exhibited greater WUE than late types. This was attributed to more efficient partitioning of photosynthates by early flowering genotypes. Observations on plant water status, NPK concentrations as well as Nitrogen Harvest Index (NHI) and Nitrogen Use Efficiency (NUE) also showed superiority of early genotypes over late genotypes under field drought conditions (Table 2).

Table 2. Seed yield, nitrogen uptake, nitrogen harvest index and nitrogen use efficiency of moth bean genotypes under field drought condition (Vyas *et al.*, 2003)

Genotypes	Seed yield (kg ha^{-1})	N uptake (kg ha^{-1})	NHI (%)	NUE (kg seed kg^{-1}N)
Early flowering				
RMO-40	723	16.04	35.2	45.1
RMD-117	495	11.77	37.8	42.1
RMO-224	644	14.88	46.8	43.3
RMO-225	481	11.05	50.9	43.5
RMO-257	443	9.73	50.9	45.5
Mean	**557**	**12.69**	**44.3**	**43.9**

Late flowering				
Jadia	408	17.88	32.4	22.8
FMO-1023-1	447	18.74	30.5	29.2
FMO-1096-1	385	10.72	27.9	35.9
Jawala	486	15.62	37.0	31.0
JMM-DBS-1	356	11.52	27.8	30.9
FMO-1118-1	443	16.68	31.1	26.6
IPCMO-880	345	12.27	31.2	28.1
FMO-1460-1	244	12.13	24.7	20.1
JMM-259	297	12.20	22.4	24.3
Mean	**390**	**14.20**	**29.6**	**27.7**

Likewise field studies undertaken with diverse early and late maturing genotypes of clusterbean under rainfed conditions for two consecutive years (1993-low rainfall, 1994-high rainfall) revealed significant genotypic differences in seed yield, WUE, nutrient uptake and leaf metabolism (23). The seed yield of early genotypes (RGC-936 and Suvidha) was significantly higher during low rainfall year while late genotypes (Maru guar and HG-75) out yielded early types during high rainfall year. WUE was 56.8% higher in early as compared to late genotypes during drought year. This was accompanied by higher nitrogen harvest index and nitrogen use efficiency in early maturing genotypes indicating a better utilization of nitrogen (Table 3).

Table 3. Seed yield, harvest index, WUE and NUE of early and late genotypes of clusterbean under low and high rainfall conditions (Garg *et al.*, 2003)

Year	Rainfall (mm)	Seed yield (kg ha^{-1})		HI (%)		WUE (kg ha^{-1} mm^{-1})		NUE (kg seed kg^{-1}N)	
		Early	Late	Early	Late	Early	Late	Early	Late
1993	206	995	625	33.2	19.7	4.66	2.97	17.2	12.2
1994	479	660	800	11.6	12.7	1.83	2.38	11.7	10.5

Studies have further indicated that early genotypes of both these legumes maintain higher plant water status and experience less metabolic changes than late genotypes under water stress (1,26). For instance genotype CZM-32E of moth bean displayed higher drought tolerance capacity than cv. Maru moth due to maintenance of higher Ψ plant and RWC, higher photosynthetic rate, higher nitrate reductase activity and better carbohydrate metabolism under different intensities of water stress (Table 4).

Table 4. Water stress effects on water potential, net photosynthetic rate and starch in moth bean genotypes (Garg *et al.*, 2001)

Water stress (d)	Ψ plant (MPa)			Net photosynthetic rate (μ mol $m^{-2}s^{-1}$)			Starch ($mg\ g^{-1}$ dw)		
	V1	V2	V3	V1	V2	V3	V1	V2	V3
0	-0.60	-0.70	-0.70	8.0	8.4	11.8	131	127	146
3	-0.90	-0.95	-0.90	4.2	3.5	9.4	77	120	128
6	-1.25	-1.20	-1.10	2.7	2.6	5.2	60	73	77
9	-1.50	-1.35	-1.30	1.5	1.7	4.2	55	57	71

V1=RMO-40; V2=Maru Moth; V3=CZM-32E

Water stress had maximum adverse effect on seed yield at the flowering stage in genotypes Maru guar and FS-277 but at the pod development stage in clusterbean genotypes Suvidha, HFG-182 and JGC-19 (15). Different physiological and biochemical parameters also displayed higher metabolic derangements under water stress at the flowering stage in all the genotypes.

In moth bean early flowering genotypes (RMO-40, CZM-99) consistently displayed less metabolic alterations than late genotypes (Maru moth, CAZRI moth-1) at vegetative and flowering stages but flowering stage was more sensitive to water stress. The maintenance of higher photosynthetic rate and better metabolic efficiency in early genotypes under water stress led to significantly less reduction in their seed yield and DMP as compared to late flowering genotypes (24)

Results of various studies on clusterbean, mung bean and moth bean indicated stress mediated changes in various leaf metabolites and enzyme activities that modulated the growth and yield of the legumes (2,3). It was found that under increasing water stress, normal GS-GOGAT system becomes less effective but GDH activity increases many fold indicating a shift in the pathway of ammonia assimilation which is an adaptive mechanism to detoxify the accumulated ammonia under acute water stress (15,27).

Evaluation of relative drought tolerance of arid legumes at vegetative and flowering stages revealed that moth bean displayed higher drought tolerance than clusterbean, mung bean and cowpea (16). Mung bean at the vegetative stage and cowpea at the flowering stage recorded maximum reduction in seed yield crops water stress due to more alterations in various physiological and biochemical parameters. Maintenance of high leaf turgidity, net photosynthetic rate and stomatal conductance at almost similar stress levels (Ψ_{plant}) along with less alterations in leaf metabolites such as total chlorophyll, starch, reducing sugars, soluble protein and free amino acids in moth bean as compared to clusterbean and other legumes was reflected in its yield stability under water stress (Table 5). Further, high plant water potential, leaf turgiscence and high photosynthetic rate on restoration of water supply to stressed plants showed high capacity of recovery in all the legumes, but more in case of moth bean.

Table 5. Influence of drought at flowering stage on comparative performance, water status, photosynthesis and leaf metabolism of clusterbean, moth bean, mung bean and cowpea (Garg *et al.*, 2005).

Parameter	Treatment	Clusterbean	Moth bean	Mung bean	Cowpea
Seed yield (g $plant^{-1}$)	C	7.60	4.13	7.64	10.03
	D	5.10	3.17	4.55	5.35
ψ plant (-MPa)	C	0.87	0.83	0.80	0.80
	D	1.90	1.93	1.87	1.77
RWC (%)	C	94.6	95.6	83.3	90.4
	D	65.4	77.8	75.9	81.4
Net photosynthesis (μ mol m^{-2} s^{-1})	C	9.10	6.50	6.50	5.50
	D	4.48	4.80	2.51	2.61
Chlorophyll (mg g^{-1} dw)	C	9.79	9.96	9.77	9.97
	D	7.93	8.87	7.22	7.55
Reducing sugars (mg g^{-1} dw)	C	17.12	26.61	20.64	20.95
	D	33.04	27.94	22.75	24.68
Starch (mg g^{-1} dw)	C	18.06	219.2	308.4	210.1
	D	142.4	179.3	190.6	147.2
Soluble protein (mg g^{-1} dw)	C	89.6	61.8	74.1	75.0
	D	85.6	60.7	71.9	69.5

C- Control Plants, D-Droughted plants

Osmotic adjustment is the active accumulation of solutes within the plant tissues in response to lowering of soil water potential (19). The degree of osmotic adjustment varies with species and genotypic variations within a species have been demonstrated in a number of crops including several legumes (5). Osmotic adjustment has been shown to enable photosynthesis to be maintained at low leaf water potential, to delay leaf senescence, to improve floral survival and to improve yield under water-limited conditions. A wide variety of organic solutes in plant tissues during water stress contribute to osmotic adjustment. Stress induced accumulation of free proline, free amino acids and reducing sugars has been observed in clusterbean and moth bean (15,16,26,27) and other crops. However, clusterbean and pearl millet accumulate high levels of free proline as compared to moth been and sesame plants under water deficits (Table 6). Over production of proline and glycine betaine that act as osmo protectants and their correlation with stress tolerance has been well emphasized (28). These studies indicated the possibility of genetic manipulation of proline over-production in drought resistant plants and control of osmoregulation in others.

Table 6. Influence of drought intensity on accumulation of free proline in certain arid zone crop plants (Garg, 2003)

Water stress levels	Proline concentration (mg g^{-1} dw)			
	Pearl millet	Clusterbean	Moth bean	Sesame
Control	0.75	0.47	0.38	0.45
Mild	3.26	3.91	0.71	0.62
Moderate	12.31	8.53	1.71	1.15
Severe	15.34	12.91	1.82	1.15

Most of the adaptations to drought have some limitations. A genotype of short duration usually yields less compared to that of normal duration under better moisture availability (1). The mechanisms that confer drought resistance by reducing water loss (such as stomatal closure and reduced leaf area) generally result in reduced assimilation of carbon dioxide i.e. reduced photosynthetic efficiency. Osmotic adjustment increases drought resistance by maintaining plant vigour but the increased solute concentrations responsible for osmotic adjustment may have additional energy requirements at the cost of plant growth (5,7).

Yield Improvement under Drought

Crop yields under drought can be improved by laying emphasis on integrated traits like seedling establishment, early growth vigour, developmental plasticity, leaf area maintenance, osmotic adjustment, high transpiration efficiency and mobilization of pre-anthesis stored reserves. In addition better crop management including appropriate planting dates, row spacing and plant populations, optimum fertilizer use and adoption of drought ameliorative measures may greatly help in alleviation of drought effects. In recent years several attempts have been made to minimise the detrimental effects of drought in clusterbean and moth bean which are briefly outlined here.

In clusterbean and moth bean seed yields were significantly more in early flowering/maturing genotypes under low rainfall situation. However, late genotypes provided higher yields under good rainfall conditions. For stability of yield, planting of early and late genotypes of these legumes in 1:1 ratio was found most appropriate under low as well as higher rainfall condition (Table 7).

The improved performance of mixed crop, particularly under low rainfall situation could be attributed to more efficient use of water, low initial canopy development and higher photosynthetic rates (29,30).

Table 7. Yield stability in moth bean and clusterbean through varietal intercrops (Kathju et al., 2003; Vyas et al., 2003)

Planting pattern	Seed yield (kg ha^{-1})		WUE (kg ha^{-1} mm^{-1})		Net Photosynthesis (μ mol m^{-2} s^{-1})	
	1995	1996	1995	1996	1995	1996
MOTH BEAN						
Early (RMO-40)	744	825	2.66	2.29	11.3	19.2
Late (Maru moth)	528	1011	1.88	2.80	9.7	20.5
E + L	755	1188	2.72	3.32	12.3	22.0
CLUSTERBEAN						
Early (Suvidha)	833	986	2.89	2.66	22.7	26.8
Late (Maru guar)	717	1122	2.54	2.93	21.0	26.0
E + L	928	1172	3.20	3.16	23.6	27.8

1995: Low rainfall year (267 mm); 1996: High rainfall year (424 mm)

Optimum plant population or row spacing is one of the major management variables under dry land conditions. In water deficit regions large canopy growth may be disadvantageous as it may exhaust the available soil moisture more rapidly from the root zone during drought. Studies with clusterbean genotypes have shown that 40-45 cm row spacing was significantly better for achieving higher seed yields, WUE and NUE under arid conditions (31). In another study moth bean genotypes planted at 30 x 15 cm provided significantly higher seed yields, DMP and higher WUE as compared to 45 x 10 cm spacing (32).

Addressing the soil fertility constraints can also alleviate drought stress effects on crop productivity in arid environments (33). The supply of phosphorus is particularly critical as declining water availability causes reduced P availability and absorption by moth bean and clusterbean under dry land conditions. Phosphorus application significantly ameliorated the detrimental effects of drought on water status, rates of net photosynthesis, contents of chlorophyll, starch, soluble protein and nitrate reductase activity in moth bean genotypes (34). Consequently seed yield was significantly improved despite drought condition. Similarly application of potassium under soil moisture stress has been found to play a major role through its favourable influence on plant water relations, physiological processes like net photosynthesis and stomatal conductance and seed yields (35). Maintenance of adequate potassium in the soil significantly improved plant water status, net photosynthesis, and yield of clusterbean (Fig 1.) under drought imposed at various developmental stages (36). Similarly seed treatment with thiourea (500 ppm) followed by foliar sprays (1000 ppm) at vegetative and flowering stages significantly mitigated the detrimental effects of field drought in clusterbean (37) through enhancement of photosynthetic efficiency and nitrogen metabolism (Table 8).

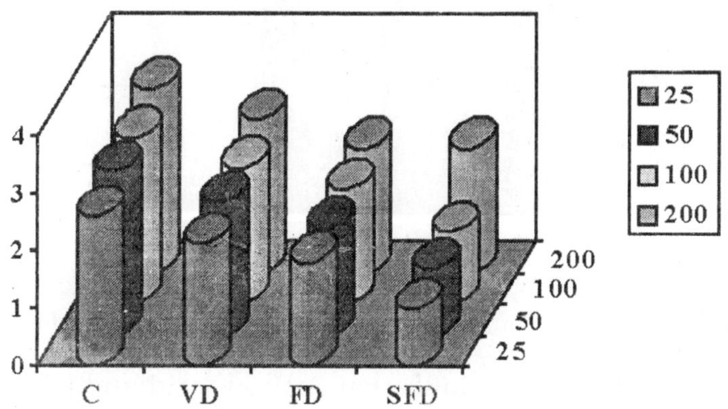

Figure 1. Seed yield of clusterbean under drought at vegetative (VD), flowering (FD) and seed formation (SF) growth stages under varying levels of potassium

Table 8. Influence of thiourea on clusterbean (Garg *et al.*, 2006)

Treatment	Seed yield (kg ha^{-1})	DMP (kg ha^{-1})	Net Photosynthesis (μ mol m^{-2} s^{-1})	NRA (μg NO_2 g^{-1} dwh^{-1})
Control	295	1172	10.3	348
Seed treatment (Thiourea)	382	1409	13.1	512
Seed treatment + spray (Thiourea)	410	1512	13.5	581

Data average of two years (1999 and 2000)

Likewise, kinetin (5 ppm) either as pre-sowing seed treatment or as foliar spray or both (at 25 and 40 DAS) significantly improved the growth, DMP, seed yield and WUE of clusterbean under field drought condition (38). Combined application of phosphorus and thiourea has been found to act synergistically and improve growth and yield of clusterbean under simulated drought conditions (39). This is mediated due to their favourable and additive effects on photosynthetic efficiency, leaf area development and nitrogen metabolism.

The usefulness of thiourea application has also been demonstrated at farmers' fields (40) where seed treatment with 500 ppm thiourea followed by a single spray of thiourea (1000 ppm) at the pre-flowering stage significantly improved the growth and seed yield of clusterbean, moth bean and mung bean (Table 9 & Fig 2.). It has been concluded that thiourea treatment can significantly enhance the productivity of all these legumes under arid conditions (41).

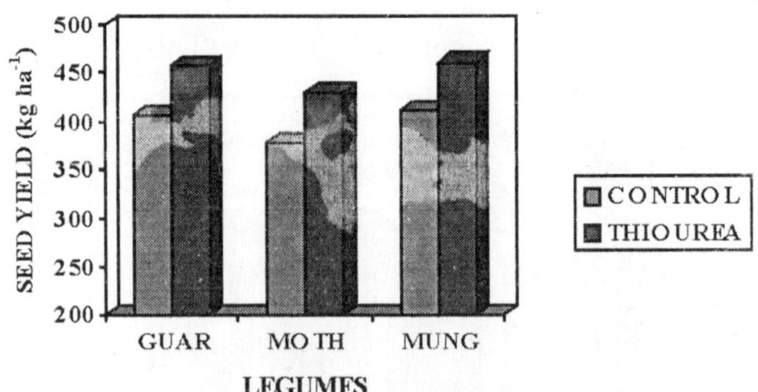

Figure 2. Influence of thiourea on seed yield of arid legumes at farmers field

Table 9. Influence of thiourea application on seed yield and DMP of clusterbean and moth bean genotypes at farmer's field (Garg and Burman, 2006)

Legumes	Seed yield (kg ha^{-1})		Dry matter production (kg ha^{-1})	
	Control	Thiourea treated	Control	Thiourea treated
Moth bean				
RMO-40	507	569	1218	1320
CZM-99	450	521	1125	1254
Clusterbean				
RGC-936	405	467	1416	1602
RGC-1003	438	543	1377	1994

Improvement in performance of a crop under water shortage can arise both from genetic changes to the crop that enable it to perform better in drought-prone environments and from agronomic changes that enable it to perform better under these constraints. It is evident that physiological manipulation along with good crop management is required for the most efficient use of limited rain water by different arid legumes. Thus, to face drought stress and to achieve sufficient and sustainable seed yield in the future, requires a multidisciplinary approach based on plant genomic, physiology, agronomy and modelling.

REFERENCES

1. Garg, B.K. 2003. Physiological aspects of abiotic stress tolerance in arid legumes (in) *Advances in Arid Legumes Research* pp: 347-354. A Henry, D Kumar and N B Singh (Eds) Scientific Publishers, Jodhpur.
2. Garg, B. K. and Burman, U. 2002. Physiology (in) *Guar in India* pp. 63-83 D.
3. Garg, B. K. and Burman, U. 2002. Moth bean physiology (in) *Moth bean in India* pp. 87-102, D. Kumar and N. B. Singh (Eds) Scientific Publishers, Jodhpur.
4. Arnon, I. 1992. Agriculture in Drylands – Principles and Practice. Elesiver, Amsterdam
5. Turner, N. C. 1997. Further Progress in crop water relations. *Advances in Agronomy* 39: 1-51.
6. Garg, B.K. 2002. Mechanisms of drought tolerance and strategies for yield improvement under drought. *Current Agriculture.* 26:15-21.
7. Turner, N. C. 1986. Adaptations to water deficits. *Australian Journal of Plant Physiology* 13: 175-190.
8. Levitt, J. 1980. *Responses of Plants to Environmental Stresses* Vol 2. Academic Press, New York
9. Voltas, J., Lopez, Coircoles, H. and Borras, G. 2005. Use of bipolar analysis and factorial regression for the investigation of superior genotypes in multi-environment trials. *European Journal of Agronomy*. 22:309-324.
10. Garg, B.K. 2008. Physiology of drought tolerance of clusterbean and moth bean. *Journal of Arid Legumes*. 5:8-14.
11. Idso, S.B., Reginado, R., Reicosky, D. and Hatfield, J. 1981. Determining soil induced plant water potential depression in alfalfa by means of infrared thermometer. *Agronomy Journal*. 73:826-830.
12. Karamanos, A.J. and Papatheohari, A.Y. 1999. Assessment of drought resistance of crop genotypes by means of Water Potential Index. *Crop Science*. 39:1792-1797.
13. Rizza, F., Badeck, F.W., Cattivelli, I. and Staca, A.M. 2004. Use of a water stress index to identify barley genotypes adapted to rainfed and irrigated conditions. *Crop Science*. 44:2127-2137.
14. Jones, H.G. 2007. Monitoring plant and soil water status: established and novel methods revisited and their relevance to studies of drought tolerance. *Journal of Experimental Botany*. 58:119-130.
15. Garg, B.K., Kathju, S., Vyas, S.P. and Lahiri, A.N. 1998. Influence of water deficit stress at various growth stages on some enzymes of nitrogen metabolism and yield in clusterbean genotypes. *Indian Journal of Plant Physiology* 3: 214-218
16. Garg, B.K., Burman, U. and Kathju, S. 2005. Comparative water relations, photosynthesis and nitrogen metabolism of arid legumes under water stress. *Journal of Plant Biology* 32: 83-93.

17. Fisher, R.A. Rees, D., Sayrre, K.D. Lu, Z.M. and Candon, A.G. 1998. Wheat yield progress associated with higher stomatal conductance and photosynthetic rate and cooler canopies. *Crop Science* 38:1467-1475.

18. Campos, H., Cooper, M., Habben, J.E., Edmeades, G.O. and Schussler, J.R. 2004. Improving drought tolerance in maize: a view from industry. *Field Crop Research*. 90: 19-34.

19. Morgan, J.M. 1984. Osmoregulation and water stress in higher plants. *Annual Review of Plant Physiology* 35: 292-319.

20. Moinuddin, A., Fisher, R.A. Sayre, K.D. and Reynolds, M.P. 2005. Osmotic adjustment in wheat in relation to grain yield under water-deficit environments. *Agronomy Journal*. 97:1062-1071.

21. Farquhar, G.D. and Richards, R.A. 1984. Isotopic composition of plant carbon correlates with water-use efficiency of wheat genotypes. *Australian Journal of Plant Physiology*. 11: 39-52.

22. Blum, A. 1988. Improving wheat grain filling under stress by stem reserve mobilization. *Euphytica*. 100: 77-83.

23. Garg, B.K., Kathju, S., Vyas, S.P. and Lahiri, A.N. 2003. Relative performance, evapotranspiration, water and nitrogen use efficiency and nitrogen metabolism of clusterbean genotypes under arid environment. (in): *Advances in Arid Legumes Research*. pp 212-220 A. Henry, D. Kumar and N.B. Singh (Eds) Scientific Publishers, Jodhpur

24. Garg, B. K., Burman, U., and Kathju, S. 2004a. Effect of water stress on moth bean (*Vigna aconitifolia*) genotypes. *Indian Journal of Plant Physiology* 9: 29-35.

25. Vyas, S.P., Garg, B.K., Kathju, S. and Lahiri, A.N. 2003a. Relative drought tolerance of moth bean genotypes (in) *Advances in Arid Legumes Research* pp. 394-397. A Henry, D Kumar and N B Singh (Eds) Scientific Publishers, Jodhpur.

26. Garg, B. K., Kathju, S. and Burman, U. 2001. Influence of water stress on water relations, photosynthetic rate and nitrogen metabolism of moth bean genotypes. *Biologia Plantarum* 44: 289-292.

27. Vyas, S. P., Kathju, S., Garg, B.K. and Lahiri, A.N. 1996. Activities of nitrate reductase and ammonia assimilating enzymes of moth bean under water stress. *Science and Culture* 62: 213-214.

28. Kishore, P.B.R., Hong-Zong, L., Verma, P.A.S. and Miiao, C.H. 1995. Over production of DELTA-pyrroline S-carbohylase synthetase increases proline production and confers osmotolerance in transgenic plants. *Plant Physiology* 108: 1387-1394. Kumar and N B Singh (Eds.) Scientific Publishers, Jodhpur.

29. Kathju, S., Garg, B.K., Vyas, S.P. and Lahiri, A.N. 2003. Sustainable production of moth bean through genotype management under arid environments. *Journal of Arid Environments* 53: 137-143.

30. Vyas, S.P., Garg, B.K., Kathju, S. and Lahiri, A.N. 2003b. Genotypic management of clusterbean for sustainable production under arid environment. *Indian Journal of Pulse Research* 16: 104-107.

31. Garg, B.K., Kathju, S., Vyas, S.P. and Lahiri, A.N. 2002. Influence of spacings on performance, water use, nutrient uptake and nitrogen use efficiency of clusterbean under diverse rainfall conditions. (in) *Human Impact on Desert Environment* pp. 294-304. P. Narain, S. Kathju, Amal Kar and M.P. Singh (Eds) Scientific Publishers, Jodhpur.
32. Burman, U., Garg, B.K. and Kathju, S. 2002. Effect of spacing on seed yield and physiological traits in moth bean (*Vigna aconitifolia*). *Indian Journal of Agricultural Sciences* 72: 76-79.
33. Lahiri, A.N. 1980. Interaction of water stress and mineral nutrition on growth and yield (in): *Adaptations of Plants to Water and High Temperature Stress* pp. 341-362. N C Turner and P J Kramer (Eds) John Wiley, New York
34. Garg, B.K., Burman, U. and Kathju, S. 2004b. The influence of phosphorus nutrition the physiological response of moth bean genotypes to drought. *Journal of Plant Nutrition and Soil Science* 167: 503 – 508.
35. Mengel, K and Kirkby, E.A. 1980. Potassium in crop production. *Advances in Agronomy* 33: 59-110.
36. Vyas, S.P., Garg, B.K., Kathju, S. and Lahiri, A.N. 2001. Influence of potassium on water relations, photosynthesis, nitrogen metabolism and yield of clusterbean under soil moisture deficit stress. *Indian Journal of Plant Physiology* 6: 30-37.
37. Garg, B.K., Burman, U. and Kathju, S. 2006. Influence of thiourea on photosynthesis, nitrogen metabolism and yield of clusterbean (*Cyamopsis tetragonoloba* (L.) Taub.) under rainfed conditions of Indian arid zone. *Plant Growth Regulation* 48:237-245
38. Burman, U., Garg, B.K. and Kathju, S. 2003. Influence of kinetin on photosynthesis, nitrogen metabolism and yield of clusterbean under moisture deficit conditions. *Indian Journal of Plant Physiology* 8: 287-291.
39. Burman, U., Garg, B.K. and Kathju, S. 2004. Interactive effects of thiourea and phosphorus under simulated water stress. *Biologia Plantarum* 48: 61-65.
40. Garg, B.K. and Burman, U. 2006. Yield improvement of clusterbean and moth bean by thiourea application under arid conditions. *Journal of Arid Legumes* 3:51-54.
41. Garg, B.K. 2007. Role of thiourea in improving the productivity of arid legumes. *Journal of Arid Legumes* 4: 1-5.

Performance of promising moth bean genotypes in extreme arid environment

R.S. Mertia, B.K. Kandpal*, R.N. Kumawat and S.S. Mahajan

Central Arid Zone Research Institute, Regional Research Station, Jaisalmer-345001, Rajasthan

*Corresponding author (Email: basantkandpal@yahoo.com)

ABSTRACT

A varietal trial on promising moth bean (*Vigna aconitifolia*) genotypes was conducted during *Kharif* 2006 to evaluate their growth and yield potential in extreme arid environment of Jaisalmer. Total eleven genotypes including two national check varieties (CZM-1 and RMO- 225) were sown in randomized block design with four replications at Chandan farm of Jaisalmer. Among the genotypes comparatively late 50 % flowering was observed in GMO-0111 (42 days), GMO-03-06 (40 days), GMO -03-12 (38 days) and RMO 140 (37 days) in comparison to 33-34 days in check varieties. However, total maturity period of all the genotypes remain almost same. Maximum plant height and number of branches per plants were recorded in RMO-140 followed by GMO-0111 and RMO-06-04. The number of pods per plant was highest in RMO-225 followed by HM-61. Seeds per pod were statistically at par in all the genotypes while 100 seed weight was maximum in GMO-0111 but remained at par with GMO-03-12, CZM-1 and RMO-06-01. Over all GMO-03-06 produced significantly higher seed yield (810 kg/ha) than both check varieties but remained at par with RMO 06-01, RMO 140 and GMO-0111. Almost similar trend was recorded for straw yield as well. The overall findings reveal superiority of dual purpose genotypes viz. GMO-03-06, RMO-06-01 and RMO-140 under extreme arid environment of Jaisalmer.

Key words: Moth bean, genotypes, performance, yield.

INTRODUCTION

Moth bean [*Vigna aconitifolia* (Jacq.) Marechall] is one of the most important drought hardy and high temperature tolerant *Kharif* crop among the arid legumes cultivated in thar desert of India and Pakistan. It serves as a major source of dietary protein along with cowpea and green gram in this region (1). Dry seeds are also used for a number of delicious confectionary items such as *Papad, Mangon, Mogar, Namkins* and popularly known salted snacks (2) called *Bikanery-Bhujia* (3). High demand from confectionary as well as household resulted into sufficient area under moth bean in western Rajasthan. In spite of high demands the yield of moth bean is very low. The productivity of moth bean depends upon the selection of suitable germplasm besides adoption of soil-site specific agronomic practices. The germplasm must be capable to survive intermittent dry spells, recoup quickly with subsequent rains, matures before the terminal drought and makes non-significant yield losses. The continuous research efforts under National Network Research Project on Arid Legumes have resulted into identification and release of several desired genotypes. However, information to their adaptability to a soil-site is essentially required before introduction to an area. Keeping this hypothesis a study was conducted during *Kharif* 2006 at Chandan Farm of Regional Research Station, Jaisalmer to evaluate recently identified moth bean germplasm.

MATERIALS AND METHODS

The Climate

The agro-climatic situation of Jaisalmer district is characterized by low and erratic rainfall, high wind speed, extreme of temperatures, high evapotranspiration rate, serve wind erosion and thus, frequent droughts. The over all climate is categorized under Marusthali, hot, hyper arid eco-sub-region (4).

During *Kharif*, 2006, the first rain of monsoon was recorded on June 28. It was followed by a dry spell throughout July. The monsoon was re-activated on August 2. Out of total 223 mm rainfall 221 mm rainfall in 11 rainy days was recorded during August month. The monsoon withdrew after September 7 (2.00 mm). Overall 150.5 mm rainfall was received during the crop growing period against 439.0 mm potential evapotranspiration. The mean temperature throughout the growing period remained around 30°C, while average wind speed was 7.26 km hr^{-1}.

The Soils

The experiment was laid out in almost level (0-1% slope) aeolian plain. The soils of the sites were classified as deep, mixed, hyperthermic, calcareous family of Typic Torripsamments (Table 1). The soil was low in organic carbon (0.12 %) and available N content (207 kg $KMnO_4$-N/ha), medium in P (14 kg P/ha) and K content (138 kg K/ha). These almost level, well drained, very deep sandy loam soils with 4.0 to 6.2% free $CaCO_3$ content provide sufficient soil volume and condition for *in situ* rain water conservation and penetration of moth bean roots deeper into sub surface soils. However, alkaline condition of the soils (pH> 8.0) imposed slight limitation for successful rainfed cultivation of moth bean.

Table 1. Physio-chemical properties of the soils of the experimental site

Horizon	Depth (cm)	Particle Size Distribution			$CaCO_3$ (%)	pHe	ECe dSm^{-1}	Organic Carbon (%)	CEC (Cmol (+p)/kg)
		Sand (%)	Clay (%)	Texture					
Mixed, hyperthermic, calcareous family of Typic Torripsamments									
Ap	0-20	77.4	6.1	ls	5.4	8.7	0.8	0.12	3.5
C1	20-55	83.2	6.8	ls	4.0	8.6	0.5	0.08	2.8
C2	55-108	86.1	6.5	ls	5.2	8.6	0.3	0.06	2.4
C2	108-137	86.7	6.7	ls	6.2	8.7	0.4	0.06	2.4

ls= loamy sand

The Experimental Details

Eleven improved genotypes/varieties were sown on August 9, 2006 in complete randomized block design with 4 replications. Genotypes sown include HM 61, RMO- 140, RMO-06-01, RMO-06-04, MH-52, CZM-105, GMO-03-06, GMO-03-12, GMO-0111 and 2 checks (CZM-1 and RMO-225). The genotypes were sown in lines at 30 cm spacing. Thinning/gap filling of the experimental plots was done 10 days after sowing to maintain the optimum plant population. At maturity 5 plants per plot were randomly harvested from second and third rows from the western side of the each plot to estimate mean plant height (cm), number of branches per plant, number of pods per plant, number of seeds per pods and 100 seed weight (g). The net plot was demarcated and harvested to estimate seed and biological yield (kg/ha). Analysis of variance (ANOVA) technique was used to compare the means of different growth and yield attributes to draw logical conclusions.

RESULTS AND DISCUSSION

Growth and Yield attributes

Under extreme arid conditions availability of moisture (rainfall) is the most critical factor besides optimum temperature and soil health for initial establishment, growth and development of a crop. Frequent rains in sufficient quantity during August month effectuated all the genotypes to achieve maximum growth potential and initiation of flowering. Thus, the genotypes showed significant variation in plant height (15.5-20.2 cm), branches per plant (0.5-2.9), straw weight (6.15-13.49 g plant^{-1}), pods per plant (22.8-34.9), 100-seed weight (2.85-3.44 g), seed weight per plant (3.16-4.53 g) and total biomass per plant (9.31-18.03 g). The detailed data is presented in Table 2 and 3.

Genotype GMO-03-06 attained significantly lower height than both check varieties. Height of rest of the varieties was at par with checks except RMO-140 whose height was significantly more than the check CZM-1 only. RMO-140 produced significantly higher number of branches per plant than rest of the genotypes including both checks except GMO-0111 and RMO-06-01, which remained at par. Genotype CZM-105 recorded lowest straw weight and total biomass per plant however remained at par with rest of the genotypes except RMO-06-01, RMO-06-04, RMO-0140 and GMO-0111. Overall RMO-06-01 produced significantly more straw and total biomass per plant than rest of the genotypes except RMO-06-04 which is statistically at par. Frequent rains up to vegetative stage of the genotypes made them attain height, branches per plant and overall biological frame as per their genetic potentials.

Days to 50% flowering shows conspicuous division of genotypes in two groups- early flowering genotypes (mean 33.6 days) and late flowering genotypes (mean 39.4 days). Genotypes GMO-0111, GMO-03-06, GMO-03-12 and RMO-140 took more than 37 days to 50% flowering, while rest genotypes flowered early. Genotype GMO-03-12 took significantly maximum days to maturity (82.3 days) than rest of the genotypes except GMO-0111 and RMO-140, which were at par. Genotypes CZM-1 matured in shortest duration but remained at par with RMO-06-04, HM-61 and RMO-225. Variety RMO-225 produced significantly maximum number of pods per plant but remain at par with HM-61, RMO-140, MH-51 and GMO-03-06. It is mainly due to the overall larger biological frame of these genotypes on which more flowers- and eventually more pods-can develop. Several workers have reported similar trend (5, 6, 7 and 8). Pod length and number of seeds per pod did not show any significant variation among the genotypes. It may be due to experiencing almost similar moisture stress situation during pod formation and elongation stages. 100-seed weight was found significantly maximum in genotype GMO-0111 but remained at par with GMO-03-12, CZM-1 and RMO-06-01. 100-seed weight is the first benchmark for the photosynthates translocation capacity of the genotypes from source to sink. These genotypes thus, showed efficient translocation capacity under moisture stress conditions.

Seed weight per plant is the function of pods per plant, seeds per pod and 100-seed weight. Thus, their cumulative impact has produced significant variation in seed weight per plant. Genotype RMO-06-01 produced significantly maximum seed weight per plant (4.53 g) but remained at par with RMO-225, HM-61, RMO-06-04, GMO-03-06 and GMO-0111. Several workers have reported similar findings (5, 7).

Total biological and seed yield

At constant plant density, biological and seed yield of moth bean are the function of overall growth and yield attributes per plant (5, 8). The significantly larger biological frame (plant height and number of branches per plant) to produce more number of flowers and thus, more pods per plant and 100-seed weight resulted into significant variation in straw yield (1292-2571 kg/ha) and seed yield (412-810 kg/ha) (Table 3). Genotype GMO-03-06 produced significantly higher straw and seed yield than other genotypes but remained at par with RMO-06-01 and RMO-140. Variety CZM-1 produced minimum straw and seed yield.

Table 2. Growth parameters of various moth bean genotypes at extreme arid conditions

Variety	Plant Height (cm)	Branches per plant	Straw weight (g/plant)	Total biomass (g/plant)	Days to 50% flowering	Days of maturity
HM-61	16.8	1.1	8.28	12.38	34.0	75.3
RMO-140	20.2	2.9	10.04	13.78	37.3	79.8
RMO-06-01	19.0	2.0	13.49	18.03	33.5	78.0
RMO-06-04	18.7	1.8	10.70	14.72	33.8	74.5
MH-52	16.9	1.1	7.20	10.85	32.8	75.8
CZM-105	16.6	0.5	6.15	9.31	33.8	77.3
GMO-03-06	15.5	1.8	8.59	12.61	40.3	79.0
GMO-03-12	18.7	1.5	8.77	12.21	38.3	82.3
GMO-0111	19.3	2.4	9.49	13.28	41.8	80.0
CZM-1 (C)	17.6	1.4	7.06	10.33	32.8	74.0
RMO-225(C)	18.4	1.7	8.32	12.69	34.5	76.8
SEm±	*0.7*	*0.3*	*1.10*	*1.18*	*0.6*	*1.0*
CD (0.05%)	*1.9*	*0.9*	*3.19*	*3.41*	*1.8*	*2.9*
CV (%)	7.4	37.4	24.8	18.5	3.4	2.5

Sufficient moisture availability during grand growth stage of crop incited genotypes to achieve their full genetic growth potential. Therefore, semi-erect genotypes with longer vegetative period attained more growth and thus higher straw yield. Significant moisture stress at pod formation stage restricted genotypes to attain their potential seed yield. The impact of moisture stress was more in longer duration genotypes as reflected by harvest index. However, the impact was nullified to a major extent due to more sink translocation from larger photosynthetic area.

The data on harvest index suggest that early flowering genotypes had better harvest index (32.2-37.8%) in comparison to late flowering genotypes (29.6-31.5%). Among the genotypes MH-52 recorded significantly maximum harvest index (37.8%) followed by RMO-06-04 and RMO-225. Lowest harvest index (29.6%) was recorded in GMO-03-12. It thus, seems that the partitioning of photosynthates towards seed was more efficient in the early flowering genotypes. The results are in confirmation with the findings of other workers (5, 9).

Table 3. Yield attributes and yield (kg/ha) of various moth bean genotypes at extreme arid conditions.

Variety	Pod per plant	Pod length (cm)	Seeds per pod (No.)	100 seed weight (g)	Seed weight (g/plant)	Seed yield (kg/ha)	Straw yield (kg/ha)	Harvest Index (%)
HM-61	34.5	3.23	4.25	3.01	4.10	565	1753	32.4
RMO-140	32.7	3.28	4.45	3.11	3.75	704	2270	30.9
RMO-06-01	28.1	3.15	4.35	3.22	4.53	752	2310	32.7
RMO-06-04	26.9	3.28	4.58	2.88	4.02	514	1446	35.4
MH-52	30.2	3.20	4.25	3.16	3.65	514	1370	37.8
CZM-105	25.5	3.08	4.18	2.85	3.16	523	1532	33.8
GMO-03-06	32.4	3.25	4.58	2.93	4.02	810	2571	31.5
GMO-03-12	23.4	3.33	4.53	3.32	3.44	514	1742	29.6
GMO-0111	25.0	3.38	4.93	3.44	3.80	676	2202	30.8
CZM-1 (C)	22.8	3.08	4.13	3.30	3.28	412	1292	32.2
RMO-225(C)	34.9	3.28	4.55	3.18	4.38	556	1597	34.8
SEm±	*2.1*	*0.07*	*0.28*	*0.08*	*0.28*	*40*	*105*	*1.1*
CD (0.05%)	*6.2*	*NS*	*NS*	*0.23*	*0.81*	*115*	*305*	*3.3*
CV (%)	14.9	4.3	12.5	5.1	14.7	13.4	11.6	7.0

CONCLUSION

The findings presented here indicate that under rainfed extreme arid situation of Jaisalmer dual purpose moth bean genotypes viz. GMO-03-06, RMO-06-01 and RMO-140 are more water efficient and able to produce higher seed and straw yield. Since these genotypes have proved their superiority over other genotypes and check varieties, a complete package of practice must be developed for large scale cultivation if farmers' field under livestock based farming system of extreme arid thar region.

REFERENCES

1. Sihag, S.K.; Khatri, R.S. and Singh, J.V. (2004). Genetic divergence analysis in moth bean [*Vigna aconitifolia* (Jacq.) Marechall]. *Journal of Arid Legumes* 1(2):98-100.
2. Henry, A. and Singh, R.P. (1985). Moth bean. In: *Efficient management of dry land crops*. (Eds.) V. Balasubbramanian and J. Venkateshwarlu. Central Research Institute for Dryland Agriculture, Hyderabad. pp 199-207.
3. Bhansali, A.K.; Choudhary, B.R.; Kumhar, S.R. and Bhandari, M.M. (2004). Stability of seed yield and other traints in moth bean [*Vigna aconitifolia* (Jacq.) Marchall]. *Journal of Arid Legumes* 1(2):95-97.
4. Velayutham, M.; Mandal, D.K.; Mandal, C. and Sehgal, J. (1999). Agro-ecological sub-regions of India for planning and development. NBSS &LUP Pub. 35. National bureau of Soil Survey and land use planning, Nagpur, India. 372p.
5. Mertia, R.S.; Kandpal, B.K. and Kumawat, R.N. (2006). Comparative performance of moth bean (*Vigna aconitifolia*) varieties under rainfed conditions. *Journal of Arid Legumes* 3(1):15-18.
6. Kakani, R.K.; Garg, D. K.; Sharma, R.C. and Shiva Narayana (2003). Evaluation of germplasm lines of moth bean. In: *Advances in Arid Legume Research* (Eds. A. Henery, D. Kumar and N.B. Singh). Scientific Publishers (India) Jodhpur. 107-111.
7. Tikka, S.B.S.; Yadvendra, J.P.; Bordia, P.C. and Kumar, S. (1976). A correlation and path coefficient analysis of components of grain yield in moth bean. *Gentica Agriculture.* 30: 241-248.
8. Tikka, S.B.S.; Asawa, B.M. and Kumar, S. (1977). Correlation response to selection in moth bean *Gujarat Agriculture University Research Journal* 3:1-4.
9. Vyas S.P.; Garg, B.K.; Kathju, S. and Lahiri, A.N. (2003). Relative drought tolerance of moth bean genotypes. In: *Advances in Arid Legume Research* (Eds. A. Henery, D. Kumar and N.B. Singh). Scientific Publishers (India) Jodhpur. 393-397.

Direction and magnitude of association among the yield and its components characters in M_4 generation in horsegram (*Macrotyloma uniflorum* Lam. Verdc)

N.B. Patel, S.B.S. Tikka and J.B. Patel

S. D. Agricultural University, Sardarkrushinagar (Gujarat)

ABSTRACT

Effect of different doses of gamma rays (5, 10, 15, 20, 25, 30, 35 and 40) using three varieties of horsegram viz., AK-21, AK-42 and Maru-K-1 by using the common and reliable parameters under laboratory and field conditions in horsegram (*Macrotyloma uniflorum* Lam. Verdc) was studied at Main Pulses Research Station, S.D. Agricultural University, Sardarkrushinagar during summer-2004. In context to relationship of yield per plant and component characters in M_4 generation, it was observed that yield per plant showed highly significant and positive genotypic correlation with plant height, number of effective branches per plant and number of pods per plant in AK-21; with days to 50% flowering, plant height, number of effective branches per plant, number of pods per plant and number of seeds per pod in AK-42; with plant height, number of pods per plant, pod length and number of seeds per pod in MK-1 and with days to 50% flowering and days to maturity in pooled over varieties. Yield per plant also showed significant negative genotypic correlation with days to 50% flowering and pod length in AK-21; with days to 50% flowering in MK-1 and with number of effective branches per plant and test weight in pooled analysis.

Key words: Horsegram, gamma rays, M_4 generation, components, association.

INTRODUCTION

Horsegram (*Macrotyloma uniflorum* Lam. verdc, Dolichos biflorus) 2n=20, vernacularly known as Kulthi, Kalath is a traditional tropical grain legume is extensively grown in peninsular India and the cultivation extends up to 5000 feet elevation in hilly areas of Himachal Pradesh and Nepal. It is cultivated both in wet and cool seasons. In Gujarat it is a minor pulse crop grown in two thousand odd hectares in punctuated pockets in the tribal and difficult terrains of Dangs, Rajkot, Surendranagar, Narmada, Sabarkantha, Panchmahals, Dahod and Banaskantha districts. Despite the average productivity is around 526 kg/ha that is more than the national average (1).

Improvement in any crop depends upon the extent, nature and magnitude of genetic variability present in the population and the extent to which it is heritable. Yield is complex and polygenic characters and highly influenced by the environment, therefore, it is necessary for plant breeder to know the role of environment for expression of traits. To exploit variability for improving seed yield, knowledge on interrelationship between yield and its component and among the component character is essential. Further, path coefficient analysis provides precise information about the cause and effect situation and helps in determining the selection criteria in crop improvement. The literature available in horse gram has been reviewed are presented below.

MATERIALS AND METHODS

Pure, healthy and mature seeds (250g) of the three cultivars of horsegram (*Macrotyloma unifloram* Lam. Verdc) *viz.,* AK 21, AK 42 and Maru K-1 were obtained from the germplasm pool maintained at Main Pulses Research Station, Sardarkrushinagar Dantiwada Agricultural University, Sardarkrushinagar and the same were exposed to different doses of gamma rays at Bhabha Atomic Research Centre (BARC), Trombay with gamma rays intensity of 1.8 kR per minute. The doses were given *viz.,* 0, 5, 10, 15, 20, 25, 30, 35 and 40 kR to all the three varieties, therefore, making 27 treatments as mentioned below:

AK-21	:	0, 5, 10, 15, 20, 25, 30, 35 and 40 kR
AK-42	:	0, 5, 10, 15, 20, 25, 30, 35 and 40 kR
Maru-K-1	:	0, 5, 10, 15, 20, 25, 30, 35 and 40 kR

The mutated seed were grown at Main Pulses Research Station, Sardarkrushinagar Dantiwada Agricultural University, Sardarkrushinagar during *summer* 2004 to *kharif* 2005. M_1 generation was raised with perfect agronomic conditions in field with out any replication. From each treatment of M_1 generation, 25 normal appearing plants were randomly selected to provide material for M_2 generation up to 20 kR. M_2 generation was raised to assess induced polygenic variability and to score the types of macro-mutations and their frequency. M_2 and M_3 generation were raised in Compact Family Block Design with three replications. 25 seeds each from selected plants from M_1 were sown in a row for each replication. Selected 25 plants from each treatment of M_1

generation become families in M_2 generation. The row to row spacing was 45 cm and plant to plant 15 cm. Similarly M_3 generation was sown in Compact family block design. M_4 generation was raised in Randomized block design with three replications. Selected 58 lines which yielded higher than respective check in M_3 generation were selected and sown for their superiority in M_4 generation along with respective checks. The row to row spacing was 45 cm and plant to plant 15 cm. All recommended agronomical practices were followed.

The phenotypic, genotypic and environmental correlation coefficients between pairs of all characters studied were worked out. For this purpose, data were subjected to covariance analysis (2). The form of analysis used for estimating covariance components between different pair of observations was the same as that of for analysis of variance except that sum of squares and mean squares were replaced by sum of products, respectively.

RESULTS AND DISCUSSION

In M_4 generation, mutants which recorded higher yield in M_3 generation with respective check varieties are selected and evaluated for further study. Total of 58 mutants comprising of 20 mutants from AK-21, 16 from AK-42 and 22 from MK-1 were selected and evaluated along with three respective checks (AK-21, AK-42 and MK-1) for correlation of yield and yield attributing characters.

The correlation between two characters in an ordinary population should be the composite of the effect of selection, gene linkage and pleiotropy (3). If the nature of selection practiced in the treated and the control population is the same, any difference in the correlation coefficients between control and treated population will be due to the effect of mutagens on gene linkage and altered pleiotropic effect of the newly mutated gene. Hence, coefficients of correlation between all the nine characters were computed in the treatments including control for all the three varieties individually as well as pooled over three varieties.

Relationship of yield per plant and component characters in M_4 generation, it was observed that yield per plant showed highly significant and positive genotypic correlation with plant height, number of effective branches per plant and number of pods per plant in AK-21 (Table 1); with days to 50% flowering, plant height, number of effective branches per plant, number of pods per plant and number of seeds per pod in AK-42 (Table 2); with plant height, number of pods per plant, pod length and number of seeds per pod in MK-1 (Table 3) and with days to 50% flowering and days to maturity in pooled over varieties (Table 4). Yield per plant also showed significant negative genotypic correlation with days to 50% flowering and pod length in AK-21 (Table 1); with days to 50 %

flowering in MK-1 (Table 3) and with number of effective branches per plant and test weight in pooled analysis (Table 4). Perusal of data showed that yield per plant exhibited significant positive phenotypic correlation with number of pods per plant in AK-21 (Table 1); with number of pods per plant and number of seeds per pod in AK-42 (Table 2) and MK-1 (Table 3), respectively and with days to 50% flowering on pooled basis (Table 4). Similar results were obtained by Shivashankar *et al.*, (4) for plant height, number of pods per plant and number of seeds per pod, Patil and Deshmukh (5) for number of pods per plant and number of effective branches per plant, Birari *et al.* (6) for days to maturity, number of pods per plant and number of seeds per pod, Rao and Nanda (7) for days to maturity, plant height and pods per plant, Sharma (8) for pod length, number of pods per plant and number of seeds per pod, Lad *et al.*, (9) for number of pods per plant, pod length and number of seeds per pod, Nagaraja *et al.*, (10) for number of effective branches per plant, Nehru *et al.*, (11) for number of pods per plant, Prakash and Khanure (12) for number of pods per plant, number of effective branches per plant and plant height.

Table 1: Genotypic and phenotypic correlation coefficients among nine different characters in M_4 generation of AK-21

Characters	r	Days to 50% flowering	Days to maturity	Plant height	Number of effective branches per plant	Number of pods per plant	Pod length	Number of seeds per pod	Test weight
Yield per plant	r_g	-0.634**	-0.394	3.149**	1.395**	1.145**	-1.196**	0.140	-0.031
	r_p	0.227	-0.215	0.075	-0.083	0.779**	-0.110	0.334	-0.007
Days to 50% flowering	r_g		-0.092	26.589**	2.084**	-0.285	0.067	-0.349	-0.366
	r_p		-0.101	0.218	0.216	0.160	-0.073	-0.032	-0.240
Days to maturity	r_g			-31.587**	-2.166**	-0.377	-1.098**	-0.905**	-0.072
	r_p			0.207	-0.091	-0.039	-0.165	-0.327	0.031
Plant height	r_g				23.193**	7.430**	-64.098**	44.332**	39.672**
	r_p				0.020	0.080	-0.198	-0.012	0.100
Number of effective branches per plant	r_g					1.382**	-0.174	-0.745**	-0.874**
	r_p					-0.121	-0.004	0.035	-0.197
Number of pods per plant	r_g						-0.686**	0.063	-0.160
	r_p						-0.114	-0.069	-0.026
Pod length	r_g							-0.018	0.389
	r_p							0.026	0.138
Number of seeds per pod	r_g								0.334
	r_p								0.146

Negative correlation coefficient for days to 50% flowering with yield per plant in AK-21 and MK-1 were important for selecting early type of plants with higher yield. Similar result obtained by Rao and Nanda (7) and Sood *et al.* (13) in horsegram. Number of pods per plant exhibited positive significant genotypic correlation with number of seeds per pod in AK-42 (Table 2) and with pod length and number of seeds per pod in MK-1 (Table 3). This trait possessed significant positive and negative correlation with test weight and pod length on pooled basis (Table 4). Pod length exhibited positive significant genotypic correlation with number of seeds per pod in MK-1 only (Table 3).

Table 2: Genotypic and phenotypic correlation coefficients among nine different characters in M_4 generation of AK-42

Characters	r	Days to 50% flowering	Days to maturity	Plant height	Number of effective branches per plant	Number of pods per plant	Pod length	Number of seeds per pod	Test weight
Yield per plant	r_g	1.230**	-0.047	0.711**	1.367**	0.847**	-0.264	0.795**	-0.373
	r_p	-0.033	-0.033	0.217	0.190	0.792**	0.037	0.540*	-0.073
Days to 50% flowering	r_g		1.888**	0.718**	-0.959**	0.522*	-2.051**	0.969**	0.607**
	r_p		0.080	0.376	0.038	-0.092	-0.024	0.015	0.096
Days to maturity	r_g			-0.318	1.776**	0.373	-0.066	-0.287	0.301
	r_p			-0.160	-0.284	0.003	-0.160	-0.027	0.163
Plant height	r_g				-1.598**	0.205	-0.292	0.375	-0.571*
	r_p				0.115	0.046	0.033	0.274	-0.389
Number of effective branches per plant	r_g					-0.028	0.048	1.615**	4.575**
	r_p					0.191	0.133	0.160	0.126
Number of pods per plant	r_g						-0.542*	0.541*	-0.308
	r_p						-0.029	0.114	0.016
Pod length	r_g							-0.281	-0.060
	r_p							-0.119	0.100
Number of seeds per pod	r_g								-0.040
	r_p								-0.044

Table 3: Genotypic and phenotypic correlation coefficients among nine different characters in M_4 generation of MK-1

Characters	r	Days to 50% flowering	Days to maturity	Plant height	Number of effective branches per plant	Number of pods per plant	Pod length	Number of seeds per pod	Test weight
Yield per plant	r_g	-0.505*	-1.831**	0.707**	-0.129	6.628**	0.563**	0.905**	-0.285
	r_p	0.106	-0.096	0.184	0.107	0.773**	-0.086	0.571**	-0.101
Days to 50% flowering	r_g		-0.927**	0.646**	-1.814**	-17.157**	0.217	-0.295	0.213
	r_p		0.024	0.160	0.033	0.109	0.201	-0.049	0.085
Days to maturity	r_g			-1.686**	-2.827**	-0.015	-1.296**	-1.231**	-0.223
	r_p			0.002	0.045	-0.002	-0.163	-0.264	-0.165
Plant height	r_g				-3.392**	40.195**	0.734**	0.711**	0.050
	r_p				0.106	0.232	0.021	0.078	0.069
Number of effective branches per plant	r_g					0.153	0.032	-2.520**	2.599**
	r_p					0.045	-0.073	-0.009	0.124
Number of pods per plant	r_g						49.356**	64.586**	-7.754**
	r_p						-0.136	0.123	-0.002
Pod length	r_g							0.900**	0.048
	r_p							0.157	0.009
Number of seeds per pod	r_g								-0.008
	r_p								-0.038

Table 4: Pooled genotypic and phenotypic correlation coefficients among nine different characters in M_4 generation of three different varieties of horsegram

Characters	r	Days to 50% flowering	Days to maturity	Plant height	Number of effective branches per plant	Number of pods per plant	Pod length	Number of seeds per pod	Test weight
Yield per plant	r_g	0.522**	3.396**	-0.171	-0.260*	-0.171	-0.053	-0.036	-0.272*
	r_p	0.255**	0.088	0.064	0.052	-0.037	-0.018	-0.023	0.103
Days to 50% flowering	r_g		-1.647**	0.289*	-0.130	0.597**	-0.881**	-0.042	0.646**
	r_p		0.079	0.140	-0.061	0.107	0.033	-0.026	0.161
Days to maturity	r_g			1.865**	-0.238	-1.633**	-3.442**	0.505**	2.219**
	r_p			0.038	0.003	0.030	-0.073	0.009	0.071
Plant height	r_g				-0.312*	0.480**	-0.596**	-0.184	0.846**

	r_p	-0.105	0.055	-0.008	-0.004	0.776**
Number of effective branches per plant	r_g		0.213	-0.750**	0.118	-0.281**
	r_p		0.025	-0.164	0.080	-0.061
Number of pods per plant	r_g			-0.915**	0.125	0.605**
	r_p			-0.225*	0.023	0.482**
Pod length	r_g				0.001	-0.800**
	r_p				-0.004	-0.114
Number of seeds per pod	r_g					-0.197
	r_p					-0.058

If the genetic correlation is high, the two characters can be regarded as being substantially the same, and if there are no special circumstances affecting the heritability or the intensity of selection, it will make little difference in which environment the selection is carried out (14). From these inter-relationships, it can be presumed that for improving yield in horsegram, an ideal plant type would be dwarf, early in flowering and maturity with more number of pods with more pod length and higher 100-seed weight.

REFERENCES

1. Kumar, D. 2006. Horse gram in India. Scientific publishers (India), Jodhpur.
2. Al-Jibouri, H.A., Miller, P.A. and Robinson, H. 1958. Genotypic and environ-mental variance and covariance in an upland cotton cross of interspecific origin. Agronomy J., 50: 633-637.
3. Sakai, K.I. and Suzuki, A. 1964. Induced mutation and pleiotropy of genes responsible for quantitative characters in rice. *Radiation Botany*. 7 : 7-15.
4. Shivashankar, G., Jalikot, S.H., Chikkadevaiah and Shettar, B.I. 1977. Heritability and correlation studies in horse gram (*Macrotyloma uniflorus* Syn. *Dolichos biflorus*). *Mysore Journal of Agricultural Sciences*. 11 (2) : 211-214.
5. Patil, J.V. and Deshmukh, R.B. 1983. Association analysis in horse gram. *Journal of Maharashtra Agricultural Universities*. 8 (2) : 120-121.
6. Birari, S.P., Deorukhakar, N.V. and Jamadagni, B.M. 1987. Genetic parame-ters and character association in horse gram (*Dolichos biflorus* L.). *Indian Journal of Agricultural Research*. 21 (1) : 55-58.
7. Rao, S.S. and Nanda, H.C. 1994. Genetic variability and correlation studies for grain yield and its components in horse gram. *Indian Journal of Pulses Research*. 7 (1) : 59-61.
8. Sharma, R.K. 1995. Nature of variation and association among grain yield and yield components in horse gram. *Crop Improvement*. 22 (1) : 73-76.

9. Lad, D. B., Chavan, S.A. and Dumbre, A.D. 1999. Genetic variability and correlation studies in horse gram. *Journal of Maharashtra Agricultural Universities.* 23 (3) : 268-271.

10. Nagaraja, N., Nehru, S. D. and Manjunath, A. 1999. Plant type for high yield in horse gram as evidenced by path coefficients and selection indices. *Karnataka Journal of Agricultural Sciences.* 12 (1/4): 32-37.

11. Nehru, S. D., Manjunath, A. and Nagaraja, N. 2000. Identification of traits to be selected in breeding for yield in horse gram [*Macrotyloma uniflorum* (Lam.) Verdec.]. *Karnataka Journal of Agricultural Sciences.* 13 (1) : 11-15.

12. Prakash, B.G. and Khanure, S. K. 2000. Genetic parameters, correlation and path coefficient analysis in horse gram. *Karnataka Journal of Agricultural Sciences.* 13 (2) : 313-314.

13. Sood, B. C., Gartan, S. L. and Kalia, N. R. 1994. Variability, correlation and path studies in horse gram. *Indian Journal of Pulses Research.* 7 (1) : 68-69.

14. Falconer, D.S. 1960. Introduction to quantitative genetics. Ronald Press Co., New York.

Study of the nature and extent of induced quantitative variability for different agronomic characters including grain yield in horsegram (*Macrotyloma uniflorum* Lam. Verdc)

N.B. Patel, S.B.S. Tikka and J.B. Patel

S. D. Agricultural University, Sardarkrushinagar

ABSTRACT

Analysis of the nature and extent of induced quantitative variability for different agronomic characters including grain yield in horsegram (*Macrotyloma uniflorum* Lam. Verdc) revealed that in M_2 generation, radiation treatment induced variation for yield per plant and other characters was bidirectional in nature in most of the treatments. Gamma rays did not change treatment mean in yield per plant in all the varieties. Increase in mean values as compared to respective control was recorded for majority of characters studied except days to flowering, days to maturity and number of effective branches per plant, indicated accumulation of positive genes. The phenotypic and genotypic coefficients of variation increased in all the gamma rays treatments for all the characters except number of effective branches per plant in MK-1 and number of seeds per pod in AK-21 and AK-42. In all the varieties, genetic parameters such as heritability and genetic advance also increased in many fold in majority of characters. In M_3 generation, gamma rays treated progenies showed high proportion of GCV and PCV in comparison to respective controls. The magnitude of GCV, PCV and genetic advance were lower in characters like days to 50% flowering, days to maturity and pod length, while moderate for remaining characters. High heritability was observed for all the characters studied. The incredible expanded range of variation observed in positive direction for all the characters in all three varieties in the progenies of M_2 and M_3 generation could be attributed to substantial changes in the genetic background comprising modifier complexes during mutation process. In M_4 generation, lowest treatment mean with respect to days to 50% flowering and days to maturity was recorded in AK-42 and for plant height in AK-21. Similarly, highest mean values were recorded for different characters such as number of effective branches per plant in MK-1, number of pods per plant, test weight and yield per plant in AK-21 and pod length and number of seeds per pod in AK-21. A wide range of phenotypic variation was observed for plant height (AK-42) and number of pods per plant (AK-21). The moderate range of phenotypic variation was found for days to 50% flowering, days to maturity, number of effective branches per plant, pod length, number of seeds per pod, test weight and yield per plant in M_4 generation. Existence of low genotypic and phenotypic components of variability for all the characters suggested that these characters are more influenced by environment and therefore, the selection of these characters will not be much effective. Highest heritability was recorded for test weight in all three varieties.

Key words: Horsegram, gamma rays, M_2, M_3, M_4 generations, induced variability, green yield.

INTRODUCTION

The yield level of horse gram (*Macrotyloma uniflorum* Lam. Verdc) is inherently low that could be attributed to relatively narrow genetic variability that constitutes the corner stone of plant improvement activities. This variability could be enhanced through hybridization that leads to busting of undesirable linkages and consequently recombination of genes. However, crossing in horse gram is tedious and time/labour intensive. Alternatively, it has been demonstrated that the mutagenesis is equally effective in breaking gene castellations/developing new recombination and there by making immense genetic variability available to the breeders for effective selection of desirable types (1). Quite a good number of studies on induced quantitative variability in different pulse crops are available in the literature (2, 3, 4 and 5). However, such information is scanty in horse gram.

It is well recognized that quantitative characters are controlled by polygenic system. Since most of the characters of economic importance in the crop plants are determined by such multiple genes, their study bears a direct impact upon breeding. The extensive research on drosophila by Buzzati-Traverso and Scossiroli (6) and Scossiroli (7) elucidated that radiations can induce polygenic mutation and consequently genetic variability for quantitative characters.

Quantitative variability induced in M_2 generation revealed efficiency and effectiveness of mutagenic treatment doses. Increased proportions of effective variability, polygenic in nature to be identified by visual sight. The same one, however, expressed through statistical procedures only. The mean, range and coefficient of variation (CV) etc., are the simple parameters of measuring the rate/ proportion of induced quantitative variants in M_2 and M_3 generations as well.

MATERIALS AND METHODS

Pure, healthy and mature seeds (250g) of the three cultivars of horsegram (*Macrotyloma unifloram* Lam. Verdc) *viz.,* AK 21, AK 42 and Maru K-1 were obtained from the germplasm pool maintained at Main Pulses Research Station, Sardarkrushinagar Dantiwada Agricultural University, Sardarkrushinagar and the same were exposed to different doses of gamma rays at Bhabha Atomic Research Centre (BARC), Trombay with gamma rays intensity of 1.8 kR per minute. The doses were given *viz.,* 0, 5, 10, 15, 20, 25, 30, 35 and 40 kR to all the three varieties, therefore, making 27 treatments as mentioned below:

AK-21	:	0, 5, 10, 15, 20, 25, 30, 35 and 40 kR
AK-42	:	0, 5, 10, 15, 20, 25, 30, 35 and 40 kR
Maru-K-1	:	0, 5, 10, 15, 20, 25, 30, 35 and 40 kR

The mutated seed were grown at Main Pulses Research Station, Sardarkrushinagar Dantiwada Agricultural University, Sardarkrushinagar during *summer* 2004 to *kharif* 2005. M_1 generation was raised with perfect agronomic conditions in field with out any replication. From each treatment of M_1 generation, 25 normal appearing plants were randomly selected to provide material for M_2 generation up to 20 kR. M_2 generation was raised to assess induced polygenic variability and to score the types of macro-mutations and their frequency. M_2 and M_3 generation were raised in Compact Family Block Design with three replications. 25 seeds each from selected plants from M_1 were sown in a row for each replication. Selected 25 plants from each treatment of M_1 generation become families in M_2 generation. The row to row spacing was 45 cm and plant to plant 15 cm. Similarly M_3 generation was sown in Compact family block design. M_4 generation was raised in Randomized block design with three replications. Selected 58 lines which yielded higher than respective check in M_3 generation were selected and sown for their superiority in M_4 generation along with respective checks. The row to row spacing was 45 cm and plant to plant 15 cm. All recommended agronomical practices were followed.

The mean values for each character were analyzed using Compact Family Block design as suggested by Panse and Sukhatme (8). The phenotypic and genotypic variance and genetic advance (9), phenotypic and genotypic coefficient of variability (10) and heritability in broad sense (11) were estimated for grain yield per plant and its component traits.

RESULTS AND DISCUSSION

The range of variation induced either towards both the direction or towards negative direction in different mutagenic treatments in M_2 generations. Negative range was observed for days to 50% flowering and days to maturity in all the three varieties and number of effective branches per plant and number of pods per plant in MK-1 and number of seeds per pod in AK-21 and AK-42. For remaining characters including yield per plant positive shift of range was observed for all the mutagenic treatments in all the varieties (Table 1A & 1B).

M_3 generation exhibited incredible range in positive direction for all the characters in all the varieties. Gamma rays at 5 kR exhibited maximum range for plant height, number of effective branches per plant and pod length in AK-21 and AK-42, number of pods per plant in AK-42 and MK-1 and yield per plant in AK-21; 10 kR gamma rays for days to 50% flowering in AK-21 and MK-1, number of seeds per pod in AK-42 and MK-1, days to maturity in AK-21, number of effective branches per plant in MK-1, pod length in MK-1 and yield per plant in AK-42; 15 kR gamma rays for days to maturity in MK-1 and number of seeds per pod in AK-21 and 20 kR gamma rays for days to 50% flowering and days to maturity in AK-42, plant height and yield per plant in MK-1 and number of pods per plant in AK-21 (Table 2A & 2B).

This expanded range of variation for different measurements generated in the progenies of M_2 and M_3 generation could be attributed to substantial changes in the genetic background comprising modifier complexes during mutation process. Such mutations are aptly called micromutations (12), which have been measured in suitable biometric scale in this study. Mutations involving major loci may be drastic in their effects, very often injurious. However, micromutations are of great agricultural value since they tend to adjust themselves under diverse conditions. Expanded variability towards positive as well as negative direction has also been reported by various pulse workers (13 and 14 in green gram, 15 in cowpea, and 16 in black gram). Thus, it appeared that mutagenic treatments of three varieties have substantially enlarged the spectrum of variation useful for crop improvement as envisaged by Fray (17). This widely spread phenotypic range indicated that with the proper selection methods, there is a scope for improvement in different characters.

Induced variation comprises both environmental and genetic components and of the genetic variation only the heritable and fixable portion would be practical significance. In the present study, this aspect was studied through the estimation of phenotypic and genotypic coefficients of variation, heritability and genetic advance. The phenotypic and genotypic coefficients of variation were increased in all the gamma rays treatments in M_2 generation for all the characters except number of effective branches per plant in MK-1 and number of seeds per pod in AK-21 and AK-42, which had lower GCV and PCV values than their respective controls (Table 1A & 1B). In all the varieties, genetic parameters such as heritability and genetic advance also increased in many fold in almost all the treatments except genetic advance in plant height and number of effective branches per plant in MK-1 and number of seeds per pod in AK-21 and heritability and genetic advance in pod length in AK-21. Results indicated that gamma rays treatment found effective to induce variability in all three varieties of horsegram. The magnitude of GCV, PCV and genetic advance were lower in characters like days to 50% flowering, days to maturity and number of seeds per pod, while moderate for the remaining characters. The high estimates of heritability were observed for days to 50% flowering, days to maturity, plant height, pods per plant, pod length and yield per plant, while moderate for effective branches per plant and seeds per pod (Table 1A and 1B).

Table 1A: Phenotypic range, treatment mean, phenotypic and genotypic coefficient of variation, heritability (broad sense) and genetic advance for days to 50% flowering (nos.), days to maturity (nos.), plant height (cm) and for number of effective branches per plant (nos.) in M_2 generation of three different varieties in Horsegram

Treatments	Phenotypic range	Treatment mean	GCV %	PCV%	Heritability % (B.S)	GA %	GA (% mean)	Phenotypic range	Treatment mean	GCV %	PCV %	Heritability % (B.S)	GA %	GA (% mean)
	Days to 50 % flowering (nos.)								Plant Height (cm)					
			AK-21								AK-21			
Control	54.33-60.67	57.48	1.99	3.98	25.03	1.18	2.05	43.47-66.67	51.39	9.65	10.79	80.03	9.14	17.79
5 kR	52.00-65.33	56.93	5.11	6.45	62.66	4.74	8.33	39.33-62.33	47.55	11.32	12.65	80.07	9.92	20.87
10 kR	50.33-62.00	57.41	5.46	6.34	74.22	5.57	9.70	32.00-57.00	46.17	13.90	14.94	86.58	12.30	26.64
15 kR	53.67-61.67	56.99	3.37	5.00	45.26	2.66	4.66	27.33-55.33	41.15	14.90	16.05	86.26	11.73	28.51
20 kR	54.67-64.00	58.49	3.09	6.15	25.23	1.87	3.20	24.50-47.33	38.38	13.74	19.44	49.99	7.68	20.02
			AK-42								AK-42			
Control	51.00 - 57.67	55.16	2.45	4.18	34.32	1.63	2.96	45.53-69.53	58.11	11.96	13.78	75.41	12.44	21.40
5 kR	51.33 - 65.33	56.95	4.64	6.95	44.56	3.63	6.38	32.40-53.67	43.96	13.96	15.11	85.32	11.67	26.56
10 kR	53.67 - 62.00	56.37	1.30	6.32	4.25	0.31	0.55	35.00-55.69	42.84	13.67	15.06	82.37	10.95	25.56
15 kR	54.00 - 62.33	57.72	2.71	4.28	40.26	2.05	3.55	32.33-48.00	40.15	9.02	12.08	55.77	5.57	13.88
20 kR	48.33 - 65.00	57.32	6.14	7.20	72.70	6.18	10.78	19.67-46.54	37.65	18.12	19.33	87.90	13.18	35.00
			MK-1								MK-1			
Control	52.67 - 62.00	56.95	1.99	6.46	9.46	0.72	1.26	48.67-69.33	59.37	12.15	12.74	90.92	14.17	23.86

5 kR	50.00 - 62.67	57.39	3.20	7.67	17.42	1.58	2.75	37.00-64.60	49.74	13.58	14.46	88.26	13.07	26.28
10 kR	54.00 - 64.00	57.91	2.56	6.49	15.56	1.20	2.08	28.33-49.33	41.48	11.66	13.15	78.63	8.84	21.30
15 kR	53.00 - 64.00	58.25	4.64	5.76	64.96	4.49	7.71	29.63-49.60	40.04	14.23	15.45	84.86	10.81	27.01
20 kR	50.33 - 64.67	58.16	4.56	5.75	62.87	4.33	7.45	25.90-47.67	36.70	13.36	15.21	77.19	8.87	24.18
		Days to maturity						Number of effective branches per plant						
		AK-21						AK-21						
Control	101.67-112.67	106.76	1.74	4.10	17.97	1.62	1.52	2.60 - 3.67	3.15	4.88	18.15	7.23	0.09	2.71
5 kR	100.33-117.00	109.93	3.84	4.21	83.14	7.93	7.22	2.20 - 4.37	2.89	15.87	23.74	44.70	0.63	21.86
10 kR	105.25-119.00	114.07	2.35	3.04	59.95	4.28	3.75	2.23 - 3.67	2.94	11.54	24.14	22.86	0.33	11.37
15 kR	102.67-120.00	112.63	3.97	4.55	76.20	8.04	7.14	1.73 - 3.60	2.85	12.01	22.66	28.12	0.37	13.12
20 kR	106.67-120.33	114.97	2.73	3.67	55.27	4.81	4.18	2.30 - 3.67	2.97	11.15	21.86	26.00	0.35	11.71
		AK-42						AK-42						
Control	101.00-109.67	105.20	1.08	3.13	11.98	0.81	0.77	2.63 - 3.80	3.21	7.32	14.21	26.55	0.25	7.77
5 kR	102.00-114.67	106.72	3.01	3.65	68.25	5.47	5.13	1.40 - 4.00	2.87	21.22	27.47	59.70	0.97	33.78
10 kR	102.33-113.33	106.40	2.39	3.49	45.76	3.58	3.36	2.00 - 3.63	2.72	13.67	22.28	37.61	0.47	17.27
15 kR	103.42-120.33	111.07	4.16	5.37	59.83	7.36	6.62	2.33 - 3.80	3.06	15.52	21.00	35.54	0.47	15.38
20 kR	106.33-119.67	114.32	2.58	3.37	58.93	4.67	4.09	2.30 - 3.67	3.00	10.62	22.83	21.63	0.30	10.17
		MK-1						MK-1						
Control	102.44-113.33	108.01	1.74	3.18	29.81	2.11	1.95	1.33 - 3.63	2.79	17.57	27.35	41.28	0.65	23.26
5 kR	106.33-116.67	112.01	2.73	3.39	64.96	5.08	4.54	2.23 - 3.67	3.14	9.07	20.51	19.54	0.26	8.26
10 kR	103.00-114.67	107.05	1.06	3.74	8.03	0.66	0.62	1.67 - 3.77	2.80	15.64	24.34	41.32	0.58	20.72
15 kR	104.33-120.67	111.63	3.44	4.16	68.55	6.56	5.87	2.00 - 3.67	3.01	11.99	20.34	34.75	0.44	14.56
20 kR	109.00-122.67	116.07	2.97	3.58	68.81	5.90	5.08	2.33 - 3.67	3.05	10.90	22.50	23.48	0.33	10.88

Table 1B: Phenotypic range, treatment mean, phenotypic and genotypic coefficient of variation, heritability (broad sense) and genetic advance for pods per plant (nos.), pod length (cm), seeds per pod (nos.) and yield per plant (g) in M_2 generation of three different varieties in Horsegram

Treatments	Phenotypic range	Treatment mean	GC-V%	PC-V%	Heritability % (B.S)	GA %	GA (% mean)	Phenotypic range	Treatment mean	GC-V%	PC-V%	Heritability % (B.S)	GA %	GA (% mean)
		Pods per plant (nos.)						Seeds per pod (nos.)						
		AK-21						AK-21						
Control	34.67-52.50	42.04	10.76	13.64	62.17	7.35	17.47	2.80-5.13	4.25	7.59	17.01	19.93	0.30	6.98
5 kR	30.00-54.20	41.62	16.04	18.11	78.44	12.18	29.27	4.47-5.77	5.17	5.34	11.81	20.46	0.26	4.98

10 kR	30.67-51.00	43.24	11.76	14.61	64.78	8.43	19.50	4.33-6.00	5.08	3.16	15.72	4.04	0.07	1.31
15 kR	33.80-52.77	43.64	8.91	12.14	53.85	5.88	13.47	4.07-5.57	4.93	3.93	14.99	6.87	0.10	2.12
20 kR	25.67 - 49.27	39.54	16.98	19.40	76.62	12.11	30.62	4.40-5.70	5.05	3.26	12.92	6.37	0.09	1.69
				AK-42							AK-42			
Control	33.67 - 53.33	43.89	9.30	12.70	53.57	6.15	14.01	3.20-5.77	4.45	7.16	17.91	16.01	0.26	5.91
5 kR	26.67 - 57.93	41.58	19.59	21.59	82.34	15.23	36.62	4.33-5.77	5.24	6.68	11.13	36.02	0.43	8.26
10 kR	16.00 - 53.47	41.41	20.50	21.94	87.30	16.34	39.45	4.47-5.80	5.20	5.78	10.72	29.08	0.33	6.42
15 kR	36.33 - 49.33	43.58	6.31	9.43	44.72	3.79	8.69	4.13-5.87	4.87	6.81	11.55	34.73	0.40	8.26
20 kR	24.33 - 48.87	39.81	16.14	18.31	77.72	11.67	29.32	3.73-5.67	4.95	5.82	14.50	16.13	0.24	4.82
				MK-1							MK-1			
Control	21.11 - 51.01	39.49	19.80	22.60	76.77	14.11	35.74	4.00-5.57	4.65	5.88	11.26	27.32	0.29	6.34
5 kR	24.67 - 49.67	37.84	17.62	20.03	77.32	12.07	31.91	4.10-5.93	5.02	7.96	12.82	38.58	0.51	10.19
10 kR	25.33 - 50.38	42.62	14.06	16.42	73.24	10.56	24.78	4.27-5.97	5.10	4.40	13.39	10.80	0.15	2.98
15 kR	18.33 - 47.93	36.32	20.73	22.81	82.59	14.10	38.81	3.57-5.90	4.68	11.17	15.07	54.92	0.80	17.05
20 kR	18.33 - 46.33	33.75	22.83	25.12	82.61	14.43	42.75	4.00-5.87	5.03	6.00	14.39	17.37	0.26	5.15
				Pod length (cm.)							Yield per plant (g)			
				AK-21							AK-21			
Control	2.93 - 4.93	3.72	13.18	14.66	80.89	0.91	24.42	5.27-8.63	7.31	12.01	15.87	57.25	1.37	18.71
5 kR	3.30 - 5.53	4.06	8.64	16.16	28.59	0.39	9.52	4.49-11.7	7.86	22.78	29.33	60.33	2.87	36.46
10 kR	2.72 - 5.32	3.91	13.35	18.43	52.41	0.78	19.90	5.32-10.8	8.46	14.40	21.72	43.95	1.66	19.67
15 kR	3.24 - 4.68	3.91	10.35	12.14	72.70	0.71	18.19	7.01-11.2	8.86	9.73	16.10	36.56	1.07	12.12
20 kR	3.40 - 4.50	3.93	9.45	11.98	62.20	0.60	15.36	4.13-10.3	7.31	19.28	24.32	62.82	2.30	31.48
				AK-42							AK-42			
Control	3.43 - 5.43	4.13	10.58	12.45	72.18	0.76	18.52	5.63-9.73	7.53	11.41	19.36	34.71	1.04	13.84
5 kR	3.47 - 5.30	4.10	7.19	12.45	33.35	0.35	8.55	5.75-11.8	8.49	21.04	25.39	68.66	3.05	35.92
10 kR	3.14 - 6.13	4.17	13.80	15.85	75.83	1.03	24.76	4.74-11.8	8.41	17.65	23.01	58.81	2.34	27.88
15 kR	3.31 - 4.63	3.95	10.19	12.23	69.38	0.69	17.48	5.11-10.2	8.33	14.34	18.65	59.18	1.89	22.73
20 kR	3.48 - 5.59	4.28	11.83	13.65	75.13	0.90	21.13	4.52-8.90	6.77	19.29	23.44	67.78	2.21	32.72
				MK-1							MK-1			
Control	3.53 - 4.67	4.12	6.63	9.21	51.83	0.40	9.83	5.96-8.52	7.48	6.35	11.73	29.26	0.53	7.07
5 kR	3.11 - 5.56	4.01	9.26	15.04	37.87	0.47	11.73	4.55-11.0	7.25	19.06	24.43	60.86	2.22	30.63
10 kR	3.21 - 4.60	3.80	8.74	12.14	51.85	0.49	12.97	5.09-11.3	8.74	12.16	19.40	39.29	1.37	15.70
15 kR	3.28 - 5.90	4.05	15.60	16.59	88.38	1.22	30.21	3.93-11.1	6.96	28.23	30.79	84:06	3.71	53.31
20 kR	3.01 - 4.79	3.99	12.02	13.53	78.97	0.88	22.01	3.91-10.4	6.62	28.83	31.45	84.02	3.60	54.44

In M_3 generation, gamma rays treated progenies showed high proportion of GCV and PCV in comparison to respective controls (Table 2A & 2B). Gamma rays 20 kR exhibited highest GCV and PCV in days to maturity in all three varieties; days to 50% flowering and plant height in AK-42 and yield per plant in MK-1; 15 kR gamma rays for number of pods per plant in MK-1 and number of seeds per pod in AK-21 and AK-42; 10 kR gamma rays for days to 50% flowering in AK-21 and MK-1, number of effective branches per plant, pod length and number of seeds per pod in MK-1; number of pods per plant in AK-21 and yield per plant in AK-42, 5 kR gamma rays for pod length in AK-21 and AK-42, plant height and number of effective branches per plant in AK-21. The magnitude of GCV, PCV and genetic advance were lower in characters like days to 50% flowering, days to maturity and pod length, while moderate for remaining characters. High heritability was observed for all the characters under study.

Table 2A: Phenotypic range, treatment mean, phenotypic and genotypic coefficient of variation, heritability (broad sense) and genetic advance for days to 50% flowering (nos.), days to maturity (nos.), plant height (cm) and for number of effective branches per plant (nos.) in M_3 generation of three different varieties in Horsegram

Treat-me-nts	Phenotypic range	Treat-ment mean	GCV%	PCV %	Herit-ability % (B.S)	GA %	GA (% mean)	Phenotypic range	Treat-ment mean	GCV%	PCV%	Herita-bility % (B.S)	GA %	GA (% mean)
	Days to 50% flowering (nos.)							Plant height (cm)						
	AK-21							AK-21						
Control	55.00-62.00	57.92	1.94	5.16	14.19	0.87	1.51	47.00 -58.00	52.72	4.23	6.08	48.29	3.19	6.05
5 kR	52.67-63.67	58.23	5.56	6.89	65.14	5.38	9.25	21.77 - 58.73	45.88	19.91	20.35	95.73	18.41	40.14
10 kR	49.67-67.00	57.69	5.98	6.94	74.08	6.11	10.60	34.33 - 60.60	48.58	15.59	16.88	85.29	14.40	29.65
15 kR	55.00-66.33	58.37	4.36	5.83	55.87	3.92	6.71	24.30 - 55.00	42.70	15.89	17.63	81.22	12.60	29.50
20 kR	54.00-63.67	58.32	4.44	5.43	66.62	4.35	7.46	24.33 - 53.00	39.65	14.46	16.74	74.60	10.20	25.72
	AK-42							AK-42						
Control	53.67-58.33	55.89	1.64	3.03	29.11	1.02	1.82	47.00 - 57.00	53.82	2.49	6.68	13.87	1.03	1.91
5 kR	51.67-64.33	57.21	3.80	6.49	34.19	2.62	4.57	30.67 - 60.73	46.33	16.09	19.50	68.08	12.67	27.35
10 kR	52.00-62.33	57.23	3.86	5.79	44.36	3.03	5.29	29.30 - 57.60	42.59	16.22	17.97	81.43	12.84	30.15
15 kR	55.00-63.67	58.12	2.15	4.01	28.91	1.39	2.39	26.00 - 51.74	39.78	15.50	19.39	63.93	10.16	25.53
20 kR	50.00-65.00	57.23	5.49	6.46	72.33	5.51	9.63	26.33 - 55.87	40.01	18.18	19.93	83.20	13.67	34.16
	MK-1							MK-1						
Control	54.33-63.67	57.47	2.81	4.94	32.46	1.90	3.30	53.00 - 57.67	55.71	1.22	4.43	7.56	0.38	0.69
5 kR	50.67-62.00	56.32	4.83	6.51	54.90	4.15	7.37	41.67 - 61.00	51.08	12.07	12.94	87.07	11.85	23.20
10 kR	49.67-63.67	57.65	5.40	6.51	68.86	5.33	9.24	32.63 - 57.07	44.94	15.46	16.57	87.04	13.35	29.71
15 kR	53.00-64.67	58.36	4.20	5.20	65.12	4.07	6.98	31.33 - 57.54	42.62	16.94	19.01	79.41	13.26	31.10
20 kR	54.67-61.67	58.51	2.41	4.08	34.96	1.72	2.94	30.77 - 57.47	45.53	14.79	20.04	54.50	10.24	22.50

	Days to maturity (nos.)							Number of effective branches per plant (nos.)						
	AK-21							AK-21						
Control	106.0-111.0	108.6	0.71	1.69	17.66	0.67	0.62	2.00 - 3.80	2.90	11.28	19.24	34.36	0.39	13.62
5 kR	102.8-116.7	108.6	3.27	4.06	64.87	5.88	5.42	1.87 - 3.87	3.11	16.11	21.31	57.13	0.78	25.08
10 kR	102.3-117.0	109.8	3.47	4.08	72.59	6.69	6.09	2.33 - 3.73	2.94	7.28	17.93	16.47	0.18	6.08
15 kR	102.8-117.0	109.0	3.37	3.91	74.31	6.52	5.98	2.10 - 3.83	3.01	12.84	20.19	40.41	0.51	16.81
20 kR	102.4-117.0	109.6	4.66	5.18	81.01	9.48	8.65	2.17 - 3.80	2.82	11.37	20.30	31.40	0.37	13.13
	AK-42							AK-42						
Control	100.3-107.3	104.4	1.49	2.33	40.99	2.06	1.97	2.00 - 3.77	2.85	14.13	21.75	42.20	0.54	18.91
5 kR	102.3-116.3	106.5	3.32	4.36	57.96	5.54	5.20	2.00 - 4.00	3.08	17.92	22.84	61.57	0.89	28.97
10 kR	104.3-117.7	109.4	3.31	3.75	78.26	6.61	6.04	2.00 - 3.77	2.84	14.17	22.89	38.31	0.51	18.07
15 kR	100.3-112.0	105.9	2.52	3.50	51.86	3.95	3.73	2.30 - 3.57	2.82	9.97	19.33	26.60	0.30	10.59
20 kR	101.1-118.3	108.9	4.52	5.03	80.61	9.10	8.36	2.07 - 3.93	2.83	12.05	19.49	38.22	0.43	15.35
	MK-1							MK-1						
Control	105.3-114.3	107.9	1.75	2.48	49.62	2.74	2.53	2.23 - 3.73	2.93	13.63	19.16	50.59	0.58	19.97
5 kR	101.0-115.7	107.5	3.78	4.42	73.11	7.15	6.66	2.33 - 3.83	2.96	9.61	19.17	25.13	0.29	9.93
10 kR	103.0-116.3	108.9	2.99	3.70	64.94	5.40	4.96	2.17 - 3.93	3.04	14.65	21.32	47.22	0.63	20.74
15 kR	100.0-116.3	106.1	3.17	3.79	69.84	5.80	5.46	2.30 - 4.00	2.91	11.94	21.26	31.56	0.40	13.82
20 kR	102.7-115.7	108.9	3.90	4.54	73.93	7.52	6.91	2.20 - 3.87	2.89	12.59	20.04	39.49	0.47	16.30

Table 2B: Phenotypic range, treatment mean, phenotypic and genotypic coefficient of variation, heritability (broad sense) and genetic advance for pods per plant (nos.), pod length (cm), seeds per pod (nos.) and yield per plant (g) in M_3 generation of three different varieties in Horsegram

Treatm-ents	Phenotypic range	Treat-ment mean	GCV %	PCV %	Herita-bility % (B.S)	GA %	GA (% mean)	Pheno-typic range	Treat-ment mean	GCV %	PCV %	Herita-bility % (B.S)	GA %	GA (% mean)
	Pods per plant (nos.)							Seeds per pod (nos.)						
	AK-21							AK-21						
Control	31.33-42.00	35.93	3.56	9.25	14.80	1.01	2.82	4.27 - 5.43	4.81	2.65	9.60	7.59	0.07	1.50
5 kR	22.33-52.00	38.51	20.89	22.48	86.38	15.41	40.00	3.70 - 5.77	5.00	9.20	12.30	55.99	0.71	14.18
10 kR	24.00-54.60	37.74	24.82	26.14	90.18	18.32	48.56	3.63 - 6.13	4.82	11.81	14.41	67.15	0.96	19.93
15 kR	22.33-47.07	38.37	18.75	20.66	82.43	13.46	35.08	3.27 - 5.97	4.50	14.32	16.96	71.25	1.12	24.90
20 kR	18.67-49.60	34.32	22.88	25.05	83.44	14.78	43.06	3.40 - 5.67	4.72	10.51	13.56	60.07	0.79	16.79
	AK-42							AK-42						
Control	26.33-46.27	39.07	13.20	16.52	63.82	8.49	21.72	4.57 - 5.67	5.02	2.84	9.45	9.03	0.09	1.76
5 kR	22.67-54.67	36.64	22.02	23.41	88.46	15.63	42.67	4.27 - 5.90	4.94	8.47	12.09	49.12	0.60	12.23
10 kR	22.67-50.27	34.62	21.05	23.99	76.98	13.17	38.05	3.20 - 5.67	4.66	11.75	14.74	63.56	0.90	19.29

15 kR	22.67-48.40	36.30	17.17	19.48	77.69	11.32	31.18	3.47 - 5.80	4.50	12.79	16.42	60.71	0.92	20.53
20 kR	23.67-49.73	34.89	19.63	22.19	78.25	12.48	35.77	3.60 - 5.87	4.65	8.93	13.52	43.64	0.57	12.16
	MK-1							MK-1						
Control	29.67-44.27	36.46	7.32	11.78	38.59	3.41	9.36	4.17 - 5.13	4.72	3.92	9.50	17.01	0.16	3.33
5 kR	21.67-48.71	38.94	15.88	18.41	74.35	10.98	28.20	4.33 - 5.93	4.99	9.19	13.25	48.10	0.66	13.13
10 kR	26.00-46.67	38.63	14.66	17.95	66.64	9.52	24.64	3.27 - 6.00	4.55	12.83	16.16	63.09	0.95	21.00
15 kR	25.00-49.00	37.02	20.98	22.81	84.62	14.72	39.76	3.97 - 6.10	5.01	10.27	13.57	57.32	0.80	16.02
20 kR	26.00-52.13	36.84	16.55	19.34	73.27	10.75	29.19	3.70 - 5.80	4.64	10.86	13.64	63.41	0.83	17.82
	Pod length (cm.)							Yield per plant (g)						
	AK-21							AK-21						
Control	3.40-4.11	3.81	4.41	5.87	56.60	0.26	6.84	5.84 - 9.71	7.85	7.47	17.54	18.11	0.51	6.55
5 kR	3.00-4.88	3.78	10.11	11.73	74.19	0.68	17.93	5.00=10.88	7.49	16.73	20.75	64.97	2.08	27.78
10 kR	3.50-4.27	3.89	5.35	6.05	78.29	0.38	9.75	3.72 - 9.57	7.32	14.78	25.81	32.81	1.28	17.44
15 kR	2.89-4.53	3.77	8.53	9.36	83.09	0.60	16.02	5.22-10.95	7.22	12.71	26.97	22.21	0.89	12.34
20 kR	3.20-4.20	3.75	6.16	7.28	71.69	0.40	10.75	4.73-10.01	6.82	7.21	32.15	5.03	0.23	3.33
	AK-42							AK-42						
Control	3.36-4.03	3.79	4.26	5.43	61.72	0.26	6.90	6.13-10.87	7.97	10.52	23.18	20.61	0.78	9.84
5 kR	3.12-4.18	3.68	7.45	9.20	65.60	0.46	12.43	5.39-10.11	7.19	16.47	23.16	50.58	1.73	24.13
10 kR	3.30-4.23	3.80	6.49	7.29	79.30	0.45	11.91	3.60-11.60	6.58	19.39	32.89	34.76	1.55	23.55
15 kR	3.53-4.27	3.92	4.08	5.43	56.53	0.25	6.33	5.97-10.30	7.34	12.65	17.08	54.84	1.42	19.29
20 kR	3.43-4.44	3.93	6.17	7.44	68.72	0.41	10.54	4.99-10.13	7.43	9.68	22.26	18.92	0.64	8.68
	MK-1							MK-1						
Control	3.41-4.03	3.73	3.41	5.62	36.69	0.16	4.25	5.45 - 8.60	7.62	8.11	13.61	35.51	0.76	9.96
5 kR	3.53-4.80	3.86	7.00	8.03	75.92	0.49	12.56	5.36-10.38	7.37	12.29	27.19	20.44	0.84	11.45
10 kR	3.18-4.54	3.82	8.67	9.42	84.76	0.63	16.45	5.36-9.74	7.08	13.32	18.91	49.62	1.37	19.33
15 kR	3.47-4.43	3.86	5.36	6.33	71.69	0.36	9.35	5.86-10.06	7.46	12.62	22.00	32.90	1.11	14.91
20 kR	3.23-4.27	3.84	6.37	7.50	72.19	0.43	11.16	3.51-10.61	6.94	22.96	28.46	65.07	2.65	38.15

The results thus, indicated that magnitude of PCV, GCV, heritability and genetic advance varied according to mutagenic treatment employed, characters studied and the genotypes used. In general, gamma rays proved to be efficient in generating the variability in most of the characters. The present findings are in agreement with the result of Ambarkar (18) in chickpea, Mahadevu *et al.* (15) and Mohanasundaram *et al.* (19) in cowpea, and Ahmed and Yaqoob (20) in green gram.

The increase in genetic variability and heritability estimates indicated the induction of micromutations governing the quantitative characters. High genotypic coefficient of variation and high heritability associated coupled with high genetic advance suggested involvement of additive type of gene action in the inheritance of all the characters under study. Thus, results of the present study revealed that significant gain could be expected from selection for most of the characters. Robinson (21) categorized heritability estimates as low (5 to 10 per cent), medium (10 to 30 per cent) and high (30 to 60 per cent). He further emphasized that this classification is only approximate and denotes the average of heritability estimates. The estimates of heritability were moderate to high for all the characters under study. Lower heritability estimates are expected for a character in a population which has been subjected to selection procedure, since the additive genetic variance is reduced (21). However, in the present studies, generation has not been subjected to any kind of artificial selection. In addition broad sense heritability estimates are expected to be high because total genetic variance on which these estimates are based is made up of two parts *viz.*, additive genetic variance, non-additive genetic variance due to dominance and non-allelic gene interactions, while narrow sense heritability estimates takes in to account only additive or additive x additive genetic variance.

In M_4 generation, mutants which recorded higher yield in M_3 generation with respective check varieties are selected and evaluated for further study. Total of 58 mutants comprising of 20 mutants from AK-21, 16 from AK-42 and 22 from MK-1 were selected and evaluated along with three respective checks (AK-21, AK-42 and MK-1) for variability parameters, heritability, genetic advance, correlation and divergence analysis for yield and yield attributing characters. Existence of wide range of phenotypic variability for all the characters indicated considerable amount of variability present in different characters studied (Table 3). Lowest treatment mean with respect to days to 50% flowering and days to maturity was recorded in AK-42 (56.44 days and 105.51 days, respectively) and for plant height in AK-21 (58.01 cm). Similarly, highest mean values were recorded for different characters such as number of effective branches per plant in MK-1 (3.09), number of pods per plant in AK-21 (38.19), pod length in AK-42 (4.06 cm), number of seeds per pod in AK-42 (4.81), test weight in AK-21 (3.59 g) and yield per plant in AK-21 (7.21 g). Similar results were observed by various workers; in green gram, Kwon and Oh (22) for days to 50% flowering and yield per plant; Malik (23) for days to maturity and yield per plant and Patil *et al.* (24) for number of pods per plant, yield per plant and test weight, in frenchbean, Swarup and Gill (25) for number of pods per plant, number of seeds per pod and test weight, in urdbean, Singh and Raghuvanshi (26) for plant height and yield per plant, in cowpea, Ojomo and Cheddar (27) for yield per plant and in chickpea, Shamsuzzaman and Shaikh (28) for days to maturity, yield per plant and test weight.

A wide range of phenotypic variation was observed for plant height and number of pods per plant. In plant height, maximum range was observed in AK-42 (25.7), while for number of pods per plant maximum range was noticed in AK-21 (16.99). These findings are in agreement with the results of Rao and Nanda (29); Lad *et al.* (30) and Nehru *et al.* (31) in horse gram. The moderate range of phenotypic variation was found for days to 50% flowering, days to maturity, number of effective branches per plant, pod length, number of seeds per pod, test weight and yield per plant. These results are in conformity with the findings of Nehru *et al.* (31) and Sharma (32) in horsegram. The phenotypic range of variation is not the precise criterion to estimate the amount of genetic variability present in a breeding population. The genotypic and phenotypic components of variability were low for all the characters suggested that these characters are more influenced by environment and therefore, the selection of these characters will not be much effective. However, they can be used in hybridization programme. However, the highest GCV was observed for yield per plant in AK-42 (10.57%) followed by number of pods per plant in AK-21 (10.06%), yield per plant in MK-1 (9.39) and plant height in AK-42 (9.05), indicated that selection of these characters will be effective for the improvement in horsegram. These findings are in conformity with the findings of (30, 31 and 33 in horsegram). Burton (10) described that the genetic coefficient of variation is a better index for measuring the genetic variability present in the population.

Table 3 Phenotypic range, treatment mean, phenotypic and genotypic coefficient of variation, heritability (broad sense) and genetic advance for different characters in M_4 generation of three different varieties in Horsegram

Varieties	Phenotypic range	Treatment mean	GCV%	PCV%	Heritability % (B.S)	GA %	GA (% mean)
		Days to 50% flowering (Nos)					
AK-21	53.56 – 60.22	56.72	2.95	4.71	39.14	2.16	3.80
AK-42	52.78 – 58.67	56.44	0.99	4.45	4.97	0.26	0.46
MK-1	52.56 – 61.44	56.89	2.15	5.47	15.38	0.99	1.73
		Days to maturity (Nos.)					
AK-21	102.56 – 113.00	106.30	1.39	4.22	10.80	1.00	0.94
AK-42	101.33 – 110.67	105.51	1.48	3.27	20.64	1.47	1.39
MK-1	102.78 – 112.00	106.16	1.31	3.58	13.44	1.05	0.99

			Plant height (cm)				
AK-21	50.88 – 65.77	58.01	2.14	12.13	3.11	0.45	0.78
AK-42	46.61 – 72.31	58.38	9.05	14.08	41.29	6.99	11.97
MK-1	46.88 – 70.15	58.23	5.39	15.06	12.81	2.31	3.97
			Number of effective branches per plant (Nos.)				
AK-21	2.12 – 3.42	3.00	4.29	15.71	7.44	0.07	2.41
AK-42	2.37 – 3.26	2.98	2.50	10.06	6.18	0.04	1.28
MK-1	2.61 – 3.67	3.09	3.71	11.52	10.40	0.08	2.47
			Number of pods per plant (Nos.)				
AK-21	29.42 – 46.41	38.19	10.06	18.43	29.78	4.32	11.31
AK-42	31.13 – 47.55	37.96	6.86	18.65	13.51	1.97	5.19
MK-1	29.82 – 45.69	38.12	3.57	17.40	4.20	0.57	1.51
			Pods per plant (Nos.)				
AK-21	3.38 – 4.82	4.04	5.19	14.13	13.50	0.16	3.93
AK-42	3.63 – 5.00	4.06	7.60	12.99	34.27	0.37	9.17
MK-1	3.52 – 4.88	4.04	4.71	12.28	14.73	0.15	3.73
			Number of seeds per pod (Nos.)				
AK-21	3.78 – 5.44	4.77	6.46	12.79	25.51	0.32	6.72
AK-42	3.93 – 5.46	4.81	5.79	11.67	24.63	0.28	5.92
MK-1	4.06 – 5.56	4.72	6.39	12.54	25.94	0.32	6.70
			Test weight (g)				
AK-21	3.33 – 4.02	3.59	4.80	5.19	85.74	0.33	9.16
AK-42	3.11 - 3.71	3.51	3.41	4.09	69.53	0.21	5.86
MK-1	3.14 – 3.83	3.53	4.60	5.12	80.91	0.30	8.53
			Yield per plant (g)				
AK-21	5.56 – 9.01	7.21	8.94	21.12	17.95	0.56	7.81
AK-42	5.20 – 9.59	7.15	10.57	23.47	20.29	0.70	9.81
MK-1	5.32 – 9.84	7.20	9.39	24.24	15.03	0.54	7.50

Heritability estimates were observed moderate for almost all the characters. Highest heritability was recorded for test weight in all three varieties. Moderate estimates of heritability were examined for days to 50% flowering in AK-21; days to maturity in AK-42; plant height in AK-42; number of pods per plant in MK-1; pod length in AK-42; yield per plant in AK-42 and number of seeds per plant in all three varieties. The higher estimates of heritability indicated that these characters were less affected by the environment.

Shift in gene frequency towards superior side selection pressure is termed as genetic advance and is generally expressed as percentage of mean (genotypic gain). Johnson *et al.* (9) reported that it is more useful to estimate heritability values together with genetic advance in predicting the ultimate choice of the best genotype through selection. High genetic gain along with high heritability showed most effective condition for selection. In the present investigation, none of the genotypes showed high heritability coupled with high genetic advance but moderate heritability coupled with moderate genetic advance were recorded for plant height in AK-42 and number of pods per plant in AK-21 indicated that these characters are govern by additive gene action and therefore, direct selection would be effective. High heritability with low genetic advance was recorded for test weight in all three varieties indicated that heritability is mainly due to the non-additive genetic effects (dominance and epistasis). Moderate heritability coupled with low genetic advance was observed for days to 50% flowering in AK-21 and days to maturity, pod length and yield per plant in AK-42. These findings are also supported by Nehru *et al.* (31).

REFERENCES

1. Virapanashappa, K., Mahishi, D.M. and Shivashankar, G. 1980. Variation in cowpea following hybridization and mutagenesis. *Indian J. Genet.* 40 : 396-398.
2. Singh, V.P. and Chaturvedi, S.N. 1980. Mutagenic efficiency of EMS, NMU, gamma rays and their combined treatments in *Vigna radiata* (L.) Wilczek. *Genetic Agrari.*, 34 (3-4) : 331-348.
3. Shaikh, M.A.Q., Ahmed, Z.U., Majid, M.A., Bhuiya, A.D. Kaul, A.K. and Mia, M.M. 1980. Development of a high yielding chickpea mutant. Nutation breeding Newsletter 16 : 1-2.
4. Singh, D.P., Sharma, B.L. and Kunadia, B.A. 1980. Induced variability of handling M_2 populations. Tropical Grain Legume Bulletin. 19 : 30-34.
5. Tikka, S.B.S. 1982. Mutagenesis in Finger Millet (*Elusine coracana* Gaertn.) Ph.D. Thesis (Unpublished) submitted to Gujarat Agricultural University, Sardarkrushinagar.
6. Buzzati-Traverso, A.A. and Scossiroli, R.E. 1956. X-ray induced mutations in polygenic system. Proceeding 2nd UN International Conference, PUAEA. 22 : 293-97.
7. Scossiroli, R.E. 1959. Selezione artificiale per the carattera quantitative in populazioni di *Drosophila melanogaster* irradiate contraggi X. *Comitato Nazionale* per li Ricerche Nucleari, Bulletin. 4 : 132.
8. Panse, V.G. and Sukhatme, P.V. 1978. "Statistical Methods for Agricultural Workers". I.C.A.R., New Delhi. 3rd Ed.
9. Johnson, H.W., Robinson, H.F. and Comstock, P.F. 1955. Estimation of genetic and environmental variability in sorghum. *Agronomy Journal*, 47 : 314-318.

10. Burton, G.W. 1952. Quantitative inheritance in grasses. Proc. 6th Int. Grassland. Conqr., 1 : 277-283.

11. Burton, G.W. and De Vane, E.H. 1953. Estimating heritability in tall fescue (*Festuca arundinacese*) from replicated clonal material. *Agronomy Journal*, 45 : 478-481.

12. Gaul, H. 1958. Present aspect of induced mutations in Plant Breeding. *Euphytica*. 7 : 275-279.

13. Kumar, Y. and Mishra, V.K. 1999. Effect of gamma ray and diethyl sulphate on germination, growth and fertility and yield in green gram (*Vigna radiata* (L.) Wilckzek). *Annals of Agricultural Research*. 20 (2) : 144-147.

14. Chevan, A.A., Patil, V.D. and Pawar, R.B. 2000. Induced mutations in mungbean-variety BM-4. *Proceding of the DAE-BRNS Symposium on the use of nuclear and molecular techniques in crop improvement*. Division of nuclear Techniques in food and Agriculture, Vienna, Austria. December 6-8, 2000, Mumbai. Pp. 149-155.

15. Mahadevu, P., Thinmaiah, S. K. and Kulkarni, R. S. 1998. A *Pythium* rot resistant mutant of cowpea (*Vigna unguiculata* (L.) Walp) cv. KBC-1 (in) DAE Symposium on induced mutations and molecular techniques in improving crop productivity and quality. 50 Chennai.

16. Khan, N. M. and Tak, M. G. 2000. Induced mutations in black gram (*Vigna mungo* (L.) Hepper). DAE- BRNS Symposium, Mumbai, 131-140.

17. Frey, K.J. 1965. Mutation Breeding for Quantitative Attributes. In : Use of Induced Mutations in Cross Breeding. Proceeding Advisory Group, Vienna, 1975. IAEA, Vienna. : 95-111.

18. Ambalkar, A.S. 1997. Mutagenesis studies in chickpea (*Cicer arietinum* L.). A M.Sc. (Agri.) thesis submitted of Mahatma Phule Krushi Vidyapeeth, Rahuri.

19. Mohanasundaram, M. Thamburaj, S. and Natrajan, S. 1998. Observations on chlorophyll mutant in vegetable cowpea (*Vigna unguiculata* (L.) Walp) induced by gamma-rays. *South Indian Horticulture*. 46 (3-6) 2435-246.

20. Ahmad, B and Yaqoob, M. 1993. Radiation for induced mutation in mungbean (*Vigna radiata* (L.) Wilczek). *Sarhad Journal of Agriculture*. 9 (5) : 423-427.

21. Robinson, H.F. 1966. Quantitative genetics in relation to breeding on the centennial of Mendelism. *Indian J. Genet.*, 26A : 171-187.

22. Kwon, S.H. and Oh, J.H. 1983. Mungbean mutation breeding in Korea. *Mutation Breeding Newsletter*. 21 : 4.

23. Malik, I.A. 1988. High yielding and early maturing mutants in mungbean (*Vigna radiata* (L.) Wilczek). *Mutation Breeding Newsletter*. 32 : 7-8.

24. Patil, R.A., Jahagirdar, J.E,. Shinde, V.S. and Ghodke, M.K. 1993. Mungbean variety 'BM-4' is suitable for central zone. *Indian Farming*. 42 : 12.

25. Swarup, V. and Gill, N.S. 1968. X-ray induced mutations in french bean. *India J. Genet.* 28 : 44-58.

26. Singh, R.K. and Raghuvanshi, S.S. 1987. A higher yielding mutant of black gram with improve nodule formation. *Mutation Breeding Newsletter*. 30 : 18-17.

27. Ojomo, O.A. and Chedda, H.R.M. 1972. The variation for some characters in cowpea (*Vigna unguiculata* (L.) walp). following ionizing radiation. *Radiation Botany*. 12: 273-380.

28. Shamsuzzaman, K.M. and Shaikh, M.A.Q. 1991. Early maturing and higher seed yielding chickpea mutants. *Mutation Breeding Newsletter*. 37 : 4-5.

29. Rao, S.S. and Nanda, H.C. 1994. Genetic variability and correlation studies for grain yield and its components in horse gram. *Indian Journal of Pulses Research*. 7 (1) : 59-61.

30. Lad, D.B., Chavan, S.A. and Dumbre, A.D. 1999. Genetic variability and correlation studies in horse gram. *Journal of Maharashtra Agricultural Universities*. 23 (3) : 268-271.

31. Nehru, S.D., Manjunath, A. and Nagaraja, N. 2000. Identification of traits to be selected in breeding for yield in horse gram [*Macrotyloma uniflorum* (Lam.) Verdec.]. *Karnataka Journal of Agricultural Sciences*. 13 (1) : 11-15.

32. Sharma, R.K. 1995. Nature of variation and association among grain yield and yield components in horse gram. *Crop Improvement*. 22 (1) : 73-76.

33. Prakash, B.G. and Khanure, S.K. 2000. Genetic parameters, correlation and path coefficient analysis in horse gram. *Karnataka Journal of Agricultural Sciences*. 13 (2) : 313-314.

Gamma ray and pH sensitivity of cowpea in respect of seedling height

Deepa Maindiratta and D. Kumar

Central Arid Zone Research Institute, Jodhpur.

ABSTRACT

Assessment of radio sensitivity of mutagenic agents was first step in selecting the appropriate agents, their combinations, doses and the concentration. Appropriate doses and their combination lead to increase mutation frequency and directed mutagenesis. In view of it fixed quantity seeds of GC-3 cow pea variety was exposed to different doses of gamma rays (10, 20 & 30kR), pH levels (4, 7 and 9.2) and their possible combinations including control. We report here the effect of these treatments on seedling height of cowpea at three intervals leading third, sixth and ninth day in laboratory at usual temperature and conditions. There was steady decrease in seedling height with rise in gamma rays doses. For instance, it reduces to the extent of 69.74, 71.71 and 70.27 % at 3, 6 and 9th day respectively. The effect of pH was reverse to that of gamma ray. For instance, the reduction over control of seedling height at 3rd day was 17.11% at acidic pH of 4 while in the neutral pH the value was 11.84% at 3rd day and there was an increase over control in seedling height by 4.61% and 2.70% at 6 and 9th day respectively were a maximum value of seedling at the respective interval and increase in seedling height occur 5.3 and 2.7 at 6 and 9th day. The maximum value of reduction over control was observed on 3rd day at 9.2pH i.e, 7.89%. However on 6th and 9th day there was an increase in height over control at 9.2 pH i.e, 5.26% and 7.57% respectively. The maximum reduction in seedling height was observed when the interaction between two was examined at 7pH + 30kR at 9th day of interval from these results it may therefore concluded that LD_{50} in respect of gamma rays, pH and their combinations was almost 51.32% at 3rd day.

Key words: Gamma ray, pH sensitivity, cowpea, seedling height.

INTRODUCTION

Cowpea commonly known as loobia is an important drought hardy (1), deep rooted, annual legume of great significance. The crop is characterized for multi use purpose: food, feed, fodder, green vegetable etc. The crop is also known for a variety of variability for stature plant type, maturity period, grain colour, grain shape and nutritative values. Inspite of referred virtues associated with this arid legume there are certain bottle neck also. For instance, growth period is generally extended, reproductive phase is shorter, asymmetrical or unsynchronized growth behavior, poor conversion leading to in efficient source-sink relationship consequently in poor productivity. There is therefore, scope for its genetic fillip by using appropriate breeding tools. We intend to use mutation breeding for its genetic acceleration. In mutation breeding however, the first and the initial step towards assessing the relative response of the concern crop towards mutagenic agent. With a desire to assess the sensitivity of cowpea (GC-3) towards gamma-ray doses and different pH levels to be judge by initial growth traits.

MATERIALS AND METHODS

Present study on cowpea GC-3 cultivar was initiated at this institute during the month of April 2008. Pure seeds of GC-3 were obtained from S.D Agricultural University, S.K Nagar. Adequate quantity of seeds treated with three pH levels (4, 7 & 9.2) for 15 minutes. Treated seeds were thereafter expose to different doses of CO^{60} gamma-ray (10, 20 & 30kR) at DRDO, Jodhpur with an intensity of 2.6kR/min. There were total 16 combinations Ten seeds from each of the treatment were used for such studies to assess the radiobiological responses at the pH and gamma ray level. Seeds were plated in moisten blotting papers in petri plates at 35°C ± 1. The same were watered at regular intervals. The data on seedling height were recorded at 3, 6 and 9th day of planting.

RESULT AND DISCUSSION

In present study seedling height and its response to gamma-ray and pH treatments are presented in table-1. A critical view of the figure provided clearly revealed that in general seedling growth increased steadily with rise in gamma ray doses, but the trend for seedling height when compare to gamma-ray doses was some what reverse at the pH levels. For instance, seedling height at 3,6and 9th day was 6.1, 11.8 and 13.8 cm. at 10 kR indicating 19.74, 22.37 and 25.41% reduction over control. At 20 kR there was more reduction in seedling height as judge by % reduction. The value increased to -51.3, -32.8 and -46.4 % over control. The same figure at 30 kR was to the twin of -69.7, -71.7 and 70.27. The data clearly reflect that seedling height decreased dramatically with rise in gamma-ray doses and growth interval both. On the contrary, the same figure at 4 pH was -17.1, -3.2 and -5.9, respectively. At 7 pH the same figure were -11.8, +4.61 and +2.70. At 9.2 pH the decreased at three respective interval in seedling height was -7.98, +5.26 and+7.57. These data revealed that the seedling height was reduced up to 3rd day only with rise in pH level and at 6 and 9th day seedling height was increased over control. In view to study the combine effect of pH level and gamma-ray doses till seedling height depression with growth can be judge at 9th day for convenience. If we see the % increase and decrease value at 9th day data clearly reveals that the maximum seedling height was reduced (-72.9%) at 7pH + 30 kR followed by 30 kR and 7 pH +20 kR (-8.92 %). On the contrary, 9.2 pH gave accumulatry effect in growth at 6 and 9th day. Thus, 7 pH with 30 kR gamma-ray and only 30 kR proved most sensitive doses for GC-3. In other words, 30 kR could be enough without pH level. pH level either 4 or 7 or9.2 is not adding additional sensitivity that's gamma-ray itself induced. It is because of differential pH level proving beneficial for growth [2].

Table 1: Seedling height and +/- over control at different intervals

S.No	Treatment	3rd day		6th day		9th day	
		Height (cm.)	+/- over control (%)	Height (cm.	+/- over control (%)	Height (cm.	+/- over control (%)
1	GC-3 (control)	7.6	-	15.2	-	18.5	-
2	GC-3 + 10kR	6.1	-19.74	11.8	-22.37	13.8	-25.41
3	GC-3 + 20kR	3.7	-51.32	10.2	-32.89	9.9	-46.49
4	GC-3 + 30kR	2.3	-69.74	4.3	-71.71	5.5	-70.27
5	GC-3 + 4pH	6.3	-17.11	14.7	-3.29	17.4	-5.95
6	GC-3 + 4pH+10kR	6.6	-13.16	16.5	+8.55	16.3	-11.89
7	GC-3 + 4pH + 20kR	4.8	-36.84	10	-34.21	10.3	-44.32
8	GC-3 + 4pH + 30kR	4.4	-42.11	7.5	-50.66	7.8	-57.84
9	GC-3 + 7pH	6.7	-11.84	15.9	+4.61	19	+2.70
10	GC-3 + 7pH + 10kR	5.7	-25.00	10.6	-30.26	11.7	-36.76
11	GC-3 + 7pH + 20kR	3.0	-60.53	7.3	-51.97	7.6	-58.92
12	GC-3 + 7pH + 30kR	3.2	-57.89	4.9	-67.76	5	-72.97
13	GC-3 + 9.2pH	7.0	-7.89	16	+5.26	19.9	+7.50
14	GC-3 + 9.2pH + 10kR	5.0	-34.21	15	-1.32	16.2	-12.43
15	GC-3 + 9.2pH + 20kR	5.1	-32.89	9.6	-36.84	10	-45.95
16	GC-3 + 9.2pH + 30kR	5.4	-28.95	11.2	-26.32	12	-35.14
	Mean	5.2		11.294		12.556	

REFERENCES

1. Singh, B.B. & Sharma. B. (1996). Restructuring cowpea for higher yield. *Indian Journal of Genetics*. 56(4): 389-405.
2. West, D.W and Francois, L.E (1982) Effects of salinity on germinations, growth and yield of cowpea. *Irrigation Science*. 3: 169-175.

Genetic analysis of yield and its components in mungbean (*Vigna radiata* (L.) Wilczek.

C.G. Intwala, P.S. Vashi, V.S. Patel and S.R. Patel

Department of Agricultural Botany, N. M. College of Agriculture, Navsari Agricultural University, Navsari – 396 450.

ABSTRACT

Combining ability studies were carried out through line x tester analysis for nine quantitative characters. Both, general and specific combining ability were significant for seed yield and its related traits, indicating the importance of additive as well as non-additive variances, former being pre-dominant. The magnitude of gca variance due to males for branches per plant, clusters per plant, pods per plant, seeds per pod, seed yield per plant and 100-seed weight were comparatively higher than that due to females which indicated greater diversity among males for these components. Parents GM-4, VRMGg-1, SM-52, SM-56, Pusa bold-1 and Pusa 9632 were found to be good general combiner for seed yield and its related traits. On the basis of *per se* performance, sca effects and heterobeltiosis, the best crosses for seed yield per plant were VRMGg-1 x SM-52, VRMGg-1 x SM-7 and GM-4 x Pusa bold-2.

Key words: Mungbean, gene action, line x tester analysis, combining ability

INTRODUCTION

Mungbean (*Vigna radiata* (L.) Wilczek) is an important pulse crop in India. Selection of parents in any breeding programme depends on their ability to combine well and also on the potentiality to throw transgressive segregants. The knowledge of combining ability provides a useful criterion for selection of desirable parents and hybrids for further exploitation in breeding programme. Therefore, the present investigation has been taken up to derive information about general combining ability of the varieties and gene action involved in the inheritance of yield and its components.

MATERIALS AND METHODS

The parental material for the present study comprised of four diverse lines Co-4, VRMGg-1, GM-3 and GM-4 and ten testers K-851, SM-7, SM-47, SM-48, SM-49, SM-52, SM-56, Pusa bold-1, Pusa bold-2 and Pusa 9632. Forty crosses were made in a line x tester fashion using above four lines and ten testers during *rabi* 2001. The F_1s and parents were grown in contiguous blocks in randomized block design with three replications at three different locations during summer-2001. The parents and F_1s were represented by a three row plot of 30 plants, placed at 45 x 15 cm. The observations were recorded on branches per plant, clusters per plant, pods per cluster, pods per plant, pod length (cm), seeds per pod, seed yield per plant (g), 100-seed weight (g) and harvest index (%). The mean values were subjected to combining ability analysis as per the line x tester mating design [1].

RESULTS AND DISCUSSION

The analysis of variance revealed that considerable amount of variability existed in the experimental material for all the traits studied (Table 1). Parents and hybrids both interacted significantly with change in location for all the traits indicating the presence of substantial heterosis in crosses. The gca and sca variances were highly significant for majority of the characters (Table 2). This suggested that both additive and non-additive variances were important in the expression of traits. Significance of both the variances has been reported by several workers [2, 3, 4, 5] for seed yield and its important traits. Preponderance of additive type of gene action was observed in the inheritance of branches per plant, pods per cluster, clusters per plant, pods per plant, seeds per pod and seed yield per plant. These findings were in agreement with various researchers [3, 4, 5, 6, 7]. In case of pod length, the ratio of variance sca/gca was 2.74 (Table 1) which suggested that both additive and non-additive type of variances offer good opportunities for selection and accumulation of desired genes for the improvement of this trait in mungbean.

The magnitude of gca variance due to males for branches per plant, clusters per plant, pods per plant, seeds per pod, seed yield per plant and 100-seed weight was comparatively higher than that due to females (Table 1) which indicated greater diversity among males for these components. Among parents, female lines confirmed greater stability of performance than males. The ratio $\sigma^2_{sca}/\sigma^2_{gca}$ vs σ^2_{gca} x location /σ^2_{sca} x location revealed greater influence of environment on estimates of sca variance for majority of characters (Table 1), suggesting need for evaluation of hybrids in multilocational trials.

Table 1: Analysis of variance for combining ability for nine characters of mungbean

Source of Variation	df	Branches plant⁻¹	Pods cluster⁻¹	Clusters plant⁻¹	Pods plant⁻¹	Pod Length	Seeds pod⁻¹	Harvest Index	Seed yield plant⁻¹	100- Seed weight
Locations	2	33.482	7.570	73.018	2491.781	3.315	4.032	953.129	128.125	30.868
Replications/ Locations	6	0.048	0.141	0.068	1.779	0.413	0.088	3.019	0.178	0.143
Females	3	9.189**	10.393**	57.996**	2329.976**	3.011*	2.296*	5198.093**	201.073**	3.762
Males	9	10.645**	11.234**	50.987**	1396.038**	0.233	70.067**	966.884**	155.842**	13.510
Females x Males	27	0.516	0.921	2.026**	65.313**	0.305**	1.582**	118.707**	5.010*	1.432
Females x Locations	6	2.233	1.165	8.463**	68.086**	0.447**	2.382**	86.479*	7.220*	3.716
Males x Locations	18	0.281	4.478**	7.101**	45.739**	0.204	0.901	65.211*	5.395*	0.740
Females x Males x Locations	54	0.545**	0.846**	1.106**	20.554**	0.155	0.876**	40.745**	3.363**	0.460
Error	234@	0.0685	0.123	0.136	0.477	0.140	0.160	4.185	0.101	0.0525
σ^2_{gca}		0.137	0.121	0.726	27.989	0.0182	0.524	46.756	2.771	0.0782
σ^2_{sca}		-0.01	0.03	0.310	14.250	0.050	0.240	25.99	-0.330	0.320
$\sigma^2_{gca} / \sigma^2_{sca}$		-0.073	0.247	0.426	0.509	2.747	0.458	0.555	-0.119	4.092
σ^2_{gca} x Loc.		0.034	0.106	0.321	1.731	0.008	0.036	0.862	0.0723	0.0818
σ^2_{sca} x Loc.		0.160	0.240	0.320	6.690	0.01	0.240	12.19	1.090	0.140
σ^2_{gca} x Loc. / σ^2_{sca} x Loc.		4.719	2.264	1.006	3.865	1.232	6.575	14.142	15.076	1.711

*, ** significant at 5 and 1 per cent probability levels, respectively

@ for individual location, d.f. for error and replication were 78 and 2, respectivel

The estimates of gca effects (Table 2) revealed that, among females VRMGg-1 was the best general combiner for seed yield per plant, branches per plant, clusters per plant and pods per plant. GM-4 displayed highly significant positive gca effects for pods per cluster, clusters per plant, pods per plant, harvest index, 100-seed weight and seed yield per plant. Among males, SM-56, Pusa 9632, SM-52 and Pusa bold-1 were good general combiners for seed yield per plant and some of its direct components. Female VRMGg-1 depicted high gca effect for seed yield and it was associated with high positive gca effects for branches per plant, clusters per plant, pods per cluster and harvest index. High positive gca effect for grain yield in GM-4 was related with high gca effects for pods per cluster, clusters per plant, pods per plant, harvest index and 100-seed weight. Similarly in SM-56 and Pusa 9632 the gca effects for seed yield per plant was associated with pods per cluster, clusters per plant, pods per plant, harvest index and 100-seed weight. In SM-52 and Pusa bold-1 the gca effects for branches per plant, clusters per plant, pods per cluster, harvest index and 100-seed weight were found to be associated with high gca effects for seed yield per plant. Similar results have been reported by earlier workers [4, 8, 9]. These parents may be intensively used in the future hybridization programme aimed at amelioration of seed yield.

Table 2. Estimates of general combining ability effects of parents and specific combining ability effects of crosses in mungbean

Parents	Branches plant^{-1}	Pods cluster^{-1}	Clusters plant^{-1}	Pods plant^{-1}	Pod Length	Seeds pod^{-1}	Harvest Index	Seed yield plant^{-1}	100- Seed weight
					Females				
Co-4	0.02	-0.43**	-1.16**	-7.54**	0.21**	-0.03	-11.21**	-2.21**	0.04*
VRMGg-1	0.40**	-0.09**	0.69**	3.58**	-0.22**	-0.05	1.87**	1.04**	-0.23**
GM-3	-0.04*	0.36**	0.17**	2.20**	-0.06*	0.13**	4.12**	0.45**	0.01
GM-4	-0.38**	0.16**	0.29**	1.77**	0.07*	-0.06*	5.22**	0.72**	0.18**
SE (g_i) ±	0.02	0.03	0.03	0.05	0.03	0.03	0.16	0.03	0.02
					Males				
K-851	-0.44**	-0.65**	-0.82**	-3.81**	-0.09	-0.59**	-0.61*	-1.01**	-0.53**
SM-7	-0.44**	-0.29**	-0.85**	-6.08**	0.03	1.17**	-4.44**	-1.68**	-0.62**
SM-47	-0.56**	0.03	-0.90**	-5.81**	-0.05	-1.17**	-4.33**	-1.53**	-0.73**
SM-48	-0.53**	-0.90**	-1.09**	-5.23**	0.05	1.32**	-4.56**	-1.98**	-0.33**
SM-49	-0.27**	0.18**	-0.47**	-3.82**	-0.04	1.72**	-6.42**	-2.04**	-0.30**
SM-52	0.39**	-0.51**	1.08**	6.68**	0.16**	-1.43**	5.18**	2.02**	0.15**
SM-56	0.60**	0.70**	1.81**	9.39**	-0.10*	-1.70**	7.68**	3.20**	0.14**
Pusa bold-1	0.53**	0.41**	1.22**	5.34**	-0.02	0.65**	5.81**	2.01**	0.73**
Pusa bold-2	0.88**	0.42**	-1.26**	-3.54**	-0.03	1.49**	-2.31**	-1.22**	1.13**
Pusa 9632	-0.16	0.61**	1.28**	6.91**	0.08	-1.46**	4.00**	2.22**	0.37**
SE (g_j) ±	0.04	0.05	0.05	0.09	0.05	0.05	0.28	0.04	0.03

*, ** significant at 5 and 1 per cent probability levels, respectively

Table 3: Top three cross combinations based on sca effects and mean performance for seed yield per plant and its important traits in mungbean (data pooled over locations)

Characters	Best specific combinations	sca effect	Mean performance	gca effect of parent
Seed yield per plant	VRMGg-1 x SM-52	1.16**	14.18	G x G
	VRMGg-1 x SM-7	0.86**	10.17	G x P
	GM-4 x Pusa bold-2	0.54**	10.00	G x P
Branches per plant	GM-4 x Pusa bold-1	0.43**	3.56	A x G
	GM-3 x SM-7	0.40**	2.90	P x P
	GM-4 x Pusa bold-2	0.36**	3.84	A x G
Pods per cluster	GM-4 x SM-52	0.54**	4.64	G x P
	VRMGg-1 x SM-48	0.49**	3.96	P x P
	VRMGg-1 x SM-7	0.41*	4.48	P x P
Clusters per plant	GM-4 x SM-56	0.77**	8.87	G x G
	VRMGg-1 x SM-52	0.67**	8.43	G x G
	Co-4 x K-851	0.60**	4.62	P x P
Pods per plant	VRMGg-1 x SM-52	4.61**	39.42	G x G
	VRMGg-1 x SM-56	3.57**	41.09	G x G
	Co-4 x SM-49	3.18**	16.37	P x P
Pod length	GM-4 x SM-49	0.27**	7.04	G x A
	Co-4 x SM-56	0.25**	7.10	G x P
	VRMGg-1 x SM-56	0.22**	6.64	P x P
Seeds per pod	Co-4 x K-851	0.89**	10.48	A x P
	VRMGg-1 x Pusa bold-2	0.47**	12.12	A x G
	VRMGg-1 x SM-48	0.46**	11.93	A x G
Harvest index	GM-4 x SM-49	7.09**	49.16	G X P
	VRMGg-1 x Pusa-9632	5.59**	54.72	G X G
	Co-4 x SM-56	4.51**	44.24	P x G
100-seed weight	VRMGg-1 x K-851	0.61**	4.88	P x P
	Co-4 x SM-47	0.52**	4.87	G x P
	GM-3 x Pusa bold-2	0.50**	6.69	A X G

*, ** significant at 5 and 1 per cent probability levels, respectively

The comparative study (Table 3) of best three crosses on the basis of sca effects and based on mean performance for seed yield and its related characters revealed that the sca effects and *per se* performance were not positively associated in all the cases. For example cross VRMGg-1 x SM-52 gave the highest sca effect for seed yield but is *per se* performance was not in accordance. Thus, crosses which would give the highest sca effects would not necessarily give the highest mean values in all the cases. The sca effects of top crosses and gca effects of their parents indicated that majority of these good specific cross combinations were the result of good x good or poor x poor combination. High sca effects due to good x good combiner reflect additive x additive type of gene interaction and superiority of favourable genes contributed by their parents, while these involving good x poor or poor x poor combiners indicated the interactions of additive x dominance or dominance x dominance, respectively. Bi-parental progeny selection might be useful to get some transgressive segregants from crosses involving good x good and good x poor combinations. It is therefore suggested that the crosses showing high sca effects and involved both or one high general combiner could be successfully exploited for further amelioration of seed yield as they are expected to throw stable performing transgressive segregants carrying fixable gene effects in mungbean. Similar findings have also been reported by earlier workers [4, 10].

REFERENCES

1. Kempthorne, O. (1957). An introduction to genetic statistics. John Wiley and Sons., Inc., New York pp. 468-471.
2. Naidu, N. V. and Satyanarayana, A. (1993 a). Heterosis and combining ability in mungbean (*Vigna radiata* (L.) Wilczek). *Indian J. Pulses Res.*, **6**(1) : 38-44.
3. Dasgupta, T.; Banik, A. and Das S. (1998). Combining ability in mungbean. *Indian J. Pulses Res.*, **11**(1) : 28-32.
4. Aher, R. P.; Dahat, D. V. and Surve, P. P. (2001). Diallel analysis for yield contributing characters in mungbean. *Legume Res.*, **24**(2) : 124-126.
5. Pithia, M. S. and Borad, P. K. (2003). Combining ability analysis in mungbean. National Symposium on Pulses for Crop Diversification and Natural Resource Management, 20-22 December, IIPR, Kanpur, pp. 89.
6. Kute, N. S.; Deshmukh, R. B.; Sarode, N. D. and Mamjal, M. R. (1999). Combining ability for yield and its components in mungbean (*Vigna radiata* (L.) Wilczek). *Madras Agric. J.*, **86** (4-6) : 210-212.
7. Khattak, G. S. S.; Haq, M. A.; Ashraf, M. and Srinives, P. (2001). Combining ability in mungbean. *Korean J. Crop Sci.*, **46**(5) : 424-427.
8. Manivannan, N. (2002). Line x tester analysis in *Kharif* green gram. *Legume Res.*, **26**(2) : 127-130.

9. Anbumalarmathi, J.; Rangasamy, P. and Babu, S. (2004). Studies on combining ability and heterosis for yield and yield components in green gram (*Vigna radiata* (L.) Wilczek). *Madras Agric. J.*, **91** (1-3) : 79-82.

10. Singh, B. B. and Dikshit, H. K. (2003). Combining ability studies for yield and architectural traits in mungbean (*Vigna radiata* (L.) Wilczek). *Indian J. Genet.*, **63** (4) : 351-352.

Stability analysis for seed yield and its components in mungbean (*Vigna radiata* (L.) Wilczek)

C.G. Intwala, P.S. Vashi, V.S. Patel and V. Khandelwal

Department of Agricultural Botany, N.M. College of Agriculture, Navsari Agricultural University, Navsari – 396 450.

ABSTRACT

Fifty four genotypes of mungbean were studied under three different environments (locations) to characterize stability for seed yield and its components. The variance due to G x E interaction was highly significant for majority of the traits, indicating wide differences between environments and differential behaviour of genotypes under different environments. The significance of G x E (linear) for seed yield per plant and its related traits suggested importance of both linear and non-linear components in building up total G x E interaction. Female VRMGg-1 was found average stable for seed yield per plant while male Pusa 9632 was found suitable for unfavorable environment. Among the high yielding hybrids, GM-4 x K-851 and GM-3 x SM-52 were found stable for seed yield while VRMGg-1 x SM-56, VRMGg-1 x SM-52, GM-3 x SM-52 and GM-4 x Pusa 9632 were found suitable for favorable environment for seed yield and one or two yield components.

INTRODUCTION

Breeding genotypes with wider adaptability has been an ultimate aim of plant breeders. Study of genotype x environment interaction not only helps in planning the breeding programme but also enables the identification of highly responsive and high yielding genotypes suitable for production in targeted production conditions, where genetic potential of the genotypes can be fully exploited. In mungbean, stability for morphological traits has been found to play a valuable role in imparting stability for grain yield. Therefore, it can be said that stability of yield in a genotype is determined to a greater extent by the relative stability of performance for different components. Since in mungbean, the literature on stability analysis revealed meager information on genotype x environment interaction for yield and its components, therefore the present investigation was taken into account for generating the information on G x E interaction yield and its components.

MATERIALS AND METHODS

Experimental material comprising of 40 F_1s, 4 females and 10 males were evaluated during summer-2001 in randomized block design (RBD) with three row plot of 30 plants, planted at 45 x 15 cm in three replications under three diverse environments (locations) i.e., Navsari, Achhalia and Vanarasi. Five competitive plants were randomly selected from each plot in each replication to record the observations for the characters branches per plant, pods per cluster, clusters per plant, pods per plant, seeds per pod, seed yield per plant (g) and 100-seed weight (g). The data were analysed for stability parameters following model [1]. A genotype having unit regression co-efficient (b_i = 1) and non-significant deviation from regression ($S^2d_i = 0$) was considered as stable.

RESULTS AND DISCUSSION

The pooled analysis of variance (Table 1) showed that mean squares due to genotype and environment were highly significant indicating presence of variability among the genotype and environment. The variances due to genotype x environment interaction were found to be significant for all the characters. These results are in agreement with those reported by earlier workers [2, 3, 4, 5]. The environment (linear) component was also found highly significant for all the traits indicating the influence of environment on the expression of characters. [2,3,6] in greengram had also reported similar findings. For the expression branches per plant, clusters per plant and pods per plant preponderance of non-linear components was noticed. These results are in agreement with those reported by various researchers [2, 4, 7]. The non-linear component (pooled deviation) was also highly significant for all the characters except seeds per pod in all the environments indicating that this part of the variation in the performance of the genotype is unpredictable and both linear regression and deviations from linearity were the major components for difference in stability for the characters. These results are in agreement with those reported by earlier workers [7, 8].

Table 1 : Stability analysis of variance for different characters in mungbean

Source of variation	d.f.	Branches per plant	Pods per cluster	Clusters per plant	Pods per plant	Seeds per pod	Seed yield per plant(g)	100-seed weight (g)
Genotype (G)	53	$_{++}$1.160**	$_{++}$1.269**	$_{++}$5.761**	$_{++}$180.158**	$_{++}$4.957**	$_{++}$18.794**	$_{++}$1.274**
Environment (E)	2	$_{++}$11.141**	$_{+}$1.776**	$_{++}$24.778**	$_{++}$839.553**	$_{++}$1.725**	$_{++}$28.538**	$_{++}$13.851**
G x E	106	0.204**	0.525**	0.983**	11.039**	0.311**	1.380**	0.247**

Environment (linear)	1	++22.283**	++3.553**	++49.556**	++1679.121**	++3.446**	++57.072**	++27.707**
G x E (linear)	53	0.180**	+0.611**	0.620**	11.617**	++0.484**	++1.689**	++0.321**
Pooled deviation	54	0.224**	0.431**	1.322**	10.268**	0.135	1.051**	0.169**
Pooled error	318	0.0658	0.108	0.119	0.421	0.148	0.0971	0.0465

*, ** Significant at 5 and 1 per cent probability levels against pooled error

+, ++ Significant at 5 and 1 per cent probability levels against pooled deviation

Estimates of stability parameters for the characters under study are given in Table 2. Genotypes were considered as stable having high mean performance (x), regression coefficient close to unity ($b_i = 1$) and deviation from regression approaching to zero ($S^2d_i = 0$). The genotype having $b_i > 1$ and significant S^2d_i was considered as suitable for favourable environment and the genotype with $b_i < 1$ and non significant S^2d_i was treated as good for poor environment (1). The environmental indices (Table-3) for all the characters for which stability parameters were worked out indicated that E_2 (Achhalia) was the most favourable environment for seed yield per plant (0.578) and its important related traits like branches per plant (0.287), clusters per plant (0.415), seeds per pod (0.118), 100-seed weight (0.469) and harvest index (2.778). The genotypes having significant S^2d_i were treated as unstable.

Table 2 : Stability parameters for different characters of Mungbean

Genotype	Branches per plant			Pods per cluster			Clusters per plant			Pods per plant			Seeds per pod			Seed yield per plant (g)			100-seed weight (g)		
	Mean	bi	S²di	Mean	bi	S²di	Mean	bi	S²di	Mean	bi	S²di	Mean	bi	S²di	Mean	bi	S²di	Mean	bi	S²di
Co-4	2.38	0.06**	0.00	4.02	2.10**	0.33**	4.00	0.56*	0.32**	13.59	0.03**	1.22**	0.99	4.52*	0.16*	5.52	-0.03	-0.03	4.23	0.25**	-0.01
VRM Gg-1	2.80	0.58	0.27**	4.03	4.87**	0.19**	4.84	0.29	1.01**	19.39	0.51	2.76**	8.81	2.61**	0.02	7.39	0.78	-0.02	4.53	0.45*	0.02
GM-3	2.33	0.63*	-0.02	3.69	2.39**	-0.02	4.59	0.16	-0.04	16.23	0.38	9.01**	9.44	0.84	-0.05	6.93	0.54	0.76**	4.97	0.45	0.12**
GM-4	2.50	0.95	-0.02	4.00	2.62**	0.09	4.42	0.60	0.88**	18.96	1.14	5.82**	9.58	-1.27	0.20**	8.87	0.68	0.39**	5.26	0.35*	0.03
K-851	2.20	0.63*	-0.02	3.50	1.22**	0.15**	4.64	0.52**	0.07	15.30	0.34	9.90**	9.28	-1.76	0.44**	6.42	0.59**	-0.02	5.43	2.63	0.51**
SM-7	2.19	0.95	0.79**	3.93	2.01	0.22**	5.06	0.63**	-0.03	13.54	0.13*	4.42**	10.91	5.07	-0.05	5.76	0.24	1.76**	5.29	2.45	0.69**
SM-47	2.11	0.64	0.10**	3.70	2.51	0.96**	3.99	0.40	0.21**	12.28	0.40**	-0.13	9.40	-0.80	-0.05	5.56	0.06	0.38**	4.92	1.84**	0.03
SM-48	1.80	0.58	0.19**	3:54	0.98**	0.16**	4.02	1.00	0.22**	12.69	0.21**	0.01	10.59	1.28	-0.03	6.51	1.11	1.48**	5.09	1.33	0.07**
SM-49	1.84	0.28	0.23**	4.68	2.43**	0.01	4.10	0.49	0.06	15.80	0.58**	0.30	10.93	3.99**	0.01	6.23	0.79	0.41**	5.03	1.31	0.07**
SM-52	3.07	0.67	0.07*	3.93	0.15**	0.05	5.97	1.64	0.12*	23.41	1.36	16.06**	8.67	0.83	-0.05	8.97	-1.44	1.57**	5.43	1.12	-0.01
SM-56	2.96	0.96	0.03	4.20	1.94	-0.04	6.28	1.74	1.09**	23.26	1.58	41.24**	9.06	0.22	-0.02	10.42	-0.73	1.58**	5.56	0.51	0.25**
Pusa bold-1	2.94	1.16	-0.02	3.66	0.19**	-0.02	5.80	1.66	4.88**	20.27	1.21	13.88**	9.60	3.04	0.46**	9.12	-0.71	1.20**	6.02	0.48**	-0.01
Pusa bold-2	2.94	0.39	0.32**	3.27	1.48	0.07	3.46	0.37	0.53**	11.52	0.15**	1.41**	11.07	2.19	0.03	6.49	2.20**	1.13**	6.13	0.27**	0.00

Pusa-9632	2.20	-0.84	0.36**	4.34	3.67*	0.88**	5.39	0.19	5.25**	19.86	0.87	21.05**	8.47	1.65**	-0.04	9.26	1.37**	-0.02	5.49	0.62	0.19**
Co-4 x K-851	2.59	0.45	0.02	3.53	-1.77	-0.04	4.62	1.01**	0.17**	14.06	0.34**	0.03	10.48	0.09*	-0.04	6.11	1.55	3.35**	4.57	1.67	0.09**
Co-4 x SM-7	2.38	0.21	-0.02	3.88	1.48	1.29**	4.13	1.00	0.08	13.21	0.33**	0.17	11.60	0.67	-0.04	5.17	1.12	0.40**	4.91	1.93*	0.09**
Co-4 x SM-47	2.64	0.07**	-0.01	4.07	2.53	0.51	4.20	0.18	0.79**	14.26	0.36	3.59**	9.39	4.38**	-0.02	6.38	3.49**	-0.01	4.87	2.28**	0.08**
Co-4 x SM-48	2.56	0.48	0.28**	2.87	2.52**	-0.03	3.96	0.43	0.90**	13.68	0.67*	0.38	11.62	3.66	-0.05	5.93	2.99**	0.26**	5.14	2.24**	0.01
Co-4 x SM-49	2.96	1.44**	-0.02	4.30	2.67**	0.11*	4.81	1.21	1.98**	16.37	1.30	4.54**	11.50	4.10**	-0.01	5.76	2.81*	0.35**	4.78	1.85	0.16**
Co-4 x SM-52	3.40	0.99	0.04	3.41	3.49**	0.00	5.80	0.31	0.20**	20.89	1.22**	-0.10	8.51	4.61**	-0.03	9.57	0.81	1.18**	4.87	0.99	0.08**
Co-4 x SM-56	3.53	1.00	0.00	4.52	3.16	0.83**	5.96	0.39	1.09**	22.23	1.38**	-0.06	8.44	0.22	-0.05	11.47	1.07	0.26**	5.01	0.96	0.29**
Co-4 x Pusa bold-1	3.17	0.99	0.04	4.48	1.40**	0.28**	5.29	0.13	6.50**	20.01	1.26	2.28**	10.68	0.80**	-0.02	10.20	-0.85	3.92**	5.38	1.04	0.47**
Co-4 x Pusa bold-2	3.80	1.58	0.13**	4.33	0.96	0.00	3.68	0.76	0.16**	14.69	0.73**	0.06	10.84	2.46**	-0.04	6.74	0.13**	-0.02	5.91	1.98	0.47**
Co-4 x Pusa-9632	2.89	1.23	1.69**	4.81	0.63	-0.01	5.97	0.48	1.59**	20.66	1.16*	0.00	8.67	-1.93	-0.05	10.07	2.61	5.43**	5.35	1.79	0.21**
VRM Gg-1 x K-851	2.89	1.54	0.57**	3.69	0.25	-0.03	4.87	1.56	0.08	24.48	1.03	1.95**	8.52	0.75	-0.05	10.39	2.52**	0.08	4.88	2.51	-0.01
VRM Gg-1 x SM-7	3.02	1.58	-0.02	4.48	4.69*	1.78**	5.58	2.52	6.47**	21.49	1.59	35.18**	11.10	1.68	0.10	10.17	2.54	0.96**	3.98	1.24	0.04
VRM Gg-1 x SM-47	2.60	1.52**	-0.02	4.61	8.33**	0.85**	5.69	1.65	-0.04	18.71	0.70**	0.21	8.59	-0.85	-0.05	9.09	1.33	1.14**	3.61	0.86	0.00
VRM Gg-1 x SM-48	2.78	1.66	0.13**	3.96	-1.50	1.19**	5.48	1.29	3.98**	20.97	0.86	7.46**	11.93	6.88**	0.06	8.72	0.83	0.15**	4.84	2.59**	0.00
VRM Gg-1 x SM-49	3.02	1.79	0.09**	4.58	6.67	-0.04	6.66	2.07	1.53**	22.88	1.90	36.66**	12.14	4.72**	0.032	8.52	1.17	0.27**	4.86	1.75*	0.03
VRM Gg-1 x SM-52	3.96	1.72*	0.02	3.54	3.21**	0.11*	8.43	1.71	-0.04	39.42	1.99	58.24**	8.97	2.26	-0.05	14.18	2.13**	0.02	4.96	1.26	-0.01
VRM Gg-1 x SM-56	4.10	1.68	0.10**	5.39	0.66	0.15**	8.44	1.92	0.19**	41.09	2.15	57.21**	8.60	0.04	-0.03	14.24	3.14**	-0.03	5.13	1.39	0.11**

VRM Gg-1 x Pusa bold-1	3.83	2.26**	0.01	4.44	1.48	0.07	8.26	1.87	0.41**	33.32	1.66	15.62**	10.93	3.63	0.09	12.53	1.09	4.67**	5.43	0.16	0.51**
VRM Gg-1 x Pusa bold-2	4.09	2.11**	0.03	4.36	0.84**	-0.02	5.78	2.13*	0.16**	22.71	1.22	5.10**	12.12	3.09	0.08	9.27	1.33	0.01	5.47	0.95	0.86**
VRM Gg-1 x Pusa-9632	3.47	1.54	1.20**	4.54	3.08**	0.13**	7.73	1.39	0.12*	36.23	2.46**	-0.04	8.67	1.10	0.16*	12.22	1.89	0.24**	4.95	1.26	0.15**
GM-3 x K-851	2.47	0.42	0.07**	4.09	4.17**	0.12**	5.27	0.53	2.60**	20.08	0.98	1.07**	10.17	9.18**	-0.02	9.19	1.57	0.17**	4.19	1.18	0.07**
GM-3 x SM-7	2.90	0.06	-0.02	4.59	1.36	2.92**	5.68	2.93	2.40**	19.54	1.01	7.16**	11.48	3.39	0.07	8.88	0.96	0.11*	4.22	0.80	0.07**
GM-3 x SM-47	2.42	1.62	0.10**	4.98	-0.24	0.89**	5.37	2.07	1.15	20.18	0.76	1.27**	9.03	3.11**	-0.04	9.10	-0.66	1.57**	4.14	0.30	0.32**
GM-3 x SM-48	2.33	1.59	0.24**	3.71	-1.17**	0.11	4.93	1.44	1.50**	22.46	0.01	8.59**	11.16	-2.23**	0.01	8.65	1.78	0.25**	3.97	0.45	0.04
GM-3 x SM-49	2.80	-0.48**	0.02	4.83	4.25**	0.61**	5.54	2.06	1.60**	19.09	-0.02**	0.25	12.32	1.69	-0.03	8.31	-0.88**	0.01	4.83	0.11	0.24**
GM-3 x SM-52	3.07	0.09	0.09**	4.17	-5.60**	0.92**	6.91	0.21*	0.05	33.03	0.39**	-0.13	8.80	2.86	0.93**	12.17	0.70	0.06	5.69	0.52	0.36**
GM-3 x SM-56	3.31	0.00	0.42**	5.17	5.84**	-0.03	7.97	0.65**	-0.03	36.91	1.58**	-0.08	8.34	-2.79	0.68**	13.61	1.42**	-0.02	5.01	0.11	0.17**
GM-3 x Pusa bold-1	3.46	1.13	0.21**	5.34	2.61	0.71**	7.40	0.69	1.68**	34.34	1.18	19.63**	10.97	2.40	-0.04	12.30	1.30	1.59**	6.05	0.65*	-0.01
GM-3 x Pusa bold-2	3.70	1.25	0.47**	5.43	1.24	2.08**	4.84	1.35	0.03	25.29	1.36**	-0.12	12.20	2.22	0.12	8.94	1.23	0.53**	6.69	0.50	-0.01
GM-3 x Pusa-9632	2.89	1.65	2.10**	5.75	1.39**	-0.03	7.78	0.30	2.59**	36.56	-1.91	39.77**	6.90	1.96	0.37**	12.85	1.06	0.99**	5.78	-1.09	-0.01
GM-4 x K-851	2.18	1.34	0.00	3.87	0.69	-0.01	5.96	1.50	0.05	24.34	1.39**	0.07	9.30	-1.46	0.30**	10.10	2.62	-0.03	4.39	0.42**	-0.01
GM-4 x SM-7	1.84	1.39	0.02	3.69	1.98	0.38**	5.22	1.23*	-0.03	19.64	0.96	-0.03	11.33	-1.29	-0.05	8.85	-0.04**	-0.02	4.60	0.71	0.02
GM-4 x SM-47	1.98	1.50	-0.02	4.25	-0.44	0.77**	5.13	1.82	1.05**	21.81	1.97*	4.62**	9.11	1.48	-0.04	9.13	1.04	0.15**	4.63	0.47	0.10**
GM-4 x SM-48	2.11	1.34	0.00	3.67	1.48	0.07	5.28	1.28	0.00	20.17	1.60	20.08**	11.37	-1.74	0.50**	8.61	1.56	0.04	4.90	0.80	0.11**
GM-4 x SM-49	2.04	1.65	0.06*	4.79	5.93**	0.07	5.10	1.40	1.69**	24.59	1.33	19.38**	11.73	1.24	-0.01	9.09	0.48	3.80**	4.48	0.43	-0.01
GM-4 x	3.04	1.02	0.12**	4.64	5.00**	0.11	7.16	0.05	0.95**	31.57	0.81	13.59**	8.81	2.40**	0.02	11.97	1.40	2.10**	5.26	0.03	0.35**

SM-52																					
GM-4 x SM-56	3.36	1.25	0.23**	5.52	6.74**	0.01	8.87	2.20**	0.06	35.54	1.11	4.01**	8.62	-2.15	-0.05	13.28	1.97	1.15**	5.60	1.17	0.37**
GM-4 x Pusa bold-1	3.56	1.33*	-0.02	5.16	-0.05	0.26**	7.91	0.93	6.25**	31.90	1.27	39.06**	10.83	1.34	0.13	12.81	2.27	5.38**	6.22	1.03	0.31**
GM-4 x Pusa bold-2	3.84	1.66	0.13**	5.33	0.03	1.07**	4.64	0.61	0.48**	21.22	1.11	12.68**	11.62	-1.12	0.81**	10.00	0.28	3.72**	6.61	1.15	0.29**
GM-4 x Pusa-9632	2.00	0.85	0.12**	5.13	5.36**	0.10	7.63	0.68	4.34**	32.38	1.30*	0.52**	8.74	3.14**	-0.02	12.94	3.05**	0.23	5.56	-0.25	0.02**
Parental mean	2.45	-	-	3.89	-	-	4.75	-	-	16.87	-	-	9.70	-	-	7.39	-	-	5.24	-	-
Hybrid mean	2.97	-	-	4.45	-	-	6.00	-	-	24.55	-	-	10.20	-	-	9.95	-	-	5.04	-	-
S.Em ±	0.335	0.737	-	0.464	0.256	-	0.813	0.120	-	0.227	0.575	-	0.260	0.145	-	0.725	0.997	-	0.291	0.575	-

*, ** significant at 5 per cent and 1 per cent of probability levels, respectively

In the present study by considering these parameters of stability, it was observed that, among all the genotypes female VRMGg-1 (x =7.39 g) produced higher seed yield with b_i value near to unity (b_i = 0.78) and non-significant S^2d_i value, this indicates that this genotype was most stable across the environment for seed yield. Female VRMGg-1 also exhibited above average stable for 100-seed weight. Whereas, among hybrids GM-4 x K-851 (b_i = 2.62, S^2d_i = -0.03) and GM-3 x SM-52 (b_i = 0.70, S^2d_i = 0.06) was found to be stable for seed yield per plant owing to non-significant regression co-efficient (b_i) and deviation from regression (S^2d_i) with high mean performance compared to hybrid mean. GM-4 x K-851, in addition to seed yield per plant depicted stability for clusters per plant and above average stability for 100-seed weight. In case of GM-3 x SM-52, it also exhibited stability for clusters per plant and pods per cluster. Seed yield per plant was found to be better in favourable environments in VRMGg-1 x SM-56, VRMGg-1 x SM-52, GM-3 x SM-56 and GM-4 x Pusa 9632. None of the hybrids displayed above average stability, suitable for poor environment conditions. This revealed that hybrids found stable for seed yield per plant also manifested stability for two or three component traits which indicated that stability of various component traits must be responsible for the observed stability of hybrids for seed yield. It can therefore be suggested that while making selection attention should be paid to phenotypic stability of the characters associated with seed yield and genotypes having average stability for different characters could be used in developing stable hybrids. The chances for selection of stable hybrids could be strengthen by selection in favour of stability for individual component. (9) also suggested that stability for seed yield might be due to the stability of various yield components.

Table 3: Estimates of environmental indices (Ij) for various characters under different environments

Characters	Environments		
	E-1	E-2	E-3
Branches per plant	-0.519	0.287	0.232
Pods per cluster	0.131	0.076	-0.207
Clusters per plant	-0.773	0.415	0.358
Pods per plant	-4.445	1.611	2.833
Seeds per pod	-0.214	0.118	0.097
Seed yield per plant (g)	-0.646	0.578	0.069
100-seed weight (g)	-0.545	0.469	0.076

E-1 = Navsari E-2 = Achhalia E-3 = Vanarasi

The stability parameters for various characters revealed that as many as 15 genotypes turned out to be stable for seeds per pod, 12 for pods per cluster, 11 for branches per plant and 3 for seed weight were found to be stable i.e. they had non-significant b_i and S^2d_i values. The stability of genotypes revealed that none of the parents and hybrids was found to be ideal for better as well as poor management condition for all the characters studied. Average stability of genotypes for seed yield and some of its related traits have also been reported by earlier research workers [2, 3, 4, 8] in greengram.

REFERENCES

1. Eberhart, S.A. and Russell, W.A. (1966). Stability parameters for comparing the varieties. *Crop Sci.*, **6** : 36-40.
2. Manivannan, N. (1999). Association between stability parameters in green gram. *Legume Res.*, **22** (3) : 167-171.
3. Raje, R.S. and Rao, S.K. (2001). Stability analysis for hundred seed weight in mungbean (*Vigna radiata* (L.) Wilczek). *Legume Res.*, **24** (4) : 222-230.
4. Singh, K.N.; Sharma, P.R.; Singh, M.R.K. (2003). Stability analysis in mungbean (*Vigna radiata* (L.) Wilczek). *Research on Crops*, **4** (1) : 97-103.
5. Asrat Asfaw; Yadav, O.P. and Tomer Y.S. (2004). Stability for grain yield in black gram. *Legume Res.*, **27** (1) : 42-45.
6. Manivannan, N. (2003). Stability analysis for seed yield in green gram. *Legume Res.*, **26** (2): 143-145.
7. Abdalla, A.I. and Singh, D.P. (2004). Stability analysis of yield and yield components in early maturing lines of mungbean. *Indian J. Pulses Res.*, **17** (2) : 121-124.
8. Appala Swamy, A. and Reddy, G.L.K. (2004). Stability analysis of yield in mungbean (*Vigna radiata* (L.) Wilczek). *Legume Res.*, **27** (2) : 107-110.
9. Graffius, J.E. (1956). Components of yield in oats, a geometrical interpretation. *Agron. J.*, **51** : 515-554.

Variability in seed yield, gum and protein content in clusterbean [*Cyamopsis tetragonoloba* (L.) Taub.]

Rakesh Pathak, Manjit Singh and A. Henry

Central Arid Zone Research Institute, Jodhpur-342 003, Rajasthan, India

ABSTRACT

Variability present in Clusterbean [*Cyamopsis tetragonoloba* (L.) Taub.] varieties in relation to seed yield, gum and protein content was studied along with correlation coefficient. Varieties RGC 1002 and RGC 1077 gave highest seed yield, whereas, GAUG 9808 and HGS 884 gave highest gum content during kharif 2005 and kharif 2006 respectively, while, highest crude protein was reported with RGC 1038 during kharif 2006. High positive correlation was observed between gum and endosperm percentage. However there was no apparent association between gum content and seed yield. A negative correlation was found between crude protein and gum content.

Key words: Clusterbean, correlation, variability, endosperm, gum, crude protein, seed yield.

INTRODUCTION

Clusterbean [*Cyamopsis tetragonoloba* (L.) Taub.] is an important kharif legume grown extensively in Rajasthan. The seeds of this crop are rich in protein and processed to make clusterbean flour which is a potential source of galactomannan gum. The suitability of this gum in paper, tobacco, petroleum mining, cosmetics, food and pharmaceutical industries has been highly recognized and protein of seed is the best nutritional supplement for animals. The gum is found in the endosperm whereas, seed coat and cotyledons are rich source of protein. In this way clusterbean has got a special place in the commercial scene and therefore, has a great potential in uplifting the economy of the arid and semi arid regions (1).

In the present investigation the promising varieties of clusterbean were evaluated for seed yield, gum and protein content under rainfed conditions.

MATERIALS AND METHODS

Forty promising varieties of clusterbean were raised to study the variability of gum and protein along with seed yield per plant under rainfed conditions during kharif 2005 and 2006 at CR Farm of Central Arid Zone Research Institute, Jodhpur. The experiment was laid out in randomized block design. The plot size consisted of 3 rows of 4 m long with row to row and plant to plant distance of 30 cm and 10 cm respectively. Seed was collected replication wise to study the above parameters. Seeds were analyzed for endosperm, gum and crude protein content (as per cent of seed dry weight). Determination of gum content was done by the colorimetrically method suggested by (2) and modified by (3) whereas, crude protein was estimated by Kjeltec method (4).

RESULTS AND DISCUSSION

Gum and endosperm ranged from 23.8 to 36.0 and 28.0 to 50.9 per cent respectively during 2005 whereas these were 20.2 to 35.1 and 27.2 to 46.9 per cent respectively during kharif 2006. Varieties GAUG 9808 gave the highest gum content (36%) during 2005, whereas HGS 884 (35.1%) ranked first during 2006. The mean crude protein in 40 varieties was 29.7 and it ranged from 26.4 to 32.2 per cent during 2006. Varieties RGC 1038 (32.2%) followed by HGS 02-20 (32%) gave high value of protein. HGS 02-29 and RGC 1078 showed highest endosperm during kharif 2005 and 2006 respectively. Seed yield per plant was higher for varieties RGC 1002 (15.3 g) and RGC 1077 (12.6 g) during kharif 2005 and 2006 respectively (Table 1).

The correlation coefficient were worked out among the characters viz., days to 50% flowering, 100-seed weight, seed yield per plant, endosperm, gum and protein content for kharif 2005 and 2006 and are presented in Table 2. In general 100-seed weight had significantly positive correlation with seed yield during kharif 2005 while it showed low positive correlation during 2006. There was low negative association between 100-seed weight and days to 50% flowering. Seed yield per plant gave low positive correlation with all the traits except 100-seed weight studied during kharif 2005 and 2006. Endosperm had significantly positive correlation with gum content during the period under study, whereas, crude protein showed negative correlation with gum content.

High positive correlation was observed between gum content and endosperm indicated that with the increase in the endosperm in the seed, there would be relative increase in the gum content. This observation has also been reported by (5).

The correlation coefficient between seed yield and gum content revealed a low positive correlation indicating that increase in the yield potential of the crop would not adversely affect the gum content of the seed. Similar findings were observed by (6).

The correlation between seed yield and crude protein revealed a low positive association. Such type of observation was also reported by (7, 8). The trend is indicating that any seed yield improvement programme would not adversely affect the protein content of the seed.

Gum and protein content had a low negative association. The similar association was also observed by (6, 7). The results revealed that increase in gum content may be adversely affected by protein content or vice-versa. However, there was no apparent association between gum content and seed yield.

The early flowering associated with the reduced seed size as indicated by low negative association between 100-seed weight and days to 50% flowering, though the association of flowering with other traits were positive but not significant. This indicated that early flowering did not conform the advantage in the respect of quality traits over the medium maturing varieties.

The present investigation revealed that among all the varieties GAUG 9808 and HGS 884 is best for gum content, RGC 1038 and HGS 02-20 gave high protein while RGC 1002 and RGC 1077 were high yielding varieties in the rainfed condition of arid and semi-arid regions..

Table 1. Mean values for days to 50% flowering, 100 seed weight, seed yield per plant, endosperm, gum content and crude protein in clusterbean varieties

Varieties	Days to 50% flowering		100-seed wt (g)		Seed yield/ plant (g)		Endosperm (%)		Gum content (%)		Crude protein (%)	
	2005	2006	2005	2006	2005	2006	2005	2006	2005	2006	2005	2006
RGC 1088	30.3	28.3	3.3	2.8	8.8	6.5	41.3	40.2	31.5	28.5	NA	29.9
HGS 365	30.0	28.0	3.0	2.9	6.8	6.7	44.3	42.4	32.2	33.4	NA	30.9
HGS 02-29	32.0	29.0	2.9	3.0	5.0	6.8	50.9	44.0	29.3	32.0	NA	27.4
GAUG 0013	33.0	29.0	2.9	2.9	8.4	6.9	37.1	39.3	27.8	30.4	NA	31.1
HG 155-156	33.3	29.3	3.2	2.9	10.1	7.2	42.4	40.2	33.1	30.9	NA	26.6
VIKAS 35	43.0	29.0	3.2	2.9	10.9	10.1	46.8	43.1	34.0	31.8	NA	28.5
WSP 50	37.7	29.0	3.0	2.9	8.9	7.6	34.1	36.2	26.0	29.8	NA	29.6
RGC 1092	32.7	29.0	3.0	2.8	7.5	6.3	36.7	38.2	32.2	34.1	NA	29.2

GAUG 0001	37.7	29.0	2.9	2.8	7.9	6.8	40.4	37.6	29.7	26.7	NA	31.3
GAUG 9808	34.0	30.3	3.1	2.9	7.2	5.5	38.4	37.0	36.0	33.7	NA	29.9
HGS 26-01	30.3	28.3	3.0	2.8	6.9	4.7	47.4	42.0	29.4	27.4	NA	30.6
GAUG 9703	37.3	29.3	2.8	2.8	6.3	4.8	34.8	34.2	25.4	27.4	NA	28.2
RGC 1033	37.7	29.0	2.9	2.8	6.7	4.2	43.7	44.2	26.5	24.2	NA	31.0
HGS 891	30.7	28.7	3.2	3.0	9.7	7.6	42.7	42.2	34.0	34.2	NA	27.7
RGC 1038	30.7	28.3	2.9	2.8	6.9	6.9	41.5	40.5	25.3	21.7	NA	32.2
RGC 1078	40.7	28.7	3.2	3.0	6.3	5.4	46.4	46.9	29.5	30.0	NA	29.2
RGC 1077	29.7	28.7	3.2	2.8	9.3	12.6	38.9	43.7	24.8	27.4	NA	30.3
RGC 1076	34.0	28.7	2.5	2.6	8.2	8.2	30.3	38.7	24.2	30.9	NA	28.6
HGS 02-20	30.7	29.0	2.8	2.8	6.3	4.7	40.2	40.2	28.9	31.1	NA	32.0
HGS 02-1	30.7	28.7	2.9	2.8	12.9	8.4	45.3	40.9	34.8	30.8	NA	29.0
GAUG 0101	31.3	28.3	2.9	2.8	8.2	4.9	43.4	40.5	30.9	27.1	NA	29.4
RGC 2021	37.7	29.0	2.9	2.9	6.9	3.3	43.6	39.1	28.3	24.1	NA	28.5
CLBH 201	37.3	28.3	3.1	3.0	13.7	8.7	28.0	27.2	23.8	23.8	NA	29.6
AVKG 73	30.7	28.0	3.3	3.1	9.9	6.8	30.3	30.8	27.5	27.7	NA	30.2
CAZG 6	30.7	28.0	3.2	3.1	12.2	5.4	37.0	35.7	25.1	21.3	NA	29.3
CAZG 50	34.0	28.7	3.1	2.8	12.7	6.0	42.4	43.5	25.4	27.3	NA	28.2
RGC 1030	37.0	28.7	3.3	2.8	9.1	5.6	45.1	46.2	29.1	30.2	NA	26.4
HGS 870	37.0	28.7	3.0	2.7	9.4	5.7	37.2	38.9	29.5	25.8	NA	30.9
HGS 880	36.3	29.0	3.5	3.0	11.3	5.9	36.8	35.5	26.3	20.2	NA	29.5
RGM 114	30.7	28.0	3.3	2.7	12.9	5.9	41.3	43.0	26.6	32.0	NA	29.8
RGM115	36.7	29.0	3.2	2.6	8.4	4.0	37.1	37.9	31.2	34.3	NA	28.0
HGS 884	39.7	28.7	3.1	2.6	13.0	8.2	38.4	41.6	29.1	35.1	NA	32.0
RGC 1059	30.7	28.0	3.1	2.8	7.9	4.0	41.6	39.9	33.4	32.6	NA	30.5
RGC 1055	39.3	29.3	3.0	2.7	11.0	6.5	36.9	34.8	31.8	31.6	NA	30.4
RGC 1029	36.7	28.7	3.1	2.8	12.8	7.0	41.3	44.2	27.8	30.7	NA	29.9
GAUG 003	37.0	28.7	3.3	2.9	14.4	9.4	46.6	42.7	34.1	30.8	NA	29.7
GAUG 004	40.7	31.0	3.0	2.8	12.3	11.8	42.1	39.1	30.9	31.1	NA	31.5
GAUG 005	30.7	28.3	3.6	3.1	15.1	8.2	43.7	39.8	30.7	29.6	NA	30.3
RGC 1002	40.3	30.0	3.4	3.0	15.3	11.9	45.5	40.3	34.9	30.9	NA	31.3
RGC 986	40.3	31.7	3.1	2.8	11.5	6.7	48.9	42.9	35.5	30.1	NA	29.9
Overall mean	34.8	28.9	3.1	2.9	9.7	6.9	40.8	39.9	29.7	29.3	NA	29.7

Table 2. Correlation coefficients among days to 50% flowering, 100 seed weight, seed yield per plant, endosperm, gum content and crude protein in clusterbean varieties

Name of characters	Days to 50% Flowering		100-Seed Wt (g)		Seed yield/ plant (g)		Endosperm (%)		Gum content (%)		Crude Protein (%)	
	2005	2006	2005	2006	2005	2006	2005	2006	2005	2006	2005	2006
Days to 50% Flowering	1.00	1.00	0.01	-0.11	0.22	0.22	0.05	0.03	0.12	0.19	NA	0.04
100 Seed Wt			1.00	1.00	0.55**	0.11	0.20	-0.20	0.23	-0.23	NA	-0.14
Seed yield/plant (g)					1.00	1.00	-0.03	0.04	0.12	0.17	NA	0.17
Endosperm (%)							1.00	1.00	0.53**	0.31*	NA	-0.11
Gum content (%)									1.00	1.00	NA	-0.13
Crude Protein (%)											NA	1.00

*, ** Significance at 5% and 1% respectively

REFERENCES

1. Arora, S.S. and Gupta, B.S. 1981. Genetic variability correlation and path coefficient analysis in guar (*Cyamopsis tetragonoloba* (L.) Taub.). *Curr. Agric.* Vol. 5, No. 3-4 pp. 116-122.
2. Dubois, M.K.A., Gilles, J.K., Hamilton, P.A. Rebers and Smith, F. 1956. Colorimetric method for determination of sugars and related substances. *Anal. Chem.* 28:350-356.
3. Gupta, S.L. and Bhatia, R. K. S. 1970. Hysteresis in water vapoursroption and desorption on guar gum. *J. Indian Chem. Soc.* 47:1203-1204.
4. A.O.A.C. 1984. Office methods of the analysis of the association of office analytical chemists. 14th edn., published by the *AOAC*. Inc. North nineteenth street, Suit 210 Arlington, Virginia 22209, USA.
5. Menon, Uma, Dubey, M. M. and Bhargava, P.D. 1970. Gum content variation in gaur (*Cyamopsis tetragonoloba* (L.) Taub.) *Indian J. Hered.* 2:55-58.
6. Henry, A. and Mathur, B.K. 2008. Correlation studies of quality parameters, seed yield and maturity period in cluster bean. *J. Arid Legumes* 5 (1): 70-74.
7. Henry, A. and Mathur, B.K. 2006. Genotype x Environment interactions for seed yield and quality parameters in clusterbean. *J. Arid Legumes* 3 (1): 112-114.
8. Henry, A. and Mathur, B.K. 2005. Genetic diversity and performances of cluster bean varieties for quality and quantitative characters in arid regions. *J. Arid Legumes* 2 (1): 145-148.

Genetic diversity in clusterbean [*Cyamopsis tetragonoloba* (L.) Taub.] genotypes using RAPD markers

Rakesh Pathak, Manjit Singh, A. Henry and S. K. Singh

Central Arid Zone Research Institute, Jodhpur-342 003, Rajasthan, India

ABSTRACT

Genetic diversity in Clusterbean [*Cyamopsis tetragonoloba* (L.) Taub.] genotypes was determined using RAPD markers. Out of 12 primers tested nine were highly reproducible and generated 58 RAPD bands with 87.8% polymorphism. Clustering of nine genotypes done using UPGMA revealed that the most distant pairs of clusterbean genotypes were HGS 884 and RGC 1002, GAUG 005 and RGC 005 and RGC 1029 and GAUG 003. These genotypes exhibiting genetic distances can be utilized in future breeding programmes as parents to develop diverse recombinants.

Key words: Clusterbean, genetic diversity, RAPD, polymorphism, cluster analysis

INTRODUCTION

Clusterbean [*Cyamopsis tetragonoloba* (L.) Taub.] is a commercially important arid legume extensively grown in Rajasthan, Haryana, Gujarat and Punjab. The crop has got a status of an industrial crop due to galactomannan gum, which is a water soluble high viscosity natural polysaccharide. Guar meal, a by-product of guar gum industry is used as potential source of protein in animal feeds (1).

Morphological traits have been used for the assessment of genetic diversity (2 & 3). Such traits are influenced by environmental factors as well as developmental stage of the plant, so the results do not provide a true assessment of genetic diversity. There are few reports of use of isoenzymes for study of diversity (4 & 5). The Expressions of isoenzymes is also partially affected by the stage of development. Molecular markers offer several advantages over the conventional morphological markers. Random Amplified Polymorphic DNA (RAPD), that uses arbitrary 10-base primers to amplify the random portion of the genomic DNA (6). This technique produces higher level of polymorphism which can be easily detected in the amplified products. RAPD has been used for assessment of genetic diversity in many crops but a meager knowledge is available in clusterbean (7). Studies were undertaken to find out the genetic diversity among the nine genotypes of clusterbean using 12 primers.

MATERIALS AND METHODS

Nine promising varieties of clusterbean viz., RGM 114, RGM 115, RGC 1059, RGC 1029, RGC 1002, GAUG 003, GAUG 004, GAUG 005 and HGS 884 were sown in randomized block design with 3 replication during kharif 2007 at CR Farm of Central Arid Zone Research Institute, Jodhpur. Three gram of young leaves for each varieties at early vegetative phase were collected from each replication and grounded in liquid nitrogen for DNA extraction using CTAB method described by (8) and treated with RNase to eliminate RNA. DNA concentration was measured by UV absorbance method using a spectrophotometer at 260 nm and its purity was checked by calculating the ratio of absorbance at 260 and 280 nm. The quality and quantity of DNA was also determined by 0.8% agarose gel based on the intensities of bands using λ DNA marker.

Twelve decamer primers from OPA and OPB series (Integrated DNA Technologies Inc.) were used for RAPD assays to identify highly polymorphic well resolved specific bands. The primers did not amplify any fragments during screening were not considered for study (Table 1).

PCR was performed in a PCR plate having 96 reaction tubes of 0.2 ml. Each 25 μl reaction mixture consisted of 2.5 μl of 10X PCR buffer, 1.5 mM $MgCl_2$, 0.2 μl of dNTPs mix, 20 pM of primer, 1 U of taq DNA polymerase and 12.5 ng of DNA. PCR reaction was performed in a Corbett Research, USA, thermal cycler programmed for 44 cycles of standardized cycling conditions. The reaction involved repeated cycles each consisted of denaturation at 92°C for one minute, primer annealing at 35°C for one minute, primer extension at 72°C for one minutes and final extension at 72°C for seven minutes. Amplified DNA products were separated on a 1.2% agarose gel using 6 x loading dye, in 1X TBE buffer and stained with ethidium bromide. The amplification products were visualized under UV light and a photograph was taken using Bio Imaging System.

RAPD bands were designated on the basis of their molecular size. 100 bp ladder was used as reference to estimate the molecular size. The presence of each band was scored as '1' and its absence as '0' for each primer variety combination and used to compute the measure of genetic distance for all the pairs of individuals. Faintly visible bands were not scored but major bands corresponding to faint bands were considered for scoring. The data entry was done into binary data matrix as discrete variables. The statistical analysis was carried out using NTSYS software (version 2.02) (9). Jaccards similarity coefficients were used to generate dendrogram using Unweighted Pair Group Method with Arithmetic Mean (UPGMA) and relationship between varieties were visualized as dendrogram.

RESULTS AND DISCUSSION

The nine RAPD primers generated a total of 58 fragments, 54 of them were polymorphic and 5 primers showed 100% polymorphism (Table 1). Whereas, the minimum level of polymorphism was shown by OPA 4, OPA 17 and OPA 18 (66.7%) with average of 87.8 per cent per primer indicating high marker index. The similar range of polymorphism has been reported in *Ziziphus* species by (10) based on their study using RAPD and (11) on their studies using AFLP in *Ziziphus* species. The number of amplification product ranged from 3 (OPA 4, OPA 17, OPA 18) to 10 (OPA 11, OPB 12).

The genetic diversity of nine clusterbean varieties as revealed by nine decamer primers is shown in the dendrogram (Fig. 1). Maximum similarity was observed between RGM 114 and RGC 1059 (77.5%). The most distant pairs of clusterbean varieties are RGC 1002 and HGS 884 indicating that the distant pairs of varieties could be grouped through RAPD analysis and as such can be utilized with surity in the future breeding programmes as parents to get promising recombinants. Similar observation were recorded by (7) suggesting that HGS 884 is a distant variety.

The present investigation revealed that RGC 1002 and HGS 884 have come from different genetic backgrounds and could be explored as parent to get new promising genotype in background selections during backcross breeding programmes. The study also revealed that RAPD markers are able to differentiate the varieties that are phenotypically similar.

Table 1. List of 12 primers used in the RAPD analysis and the per cent polymorphism displayed among nine genotypes of Guar

Primers	Sequences 5' → 3'	Total No. of bands (a)	Total No. of polymorphic bands (b)	Polymorphism (b/a x100)
OPA 1	CAG GCC CTT C	6	6	100
OPA 4	AAT CGG GCT G	8	8	100
OPA 6	GGT CCC TGA C	3	2	66.7
OPA 11	CAA TCG CCG T	10	10	100

OPA 17	GAC CGC TTG T	3	2	66.7
OPA 18	AGG TGA CCG T	3	2	66.7
OPB 3	CAT CCC CCT G	-	-	NA
OPB 10	CTG CTG GGA C	9	9	100
OPB 11	GTA GAC CCG T	-	-	NA
OPB 12	CCT TGA CGC A	10	9	90
OPB 17	AGG GAA CGA G	-	-	NA
OPB 18	CCA CAG CAG T	6	6	100
Total		58	54	87.8

NA- denote the no amplification of primers

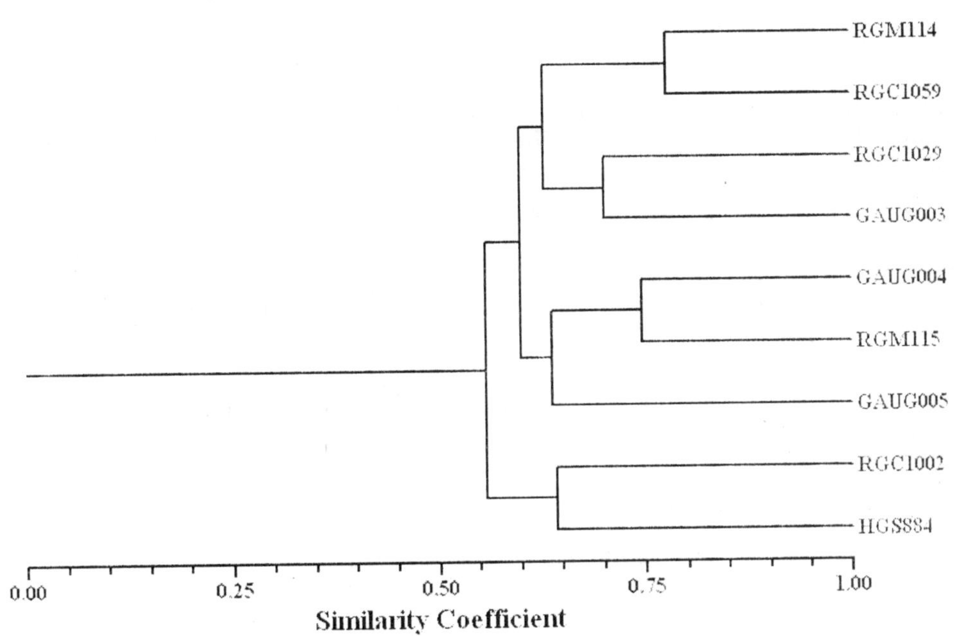

Fig 1. Dendrogram generated for clusterbean varieties using RAPD markers

REFERENCES

1. Henry, A. and Mathur, B.K. 2008. Correlation studies of quality parameters, seed yield and maturity period in cluster bean. *J. Arid Legumes* 5 (1): 70-74.
2. Henry, A. and Mathur, B.K. 2005. Genetic diversity and performance of clusterbean varieties for quality and quantitative characters in arid region. *J. Arid legumes*. 2 (1): 145-148.
3. Henry, A. and Krishna, G.V.S.R. 1990. Genetic Diversity in clusterbean. Madras Agri. J., 77(2): 116-120.

4. Nevo, E., Beiles, E. and Zoharry, D. 1986. Genetic resources of wild barley in the near East: Structure, evolution and application in breeding. *Biol. J. Linn Soc.*, 27:355-380.

5. May, B. 1992. Starch electrophoresis of allozymes. *In*: Hoelzel AR (ed.) Molecular genetic analysis of population: a practical approach IRL Press, Oxford. pp. 1-27.

6. Williams, J.G.K., Kubelik, A.R., Livak, K.J., Rafalski, J.A. and Tingey, S.V. 1990. DNA polymorphism amplified by arbitrary primers are useful as genetic markers. *Nucl. Acid Res.,* 18 (22): 6531-6535.

7. Sudhakar, P., Reddy, D.M., Srinivas, Y., Prasanthi, L. and Latha, P. 2008. Analysis of molecular diversity and detection of genotype specific fingerprints in selected lines of cluster bean (*Cyamopsis tetragonoloba* L. Taub.). *J. Arid Legumes* 5 (1): 38-41.

8. Doyle, J. J., Doyle, J. L. 1990. Isolation of plant DNA from fresh tissue. *Focus.* 12: 13-15.

9. Rohlf, F.J. 1998. NTSYS pc Numerical taxonomy and multivariate analysis system version 2.02 *F. Exeter Software, Setauket*, New York.

10. Devanshi, Singh A.K., Sharma, Pankaj, Singh Bhanumati, Singh, Rakesh and Singh, N.K. 2007. Molecular profiling and genetic relationship among ber (*Ziziphus* spp.) genotypes using RAPD markers. *Indian J. Genet.*, 67 (2): 121-127.

11. Singh, A.K., Sharma, R., Singh, N.K., Bansal, K.C., Koundal, K.R. and Mahapatra, T. 2006. Genetic diversity in ber (*Ziziphus* spp.) revealed by AFLP markers. *J. Hort. Sci. Biotech.* 81 (2): 205-210.

Success story of genetic improvement of clusterbean at agriculture research station, Durgapura, Jaipur (Rajasthan Agricultural University, Bikaner)

S.P.S. Chaudhary, O.P. Chaudhary, N.P. Singh, D.D. Saini and R.V. Singh

Department of Plant Breeding & Genetics,
Agricultural Research Station, Durgapura, Jaipur – 302 018

ABSTRACT

Guar [*Cyamopsis tetragonoloba* (L.) Taub. Syn. C. *psoraliodes* (Lamk; D.C. 2n = 2x = 14) is an important arid legume crop of driers parts of the country in *Kharif* season. It has been cultivated, particularly in Northwestern states of India. The area under the crop remained about 21 – 23 lack hectare mainly in the drier habitats of Rajasthan, Haryana, Gujarat, Punjab and to a limited extent in Uttar Pradesh and Madhya Pradesh. The old varieties cultivated in the country are like Kutch - 8, Malosan, FS - 277, Durgajaya, Durga Safed, RGC -197, RGC - 471, Durgabhar and Pusa Navbahar (Vegetable type) were late maturing and inconsistent in performance with low yield potential (6 – 8q/ha). Systematic research was then initiated for development of improved guar varieties with higher yield potential (10 –16 q/ha) and Medium maturity (110 –115 days). The Varieties viz; RGC – 986 and RGC – 1031 (Kranti) were developed from hybridization technique in the year 1999 & 2005, respectively from a cross (RGC –516 x HG –75) and (RGC – 936 x RGC – 986) for the whole guar growing areas of the country and for Rajasthan State cultivation, respectively. The varieties viz; RGC 936 (1994), RGC –1002 (1999), RGC –1003 (1999) and RGC – 1017 (2001) were developed from hybridization technique from a cross (RGC - 516 x HG –75), (NC 4/P2-1x RGC-516), (Naveen x HG –75) and (Naveen x HG –75) for the whole guar growing areas of the country, respectively. Although short duration varieties were suitable for the extreme drought prone areas of the country, but the varieties has limitation of high productivity of seed grain and fodder yields under high fertility soils & assured rainfall areas of the country.

There fore, our efforts in this direction succeeded to develop fast growing, medium tall, profusely branched, extra medium maturing , bold seeded, High grain and fodder yielding, high Gum and viscosity value and low protein value varieties RGC – 1038 (Karan) and RGC – 1055 (Uday). A variety suitable for inter and Mixed cropping with Bajra, sorghum and maize. A guar variety *viz;* RGC – 197 (1991) was developed for inter and Mixed cropping systems in the shape of improved variety of cluster bean for Rajasthan state. The variety viz; RGC – 1066 (Lathi) was developed with short duration for inter and mixed cropping systems in the year 2007 for Rajasthan state. M-83 (Durga Kranti) a vegetable variety was developed through chemical mutation from Durgabhar variety in the year 2000 for Rajasthan state. These varieties have been now become very popular among the farmers of guar growing areas of the country as reflected by the increasing demands of its breeder seed in the last 3 – 4 years. The results of FLD's in the farmers fields also indicated its liking by the farmers and it is giving about 25 – 35% higher yield over the local varieties in different areas of the country and State.

INTRODUCTION

Guar [*Cyamopsis tetragonoloba* (L.) Taub.] is an important arid legume crop, grown in the arid and semi-arid regions, particularly in Northwestern states of India: Punjab, Haryana, Rajasthan, Gujarat and Uttar Pradesh. The agro-ecological condition of India and Pakistan are ideal for guar cultivation. However, guar has been grown successfully in UA, South Africa, Brazil, Australia, Malavi, Zaire and Sudan. India accounts for nearly 75% of the world acreage and 60% of total production. The area of guar in India has now stabilized around 21-24 lakh hectares, production 10.19 lakh tonnes and productivity (428 kg/ha) Table 1. Rajasthan is the most important state as for cultivation of guar is concerned. This state accounts for about 83% of country's area (19.85 lakh ha), seventy three percent of total production and productivity (246 kg/ha) based on last 33 years the productivity of guar is, however, poorest in Rajasthan. The highest production (9.46 lakh MT) and productivity (453 kg/ha) was recorded in Rajasthan during 1990-91 (Table 1) Guar is known for versatile uses and has thus, occupied specific status among the farmers. The guar crop is largely grown in wide range of soils (Deep sandy, clay as well as in well drained clay soils) and mostly in rain fed areas. Its deep penetrating root system enables the guar plant to utilize the available moisture more efficiently and thus offers better scope for farming in rain fed areas. Its cultivation is under rain fed areas, the production and productivity of the guar crop is largely influenced by rainfall pattern. Therefore, the guar cultivation and production is directly dependent on the performance of rainfall, particularly in western districts where it is not only erratic but also uneven. Thus, the productivity of the crop is very poor in Rajasthan (246 kg/ha) and its productivity was recorded highest in Punjab (1133 kg/ha), Haryana (686 kg/ha) and Gujarat (428 kg/ha), respectively.

Guar (*Cyamopsis tetragonoloba* L. Taub.) has emerged as a new industrial crop, which helps in earning foreign exchange all time high amount of Rs 800 hundred crores of rupees on account of export of 90530 tonnes of guar products during 1998-99 for India. There has been steady increase in the exports of guar splits as well as guar powder during last seven years. Traditionally, its green pods were used as vegetable for human consumption, plant as fodder for live stock and its empty guar pods are better in forage quality than guar fodder (Arora *et al.,* 1980) and seeds for the extraction of gum, which serves as a useful source of hydrocolloids for various uses and guar churi or splits as protein rich meal feed for non ruminants. It is also providing a potential new source of protein to meet the staggering nutritional demands of increasing population. Thus, guar crop must be appraised as a potential source of vegetable protein and pods consumed, as vegetables are rich in minerals like calcium, phosphorus, iron and vitamin A and C for human food. Guar leaves has medicinal values to cure night blindness, ashed plant mixed with oil used as a poultice on cattle boils, seeds are used as a chemotherapeutic agent against small pox, and also as laxative. Boiled guar seeds are used as poultices for the plague, enlarged livers, head swellings, and on swellings due to broken bones. Guars being a leguminous crop it also play an important role in maintaining soil health and

Haryana contributes about 9% total area, 19% production and with productivity (686 kg/ha) ranking after Rajasthan. Whereas, Gujarat contributes 6% of the total area, 8% or total production and with higher productivity (428 kg/ha) than Rajasthan. The cultivation of guar in Punjab is under irrigated condition and in clay loam soils rich in organic matter. Thus, its total area is less than 1 lakh hectare, but production share for the country is 9 per cent with the highest productivity (1133 kg/ha) in India Table 2.

MATERIALS AND METHODS

Clusterbean Improvement are to a great extent dependent on the availability of genetic variability and diversity of the source material, which is the pre-requisite for continuous genetic improvement. We made great efforts to collect and assessed the greater variability in the initial breeding material to ensures better chances of producing desired cultivation land races in the shape of improved varieties of cluster bean namely: Durgajaya, Durga Safed, RGC -197, RGC – 471 and Durgabhar (Vegetable type).

These improved varieties are generally late in maturity and susceptible to major diseases of guar with higher yield potential only under long period (July to Nov.) and well distributed rainfall seasons. As guar is grown in poor fertility and thirsty soils of the country. Therefore, the research on guar improvement has shifted towards hybridizations and mutation breeding technique. To make this crop more productive, crossing programme between locally adopted types with the more then 300 hundred gremlins have been attempted, which resulted in the development of photo-thermo insensitivity, drought tolerant, disease resistant, synchrony in maturity period, short and different maturity period, dwarf plant type, high gum and protein content varieties, showing wider adaptability. These improved cultivars recently developed & released through hybridization and pedigree methods of selection are categorized in five groups. The seed yield data obtained in different Coordinated & State Multilocational trials were analyzed as suggested by Panse and Sukhatme (1).

RESULTS AND DISCUSSION

The area sown under guar has gradually increased with the development of improved varieties in the country as well as in the Rajasthan State. The production and productivity of the crop was highly inconsistent since 1970-71, range figures do not suggest any definite trend over different cultivation years for guar crop in Rajasthan. The production and productivity actually dependent on total rainfall and its distribution. Guar cultivation in Rajasthan is largely under rain fed areas due to which there are wide fluctuations in the state's production and productivity. The production and productivity figures over different year-wise apparently suggest-vulnerability of the guar crop to the vagaries of the weather. There is an urgent need not only for increasing their production and productivity but also to stabilize the same in Rajasthan Table -1.

Table 1. Area ('000 ha), production ('000 tonnes) and seed yield (kg/ha) of cluster bean in Rajasthan

Year	Area ('000 ha)	Production ('000 tonnes)	Seed yield (Kg/ha)	Year	Area	Production	Seed yield
70-71	1171.0	431.0	37	88-89	2105.0	573.0	270
71-72	1197.0	244.0	200	89.90	1971.0	561.0	230
72-73	774.0	95.0	120	90-91	2089.6	946.2	453
73-74	1338.0	415.0	310	91-92	1558.5	204.0	131
74-75	931.0	120.0	130	92-93	1822.2	583.4	320
75-76	2294.0	476.0	210	93-94	1897.1	287.2	151
76-77	2405.0	855.0	360	94-95	1959.4	708.3	361
77-78	1957.0	687.0	350	95-96	1774.7	274.2	155
78-79	2121.0	779.0	367	96-97	1819.1	739.9	407

79-80	1435.0	1440	100	97-98	1985.3	733.6	370
80-81	1965.0	315.0	160	98-99	1611.9	319.6	198
81-82	1669.0	313.0	190	99-00	2648.5	231.6	87
82-83	1996.0	282.0	140	00-01	3056.3	481.2	157
83-84	2208.0	663.0	300	01-02	2412.6	763.3	316
84-85	2038.0	360.0	180	02-03	556.5	27.9	50
85-86	1803.0	142.1	80	03-04	2278.0	1163.00	511
86-87	1751.4	131.8	80	04-05	2364.0	344.6	146
87-88	947.0	29.2	30	05-06	1845.2	651.4	353
Mean					1939.0	526.00	264

Keeping in view the area and industrial importance of cluster bean in India, emphasis on research and development of improved cluster bean varieties was taken up by the AICRP on arid legumes project involving slowly but steadily with the initial efforts put up by scientists working in a regular project located at RAU, Agricultural Research Station, Durgapura, HAU, Hisar and GAU, S.K. Nagar. The AICRP on arid legumes has witnessed exciting progress as far genetic improvement of guar crop, many promising cultivars have been developed for their commercial cultivation under different situations of arid and semi-arid regions.

Improved varieties from simple selection method from land races are generally late in maturity and susceptible to major diseases of guar with higher yield potential only under long period (July to Oct.) and well distributed rainfall seasons. To make this crop more productive, crossing programme between locally adopted types with the diverse gremlins have been attempted, which resulted in the development of photo-thermo insensitivity, drought tolerant, disease resistant, synchrony in maturity period, short and different maturity period, dwarf plant type, high gum and protein content varieties, showing wider adaptability. These improved cultivars were viz. RGC-936, RGC-986, RGC-1003, RGC-1002 & RGC-1017, A vegetable variety M –83 (Durga Kanchan), RGC-1031 (Kranti), RGC-1038 (Karan), RGC –1055 (Uday) and RGC - 1066 (Lathi)from Durgapura, RGM-112 developed at Mandore, Jodhpur. These improved cultivars have made significant contributions towards increased production and productivity of guar. The farmers and seed production agencies may have a wide choice to grow guar varieties for different purposes in specific climatic conditions of the country & State.

Synchrony of maturity in clusterbean is of immense importance for the production of good quality seed. Most of the pods at the top of plant remain green at the time of harvesting, leading to considerable loss because of indeterminate growth habit. Therefore, efforts have been started to develop new genotypes (RGC-1076) with stem determinate, such determinate plants were medium in maturity and photo insensitive. Thus, future aim is to develop a short duration photo period insensitive and determinate in growth habit, with an elevated harvest index and faster growth rate. Some of the pipeline genotypes namely; **RGC-1033, RGC-1076, RGC-1078 and RGC-1077, RGC - 1079, RGr -6, RGr - 7, RGr - 36 and RGr - 45** could be visualized as suitable ideotype.

The research efforts for multiple disease resistant varieties have been in progress and a number of varieties combining high yield and disease resistance have been successfully developed and released.

Clusterbean genetic improvement work has been successfully succeeded in tailoring the maturity duration. Maturity duration in clusterbean has been significantly reduced. Short duration varieties in clusterbean have made significant contributions towards increased production and productivity on vast thirsty and erratic rainfall areas of Rajasthan state. These short duration varieties also found suitable in *kharif* and summer cultivation in non-traditional canal areas.

Table 2. Area ('000 ha), production ('000 tonnes) and seed yield (kg/ha) of guar in India, Rajasthan, Haryana, Gujarat and Punjab

India/States	Area (lakh ha.)	Production (lakh tonnes)	Productivity (Kg/ha.)
India	23.30	10.19	428
Rajasthan	18.85 (83.0%)	7.34 (59.0%)	370
Haryana	1.47 (6.3%)	1.08 (10.59%)	740
Gujarat	1.68 (7.21 %)	1.11 (10.89%)	664
Punjab	0.80 (3.43%)	0.65 (6.37%)	800

These improved cultivars are categorized in five groups:

1. High yielding and normal maturing varieties for high fertility soils

Therefore, We have been successful in genetic improvement in clusterbean crop and by Crossing programme a variety with high grain (14-16qt./ha.) and fodder yielding, Tall, Profusely branched, Medium maturing (110-115days), Bacterial disease resistant, high endosperm value, high gum, protein and Viscosity value variety *viz.*, RGC – 986 was developed in the year 1999, Which has wider adaptability for all guar growing areas of the country Table 3.

Table 3: High yielding and normal maturing varieties for high fertility soils

Characteristics of varieties	RGC- 986	RGC- 1031(Kranti)
Year of notification	1999	2005
Areas of adoption	Whole guar growing areas of the country	Suitable for fodder, grain & green manuring for Rajasthan state
Pedigree	RGC – 516 x HG-75	RGC- 936 x RGC- 986
Days to 50% flowering (No.)	42 - 51	44 - 51
Maturity duration(No.)	110 - 115	110 - 114
Branches/plant(No.)	5.55- 9.10	5.85 – 12.10
Plant height (cm.)	90 - 130	74 - 108
Pods/plant (No.)	38 - 84	42 - 147
Seeds/pod (No.)	7.10 – 9.20	7.70 – 8.70
Gum (%)	28.19- 30.94	28.90 – 30.36
Protein (%)(%)	28.77 – 30.66	27.98 – 30.49
Endosperm (%)	33.81- 36.24	33.52 – 36.12
Carbohydrate (%)	30.32 – 35.50	30.26 – 32.60
Viscosity profile	1777 - 4079	1690 - 3960
100 Seed weight (gm.)	3.00- 3.60	3.10 – 3.40
Major diseases	Low infestation	Bacterial blight résistance
% increase over locals in FLD's	15 - 25	20 -35
Yield potential (qt./ha.)	10 – 15 qt./ha.	10.50 – 15.77 qt./ha.
Plant type	Leaves palmate, hairy, rough and serrated. Tall and profusely branched. Seeds having high gum and protein.	Leaves palmate, hairy and dark green, tripinnate & small leaf margins. Seeds bold & light gray. High viscosity value.

In this direction a guar variety RGC – 1031 (Kranti) was released in the year 2005 for guar growing areas of the Rajasthan State, which was 10 – 15% higher yielder than RGC – 986.

2. High yielding and Early maturing varieties

Four short duration (85–97 days) varieties were developed *viz;* RGC – 936, RGC – 1003, RGC – 1002 and RGC–1017, which were dwarf (70 – 90 Cm.), fast growing, branched and bushy, most suitable for low and erratic rainfall situations of the whole guar growing areas of the country. These varieties were recommended for cultivation in the year 1994, 1999, 1999 and 2001 for the whole country, respectively. These varieties were also found suitable in *kharif* and summer cultivation in non-traditional canal areas Table - 4.

Table 4: High yielding and Early Maturing varieties

Characteristics of varieties	RGC- 936	RGC- 1002	RGC- 1003	RGC- 1017
Year of notification	1994	1999	1999	2001
Areas of adoption	Whole country	Whole country	Whole country	Whole country
Pedigree	EC –516 x HG -75	NC 4/P2-1 x RGC - 516	Naveen x HG - 75	Naveen x HG - 75
Days to 50% flowering (No.)	20 - 25	30 -36	29 -35	32 - 36
Maturity duration (No.)	85 - 85	90 - 95	85 - 92	92 - 99
Plant type	Branched, hairy & busy, serrated leaf margins. White flowers, dwarf & drought prone.	Dwarf & profusely branched, deafly serrated leaf margins. Pink flower	Dark green leaf, non serrated leaf margins. vigorous growth habit. Pink flower	Dwarf , profusely branched, hairy, deep serrated leaf margins. Pink flower
Gum (%)	27.43 – 29.00	29.02 – 30.20	30.60 – 31.10	27 - 29
Protein (%)	25.20 – 26.47	28.00 – 32.10	24.29 – 25.30	29 - 33
Endosperm (%)	33.50 – 35.13	35.00- 37.00	33.40 – 35.60	32 - 37
Viscosity profile	3215	4885	3320	3850
100 Seed weight (gm.)	2.90 – 3.10	3.10 – 3.57	2.90 – 3.10	2.80 – 3.20
Major diseases	Blight susceptible	Traces	Tolerant	Tolerant
% increase over locals in FLD's	10 -15	15 - 25	20 - 35	20 - 35
Yield potential (qt./ha.)	10 - 12	10 - 13	12.00 –15.00	10 - 14

3. High yielding and Extra medium maturing varieties

Although Short duration varieties were suitable for the extreme drought prone areas of the country, but the varieties has limitation of high productivity of grain and fodder yields under high fertility soils and assured rainfall areas of the country. Therefore, our efforts in this direction succeeded to develop fast growing, medium tall (85–125 Cm.), profusely branched, extra medium maturing (100–105 days), bold seeded, High grain(14-16 qt./ha.) and fodder yielding, high Gum and viscosity value and low protein value variety **RGC – 1038 (Karan)** and **RGC – 1055 (Uday**). Variety **RGC – 1038 (Karan)** was identified in the year 2006 for the Northern states of the country. Variety **RGC – 1055 (Uday)** was released in the year 2007 for the Rajasthan State Table 5.

Table 5: Performance of high yielding, Extra Normal maturing guar varieties

Characteristics of varieties	RGC- 1038 (Karan)	RGC- 1055(Uday)
Year of notification	2007	2007
Areas of adoption	Northern –western parts of the country	Rajasthan state
Pedigree	RGC – 516 x HG -75	RGC – 1013 x RGC - 986
Days to 50% flowering (No.)	36 – 42	36 - 39
Maturity duration(No.)	97 – 105	96 – 106
Branches/plant(No.)	5.55 – 6.75	5.58 – 5.77
Plant height (cm.)	70.20 – 75.50	60.50 – 74.74
Pods/plant (No.)	50.45 – 55.78	55 –75.87
Seeds/pod (No.)	7.25 – 7.84	7.35 –7.55
Gum (%)	28.90 – 32.60	29.57 – 30.65
Protein (%)(%)	28.65 – 30.87	28.21 – 29.12
Endosperm (%)	31.60 – 36.50	35.30 – 37.23
Carbohydrate (%)	35.20 – 37.40	36.20 – 37.50
Viscosity profile	2135 - 3774	2152 -4566
100 Seed weight (gm.)	2.99 – 3.20	3.15 –3.35
Major diseases	Traces	Traces
% increase over locals in FLD's	25 – 57.58	13.64 – 45.00
Yield potential (qt./ha.)	10.21 – 29.85	10.96 – 28.81
Plant type	Medium tall, branched, hairy, serrated leaf margins, seeds dark gray, bold seeded.	Medium tall, profusely branched, hairy, deeply serrated leaf margins, light pink flower, pods medium in size, seeds light gray, bold seeded.

4. Unbranched variety for Inter and mixed cropping systems

Farmers of the state were adopting different cropping systems. Looking their demand for a variety suitable for inter and Mixed cropping with Bajra, sorghum and maize. Scientist and their team members made great efforts to collect and assessed the greater variability in the initial breeding material to develop a desired guar variety viz; RGC – 197 (1991) for inter and Mixed cropping systems in the shape of improved variety of cluster bean for Rajasthan state. The variety was deep rooted, fast growing, Normal (110 –115 days) maturing, Tall (150 – 200Cm.), unbranched and Indeterment growth habit plant type for Inter and mixed cropping systems in arid & semi arid drought situations for high grain yields (**10 –12 qt./ha).**

The variety RGC - 197 was long duration, which has problem of black seed at the time of its harvest because of some green pods left immature. Therefore, our efforts continue to develop a variety with short duration (100- 105 days) and determent growth habit plant type. Hybridization and pedigree selection method had helped us to develop a variety **RGC – 1066 (Lathi)**, which was identified and notified (2007) for Rajasthan state, deep rooted, Medium tall (100 – 125 cm), Early maturing (100-105 days), unbranched growth habit plant type and with high grain (12-14qt./ha) yields. This variety is also suitable for inter and mixed cropping systems. Plant was also Photo-Thermo insensitive so that its cultivation may be possible in ***Zaid*** and ***Kharif*** seasons Table 6.

Table 6: Performance of high yielding, unbranched guar varieties for Inter & mixed cropping systems

Characteristics of varieties	RGC- 197	RGC- 1066(Lathi)
Year of notification	1988	2007
Areas of adoption	Rajasthan state	Rajasthan state
Pedigree	Selection from local races from Naguar District.	RGC – 936 x RGC - 197
Days to 50% flowering (No.)	42 – 45	33 – 35
Maturity duration(No.)	110 – 115	95 – 100
Branches/plant(No.)	0.00	0.00
Plant height (cm.)	95.85 – 102.03	75.50- 86.67
Pods/plant (No.)	40.15 – 52.20	45.60 – 53.25
Seeds/pod (No.)	8.15 - 8.20	7.25 – 7.79
Gum (%)	27.75 – 29.95	29.90 – 31.32
Protein (%)(%)	25.85 – 27.28	28.70 – 30.00
Endosperm (%)	32.25 – 34.20	33.60 – 35.95
Carbohydrate (%)	33.40 – 36.30	31.25 – 33.70

Viscosity profile	3100 – 3500	3225 - 3660
100 Seed weight (gm.)	3.35 – 3.75	3.25 – 3.38
Major diseases	Tolerant	Tolerant
% increase over locals in FLD's	10 - 15	15 - 35
Yield potential (qt./ha.)	10.20 – 12.20	10.32 – 14.51
Plant type	Unbranched, Tall, hairy, Broad foliage, dark green , shinning and non serrated leaf margins. Pods medium in size, seeds bold and gray.	Unbranched, Medium tall, hairy, leaves with smooth margins. Light gray, round and bold. Pods are long with deep constriction in between the seeds. Extra medium in maturity.

Table 7. Impact of improved guar varieties on productivity

Name of variety	FLDS conducted	Seed yield of Improved variety (kg/ha)	Seed yield of Local variety (kg/ha)	% increase in yield over local variety
RGC-1003	97	1048.02	828.00	26.57
RGC-1002	42	1298.06	945.86	37.24
RGC-1017	8	1023.75	788.44	29.85
RGC-936	2	1565.00	1160.00	34.91
RGC- 1031	8	1229.17	915.00	34.34
RGC-1038	4	1530.00	1019.50	50.12
RGC- 1055	12	1314.40	1015.75	29.40
RGC-986	5	924.40	777.20	18.94

5. Development of vegetable variety

M-83 (Durga Kranti) was notified in the year 2000 for Rajasthan state as a vegetable variety through chemical mutation from Durgabhar variety at ARS, Durgapura. Variety was unbranched, Tall (125 –150 cm), Indeterminate growth habit, fast growing and luxuriant, fleshy, glabrous and long pods with high green pods yields (65-75 qt./ha.), early pods starts (30-35 days) and suitable for cooking as vegetable.

REFERENCES

1. Panse, V.G. and Sukhatme, P.V. 1967. Statistical methods for agricultural Workers II Ed. ICAR Publication, New Delhi.
2. Jai Vir Singh and B. S. Dahiya., 2002. Guar. Ed. Indian society of Forage Research, Hissar (INDIA).
3. Arora, S.K. and Joshi, U.N. 1980. Guar - its uses are many and expanding all the times. Presented at Fourth All India Guar Res. Dev. Workshops held at CAZRI, Jodhpur on March 10-11.

Success story of genetic improvement in cowpea [*Vigna unguiculata* (L.) Walp.] at agricultural research station, Durgapura, Jaipur

D.D. Saini And S.P.S. Chaudhary

Department of Plant Breeding and Genetics, Agricultural Research Station, Durgapura, Jaipur

ABSTRACT

Cowpea [*Vigna unguiculata* (L.) Walp.] is one of the important arid legume crops of Rajasthan for *kharif* season. It is mainly cultivated in five districts viz., Jaipur, Ajmer, Sikar, Jhunjhunu and Nagour. The area under the crop remained about 80-85 thousand hectares upto the year 1990. The old varieties cultivated in the state like RS-9, JC-5, JC-10, C-152 and FS-68 were late maturing and inconsistent in performance with low yield potential (5-6 q/ha). Systematic research was then initiated for development of improved cowpea varieties with higher yield potential and early maturity. A variety RC-19 was developed from this work in the year 1987 at ARS, Durgapura from a cross Virginia x New Era. It is an early maturing (60-65 days) variety with synchronous maturity having yield potential of about 7-9 quintals/hectare. The variety is still popular among the farmers of the state and continuing in the seed production chain of the state. The variety is having a draw-back that its seeds are fawn coloured. Further, in the year 1990-91 on the basis of feedback from farmers, a need for white seeded cowpea variety was felt in the state and thereafter research work was started at ARS, Durgapura for development of a white seeded cowpea variety. Then out of the work done at ARS, Durgapura finally a cowpea variety RC-101 was developed and released in the year 2001 from this research station. It is first white seeded cowpea variety developed in the state from a cross IT-82 D-652xFS-68. The variety has now become very popular among the farmers of cowpea growing areas of the state as reflected by the increasing demands of its breeder seed in the last 3-4 years. The results of FLDs in the farmers' fields also indicated its liking by the farmers and it is giving about 20-30 per cent higher yield over the local varieties in different areas of the state.

Key words: Maturity, synchronous, production, cowpea

INTRODUCTION

Cowpea is one of the important arid legume crops in Rajasthan and other tropical, sub-tropical regions in India. It is cultivated mainly in five districts of Rajasthan viz., Jaipur, Ajmer, Sikar, Jhunjhunu and Nagour. The area under this crop remained about 80-85 thousand hectares upto the year 1990 and old varieties like RS-9, JC-5, JC-10, C-152 and FS-68 were being cultivated. The yield potential was also very low in these varieties. Later on, the early maturing variety RC-19 was developed from a cross Virginia x New Era at ARS, Durgapura. The variety is having yield potential of 7-9 quintals/hectare and after its release, the variety became popular among the farmers resulting into increased area and production in the state depending upon the rainfall and its distribution during the crop season. Another variety in white seeded cowpea (RC-101) was developed from ARS, Durgapura in the year 2001 and now this variety has become very popular with the farmers, as reflected by the increasing demand of its seed year after year.

MATERIALS AND METHODS

Five cowpea lines viz.Virginia, New Era, IT-82D-652, FS-68 and DCP-68 were used in the crossing programme at different period to develop the three varieties which ultimately changed the scenario of cowpea production and also resulted in the increased area and productivity in the state of Rajasthan. First early maturing (60-65 days) variety RC-19 with fawn coloured seeds was developed from a cross between Virginia and New Era. The variety has yield potential of 7-9 quintals/hectare under favourable environmental and rainfall conditions. Second variety RC-101 is a white seeded cowpea cultivar developed from a cross IT-82D-652 x FS-68 in the year 2001. Its seeds are medium bold and it is the first white seeded cowpea variety developed in the state of Rajasthan from ARS, Durgapura. Lastly, the variety RC-201 (CPD-16) has been developed from this Research Station from a cross FS-68 x DCP-68 and is in the process of identification in the state of Rajasthan. The yield data obtained in different trials were analyzed as suggested by Panse and Sukhatme (1).

RESULTS AND DISCUSSION

The area sown under cowpea has gradually increased with the development of improved varieties in the state of Rajasthan. The crop was sown in 0.754 lakh hectares during 1990-91, whereas the area has increased to 1.374 lakh hectares during 2004-05 (Table 1). The production and productivity of the crop during the period 1990-91 to 2004-05 was inconsistent which may be due to favourable/ unfavourable environmental conditions in different crop seasons. The highest productivity (6.01 q/ha) and production (0.720 lakh tones) remained during 2003-04.

Table 1. Area (lakh ha), production (lakh tonnes) and productivity (q/ha) of cowpea in Rajasthan.

Year	Area	Production	Productivity
1990-91	0.754	0.273	3.62
1991-92	0.873	0.111	1.27
1992-93	0.928	0.403	4.34
1993-94	0.987	0.127	1.29
1994-95	0.840	0.278	3.31
1995-96	1.017	0.220	2.16
1996-97	1.200	0.432	3.60
1997-98	0.995	0.338	3.39
1998-99	1.052	0.234	2.22
1999-00	1.057	0.126	1.20
2000-01	1.039	0.211	2.03
2001-02	1.369	0.474	3.46
2002-03	1.110	0.106	0.96
2003-04	1.199	0.720	6.01
2004-05	1.374	0.278	2.02

The performance of the variety RC-101 was evaluated at Durgapura during *kharif*, 1994 and at two locations during *kharif*, 1995. The mean yield recorded was 10.10 q/ha in these trials (Table 2). The variety registered yield increase of 49.85 per cent over the best check variety RC-19 in these trials. The other details of the variety have been presented (3).

Table 2 : Performance of RC-101 at Durgapura and Tabiji (Ajmer) in Station Trials.

S.N.	Variety	Grain yield (q/ha)			Mean yield (q/ha)	Per cent increase over checks
		Durgapura		Tabiji		
		Kharif, 1994	*Kharif*, 1995	*Kharif*, 1995		
1.	RC-101	13.20	6.84	10.27	10.10	-
2.	FS-68 (LC)	7.05	5.00	6.18	6.34	59.30
3.	P. Phalguni (NC)	6.66	4.30	7.95	6.30	60.32
4.	RC-19 (LC)	9.73	4.75	-	6.74	49.85
	CV (%)	25.79	20.85	7.71		
	CD at 5%	2.45	1.79	0.92		

Similarly, to evaluate the performance of cowpea cultivar CPD-16, trials were conducted at three locations during the year 2003. The variety CPD-16 gave the highest average yield of 15.32 q/ha as compared to RC-101 (12.22 q/ha) and GC-3 (13.70 q/ha). The variety recorded a grain yield increase of 11.82% over the best check variety GC-3 in this trial (Table 3).

Table 3 Performance of CPD-16 in CET trial at three locations during kharif, 2003.

S. No.	Variety	Grain yield q/ha (Kh, 2003)			Average yield (q/ha)	% increase over GC-3
		Durgapura	Tabiji (Ajmer)	Fatehpur - Shekhawati		
1.	CPD-16	22.57	14.24	9.14	15.32	11.82
2.	RC-101	19.89	11.80	4.98	12.22	
3.	GC-3	22.88	12.15	6.06	13.70	
	S.Em. ±	0.42	0.64	0.33		
	CD 5%	1.22	1.89	0.98		
	CV %	4.35	10.60	7.0		

The variety CPD-16 was again evaluated at three locations during kharif, 2007 against three checks to confirm its superiority over these varieties. The highest mean yield was recorded for CPD-16 (12.21 q/ha) against the best check variety RC-101 (9.84 q/ha). The variety has given seed yield increase of 19.41 per cent over RC-101 during kharif, 2007 (Table 4).

Table 4. Performance of CPD-16 in CET Trial at 3 locations during kharif, 2007.

Sr. No.	Variety	Grain yield q/ha (Kh, 2007)			Average yield (q/ha)	-% increase over GC-3, RC-19 and RC-101
		Durgapura	Tabiji (Ajmer)	Fatehpur - Shekhawati		
1.	CPD-16	15.55	16.25	4.82	12.21	
2.	GC-3(C)	10.42	12.81	2.89	8.71	28.66
3.	RC-19 (C)	14.20	10.80	2.59	9.20	24.65
4.	RC-101 (C)	13.68	13.78	2.05	9.84	19.41
	S.Em. ±	0.80	0.55	0.16		
	CD 5%	2.48	1.71	0.49		
	CV %	11.82	8.28	10.3		

The performance of the variety CPD-16 was also tested in FLDs during kharif, 2006 and 2007 at five different locations in these years. The variety was found superior over the local check at farmers' field. The per cent increase in yield recorded during kharif, 2006 and 2007 was 44.95 and 25.00, respectively over the local variety (Table 5). These results indicated that variety RC-101 is performing well and superior over the local variety.

Table 5: Results of FLDs conducted at Ajmer and Jaipur districts of Rajasthan during kharif, 2006 and 2007.

Sr. No.	Variety	Grain yield (q/ha)			
		Kharif, 2006		Kharif, 2007	
		Improved variety	Local	Improved variety	Local
1.	RC-101	10.80	8.30	11.40	9.10
2.	RC-101	6.00	4.00	10.85	8.70
3.	RC-101	4.00	2.40	11.20	8.90
4.	RC-101	2.80	1.60	10.70	8.40
5.	RC-101	8.00	5.50	10.60	8.70
	Mean	6.32	4.36	10.95	8.76
	Per cent increase over local		44.95		25.00

REFERENCES

1. Panse, V.G. and Sukhatme, P.V. 1967. Statistical methods for agricultural Workers II Ed. ICAR Publication, New Delhi.
2. Saini, D.D. and Chaudhary, S.P.S. 2005. Breeding cowpea varieties for semi-arid agro zones. J. Arid Legumes 2(1) : 18-19.
3. Saini, D.D. 2003. Development of early maturing white seeded cowpea varieties for Rajasthan. In : Advances in arid legumes research (proceedings of the National Symposium on Arid Legumes for Food Nutrition Security and Promotion of Trade, held at CCS HAU, Hisar, 15-16 May, 2002). Published by Indian Arid Legumes Society, CAZRI, Jodhpur: pp 47-49.

Correlation studies in Indian beans (*Dolichos lablab* L.)

Shital Shewale, N.C. Desai and Sachin Chaudhari

Department of Agricultural Botany, Navsari Agricultural University, Navsari-396 450

ABSTRACT

Genotypic and phenotypic correlation coefficients were worked out for fourteen agronomic traits in beans. It was observed that association analysis between grain yield and other thirteen traits revealed that the grain yield was significant and positively correlated with plant height, pods per plant, clusters per plant, harvest index and protein content at both genotypic and phenotypic levels. The character days to 80% maturity, branches per plant, pod length and cooking time exhibited negative correlation with grain yield per plant.

Key words : *Dolichos lablab*, correlation coefficient.

Indian bean (*Dolichos lablab* L.) is an important vegetable crop of Indian origin. It occupies a unique position for vegetable purpose among the legume vegetables. Knowledge of correlation between yield and its contributing characters are essential to find out guidelines for plant selection. Keeping in view the above facts, the present investigation was undertaken to know correlation coefficient among yield and its contributing character in 60 diverse genotypes of Indian bean collected from Regional Pulse Research Station, Navsari for making effective selection in improvement of the crop.

MATERIALS AND METHODS

Sixty genotypes of Indian bean were collected from germplasm collection at the Regional Pulse Research Station, Navsari Agricultural University, Navsari and were grown in randomized block design with three replications during the pre *rabi* seasons of 2006-07. Each genotype was grown in a three-row plot of 3m long with 60cm spacing. The crop was fertilized with 20-40-0-20 NPKS kg/ha as basal dressing and irrigated as and when required. Five plants were randomly selected in each plot to record the observation on different characters and the data were subjected to statistical analysis.

RESULTS AND DISCUSSION

In general, the genetic correlation were greater than the phenotypic correlation coefficients among all the traits indicating that though there was strong association between various characters, the phenotypic expression was influenced by the environment. The correlation coefficients among different traits are listed in Table 1. The plant height, clusters per plant, pods per plant, harvest index and protein content were positively correlated with grain yield per plant. These five traits were significantly correlated with each other which indicated that selection for either of the characters which result in improvement of other character along with yield. Days to 50% flowering showed significant positive correlation with days to 80 per cent maturity, plant height, seeds per pod, pod length, protein content and cooking time. This clearly points out that flowering time is crucial indicator of maturity. Days to 80% maturity had highly significant and positive correlation with plant height, seeds per pod, pod length, protein content and cooking time. Plant height showed significant positive correlation with branches per plant, seeds per pod, pod length, 100-seed weight, harvest index, protein content, cooking time and hard seeds. Branches per plant showed significant positive correlation with clusters per plant and seeds per pod. Clusters per plant significant correlation with pods per plant, harvest index and cooking time. Pods per plant highly significant positive correlation with harvest index, seeds per pod had positive and significant correlation with pod length, 100-seed weight, harvest index, protein content, cooking time. Pod length had significant and positive correlation with 100-seed weight and hard seeds. Harvest index was highly significant and positive association with protein content. Rest of the associations were statistically non significant [1, 2].

Days to 80% maturity, branches per plant, pod length and cooking time showed highly significant and negative correlation with grain yield /plant.

From the correlation coefficient estimates it becomes obvious that branches/ plant, clusters/plant, 100-seed weight, pods per plant and seeds per pod were the main characters contributing towards the yield. The positive correlation of plant height, clusters per plant, pods per plant, harvest index and protein content with grain yield per plant Negative correlation was observed between grain yield per plant and the characters, days to 80% maturity, branches/plant, pod length and cooking time. Considering above discussion ideal plant type would be early flowering and having more pods /plant, clusters per plant, seeds per pod and increased plant height.

Table 1. Genotypic and phenotypic correlation of grain yield with other characters in 60 genotypes of Indian bean

Characters	r	DF	DM	PH	BP	CP	PP	SP	PL	TW	HI	PC	CT	HS
YLD	r_g	0.023	0.098**	0.143**	0.062**	0.195**	0.506**	-0.017	0.243**	0.053**	0.635**	0.051**	0.125**	0.144**
	r_p	0.016	0.077**	0.122**	0.077**	0.188**	0.468**	0.049**	0.199**	0.018	0.502**	0.049**	0.096**	0.116**
DF	r_g		0.993**	0.289**	0.055**	0.147**	-0.034	0.132**	0.257**	0.148**	0.115**	0.162**	0.158**	0.019
	r_p		0.855**	0.314**	0.028	0.129**	-0.015	0.135**	0.069**	-0.004	0.074**	0.111**	0.173**	-0.005
DM	r_g			0.230**	-0.040*	0.208**	0.188**	0.144**	0.412**	0.167**	0.202**	0.190**	0.055**	0.026
	r_p			0.309**	0.020	0.144**	0.124**	0.171**	0.078**	0.095**	0.072**	0.098**	0.136**	0.002
PH	r_g				0.086**	-0.023	0.144**	0.249**	0.274**	0.501**	0.098**	0.125**	0.080**	0.235**
	r_p				0.103**	-0.015	0.135**	0.202**	0.121**	0.354**	0.086**	0.084**	0.108**	0,210**
BP	r_g					0.476**	0.038**	0.177**	0.082**	0.039*	0.120**	0.063**	0.103**	0.084**
	r_p					0.444*	-0.032	0.180**	0.060**	0.042*	0.080**	0.030	0.082**	0.068**
CP	r_g						0.397**	0.076**	0.143**	0.228**	0.167**	0.102**	0.084**	0.322**
	r_p						0.391**	-0.031	0.144**	0.212**	0.160**	0.062**	0.127**	0.302**
PP	r_g							0.103**	0.337**	0.308**	0.295**	0.038**	0.008	0.345**
	r_p							0.067**	0.285**	0.262**	0.266**	-0.017	0.014	0.327**
SP	r_g								0.430**	0.180**	0.076**	0.169**	0.067**	0.046**
	r_p								0.338**	0.083**	0.081**	0.135**	0.126**	0.025
PL	r_g									0.409**	0.228**	0.088**	0.086**	0.478**
	r_p									0.521**	0.144**	0.049**	0.065**	0.378**
TW	r_g										-0.006	0.052**	0.048**	0.869**
	r_p										0.024	0.028**	0.178**	0.686**
HI	r_g											0.077**	0.104**	0.000
	r_p											0.081**	0.076**	-0.017
PC	r_g												-0.005	0.165**
	r_p												0.032	0.138**
CT	r_g													0.153**
	r_p													0.124**

DF =Days to 50% flowering	BP =Branches per plant	SP =Seeds/pod	HI =Harvest index (%)	HS =Hard seed (%)
DM =Days to 80% maturity	CP =Clusters/plant	PL =Pod length (cm)	PC =Protein content (%)	
PH =Plant height(cm)	PP =Pods/plant	TW =100 -Seed weight (g)	CT =Cooking time (min.)	

REFERENCES

1. Desai, N.C., Tikka, S.B.S. and Chauhan, R. M. (1996). Genetic variability and correlation studies in Indian bean (*Dolichos lablab* L.). *New Botanist*, XXIII **7**:197-204.
2. Pawar, R.M. (1996). Genetic variability, correlation, path coefficient and genetic divergence in Indian bean *Thesis submitted* to GAU., S.K. Nagar.

Correlation and path coefficient in cowpea (*Vigna unguiculata* L. WALP.)

Jay Lal Mahto and V. Kerketta

Department of Plant Breeding and Genetics, Birsa Agricultural University, Ranchi 834 006, Jharkhand.

ABSTRACT

In a study of 25 genotypes of cowpea (*Vigna unguiculata* L. Walp.) in three replications in randomized block design during summer and kharif 2007 at experimental area of Birsa Agricultural University, Ranchi. The observations were taken on ten representative plants from every replication. All the characters were found to be statistical significant for all the characters under study. High genetic and phenotypic coefficients of variation were found for days to flowering, number of pods per plant and 100 seed weight. High heritability along with genetic advance was found for days to flowering and number of pods per plant. Highly positive correlation between seed yield with number of pods per plant, days to 50% flowering, days to maturity and 100 seed weight were found.

Key words: Cowpea, Variability, Correlation, Path coefficient

INTRODUCTION

Cowpea (*Vigna unguiculata* L. WALP.) is pulse crop which is used as vegetable and forage also. Not much information in cowpea is available on the extent and nature of variability of characters on the basis of which the plant breeders can accelerate their selection program. Farmers of the Jharkhand state are mostly grown old traditional cultivars, which are very poor yielder. Emphasis should be given to identify location specific promising lines to enhance production. Atlin *et al.* [1] have advocated that selection exercised in the subdivided target region will be productive. It is also known that producti-vity altered with the changes in growing environment. Therefore, present investigation was undertaken to study the genetic variation of yield, yield components and their association and path coefficient.

MATERIALS AND METHODS

Twenty-five diverse lines of cowpea were evaluated in randomized block design with three replications at experimental area of Birsa Agricultural University, Ranchi, in three replications in summer and kharif 2007. Each plot consisted of six rows of 4 m length with a spacing of 45 x 15 cm between rows and plants, respectively. Observations on ten randomly selected plants in each treatment and replication were recorded on ten metric traits *viz,* plant height (cm), number of branches / plant, 100 seed weight (g), number of seeds / pod, number of pods / plant, harvest index (%) and biological yield (g), except days to 50% flowering, days to maturity and seed yield (g), which were recorded on plot basis. Analysis of variance was performed following standard procedures [2]. The data obtained during both the seasons were analyzed for variation, phenotypic coefficient of variation (PCV) and genotypic coefficient of variation (GCV), heritability and genetic advance were computed as per standard method [3, 4, 5]. Correlation was calculated as per method suggested by Johnson *et al.,* [6] and Miller *et al.* [7] and path coefficient analysis was carried out according to Deway and Lu [8] and Wright [9].

RESULTS AND DISCUSSION

The analysis of variance indicated highly significant differences among the genotypes for all the characters under studied. Higher phenotypic coefficient of variation (PCV) and genotypic coefficient of variation (GCV) were observed for number of pods/plant, 100 seed weight, days to flowering, number of seeds/pod, seed yield, biological yield, harvest index and plant height which indicated that these traits were less influenced by environment. This was also supported by narrow gap between PCV and GCV and high heritable values for these characters. Therefore these traits could be considered effective for further improvement in early maturity cowpea. High heritability coupled with high genetic advance as per cent of mean was observed for 100 seed weight, number of pods/plant, seed yield, number of seeds/pods, harvest index, days to 50% flowering, days to maturity and biological yield indicating the predominant role of additive gene effects in the expression of these traits and thus suggested that these characters could be improved by simple phenotypic selection. Similar observations have earlier been reported by Tomer *et al.*, [10] for days 50% flowering, 100 seed weight, number of pods/plant, number of seeds/pod and seed yield; Sreekumar and Nair [11] for number of seeds/pod and Rewale *et al* [12] for 100 seed weight, plant height and harvest index in cowpea. Although seed yield was positively correlated with all the traits studied but it was significantly and positively correlated with days to 50% flowering, biological yield, number of pods/plant, number of seeds/pod, 100 seed weight and harvest index (Table 1). Dahiya *et al.*, [13] had observed similar report for number of seeds/pod, number of pods/plant, 100 seed weight with seed yield/plant. Similar results were reported by other workers in cowpea [14, 15]. This suggested that selection on the basis of these components would be more effective in improvement of seed yield. This result was in agreement with the earlier findings of Tomer *et al.*, [12] for 100 seed weight and seed yield. Since days to 50% flowering, biological yield, number of pods/plant, number of seeds/ pod, 100 seed weight and harvest index exhibited significant positive correlation with seed yield also showed high heritability and genetic advance as per cent of mean. Therefore it could be concluded that selection of genotypes based on these traits could bring about improvement in seed yield in cowpea. The highest direct of days to 50% flowering (1.141) contribution to seed yield and high for harvest index (Table 2), for selection theses characters should be considered first. The residual factor at genotypic level 38.89 percent. It indicated that all characters contributed 61% variability towards total seed yield per plant.

Table 1. Genotypic correlation in cowpea between different pairs of characters.

Characters	X_2	X_3	X_4	X_5	X_6	X_7	X_8	X_9	X_{10}
Days to 50% flowering X_1	0.842**	0.106	0.821**	0.371	0.546*	0.518*	0.153	0.439	0.658**
Days to maturity X_2		0.553*	-0.122	0.493	0.578**	-0.597**	-0.306	0.247	0.705*
Plant height (cm) X_3			-0.434	0.407	0.376	0.449	0.110	-0.127	0.263
Number of branches X_4				-0.224	-0.069	-0.128	-0.264	0.409	0.342
Biological yield / plant X_5					0.604**	0.665**	0.251	0.488*	0.856*
Number of pods / plant X_6						0.968**	0.139	0.117	0.645**
Number of seeds / pod X_7							0.183	0.151	0.690**
100 grain weight (g) X_8								0.265	0.769*
Harvest index (%) X_9									0.885**
Seed yield (g) X_{10}									

*, ** Significant at 5 per cent and 1 per cent probability level, respectively.

Table 2. Direct (diagonal) and indirect (off diagonal) effect in cowpea

	X_1	X_2	X_3	X_4	X_5	X_6	X_7	X_8	X_9
X_1	**1.1410**	0.0080	-0.0160	-0.7711	-0.0585	-0.1627	0.1530	0.0292	0.3351
X_2	0.9607	**0.0095**	-0.0702	0.7639	-0.0285	-0.0352	-0.3053	-0.0829	-0.0936

X_3	0.1209	0.0044	**-0.1513**	-0.7278	-0.0126	-0.0386	0.6028	0.1500	0.0167
X_4	0.9368	-0.0077	-0.1172	**-0.9392**	0.0428	0.1755	-2.8864	-0.7711	-0.2763
X_5	0.4233	0.0017	-0.0121	0.2551	**-0.1577**	-0.0964	0.3166	0.0887	0.3182
X_6	0.6230	0.0011	-0.0196	0.5532	-0.0510	**-0.2980**	0.2663	0.0322	-0.1656
X_7	0.5911	-0.0098	-0.3087	9.1756	-0.1690	-0.2686	**0.2955**	0.0490	-0.0821
X_8	0.1746	-0.0041	-0.1190	3.7990	-0.0733	-0.0503	0.0759	**0.1906**	0.0413
X_9	0.5009	-0.0012	-0.0033	0.3400	-0.0657	0.0646	-0.0318	0.0103	**0.7633**

REFERENCES

1. Atlin, G.N.; Baker, R.J.; Mcrae, K.B. And Lu, X. 2000. *Crop Sci.* 40:17-13.
2. Panse, V.G. and Sukhatme, P.V. 1967. Stastistical Methods for Agricultural Workers. II Ed. *ICAR Publication,* New Delhi.
3. Burton, G.W. 1952. Quantitative inheritance in grasses. Proc 6th Int. Grasses Cong., 1:277-288.
4. Johnson, H.W., Robinson, H.F. and Camstock, R.E. 1955. Estimates of genetic and environmental variability in soyabean. *Agron. J.*, 47:314-318.
5. Lush, J.L. 1940. Intrasire correlation and regression of offspring on dams as method of estimating heritability of characters. *Proc. Am. Soc. Anim. Prod.* 33:293-301.
6. Johnson, H.W., Robinson, H.F. and Camstock, R.E. 1955. Genotypic and phenotypic correlation in soyabean and their implications in selection. *Agron. J.*, 47:477-485.
7. Miller, P.A.; Willioms, J.C.; Robinson, H.P. and Camstock, R.E. 1958. Estimation of genotypic and environmental variances and covariances in upland cotton and their implication in selection in selection. *Agron. J.*, 60:126-131.
8. Deway, D.R. and Lu, K.H. 1959. A correlation and Path coefficient analysis of components of crested wheat grass seed production. *Agron. J.*, 51:511-518.
9. Wright, S. 1921. Correlation and causation. *J. Agric. Res.*, 20(7):557.
10. Tomer, R.P.S.; Verma, S.R. And Karwasra, R.R. 1991. Correlation and path coefficient analysis of early testing and mature plant characters in cowpeas. *Symposium on Genetic Research and Education: Current Trends and the nest fifty years*, 504.
11. Sreekumar, K and Nair, R.R. 1996. Variability, heritability, correlation and path coefficient studies of biological nitrogen fixation traits in cowpea. *Legume Res.*, 19(3-4):171-178.
12. Rewale, A.P., Barari, S.P. and Jamadagni, B.M. 1995. Genetic variability and heritability in cowpea. *Agric. Sci. Digest Karnal*, 15(1-2):73-76.
13. Dahiya, O.P.; Singh, D. and Mishra, S.K. 2007. Correlation and path coefficient analysis in cowpea. *J. Arid Legumes.,* 4(2):127-129.

14. Oseni, T.O., Lenge, D.P. and Pal, U.R. 1992. Correlation and path coefficient analysis of yield attributes in diverse lines of cowpea. *Indian J. Agric. Sci.*, 62(6):365-368.

15. Siddique, A.K.M.A.R. and Gupta, S.N. 1991. Correlation studies in cowpea. *Annals of Bio.*, 7(2):181-185.

Genetic variability, correlation and path analysis in mungbean (*Vigna radiata* L. Wilczek)

Jay Lal Mahto and Z. A. Haider

Department of Plant Breeding and Genetics, Birsa Agricultural University, Ranchi 834 006, Jharkhand.

ABSTRACT

Forty-one genotypes of mungbean (*Vigna radiata* L Wilczek) were evaluated during summer and rainy seasons 2007. Significant differences were observed between genotypes for all the characters under study. A wide range of variation for most of the characters was observed. High genotypic coefficient of variation was observed for seed yield/ plant, 100-seed weight, number of pods/plant, number of branches/plant, plant height and days to maturity. High heritability coupled with high genetic advance was found for 100-seed weight, number of pods/plant, seed yield/plant, plant height and days to maturity. Seed yield had highly significant positive correlation with plant height, number of branches/ plant, number of pods/plant, number of seeds/pod, 100-seed weight, days to maturity at genotypic level and almost same at phenotypic level. Number of branches/plant, number of pods/plant, number of seeds/pod, days to maturity and 100 seed weight had appreciable positive direct effect on seed yield/plant.

Key words: Mungbean, Variability, Correlation, Path coefficient

INTRODUCTION

Pulses crop have important place in Indian agriculture owing to their high protein content. Pulses are primary source of protein for the poor and the vegetarian, who constitute the majority of Indian population. Pulses are unique jewels of crop husbandry and also as a good source of green manure. Indian economy is based on sustainable agriculture, particularly rainfed agriculture, where pulse crop occupy a prominent role to play. India is the largest producer and consumer of pulses in the world, accounting for about 36% of global area,

26% of global production, 27% of consumption. Seed yield, the primary objective of a plant-breeding program, is a complex character influenced by a number of interwoven component traits having a definite relationship among them as well as with seed yield. Selection of superior genotypes based on yield *per se* alone might not always be effective. Therefore, yield components should also be considered during selection for higher yield. Indirect selection becomes important also for those characters that are not easily measured and exhibit low heritability. The knowledge about associations between characters may also help identify characters that have little or no importance in the selection program. However, under a complex situation, correlations alone become insufficient to explain relationships among characters and thus, path coefficient analysis becomes useful as it resolves associations into direct and indirect effects and thus provides better understanding the nature of correlations. Hence the present investigation was carried out to study the phenotypic and genotypic associations between yield contributing characters along with path analysis for developing suitable selection criteria for seed yield improvement in mungbean.

MATERIALS AND METHODS

The experimental material comprised 41 genotypes of coordinated program were grown at Birsa Agricultural University pulses research plots during summer and rainy season 2007 following a randomized block design with three replications having spacing 25 cm and 10 cm, between rows and plants, respectively. Data were recorded on ten randomly selected plants for days to 50% flowering, number of branches per plant, number of pods per plant, plant height (cm), number of seeds per pod, days to maturity, 100 seed weight (g) and seed yield per plant (g). Analysis of variance was performed following standard procedures [1]. The data obtained during both the seasons were analyzed for variation, phenotypic coefficient of variation (PCV) and genotypic coefficient of variation (GCV), heritability and genetic advance were computed as per standard method [2, 3, 4]. Correlation was calculated as per method suggested by Johnson *et al* [5] and Miller *et al.*,[6] and path coefficient analysis was carried out according to Deway and Lu [7] and Wright [8].

RESULTS AND DISCUSSION

Significant differences were observed between genotypes for all the characters viz., days to 50% flowering, number of branches per plant, number of pods per plant, number of seeds per pod, days to maturity, plant height, 100 seed weight and seed yield per plant. A wide range of variation for most of the characters was observed. The genetic coefficient of variation together with heritability estimates give the reliable picture of the amount of genetic gain expected from selection. According to Lush [4] heritability in broad sense represents the functioning of the genotype as a whole. Burton and Devane [9] suggested that genotypic coefficient of variation, together with heritability estimates, would give reliable indication of the extent of improvement expected from selection and further explained that expected genetic advance under a particular system provides a true practical information needed by a breeder. In order to assess the proportion of heritable variability and expected genetic gain through the selection, it is essential to estimate heritability and genetic advance together. High genotypic coefficient of variation was observed for seed yield/ plant, 100-seed weight, number of pods/plant, number of branches/plant, plant height and days to maturity (Table 1). Prakash *et al.*, [12] observed similar finding for seed yield and number of pods/plant. Kumar *et al.*, [13] reported similar result for plant height. High heritability coupled with high genetic advance was found for 100-seed weight, number of pods/plant, seed yield/plant, plant height and days to maturity. Similar results were observed by

Genotypic and phenotypic correlation exhibited the same pattern as genotypic correlation was generally higher than the phenotypic ones. Seed yield had highly significant positive correlation with plant height, number of branc-hes/ plant, number of pods/plant, number of seeds/pod, 100-seed weight, days to maturity at genotypic level and almost same at phenotypic level (Table 2). Days to 50% flowering had high significantly negative correlation with seed yield/ plant. Similar result was reported earlier by Rahim *et al.*, [10], and Verma and Garg [11]. Kumar *et al.* [13] observed similar correlation between seed yield with branches and number of pods/ plant. Number of pods/plant had high correl-ation with number of branches/ plant, and days to maturity but had negative association with 100 seed weight and number of seeds / pod at genotypic and phenotypic levels.

Correlation studies simply reveal whether the characters under study are associated or nor but it is difficult to perceive that which one of them is affecting the other. Path coefficient provides better picture of correlations in terms of direct and indirect effects of each characters on seed yield. It unreveals whether the association of these characters with seed yield is due to their direct effect on seed yield, or it is a consequence of their indirect effect via some other traits. In plant breeding, selection is generally practiced on the basis of the high direct effects. Number of branches/plant, Number of pods/plant, number of seeds/pod, days to maturity and 100 seed weight had appreciable positive direct effect on seed yield/plant (Table 3) as first degree of selection and days to 50% flowering and days to maturity as second degree of selection. The residual effects are 0.3256 (genotypic) and 0.4012 (phenotypic).

Table 1. Variability in different characters *in mungbean.*

Characters	PCV (%)	GCV (%)	Heritability(%)	GA	GA (%) Mean
Plant height (cm)	25.11	24.76	96.5	46.43	49.67
Number of branches/plant	26.18	26.23	87.8	2.22	51.13
Days to 50% flowering	13.11	11.29	96.5	11.54	22.78
Number of pods/plant	29.36	29.23	98.6	44.89	60.33
Number of seeds/pod	12.12	12.01	96.7	2.78	22.68
100 seed weight (g)	27.59	26.43	97.5	47.29	57.87
Days to maturity	14.23	13.89	97.8	23.67	25.27
Seed yield/ plant (g)	39.26	38.56	98.9	30.46	80.78

GCV=Genetic coefficient of variation, PCV= Phenotypic coefficient of variation, GA=Genetic abvance

Table 2. Correlation (genotypic (G) and phenotypic (P)) among *seed yield and its traits in mungbean*

Characters		X_2	X_3	X_4	X_5	X_6	X_7	X_8
Plant height (X_1)	P	0.358*	0.625**	-0.167	0.761**	0.070	0.082	0.275
	G	0.385*	0.681**	-0.211	0.801**	0.192	0.099	0.319 *
Number of branches/	P		0.080	0.635**	0.189	0.165	0.269	0.345 *
plant (X_2)	G		0.121	0.767**	0.213	0.199	0.301	0.397**
Days to 50% flowering	P			-0.201	0.129	-0.136	-0.309 *	-0.779**
(X_3)	G			-0.256	0.166	-0.146	-0.411**	-0.811**
Number of pods/plant	P				-0.299*	-0.312 *	0.689**	0.569**
(X_4)	G				-0.323 *	-0.333 *	0.698**	0.689**
Number of seeds/pod(X_5)	P					0.675**	0.768**	0.874**
	G					0.701**	0.781**	0.899**
100 seed weight (X_6)	P						0.354 *	0.463**
	G						0.453**	0.567**
Days to maturity (X_7)	P							0.671**
	G							0.766**
Seed yield/ plant (X_8)								

***, ** Significant at 5% and 1% of probability levels, respectively.**

Table 3. Direct (diagonal) and indirect (off diagonal) effect on *seed yield in mungbean.*

Characters	X_1	X_2	X_3	X_4	X_5	X_6	X_7
X_1 P	**0.0319**	0.5959	0.4843	0.0297	-0.1336	-0.7787	0.2754
G	**0.2593**	-0.6518	-0.2316	0.1827	-0.0211	0.2489	0.4421
X_2 P	0.0327	**0.8543**	-0.6571	-0.0122	-0.0889	0.8767	-0.2221
G	-0.1886	**0.8879**	0.3481	-0.1032	0.7419	-0.2806	-0.5079
X_3 P	-0.0215	0.8333	**-0.7337**	-0.0107	0.0754	0.8739	-0.2374
G	-0.1786	0.8685	**-0.3489**	-0.0904	0.6236	0.2788	-0.4952
X_4 P	0.1001	-0.2067	0.1574	**0.0550**	-0.1636	-0.5655	0.0566
G	0.2591	-0.3475	-0.2911	**0.2794**	-0.1165	0.2996	0.2857
X_5 P	-0.0229	0.3859	-0.3341	-0.0531	**0.1818**	0.7514	-0.1647
G	0.1987	0.4938	0.1871	-0.2443	**0.3314**	-0.2649	-0.4798
X_6 P	-0.0246	0.7623	-0.6521	-0.0288	0.1388	**0.9816**	-0.2235
G	0.2086	0.8126	0.3165	-0.1817	0.0843	**0.3374**	-0.5548
X_7 P	-0.0186	0.6387	-0.5317	-0.0101	0.0937	0.7451	**-0.2976**
G	-0.1742	0.6886	0.2640	-0.1257	0.9775	-0.2563	**-0.6584**

REFERENCES

1. Panse, V.G. and Sukhatme, P.V. 1967. Stastistical Methods for Agricultural Workers. II Ed. *ICAR Publication,* New Delhi.
2. Burton, G.W. 1952. Quantitative inheritance in grasses. Proc 6th Int. Grasses Cong., 1:277-288.
3. Johnson, H.W., Robinson, H.F. and Camstock, R. E. 1955. Estimates of genetic and environmental variability in soyabean. *Agron. J.*, 47:314-318.
4. Lush, J. L. 1940. Intrasire correlation and regression of offspring on dams as method of estimating heritability of characters. *Proc. Am. Soc. Anim. Prod.* 33:293-301.
5. Johnson, H.W., Robinson, H.F. and Camstock, R.E. 1955. Genotypic and phenotypic correlation in soyabean and their implications in selection. *Agron. J.*, 47:477-485.
6. Miller, P.A.; Willioms, J.C.; Robinson, H.P. and Camstock, R.E. 1958. Estimation of genotypic and environmental variances and covariances in upland cotton and their implication in selection in selection. *Agron. J.*, 60:126-131.
7. Deway, D.R. and Lu, K.H. 1959. A correlation and Path coefficient analysis of components of crested wheat grass seed production. *Agron. J.*, 51:511-518.

8. Wright, S. 1921. Correlation and causation. *J. Agric. Res.*, 20(7):557.

9. Burton, G.W. and Devane, E.H. 1958. Estimating heritability in tall feschue (*Festuca arunelinaceia*) from replicated clonal material. *Agron. J.*, 45:478-481.

10. Jan, R.; Hassan, S.; Rashid, A.; Jan, H.; Shah, M.A. 1993. Screening of mungbean (*Vigna radiata* L.) genotypes for genetic parameters and protein content. *Sarhad J. Agril.*, 9(4):317-321.

11. Verma, P. and Garg, D.K. 2007. Correlation and path coefficient studies in mungbean (*Vigna radiata* (L) Wilczek). *J. Arid Legumes,* 4(2): 88-91.

12. Prakash, V.; Singh, R.V. and Khedar, O.P. 2007. Genetic parameters, correlation and path analysis among yield and yield characters in mungbean. *J. Arid Legumes.*, 4(1):6-8.

13. Kumar, K.; Prasad, K.D. and Verma, A.K. 2003. Genetic variability, correlation and path coefficient analysis in greengram. *J. Res. BAU.,* 15(1):97-101.

Correlation and path coefficient analysis in segregating generation under optimum and rainfed conditions in mungbean

J.B. Patel, S.H. Haibatpure and S.B.S. Tikka

S. D. Agricultural University, Sardarkrushinagar-385 506

ABSTRACT

Thirty six F_2 populations derived by growing the complete set of 36 F_1's in the poly house at the Department of Rural Engineering, S.D. Agricultural University, Sardarkrushinagar, resulted from nine diverse genotypes of mungbean [*Vigna radiata* (L.) Wilczek] following diallel mating design. Two sets of all the 36 F_2 populations were evaluated under optimum (E_1) and rainfed conditions (E_2) during *summer* 2003 and the data were subjected to correlation coefficient and path analysis for grain yield per plant and its contributing characters in both the environments. Grain yield was found to be significantly and positively correlated with clusters per plant and pods per plant under optimum moisture condition (E_1) and pods per plant and harvest index under sub-optimum moisture condition (E_2) indicating that these characters were mainly influenced the grain yield in mungbean in respective environments. Thus, it can be presumed that for improving grain yield in mungbean, an ideal plant would be with more number of pods, clusters and higher harvest index. Path coefficient analysis revealed that pods per plant had the highest positive direct effect in both the environments. Indirect association of all the characters among themselves was observed in-significant in both the environments. Overall, the emphasis can be laid on pods per plant, clusters per plant and harvest index in that order for selection in segregating generations.

Key Words : Correlation, mungbean, optimum, path coefficient, rainfed.

INTRODUCTION

Mungbean, [*Vigna radiata* (L.) Wilczek] is originated in Hindustan and central Asiatic regions. It belongs to the family leguminoceae and sub family papilionodeae with chromosome number 2n=22. Prior to any breeding programme for the improvement in crops, it is imperative to obtain information regarding the interrelationship of different characters with yield and among themselves since it facilitates a quick assessments of high yielding genotypes in selection programme. The real or true association could be known only through genotypic correlation, which eliminates the environmental influence. The simple correlation, being calculated on an overall basis, may not give a clear picture of the cause-effect system operating on the natural selection though when less number of variables are considered, correlation study alone can serve the purpose. However, when variable increases, the situation become complex then path analysis facilitates the partitioning in to direct and indirect effects of various characters. In order to achieve a clear cut picture of interrelationship of various component characters with yield, direct and indirect effects were calculated using the path coefficient analysis at genotypic level in F_2 generation under optimum and sub-optimum conditions.

MATERIALS AND METHODS

Nine varieties of mungbean [*Vigna radiata* (L.) Wilczek], were obtained from the germplasm pool maintained at Main Pulses Research Station, S.D. Agricultural University, Sardarkrushinagar. A complete set of 36 crosses developed from these parents following diallel mating design excluding reciprocals during *Kharif* 2002. Some seeds of all these 36 F_1's were grown during *rabi* 2002-03 in the poly house to get the F_2 seeds at the Department of Rural Engineering, S.D. Agricultural University, Sardarkrushinagar. Two sets of all the 36 F_2 populations were hand dibbled in *summer* 2003 in Randomized Block Design replicated thrice. One set of the experiment was grown with optimum irrigations as and when needed (E_1) whereas, other set was exposed to the moisture stress by irrigating at 30 DAS and 45 DAS (E_2). Each entry consisted of single row with 1.5 m length. The row to row and plant to plant distance measured 30 cm and 10 cm, respectively. All the recommended package of practices was adopted to raise the good crop. Observations were recorded on grain yield per plant (g), days to 50% flowering, days to maturity, plant height (cm), branches per plant, clusters per plant, pods per plant, pod length (cm), seeds per pod, 100-seed weight (g), harvest index (%), protein content (%), leaf area index (%) and dry matter efficiency (%). The data were analyzed for estimating the correlation coefficient (1) and path analysis (2) for grain yield per plant and its contributing characters under both optimum (E_1) and rainfed (E_2) conditions.

RESULTS AND DISCUSSION

Grain yield was found to be significantly and positively correlated with clusters per plant and pods per plant under optimum moisture condition (Table

1) and with pods per plant and harvest index under sub-optimum moisture condition (Table 3) indicating that these characters were mainly influenced the grain yield in mungbean in respective environments. Thus, selection practices for the improvement in one character will automatically result in the improvement in the other. Similar results were obtained by Kalpande *et al.* (3) for clusters per plant and pods per plant, Harita and Reddy (4) for harvest index, pods per plant and clusters per plant. The other characters viz., days to maturity, plant height, seeds per pod, 100-seed weight and protein content in E_1 and days to 50% flowering, plant height, clusters per plant, seeds per pod, 100-seed weight and leaf area index in E_2 had very low and positive correlation with grain yield per plant. Pods per plant showed highly significant and positive correlation with clusters per plant under optimum moisture condition (E_1) (Table 1). Similar findings were reported by Veerabadhiram and Jehangir (5) and Raje and Rao (6). Hence, selection for more clusters per plant will naturally increase number of pods per plant which in turn improve the yield.

Path coefficient analysis (Table 2 & 4) revealed that pods per plant had the highest positive direct effect in both the environments which might increase the grain yield if other variables held constant. Similar results were computed for high direct effect of pods per plant on yield by Ramaswamy and Jehangir (7). The other characters viz., harvest index, protein content, dry matter efficiency and branches per plant under optimum and harvest index, 100-seed weight, plant height, days to 50% flowering, leaf area index and clusters per plant under rainfed condition also possessed positive but low direct effect on grain yield. Indirect association of all the character among themselves was observed insignificant in both the environments. However, the magnitude of indirect influence of most the traits via pods per plant, clusters per plant and harvest index was positive indicating that these characters are very important contributors of grain yield.

Table 1. Genotypic correlation co-efficient among different characters in F_2 generation of optimum moisture condition (E_1) in mungbean.

Character	Days to 50 per cent flowering	Days to maturity	Plant height (cm)	Branches per plant	Clusters per plant	Pods per plant	Pod length (cm)	Seeds per pod	100-seed weight (g)	Harvest index (%)	Protein content (%)	Leaf area index	Dry matter efficiency (%)	Grain yield per plant (g)
Days to 50 per cent flowering	1.0000	0.0435	0.2444	0.0454	-0.1173	-0.0002	0.0260	0.0998	0.0656	0.0988	-0.1180	0.0409	-0.0542	-0.0211
Days to maturity		1.0000	0.0886	0.1642	0.0086	0.1086	-0.075	0.0430	0.1116	0.0076	0.1618	0.0496	-0.0074	0.0934
Plant height (cm)			1.0000	0.0351	-0.1034	-0.0586	0.1176	0.0296	0.0505	-0.0965	0.1211	0.1895	0.1014	-0.0679

Branches per plant				1.0000	0.2005	0.1695	0.0719	0.0517	0.0562	-0.0336	0.1381	0.0750	-0.0351	0.1602
Clusters per plant					1.0000	0.5400**	0.0984	0.0206	0.0534	-0.0122	-0.0274	0.1240	0.3127	0.4600**
Pods per plant						1.0000	0.0148	0.0489	0.0243	-0.1117	0.0858	0.0383	-0.0378	0.8895**
Pod length (cm)							1.0000	0.1335	0.1333	-0.0333	0.1798	0.0546	-0.0696	-0.0451
Seeds per pod								1.0000	0.0642	0.0056	-0.0890	0.0524	-0.1935	0.0293
100-seed weight (g)									1.0000	0.1665	-0.1026	0.0688	0.0317	0.0016
Harvest index (%)										1.0000	-0.1832	0.0737	0.0501	-0.0434
Protein content (%)											1.0000	0.1541	-0.1218	0.0990
Leaf area index												1.0000	0.1085	-0.0451
Dry matter efficiency (%)													1.0000	-0.0157

*,** Significant at 5% and 1% level of significance, respectively.

Table 2. Direct and indirect effects of different characters on grain yield per plant in F_2 generation of optimum moisture condition (E_1) in mungbean.

Character	Days to 50 per cent flowering	Days to maturity	Plant height (cm)	Branches per plant	Clusters per plant	Pods per plant	Pod length (cm)	Seeds per pod	100-seed weight (g)	Harvest index (%)	Protein content (%)	Leaf area index	Dry matter efficiency (%)	Correlation with grain yield per plant (g)
Days to 50 per cent flowering	-0.0384	0.0006	0.0063	-0.0002	0.0078	-0.0002	0.0018	0.0022	-0.0006	0.0064	-0.0051	0.0003	-0.0019	-0.0211
Days to maturity	0.0017	-0.0137	-0.0023	-0.0008	-0.0006	0.1012	0.0005	0.0010	-00011	0.0005	0.0070	0.0004	-0.0003	0.0934
Plant height (cm)	0.0094	-0.0012	-0.0257	0.0002	0.0069	-0.0546	-0.0080	0.0007	0.0005	-0.0065	0.0053	0.0014	0.0035	-0.0679
Branches per plant	0.0017	0.0023	-0.0009	0.0050	-0.0133	0.1578	0.0049	0.0012	-0.0006	-0.0022	0.0060	-0.0005	-0.0012	0.1602
Clusters per plant	0.0045	-0.0001	0.0027	0.0010	-0.0666	0.5030	0.0067	0.0005	0.0005	-0.0008	-0.0012	-0.0009	0.0108	0.4600
Pods per plant	0.0000	-0.0015	0.0015	0.0009	0.0359	0.9314	-0.0010	-0.0011	-0.0002	-0.0072	0.0037	0.0003	-0.0013	0.8895
Pod length (cm)	0.0010	-0.0001	0.0030	-0.0004	0.0065	0.0138	-0.0677	0.0030	-0.0013	-0.0021	0.0078	-0.0004	-0.0024	-0.0451
Seeds per pod	0.0038	0.0006	0.0008	-0.0003	0.0014	0.0456	0.0090	-0.0225	0.0006	0.0004	-0.0039	0.0004	-0.0067	0.0293

100-seed weight (g)	0.0025	0.0015	-0.0013	-0.0003	-0.0036	-0.0226	0.0090	-0.0014	0.0099	0.0107	-0.0045	0.0005	0.0011	0.0016
Harvest index (%)	-0.0038	-0.0001	0.0025	-0.0002	0.0008	-0.1040	0.0023	-0.0001	0.0016	0.0644	-0.0080	-0.0005	0.0017	-0.0434
Protein content (%)	-0.0045	-0.0022	-0.0031	0.0007	0.0018	0.0799	-0.0122	0.0020	-0.0010	-0.0118	0.0434	0.0011	-0.0042	0.0990
Leaf area index	0.0016	0.0007	0.0049	0.0004	-0.0083	-0.0357	-0.0037	0.0012	-0.0007	0.0047	-0.0067	-0.0073	0.0037	-0.0451
Dry mat-ter effici-ency (%)	0.0021	0.0001	-0.0026	-0.0002	-0.0208	-0.0352	0.0047	0.0044	0.0003	0.0032	-0.0053	-0.0008	0.0344	-0.0157

Figures in bold letters indicate the direct effects.

Residual effect = 0.1954

Table 3. Genotypic correlation co-efficient among different characters in F_2 generation of sub-optimum moisture condition (E_2) in mungbean.

Character	Days to 50 per cent flowering	Days to maturity	Plant height (cm)	Branches per plant	Clusters per plant	Pods per plant	Pod length (cm)	Seeds per pod	100-seed weight (g)	Harvest index (%)	Protein content (%)	Leaf area index	Dry matter efficiency (%)	Grain yield per plant (g)
Days to 50 per cent flowering	1.000	0.0044	0.1228	-0.0129	-0.1528	0.1616	0.1543	0.2287	-0.0105	0.1139	-0.1478	0.0327	0.1062	0.1568
Days to maturity		1.000	0.0397	-0.1320	-0.0353	-0.0010	-0.0643	0.1583	-0.1915	0.0142	0.0624	-0.0306	0.0926	-0.0006
Plant height (cm)			1.000	-0.1312	0.0986	0.1646	0.1276	0.1649	0.0018	-0.0084	-0.2340	-0.0270	0.0448	0.1723
Branches per plant				1.000	0.2005	-0.1190	-0.1237	-0.1110	0.1176	-0.0336	0.1272	0.0693	-0.0811	-0.1518
Clusters per plant					1.000	0.0591	0.0165	-0.0905	0.1572	-0.0122	-0.0333	0.2398	0.0182	0.0418
Pods per plant						1.000	0.0447	0.1433	0.0337	0.5392**	-0.1886	-0.0519	0.1618	0.8456**
Pod length (cm)							1.000	0.0297	0.1426	-0.0514	-0.0551	0.1165	0.0806	-0.0447
Seeds per pod								1.000	-0.0776	0.0548	-0.0297	-0.0742	0.0143	0.1105
100-seed weight (g)									1.000	0.1153	-0.2129	0.1314	0.0874	0.1012
Harvest index (%)										1.000	-0.2212	0.0929	-0.0728	0.6557**
Protein content (%)											1.000	-0.1468	0.0177	-0.2297
Leaf area index												1.000	-0.0924	0.0003

Dry matter efficiency (%)	0.0619

*,** Significant at 5% and 1% level of significance, respectively.

Table-4. Direct and indirect effects of different characters on grain yield per plant in F_2 generation of ub- optimum moisture condition (E_2) in mungbean.

Character	Days to 50 per cent flowering	Days to maturity	Plant height (cm)	Branches per plant	Clusters ·per plant	Pods per plant	Pod length (cm)	Seeds per pod	100-seed weight (g)	Harvest index (%)	Protein content (%)	Leaf area index	Dry matter efficiency (%)	Correlation with grain yield per plant (g)
Days to 50 per cent flowering	0.0305	0.0000	0.0077	0.0010	-0.0007	0.1117	-0.0143	-0.0042	-0.0007	0.0298	0.0000	0.0004	-0.0043	0.1568
Days to maturity	0.0001	-0.0021	0.0025	0.0105	-0.0002	-0.0007	0.0060	-0.0029	-0.0134	0.0037	0.0020	-0.0004	-0.0037	-0.0006
Plant height (cm)	0.0037	-0.0001	0.0630	0.0104	0.0005	0.1138	-0.0118	-0.0030	0.0001	-0.0022	0.0000	-0.0004.	-0.0018	0.1723
Branches per plant	-0.0004	0.0003	-0.0083	-0.0792	0.0010	-0.0823	0.0115	0.0021	0.0082	-0.0088	0.0000	0.0009	0.0033	-0.1518
Clusters per plant	-0.0047	0.0001	0.0062	-0.0159	0.0048	0.0409	-0.0015	0.0017	0.0110	-0.0032	0.0000	0.0032	-0.0007	0.0418
Pods per plant	0.0049	0.0000	0.0104	0.0094	0.0003	0.6913	-0.0041	-0.0026	0.0023	0.1410	0.0020	-0.0007	-0.0065	0.8456
Pod length (cm)	0.0047	0.0001	0.0080	0.0098	0.0001	0.0309	-0.0926	-0.0005	0.0099	-0.0134	0.0000	0.0015	-0.0033	-0.0447
Seeds per pod	0.0070	-0.0003	0.0104	0.0088	-0.0004	0.0990	-0.0028	-0.0185	-0.0054	0.0143	0.0000	-0.0010	-0.0006	0.1105
100-seed weight (g)	-0.0003	0.0004	0.0001	-0.0093	0.0008	0.0233	-0.0132	0.0014	0.0697	0.0301	-0.0023	0.0017	-0.0035	0.1012
Harvest index (%)	0.0035	0.0000	-0.0005	0.0027	-0.0001	0.3727	0.0048	-0.0010	0.0080	0.2615	0.0000	0.0012	0.0029	0.6557
Protein content (%)	-0.0045	-0.0001	-0.0148	-0.0101	-0.0002	-0.1304	0.0051	0.0005	-0.0148	-0.0578	0.0000	-0.0020	-0.0007	-0.2297
Leaf area index	0.0010	0.0001	-0.0017	-0.0055	0.0012	-0.0359	-0.0108	0.0014	0.0092	0.0243	0.0000	0.0133	0.0037	0.0003
Dry matter efficiency (%)	0.0032	-0.0002	0.0028	0.0064	0.0001	0.1118	-0.0075	-0.0003	0.0061	-0.0190	0.0000	-0.0012	-0.0405	0.0619

Figures in bold letters indicate the direct effects. **Residual effect = 0.2095**

REFERENCES

1. Al-Jibouri, H.A., Miller, P.A. and Robinson, H.F. 1958. Genotypic and environmental variance and covariance in an upland cotton cross of interspecific origin. *Agronomy Journal,* 50: 633-637.
2. Dewey, D.R. and Lu, K.H. 1959. A correlation and path analysis of components of crested wheat grass seed production. *Agronomy Journal,* 51: 515-518.
3. Kalpande, H.V., Patil J.V., Deshmukh R.B., Dumbre, A.D. and Kute N.S. 1997. Correlation and path coefficient analysis in green gram in different environments. *Journal of. Maharashtra Agricultural University*, 22: 281-283.
4. Harita, S. and Reddy, S.M. 2002. Correlation and path analysis in mungbean [*Vigna radiata* (L.) Wilczek] *Legume Research,* 25: 180-183.
5. Veerabadhiram, P. and Jehangir, K.S. 1995. Genetic variability, correlation and path analysis in green gram. *Madras Agricultural Journal*, 82: 365-367.
6. Raje, R.S. and Rao, S.K. 2000.Association analysis of yield and its components in mungbean [*Vigna radiata* (L.) Wilczek]. *Legume Research,* 23 : 42-48
7. Ramaswamy, A. and Jehangir, K.S. 1998. Character association in mungbean [*Vigna radiata* (L.) Wilczek]. *Madras Agricultural Journal*, 85: 371-372.

Genetic component analysis for yield and yield components under optimum and sub-optimum moisture conditions in mungbean

S.B.S. Tikka, S.H. Haibatpure and J.B. Patel

S. D. Agricultural University, Sardarkrushinagar-385 506

ABSTRACT

Nine parents alongwith their 36 F_1's and 36 F_2's was grown in randomized block design with three replications in two sets during summer 2003 to study the genetic components in mungbean [*Vigna radiata* (L.) Wilczek]. One set of the experiment was grown with optimum irrigations as and when needed (E_1) whereas, second set was exposed to the moisture stress by irrigating at 30 and 45 DAS (E_2). The estimation of D, which measures the variance due to additive genetic effect as well as H_1 and H_2, which measure the variance due to dominance genetic effect were significant for all the characters in both the environments and generations, except additive component (D) in E_2F_2 for days to flowering, in E_2F_1 for days to maturity and clusters per plant, in E_1F_2 for seeds per pod, in E_1F_1 and E_1F_2 for 100-seed weight, in E_1F_2 and E_2F_2 for leaf area index and dry matter efficiency; dominance component (H_1) in E_1F_1 and E_2F_1 for days to flowering, in E_2F_1 for pods per plant and protein content; dominance component (H_2) in E_2F_1 for days to flowering and in E_2F_2 for protein content and dominance component (H_1 and H_2) in E_1F_1, E_1F_2 and E_2F_2 for pod length, in E_2F_1 for seeds per pod, in E_2F_2 for leaf area index and in E_1F_2, E_2F_1 and E_2F_2 for dry matter efficiency, indicating that both additive and dominance components were important for the inheritance of these traits. In general, the analysis revealed that the dominance (H_1, H_2) variance components were predominant for all the traits, as they were significant and higher in magnitude indicating significant role of dominant or non-additive components in controlling the variation of the traits under study. It also suggested that superior genotypes could be isolated in the segregating generations. Similarly, the average degree of dominance $(H_1 / D)^{1/2}$ indicated over dominance for grain yield per plant and all other characters in both the environments and generations, except pod length in E_1F_1, seeds per pod and dry matter efficiency in E_2F_1 and protein content in E_2F_2 also inferred that only small fraction of genetic variance was additive in nature. Low estimates of genes differentiating between the parents might be due to asymmetrical distribution of genes with positive and negative effects among the common parents of arrays. This was further supported by invariably less than 0.25 estimates of $H_2/4H_1$. The covariance between additive and dominant effect (F component) was non-significant for most of the characters revealing unequal distribution of recessive and dominant alleles among the parents and it was supported by KD/KR ratio. Low to moderate estimates of heritability for most of the traits including grain yield per plant indicated the preponderance of non-additive variance for yield and yield attributes.

Key Words: Additive, genetic component, heritability, non-additive, optimum and sub-optimum,

INTRODUCTION

As mungbean is usually a rainfed crop, its production is linked to rainfall. The drought reduces the grain yield as well as fodder yield. Mungbean being a low input crop is generally grown in marginal lands. Therefore, a high priority should be given for breeding varieties against drought. To breed resistant varieties it is necessary to know the mechanism of resistance and its genetic control. However, very scanty information is available on genetic aspects of moisture stress in mungbean. In addition to the knowledge of combining ability, the information regarding mode of inheritance of quantitative trait is very helpful in deciding the breeding strategy.

The diallel components cross technique has long been used for obtaining information regarding type of gene action involved in inheritance of different quantitative traits and simultaneously for parental material with respect to wealth of gene influencing expression of trait. Jink and Hayman (1) and Hayman (2 and 3) highlighted the role of genetic components in this regard.

MATERIALS AND METHODS

Nine varieties of mungbean [*Vigna radiata* (L.) Wilczek], were obtained from the germplasm pool maintained at Main Pulses Research Station, S.D. Agricultural University, Sardarkrushinagar. Thirty six crosses developed from these parents following diallel mating design excluding reciprocals during *Kharif* 2002. Some seeds of all the 36 F_1's were sown during *rabi* 2002-03 in the poly house to get the F_2 seeds at the Department of Rural Engineering, S.D. Agricultural University, Sardarkrushinagar. Two sets of all the 9 parents, their 36 F_1's and 36 F_2's was grown during *summer* 2003 in randomized block design replicated thrice. One set of the experiment was grown with optimum irrigations as and when needed (E_1) and another set was exposed to the moisture stress by irrigating at 30 and 45 DAS (E_2). Each entry consisted of single row with 1.5 m length. The row to row and plant to plant distance measured 30 cm and 10 cm, respectively. All the recommended package of practices was adopted to raise the good crop. Observations were recorded on grain yield per plant (g), days to 50% flowering, days to maturity, plant height (cm), branches per plant, clusters per plant, pods per plant, pod length (cm), seeds per pod, 100-seed weight (g), harvest index (%), protein content (%), leaf area index (%) and dry matter efficiency (%). Genetic components of variance were computed by diallel cross method, suggested by Hayman (2). The mean degree of dominance in F_1 generations was calculated as $(H_1/D)^{1/2}$ following Hayman (2), whereas in F_2 generation it was calculated as one fourth of $(H_1/D)^{1/2}$, according to Verhalen *et al.* (4). Proportion of genes with positive and negative effects in the parents was calculated by the ratio of $(H_2/4H_1)^{1/2}$, following Hayman (2). Proportion of dominant and recessive genes in the parents for F_1 generation calculated following Hayman (2) and for F_2 generation following Singh and Chaudhary (5). Number of groups of genes which control the character and exhibit dominance in both the generations calculated as the ratio of h^2/H_2 following Singh and Chaudhary (5). The coefficient of correlation 'r' between the parental means and the parental order of dominance was calculated following Dabholkar (6). Heritability in narrow sense for F_1 and F_2 generation was also calculated.

RESULTS AND DISCUSSION

The t^2 value was found non significant for most of the characters in both the generations and environments which suggested the validity of hypothesis of diallel analysis (Table 1).

Table 1. Estimates of t^2 values, linear regression coefficient (b) of Wr on Vr and r values in both the environments and generations in mungbean.

Characters		t^2	b	S.E.(b)	b-o	r	S.E.(r)
Grain yield per plant (g)	E_1F_1	2.66	0.27	0.19	1.39	-0.85**	0.20
	E_1F_2	0.19	0.34	0.29	1.16	0.60**	0.30
	E_2F_1	0.0080	-0.06	0.39	-0.17	0.71**	0.27
	E_2F_2	0.14	-0.09	0.32	-0.28	0.78**	0.24
Days to 50 per cent flowering	E_1F_1	0.31	0.45	0.26	0.72	-0.07	0.38
	E_1F_2	5.40	0.04	0.17	0.26	0.78**	0.24
	E_2F_1	0.98	0.33	0.24	1.40	0.22	0.37
	E_2F_2	5.78*	0.11	0.16	0.72	-0.58**	0.31
Days to maturity	E_1F_1	4.22	0.22	0.17	1.27	-0.31	0.36
	E_1F_2	38.86**	-0.05	0.07	-0.70	0.40	0.35

	E_2F_1	1.73	-0.04	0.23	-0.21	-0.02	0.38
	E_2F_2	1.36	0.28	0.23	1.25	0.53**	0.32
Plant height (cm)	E_1F_1	16.04**	0.04	0.11	0.31	0.54**	0.32
	E_1F_2	5.16	0.24	0.16	1.50	0.45**	0.34
	E_2F_1	0.02	-0.18	0.35	-0.52	-0.20	0.37
	E_2F_2	0.67	0.78**	0.14	5.39	0.17	0.37
Branches per plant	E_1F_1	17.03**	0.11	0.10	1.08	0.46**	0.34
	E_1F_2	34.37**	0.0077	0.0813	0.10	0.36	0.35
	E_2F_1	1.48	0.50*	0.19	2.61	0.66**	0.28
	E_2F_2	1.46	0.59*	0.18	2.65	0.67**	0.28
Clusters per plant	E_1F_1	25.36**	0.14	0.09	1.54	-0.19	0.37
	E_1F_2	1.33	0.98**	0.24	0.55	0.15	0.37
	E_2F_1	40.90**	0.15	0.07	2.07	-0.14	0.37
	E_2F_2	0.16	0.10	0.32	0.34	0.21	0.37
Pods per plant	E_1F_1	2.50	0.13	0.21	0.62	-0.83**	0.21
	E_1F_2	0.02	0.39	0.36	1.08	0.55**	0.32
	E_2F_1	0.95	-0.0017	0.26	-0.01	0.60**	0.30
	E_2F_2	3.07	0.01	0.20	0.08	0.25	0.37
Pod length (cm)	E_1F_1	15.80**	0.16	0.11	1.45	0.11	0.38
	E_1F_2	1.79	0.037	0.23	0.16	0.51**	0.33
	E_2F_1	4.84	-0.19	0.17	-1.14	0.16	0.37
	E_2F_2	2.51	0.22	0.20	1.11	-0.14	0.37
Seeds per pod	E_1F_1	0.30	0.21	0.29	0.73	0.50**	0.33
	E_1F_2	3.41	0.13	0.19	0.68	0.45**	0.34
	E_2F_1	0.24	0.87*	0.26	3.25	0.42**	0.34
	E_2F_2	0.92	0.37	0.23	1.57	0.69**	0.27
100-seed weight (g)	E_1F_1	1.02	0.20	0.25	0.79	0.57**	0.31
	E_1F_2	2.99	0.16	0.2	0.82	0.38	0.35
	E_2F_1	3.59	0.04	0.19	0.24	0.48**	0.33
	E_2F_2	2.92	0.37	0.18	2.05	-0.08	0.38
Harvest index (%)	E_1F_1	32.42**	0.28**	0.07	3.68	0.41	0.34
	E_1F_2	1.67	0.80	0.47	1.68	0.73**	0.26
	E_2F_1	89.51**	0.19**	0.05	3.85	0.45**	0.34
	E_2F_2	0.01	0.16	0.35	0.47	0.23	0.37

Protein content (%)	E_1F_1	0.10	-0.26	0.41	-0.63	0.37	0.35
	E_1F_2	0.05	-0.23	0.33	-0.70	0.78**	0.24
	E_2F_1	0.31	-0.10	0.30	-0.35	0.49**	0.33
	E_2F_2	3.04	-0.03	0.20	-0.19	0.73**	0.26
Leaf area index	E_1F_1	1.13	0.60	0.48	1.25	-0.65**	0.29
	E_1F_2	0.0013	0.52	0.32	1.59	0.78**	0.24
	E_2F_1	0.88	0.70	0.43	1.61	-0.38	0.35
	E_2F_2	0.06	-0.20	0.40	-0.49	0.64**	0.29
Dry matter efficiency (%)	E_1F_1	0.56	-0.0001	0.28	0.00	-0.40	0.35
	E_1F_2	0.13	0.0090	0.01	0.67	0.13	0.37
	E_2F_1	3.19	0.0027	0.20	0.01	0.66**	0.28
	E_2F_2	0.89	-0.01	0.01	-6.76	-0.21	0.37

*,** Significant at 5% and 1% respectively

The estimation of D, which measures the variance due to additive genetic effect as well as H_1 and H_2, which measure the variance due to dominance genetic effect were significant for all the characters in both the environments and generations, except additive component (D) in E_2F_2 for days to flowering, in E_2F_1 for days to maturity and clusters per plant, in E_1F_2 for seeds per pod, in E_1F_1 and E_1F_2 for 100-seed weight, in E_1F_2 and E_2F_2 for leaf area index and dry matter efficiency; dominance component (H_1) in E_1F_1 and E_2F_1 for days to flowering, in E_2F_1 for pods per plant and protein content; dominance component (H_2) in E_2F_1 for days to flowering and in E_2F_2 for protein content and dominance component (H_1 and H_2) in E_1F_1, E_1F_2 and E_2F_2 for pod length, in E_2F_1 for seeds per pod, in E_2F_2 for leaf area index and in E_1F_2, E_2F_1 and E_2F_2 for dry matter efficiency, indicating that both additive and dominance components were important for the inheritance of these traits (Table 2). In general, the analysis revealed that the dominance (H_1, H_2) variance components were predominant for all the traits, as they were significant and higher in magnitude indicating significant role of dominant or non-additive components in controlling the variation of all the traits under study. The present findings are also in agreement with those obtained by Das *et al.* (7), Khattak *et al.* (8) and Manivannan (9).

The average degree of dominance $(H_1 / D)^{1/2}$ indicated over dominance for grain yield per plant and all other characters in both the environments and generations, except pod length in E_1F_1, seeds per pod and dry matter efficiency in E_2F_1 and protein content in E_2F_2 (Table 2). Thus, on the basis of present study, it is evident that the characters are controlled by non-additive components. The results of present finding are in agreement with results of

Baseeruddin and Nagur (10) for days to flowering and maturity; Manivannan (9) for branches per plant and pods per plant; Bhadra *et al.* (11) for pod length; Srinivas *et al.* (12) for seeds per pod; Rosaiah *et al.* (13) and Kute *et al.* (14) for seed yield per plant and Khattak *et al.* (8) for 100-seed weight and harvest index.

Table 2. Estimates of components of variance for grain yield per plant (g), days to 50 % flowering and days to maturity under optimum (E_1) and sub-optimum (E_2) moisture conditions in mungbean.

Components of variance	Grain yield per plant (g)				Days to 50 % flowering				Days to maturity			
	E_1F_1	E_1F_2	E_2F_1	E_2F_2	E_1F_1	E_1F_2	E_2F_1	E_2F_2	E_1F_1	E_1F_2	E_2F_1	E_2F_2
D	1.36 ± 0.50**	0.97 ± 0.31**	0.71 ± 0.26**	0.72 ± 0.23**	3.00 ± 0.98**	3.51 ± 6.50	2.89 ± 1.01**	2.12 ± 4.45	2.84 ± 1.01**	2.13 ± 6.34	4.86 ± 1.87**	4.10 ± 2.56
H_1	3.73 ± 1.09*	2.74 ± 0.68**	1.67 ± 0.58*	1.47 ± 0.51*	6.01 ± 2.17*	31.61 ± 14.35	4.95 ± 2.22	28.47 ± 9.81*	12.56 ± 2.24**	53.59 ± 14.00*	16.92 ± 4.12**	25.50 ± 5.65**
H_2	2.62 ± 0.94*	1.95 ± 0.59*	1.57 ± 0.50*	1.16 ± 0.44*	4.50 ± 1.86*	27.48 ±12.34	4.26 ± 1.91*	23.81 ±8.43*	10.79 ± 1.93**	32.07 ± 12.03*	13.81 ± 3.54*	18.41 ± 4.86*
h^2	5.87 ± 0.63**	6.37 ± 0.39**	0.29 ± 0.34	2.36 ± 0.29**	-0.55 ± 1.25	18.64 ± 8.26	-0.21 ± 1.28	9.60 ± 5.65	-0.55 ± 1.29	18.58 ± 8.06	1.63 ± 2.37	6.78 ± 3.25
F	0.68 ± 1.15	0.90 ± 0.72	-0.17 ± 0.62	0.10 ± 0.54	1.70 ± 2.29	3.84 ± 15.17	1.44 ± 2.35	4.88 ± 10.37	1.95 ± 2.37	7.57 ± 14.80	6.39 ± 4.35	7.77 ± 5.97
E	0.79 ± 0.16**	0.62 ± 0.10**	0.44 ± 0.08**	0.36 ± 0.07**	2.44 ± 0.31**	2.44 ± 2.06	1.36 ± 0.32**	4.45 ± 1.41*	2.34 ± 0.32**	1.86 ± 2.01	1.80 ± 0.59*	1.74 ± 0.81
$(H_1/D)^{1/2}$	166	1.68	1.53	1.43	1.42	3.00	1.31	3.66	2.10	5.02	1.87	2.34
$H_2/4H_1$	0.18	0.18	0.24	0.20	0.19	0.22	0.22	0.21	0.21	0.15	0.20	0.18
$(4DH_1)^{1/2}+F/(4DH_1)^{1/2}-F$	1.36	1.76	0.86	1.10	1.50	1.45	1.47	1.92	1.39	2.10	2.09	2.36
h^2/H_2	2.24	3.26	0.19	2.02	-0.12	0.68	-0.05	0.41	-0.05	0.58	0.12	0.37
Heritability (%)	17.96	18.36	16.47	20.39	17.57	8.55	24.43	6.91	12.45	3.83	21.51	14.23

*,** Significant at 5% and 1% respectively.

Table 2. Estimates of components of variance for Plant height (cm), branches per plant and clusters per plant under optimum (E_1) and sub-optimum (E_2) moisture conditions in mungbean.

Components of variance	Plant height (cm)				Branches per plant				Clusters per plant			
	E_1F_1	E_1F_2	E_2F_1	E_2F_2	E_1F_1	E_1F_2	E_2F_1	E_2F_2	E_1F_1	E_1F_2	E_2F_1	E_2F_2
D	24.01± 8.96**	10.53 ± 3.98**	8.35 ± 3.05**	2.85 ± 0.79*	0.18 ± 0.06**	0.23 ± 0.08**	0.14 ± 0.05**	0.16 ± 0.05**	0.77 ± 0.26**	0.08 ± 0.07	0.76 ± 0.26**	0.07 ± 0.07
H_1	60.90 ± 19.77*	34.41 ± 8.78*	21.49 ± 6.73*	4.58 ± 1.75*	0.53 ± 0.14*	0.51 ± 0.17*	0.44 ± 0.11**	0.42 ± 0.11*	2.16 ± 0.58*	0.45 ± 0.15*	2.12 ± 0.58*	0.56 ± 0.16*

H_2	50.26 ± 17.00*	33.26 ± 7.54**	14.24 ± 5.78*	5.63 ± 1.50*	0.39 ± 0.12*	0.48 ± 0.15*	0.43 ± 0.09**	0.42 ± 0.09**	1.76 ± 0.50*	0.40 ± 0.13*	1.79 ± 0.15**	0.48 ± 0.14*
h^2	3.97 ± 11.39	46.32 ± 5.05**	3.65 ± 3.87	-1.43 ± 1.01	1.47 ± 0.08**	1.06 ±0.10*	1.18 ± 0.06**	1.18 ± 0.06**	1.22 ± 0.33*	1.13 ± 0.09**	0.98 ± 0.34**	1.56 ± 0.09**
F	9.27 ± 20.90	-2.25 ± 9.28	8.17 ± 7.11	0.87 ± 1.84	0.12 ± 0.15	-0.02 ± 0.18	0.04 ± 0.11	0.02 ± 0.05	0.04 ± 0.61	0.06 ± 0.16	-0.17 ± 0.62	0.08 ± 0.17
E	6.99 ± 2.83*	7.22 ± 1.26**	4.57 ± 0.96**	4.38 ± 0.25**	0.11 ± 0.02**	0.13 ± 0.03**	0.13 ± 0.02**	0.14 ± 0.02**	0.11 ± 0.08	0.12 ± 0.02**	0.12 ± 0.08	0.12 ± 0.02**
$(H_1/D)^{1/2}$	1.59	1.81	1.60	1.27	1.72	1.49	1.77	1.62	1.67	2.37	1.67	2.83
$H_2/4H_1$	0.21	0.24	0.17	0.31	0.19	0.24	0.24	0.25	0.20	0.22	0.21	0.21
$(4DH_1)^{1/2}+F/(4DH_1)^{1/2}-F$	1.28	0.89	1.88	1.27	1.48	0.94	1.18	1.08	1.03	1.38	0.87	1.51
h^2/H_2	0.08	1.39	0.26	-0.25	3.72	2.19	2.74	2.83	0.69	2.82	0.55	3.27
Heritability (%)	23.70	14.03	20.90	11.86	17.47	17.97	13.21	14.28	23.12	8.36	21.53	6.89

*,** Significant at 5% and 1% respectively.

Table 2. Estimates of components of variance for pods per plant, pod length (cm) and seeds per pod under optimum (E_1) and sub-optimum (E_2) moisture conditions in mungbean.

Components of variance	Pods per plant				Pod length (cm)				Seeds per pod			
	E_1F_1	E_1F_2	E_2F_1	E_2F_2	E_1F_1	E_1F_2	E_2F_1	E_2F_2	E_1F_1	E_1F_2	E_2F_1	E_2F_2
D	13.48 ± 5.02**	6.72 ± 1.66**	5.86 ± 1.89**	613 ± 1.87**	0.14 ± 0.04**	0.13 ± 0.04**	0.49 ± 0.18**	0.38 ± 0.14**	0.27 ± 0.10**	0.10 ± 0.18	0.30 ± 0.09**	0.45 ± 0.15**
H_1	34.19 ± 11.08*	17.13 ± 3.66**	8.31 ± 4.18	11.50 ± 4.13*	0.05 ± 0.09	0.14 ± 0.08	1.45 ± 0.14**	0.51 ± 0.30	0.62 ± 0.22*	1.20 ± 0.40*	0.27 ± 0.20	1.19 ± 0.33*
H_2	25.69 ± 9.53*	12.47 ± 3.15**	9.40 ±3.59*	10.83 ± 3.55*	0.11 ± 0.08	0.18 ± 0.07	1.16 ± 0.34*	0.57 ± 0.26	0.55 ± 0.19*	1.05 ± 0.34*	0.26 ± 0.18	1.14 ± 0.29*
h^2	43.94 ± 6.38**	44.93 ± 2.11**	-0.86 ± 2.41	30.41 ± 2.38**	0.02 ± 0.05	0.43 ± 0.05**	0.09 ± 0.23	0.15 ± 0.17	-0.04 ± 0.13	19.17 ± 0.23**	0.44 ± 0.12*	3.30 ± 0.19**
F	2.44 ± 11.71	4.91 ± 3.87	-4.57 ± 4.42	-2.81 ± 4.36	-0.28 ± 0.10*	-0.11 ± 0.08	0.08 ± 0.42	-0.22 ± 0.32	0.10 ± 0.23	0.08 ± 0.42	-0.05 ± 0.20	-0.08 ± 0.35
E	5.42 ± 1.59*	4.28 ± 0.52**	3.32 ± 0.16**	2.72 ± 0.59**	0.18 ± 0.01**	0.10 ± 0.01**	0.23 ± 0.06*	0.26 ± 0.04**	0.18 ± 0.03**	0.19 ± 0.06*	0.33 ± 0.03**	0.30 ± 0.05**
$(H_1/D)^{1/2}$	1.59	1.60	1.19	1.37	0.60	1.04	1.72	1.16	1.52	3.46	0.95	1.63
$H_2/4H_1$	0.19	0.18	0.28	0.24	0.56	0.32	0.20	0.28	0.22	0.22	0.25	0.24
$(4DH_1)^{1/2}+F/(4DH_1)^{1/2}-F$	1.12	1.59	0.50	0.71	0.25	0.42	1.10	0.60	1.28	1.26	0.84	0.90

h^2/H_2	1.71	3.60	-0.09	2.81	0.18	2.35	0.07	0.27	-0.08	0.63	1.61	2.90
Heritability (%)	20.14	18.63	18.30	19.57	11.76	16.67	17.62	17.67	17.88	10.20	15.46	15.41

*,** Significant at 5% and 1% respectively.

Table 2. Estimates of components of variance for 100-seed weight (g), harvest index (%) and protein content (%) under optimum (E_1) and sub-optimum (E_2) moisture conditions in mungbean.

Components of variance	100-seed weight (g)				Harvest index (%)				Protein content (%)			
	E_1F_1	E_1F_2	E_2F_1	E_2F_2	E_1F_1	E_1F_2	E_2F_1	.E_2F_2	E_1F_1	E_1F_2	E_2F_1	E_2F_2
D	0.15 ± 0.04	0.15 ± 0.04	0.09 ± 0.03**	0.06 ± 0.03	3.74 ± 5.52	1.88 ± 3.17	3.77 ± 7.60	3.36 ± 3.35	1.27 ± 0.69**	2.77 ± 0.82**	3.54± 0.74**	5.76 ± 1.23**
H_1	0.52 ± 0.09**	0.53± 0.09**	0.34 ± 0.07**	0.39 ± 0.07**	35.95 ± 12.18*	27.89 ± 7.00*	44.07 ± 16.77*	20.09 ± 7.39*	5.64 ± 1.53*	5.25 ± 1.81*	3.87 ±1.64	0.64 ± 2.72
H_2	0.47 ± 0.08**	0.51 ± 0.08**	0.25 ± 0.06**	0.33 ± 0.06**	36.11 ± 10.47*	27.77 ± 6.01**	42.65 ± 14.41*	20.43 ± 6.35*	4.15 ± 1.32*	3.80 ± 1.85*	2.25 ± 1.41	2.47 ± 2.34
h^2	0.25 ± 0.05**	1.23 ± 0.05**	0.29 ± 0.04**	0.69 ± 0.04**	1.54 ± 7.01	53.65 ± 4.03**	-0.98 ± 9.66	-0.78 ± 4.26	-0.12 ± 0.88	0.52 ± 1.04	-0.43 ± 0.95	1.12 ± 1.57
F	0.04 ± 0.09	-0.05 ± 0.09	0.12 ± 0.07	-0.04 ± 0.08	-3.20 ± 12.87	1.32 ± 7.14	-8.14 ± 17.72	-0.53 ± 7.81	2.06 ± 1.62	1.77 ± 1.91	1.48 ± 1.74	0.03 ± 2.88
E	0.05 ± 0.01**	0.06 ± 0.01**	0.03 ± 0.01*	0.03 ± 0.01*	7.34 ± 1.74**	6.20 ± 1.00**	6.97 ± 2.40*	5.82 ± 1.06**	0.62 ± 0.22*	0.72 ± 0.26*	1.12 ± 0.24**	1.77 ± 0.39**
$(H_1/D)^{1/2}$	1.86	1.88	1.94	2.55	3.10	3.85	3.42	2.45	2.11	1.38	1.05	0.33
$H_2/4H_1$	0.22	0.24	0.19	0.21	0.25	0.25	0.24	0.25	0.18	0.18	0.15	0.23
$(4DH_1)^{1/2}+F/(4DH_1)^{1/2}-F$	1.15	0.84	2.04	0.77	0.76	1.20	0.52	0.94	2.25	1.60	1.50	1.02
h^2/H_2	0.54	2.40	1.15	2.10	0.04	1.93	-0.02	-0.04	-0.03	0.14	-0.90	0.46
Heritability (%)	18.07	15.46	20.93	10.20	5.17	3.53	4.49	7.10	17.32	30.33	41.54	61.27

*,** Significant at 5% and 1% respectively.

Table 2. Estimates of components of variance for leaf area index and dry matter efficiency (%) under optimum (E_1) and sub-optimum (E_2) moisture conditions in mungbean.

Components of variance	Leaf area index				Dry matter efficiency (%)			
	E1F1	E_1F_2	E_2F_1	E_2F_2	E1F1	E_1F_2	E_2F_1	E_2F_2
D	0.06 ± 0.02**	0.02± 0.01	0.03 ± 0.01**	0.03 ± 0.02	0.01 ± 0.001**	0.02 ± 0.04	0.17 ± 0.06**	0.03 ± 0.04
H_1	0.24 ± 0.04**	0.06 ± 0.02*	0.13 ± 0.03**	0.08 ± 0.04	0.03 ± 0.00*	0.05 ± 0.09	0.12 ± 0.14	0.04 ± 0.09
H_2	0.18 ± 0.04**	0.05 ± 0.02*	0.10 ± 0.03*	0.06 ± 0.03	-0.02 ± 0.00	0.06 ± 0.08	0.07 ± 0.12	0.05 ± 0.07

h^2	0.03 ± 0.02	0.00 ± 0.01	0.03 ± 0.02	0.00 ± 0.02	0.00 ± 0.00	-0.01 ± 0.05	-0.03 ±0.08	0.01 ± 0.05
F	0.05 ± 0.05	0.01 ± 0.03	0.06 ± 0.03	0.04 ± 0.04	-0.02 ± 0.00	-0.04 ± 0.10	0.09 ± 0.14	-0.04 ± 0.09
E	0.03 ± 0.01*	0.03 ± 0.00*	0.03 ± 0.00*	0.03 ± 0.01*	0.01 ± 0.00	0.02 ± 0.01	0.07 ± 0.02*	0.03 ± 0.01*
$(H_1/D)^{1/2}$	2.00	1.73	2.08	1.63	1.73	1.58	0.84	1.15
$H_2/4H_1$	0.19	0.23	0.18	0.19	0.18	0.31	0.15	0.33
$(4DH_1)^{1/2}+F/(4DH_1)^{1/2}-F$	1.53	1.34	2.85	2.38	0.27	0.23	1.92	0.27
h^2/H_2	0.15	-0.06	0.28	0.09	-0.21	-0.16	-0.35	0.12
Heritability (%)	16.21	12.04	12.60	14.68	10.00	10.52	35.41	13.04

*,** Significant at 5% and 1% respectively

The distribution of genes with positive and negative effects in the parents was not symmetrical for all the characters. The covariance between additive and dominance effect (F component) was not significant for all the characters under study except pod length in E_1F_1 (Table 2), which revealed unequal distribution of dominance and recessive alleles among the parents.

The estimates of ratio of KD/KR (Table 2) was more than unity for all the characters in all the environments and generations under study, except grain yield per plant and clusters per plant in E_2F_1, plant height and branches per plant in E_1F_2, pods per plant and seeds per pod in E_2F_1 and E_2F_2, pod length in E_1F_1, E_1F_2 and E2F2, 100-seed weight in E_1F_2 and E_2F_2, harvest index and dry matter efficiency in E_1F_1, E_2F_1 and E_2F_2 indicating high proportion of dominant genes for these traits in the parents. For most of the exceptional traits, KD/KR ratio was nearer to unity, which showed that dominant and recessive genes were present in equal proportion. The positive and negative sign of 'F' value for respective traits also confirmed the above results.

The ratio of h^2/H_2 reflects the number of genes or gene groups controlling a particular character and exhibiting dominance. It is difficult to estimate the number of gene groups exhibiting dominance effect because the ratio h^2/H_2 is underestimated if dominance effect of all genes affecting that trait are not equal in size and direction (15) or if the distribution of the genes is correlated (16). In the present study, the characters like grain yield per plant (except in E_2F_1), branches per plant, clusters per plant (except in E_1F_1 and E_2F_1), pods per plant (except in E_2F_1), plant height, pod length and harvest index (except E_1F_1, E_2F_1 and E_2F_2), seeds per pod (except in E_1F_1 and E_1F_2) and 100- seed weight (except in E_1F_1) at least one gene or gene group exhibited dominance for the above mentioned exceptional traits in respective environment and generation and for the remaining characters under study in all the environments and generations, two gene or gene groups exhibiting dominance were observed.

Low to moderate estimates of narrow sense of heritability for almost all the traits in both the environments and generations indicated the preponderance of non-additive variance for grain yield per plant and its attributes.

REFERENCES

1. Jinks, J.L. and Hayman, B.I. 1953. Analysis of diallel crosses. *Maize Genetics Crop Newsletter*, 27 : 48-54.
2. Hayman, B.I. 1954a. The analysis of variance of diallel tables. *Biometrics*, 10 : 235-244.
3. Hayman, B.I. 1954b. The theory and analysis of diallel crosses. *Genetics*, 39 : 789-809.
4. Verhalen, L.M., Walter, C.M., Baha, A., Al-Raweej Kwee-Cong Fun and Murray, J.C. 1971. A diallel analysis of several agronomic traits in upland cotton (*Gossipium hirsutum* L.), *Crop Science*, 11 : 92-96.
5. Singh, R.K. and Chaudhary, B.D. 1977. Biometrical methods in quantitative genetic analysis. Kalyani publishers, New Delhi.
6. Dabholkar, A.R. 1992. 'Elements of Biometrical Genetics'. Concept Publishing Company, New Delhi.
7. Das, S.Y., Supriyo and Chakraborty, S. 1998. Genetic variation for seed yield and its components in greengram (*Vigna radiata* (L.) Wilczek). *Advances in Plant Science*, 11 : 271-273.
8. Khattak, G.S.S., Haq, M.A., Ashraf, M. and Srinives, P. 2001. Combining ability in mungbean (*Vigna radiata* (L.) Wilczek). Korean J. *Crop Science*, 46 : 420-423.
9. Manivannan 2002. Line x tester analysis in *kharif* greengram. *Legume Research*, 25 : 127-130.
10. Baseeruddin, M.D. and Nagur, T. 1981. Combining ability for grain yield in greengram (*Vigna radiata* (L.) Wilczek). *Madras Agricultural Journal*, 68 : 582-587.
11. Bhadra, S.I., Aktar, M.I. and Quasemn, A. 1992. Chittagong University Studies (Bangladesh), Part-II, *Science*, 5: 85-89.
12. Srinivas, P., Tangbunitivong, W. and Kriffing, B. 1991. Genetic study of yield and yield components in mungbean (*Vigna radiata* (L.) Wilczek.) grown in dry and wet reasons. *Journal of National Research Council of Thailand*, 23 : 1-13.
13. Rosaiah, G., Kumari, D.S., Satyanarayana, A. and Naidu, N.V. 1994. Combining ability studies on sprout quality traits in mungbean, *Indian Journal of Pulses Research*, 7: 1-6.
14. Kute, N.S., Deshmukh, R.B., Saroda, N.D., Mamjae, M.K. and Rahuri, M.P.K.U. 1999. Combining ability for yield and its components in mungbean (*Vigna radiata* (L.) Wilczek). *Madras Agricultural Journal*, 86 : 210-212.
15. Mather, K. 1949. Biometrical Genetics, Mathuen and Co. Ltd., London.
16. Jinks, J.L. 1954. The analysis of continuous variation in a diallel cross of *Nicotiana rustica* varieties. *Genetics*., 39 : 767-788.

Effect of pre-sowing seed treatments on seed germination, plant establishment and yield of moth bean (*Vigna aconitifolia*)

M.P. Rajora, M. Patidar and S.S. Mahajan
Central Arid Zone Research Institute, Jodhpur- 342 003

ABSTRACT

An experiment was conducted during 2005 and 2006 to assess the effect of pre-sowing seed treatments on seed germination, plant establishment and seed yield of moth bean under rainfed conditions. Six pre-sowing seed treatments, viz. control, simple hydration (H-D), H-D with 2% $CaCl_2$ at room temperature (RT), H-D with 50 ppm GA_3, H-D + Thiram @ 0.25% and H-D with 0.5% KNO_3 were given to two seed lots, viz. seeds having high germination and low germination of var. Maru moth. Seeds were hydrated for 2-hr and then dried at RT <25°C. The experiment was conducted in RBD with 4 replications. Data were recorded on seed germination (lab.), seedling dry weight (lab.), field emergence, 50% flowering, plant height, days to maturity, plant stand and seed yield. Seeds of better lot showed superiority for almost all the characteristics studied. Hydration treatments were significant only during 2006 for seed germination, seedling dry weight, field emergence and days to maturity, and $CaCl_2$ treated seeds had maximum germination (83.5%) closely followed by GA_3 (82.3%). GA_3 treated seeds had maximum germination (84.0%) during 2005. Mean of 2-year showed that GA_3 treated seeds had maximum germination (83.2%) followed by $CaCl_2$ (79.9%), and $CaCl_2$ treated seeds also had more seed yield (195 kg/ha) than control (176 kg/ha). The mean data also showed that seed germination, field emergence and seed yield enhanced up to 23.3, 1.2 and 20.5% by GA_3, Thiram and $CaCl_2$, respectively over the control in low germinating seeds as compared to 9.1, 0.2 and 17.8% by $CaCl_2$, KNO_3 and KNO_3 in high germinating seeds. The beneficial effect of pre-sowing seed hydration with GA_3 and $CaCl_2$ was found for seed germination and also $CaCl_2$ was found comparatively better for seed yield.

Key words: Seed treatments, seed yield, seed yield components, *Vigna acontifolia*.

INTRODUCTION

Moth bean {*Vigna aconitifolia* (Jacq.) Marechal} is an important *Kharif* food legume of arid and semi-arid areas of the country. It is a source of food, fodder, feed, green manuring, pasture and can easily be accommodated in a multiple cropping system. It is widely used for preparation of various confectionary items, viz. papad, bhujia, mangori, vada, kheech, etc. Being tolerant to water and heat stress, it is the most important pulse crop in arid zone perspective. Rajasthan is the major moth bean growing state in the country contributing about 85% area of which 93% (12.0 lac ha) is confined to 12 arid districts and the productivity is about 200 kg/ha. The low productivity of moth bean in arid Rajasthan is mainly due to unpredicted climatic and edaphic factors. Successful establishment of seedling in the field is an important factor in maximizing moth bean production and productivity, which largely depends on seed germination and seedling vigour. Pre-sowing seed treatments or imbibitions and drying of seed is one of the methods which results in modifying physiological and biochemical nature of seed so as to get the characters that are favourable for drought resistant (1). Studies on pre-sowing hydration-dehydration treatments have indicated that the germination, seedling establishment in the field and seed yield have been increased in many crops (2, 3). Such studies are meager in arid legumes like moth bean. Keeping this fact in view, the present study was conducted to ensure better and uniform field emergence of moth bean under arid conditions of Rajasthan.

MATERIALS AND METHODS

The experiment was conducted at CAZRI, Jodhpur during *Kharif* season of 2005 and 2006. Two seed lots, viz. A_1-high vigour seed with germination percentage 5-10% above MSCS and A_2-low vigour seed lot up to/with marginally below MSCS of moth bean variety Maru moth were used for the experiment. Both the seed lots were subjected to six pre-sowing seed hydration treatments viz., simple hydration (H-D) for 2 hours (T_1), H-D with 2% $CaCl_2$ at room temperature (2 hrs) (T_2), H-D with 50 ppm GA_3 (2 hrs) (T_3), simple H-D followed by dry dressing of Thiram @ 0.25% (T_4), and H-D with 0.5% KNO_3 (2 hrs) (T_5), and untreated seeds were used as control (T_0). Hydrated seeds were dried at room temperature below 25°C. Seed germination tests were conducted in a seed germinator at alternate temperatures of 16 and 8 hours at 20 and 30°C, respectively in sand medium. The first count for seed germination was taken on 5th day and the final count was taken on 14th day. Ten randomly selected normal seedlings were dried in an oven at 85°C for 24 hours and weighed for seedling dry weight.

Field experiment was conducted in a factorial RBD with four replications adopting the recommended package of practices under rainfed conditions. Plot size was 5 m x 2 m with row spacing of 50 cm. Data were recorded on field emergence, plant height, days to 50% flowering, days to maturity, plants per plot at harvest and final seed yield per hectare. Standard statistical procedures were used to analyse the data as suggested by Panse and Sukhatme (4).

RESULTS AND DISCUSSION

Effect of seed lot: High germinating seed lot had higher mean values for seed germination, seedling dry weight, field emergence, plant height, total plant stand and seed yield, and showed earliness for 50% flowering as well as maturity in both the years as compared to the low vigour seed lot (Table 1). However, these values were significantly superior only for seed germination, field emergence and plants per plot during both the years, while seedling dry weight during 2005 and seed yield during 2006 in high germinating seed lot. This lot also showed earliness in 50% flowering and maturity significantly during 2006. Similar results were also reported in groundnut by Dhedhi *et al.* (5). The seed yield levels of the year 2006 were low than 2005. The low yield levels during 2006 might be due to reduced plant growth as the plant height was only 15 cm, which was half of the height attained during 2005.

The differences in seed germination, seedling dry weight, field emergence and seed yield could be due to differences in seed vigour of two seed lots. Among the possible reasons for differences in vigour, the major reason might be the incidence of seed ageing. The changes that might took place in seeds during ageing manifested in slower and less uniform germination, reduced tolerance to stress, decline in seedling growth and an increase in the number of abnormal seedlings produced in low vigour seed lots. Present results are in conformity with the earlier findings of Kharab and Dahiya (6) in pigeonpea.

Effect of hydration treatments: Hydration treatments in general improved the seed germination, seedling dry weight and seed yield, however their effects were not much encouraging on field emergence, days to 50% flowering, plant height, maturity and total plant stand in the field plots. Pooled mean data revealed that among the hydration treatments, the seed treated with 50 ppm GA_3 had maximum germination (83.2%) followed by 2% $CaCl_2$ (79.9%) and 0.5% KNO_3 (78.9%). Hydration treatments were significant for seed germination, seedling dry weight, field emergence and days to maturity during 2006 (Table 1). In the year 2006, 2% $CaCl_2$ treated seeds had maximum germination (83.5%) closely followed by 50 ppm GA_3 (82.3%), whereas GA_3 treated seeds had maximum germination (84.0%) during 2005. The field emergence was not influenced significantly by hydration treatments during 2005, whereas it was highest in the control (69.3%) during 2006. The correlation between seed germination and field emergence was non significant. Although the pre-sowing hydration treatments had insignificant influence on seed yield, the mean seed yield of 193.8 kg/ha due to hydration treatments was 10% higher than control (176 kg/ha). Hydration treatment with 0.5% KNO_3 had maximum mean seed yield (197 kg/ha) followed by simple hydration (196 kg/ha) and 2% $CaCl_2$ hydration (195 kg/ha). Hence, hydration with GA_3 and $CaCl_2$ had improved seed germination and also $CaCl_2$ gave more seed yield. Singh *et al.* (7) also reported highest values for standard germination in GA_3 treated seeds of pigeonpea.

Table 1. Effect of seed and hydration on seed yield and its components in moth bean.

Treat	Seed germination(%)			Dry wt. of 10 seedling (g)			Field emergence (%)			Days to 50% flowering			Plant height (cm)			Days to maturity			Plants / plot at harvest			Seed yield (kg/ha)		
	2005	2006	Mean	2005	2006	Mean	2005	2006	Mean	2005	2006	Mean	2005	2006	Mean	2005	2006	Mean	2005	2006	Mean	2005	2006	Mean
Seed Lots																								
A1	86.7	84.8	85.8	0.24	0.21	0.22	52.5	70.8	61.7	43.4	38.3	40.9	32.1	15.2	23.7	62.8	62.6	62.7	82.3	195.1	138.7	325	90	208
A2	71.2	69.8	70.5	0.21	0.20	0.21	36.0	58.4	47.2	44.0	38.8	41.4	31.8	15.0	23.4	63.3	62.9	63.1	54.5	163.8	109.2	279	69	174
CD 5%	3.7	4.4	-	0.01	NS	-	4.4	2.9	-	NS	0.30	-	NS	NS	-	NS	0.30	-	6.0	10.00	-	NS	11	-
Hydration Treatments																								
T0	76.0	73.3	74.7	0.22	0.19	0.20	44.4	69.3	56.9	43.4	38.9	41.2	32.0	15.1	23.6	62.9	62.5	62.7	69.9	191.3	130.6	268	84	176
T1	80.3	69.5	74.9	0.23	0.21	0.22	44.4	62.8	53.6	43.9	38.5	41.2	32.5	15.1	23.8	63.1	62.9	63.0	66.5	175.9	121.2	314	78	196
T2	76.3	83.5	79.9	0.22	0.21	0.21	43.3	64.8	54.1	44.1	38.5	41.3	31.5	14.7	23.1	62.9	62.9	62.9	68.9	180.5	124.7	313	76	195
T3	84.0	82.3	83.2	0.23	0.23	0.23	41.3	60.8	51.1	43.4	38.5	41.0	31.4	15.2	23.3	63.4	63.3	63.4	66.1	167.4	116.8	302	70	186
T4	78.8	76.0	77.4	0.23	0.20	0.22	47.6	65.4	56.5	43.8	38.5	41.2	31.9	15.4	23.7	62.6	62.3	62.5	69.5	184.5	127.0	309	79	194
T5	78.5	79.3	78.9	0.24	0.19	0.22	44.6	64.5	54.6	43.9	38.4	41.2	32.5	15.0	23.8	63.3	62.8	63.1	69.8	177.3	123.6	305	89	197
CD 5%	NS	7.6	-	NS	0.02	-	NS	5.0	-	NS	NS	-	NS	NS	-	NS	0.60	-	NS	NS	-	NS	NS	-

A_1-high germination seed lot, A_2-low germination seed lot; T_0-control, T_1-simple hydration (H-D), T_2-H-D with 2% $CaCl_2$, T_3-H-D with 50 ppm GA_3, T_4-simple H-D + Thiram @ 0.25%, T_5-H-D with 0.5% KNO_3.

Table 2. Effect of seed sample x hydration interaction on seed yield and its components in moth bean.

Treat	Seed germination (%)			Dry wt. of 10 seedling (g)			Field emergence (%)			Days to 50% flowering			Plant height (cm)			Days to maturity			Plants / plot at harvest			Seed yield (kg/ha)		
	2005	2006	Mean	2005	2006	Mean	2005	2006	Mean	2005	2006	Mean	2005	2006	Mean	2005	2006	Mean	2005	2006	Mean	2005	2006	Mean
A_1T_0	87.5	78.0	82.8	0.26	0.19	0.23	50.3	77.7	64.0	43.3	38.8	41.1	33.0	15.4	24.2	62.5	62.5	62.5	78.3	208.0	143.2	295	86	191
A_1T_1	85.8	79.0	82.4	0.23	0.20	0.22	56.8	68.6	62.7	43.8	38.0	40.9	32.7	15.0	23.9	62.3	62.5	62.4	84.8	190.3	137.6	359	82	221
A_1T_2	89.0	91.5	90.3	0.23	0.20	0.21	49.8	71.0	60.4	44.0	38.0	41.0	31.8	15.0	23.4	62.8	62.3	62.6	79.5	200.8	140.2	304	87	196
A_1T_3	86.5	82.0	84.3	0.24	0.25	0.24	45.8	66.2	56.0	42.8	38.3	40.6	30.1	15.7	22.9	63.5	63.3	63.4	77.0	181.0	129.0	318	84	201
A_1T_4	83.0	87.5	85.3	0.24	0.21	0.23	56.8	68.8	62.8	43.5	38.5	41.0	31.9	15.5	23.7	62.5	62.5	62.5	88.0	195.5	141.8	333	91	212
A_1T_5	88.5	90.5	89.5	0.26	0.19	0.22	55.5	72.6	64.1	43.3	38.3	40.8	33.3	14.8	24.1	63.0	62.5	62.8	86.5	195.0	140.8	341	109	225
A_2T_0	64.5	68.5	66.5	0.19	0.18	0.19	38.5	60.9	49.7	43.5	39.0	41.3	31.0	14.9	23.0	63.3	62.5	62.9	61.5	174.5	118.0	240	82	161
A_2T_1	74.8	60.0	67.4	0.22	0.21	0.22	32.0	57.1	44.6	44.0	39.0	41.5	32.3	15.2	23.8	64.0	63.3	63.7	48.3	161.5	104.9	269	74	172
A_2T_2	63.5	75.5	69.5	0.20	0.21	0.21	36.8	58.7	47.8	44.3	39.0	41.7	31.1	14.4	22.8	63.0	63.5	63.3	58.3	160.3	109.3	323	64	194
A_2T_3	81.5	82.5	82.0	0.22	0.22	0.21	36.8	55.5	46.2	44.0	38.8	41.4	32.7	14.7	23.7	63.3	63.3	63.3	55.3	153.8	104.6	286	56	171
A_2T_4	74.5	64.5	69.5	0.22	0.19	0.21	38.5	62.0	50.3	44.0	38.5	41.3	31.9	15.4	23.7	62.8	62.0	62.4	51.0	173.5	112.3	285	67	176
A_2T_5	68.5	68.0	68.3	0.21	0.19	0.20	33.8	56.5	45.2	44.5	38.5	41.5	31.8	15.3	23.6	63.5	63.0	63.3	53.0	159.5	106.3	269	70	170
CD 5%	9.03	10.7	-	0.03	NS	-	NS	NS	-	NS	NS	-	NS	NS	-	NS	0.8	-	NS	NS	-	NS	NS	-

A_1-high germination seed lot, A_2-low germination seed lot; T_0-control, T_1-simple hydration (H-D), T_2-H-D with 2% $CaCl_2$, T_3-H-D with 50 ppm GA_3, T_4-simple H-D + Thiram @ 0.25%, T_5-H-D with 0.5% KNO_3

Interaction effect of seed lot and hydration: Interaction effect of seed lot and hydration was significant for seed germination during both the years, dry weight of seedling during 2005 and days to maturity for 2006 (Table 2). Seed treatment of better lot with 2% $CaCl_2$ had 89.0% germination during 2005, 91.5% during 2006 and these values were highest among all the combinations of interaction in the respective years. Among the combinations of low germinating seeds, 50 ppm GA_3 treated seeds had maximum germination during both the years (81.5% and 82.5%) and these values were significantly higher than control (64.5% and 68.5%). Dry weight of seedlings of GA_3 treated seeds of this lot was also higher with early crop maturity (63.3 days). Pre-sowing seed treatments did not manifest any significant superiority over the control for rest of the traits. Hydration treatments have additional advantages over the untre-ated control as indicated by the improvement in seed germination, field emerg-ence and seed yield in low vigour seed lot as compared to the high germination seed lot. The mean seed germination, field emergence and seed yield enhanced up to 23.3, 1.2 and 20.5% by GA_3, H-D followed by dry dressing with Thiram and $CaCl_2$ treatments, respectively over the control in low vigour seed lot as compared to 9.1, 0.2 and 17.8% by $CaCl_2$, KNO_3 and KNO_3 hydration in high germinating seed lot. Contrarily, Mathew and Powel (8) reported that major benefits of seed treatment may lie only in highly germinable vigorous seed lots.

Thus it can be concluded that the pre-sowing seed hydration treatments with GA_3 and $CaCl_2$ resulted in higher seed germination and also $CaCl_2$ gave more seed yield, even though these higher values were statistically at par with control, except for seed germination in 2006. Clear cut trend of pre-sowing seed hydration treatments could not be observed for rest of the traits in moth bean under arid conditions.

REFERENCES

1. Henckel, P.A. 1964. Physiology of plants under drought. *Ann. Rev. Pl. Physiol.*, **15**: 363-386.
2. Basu, R.N. and Dhar, N. 1979. Seed treatment for maintaining vigour, viability and productivity of sugar beet (*Beta vulgaris*). *Seed Science and Technology*, **7**: 225-233.
3. Subbaraman, R. and Selvara, J.A. 1989. Effect of pre sowing treatment on seed yield and quality in groundnut cv. JL 24. *Seed and Farms*, **4**: 5-9.
4. Panse, V.G. and Sukhatme, P.V. 1978. *Statistical Methods for Agricutural Workers*. Indian Council of Agricultural Research, New Delhi.
5. Dhedhi, K.K., Dangaria, C.J., Parsana, G.J. and Joshi, A.K. 2006. Effect of pre-sowing seed treatment for better crop establishment in summer groundnut. *Seed Research*, **35** (1): 17-21.
6. Kharab, R.P.S. and Dahiya, B.S. 2000. Influence of natural ageing of seeds on field performance in pigeonpea (*Cajanus cajan* L. Millsp.). *Seed Research*, 28(2): 149-152.

7. Singh, O., Deepte and Kumar, A. 1999. Effect of plant growth regulators on seed quality in partially aged seeds of pigeonpea (*Cajanus cajan*). *Seed Research*, **27** (1): 54-59.

8. Mathews, S. and Powel, A.A. 1986. Environmental and physiological constraints on field performance of seeds. *Hort. Science*, **21** (5): 1125-1128.

Potential and prospects of *in vitro* methods for the improvement of mothbean (*Vigna aconitifolia* (Jacq.) Marechal) crop

R. Raj Bhansali

Division of Plant Science and Biotechnology,
Central Arid Zone Research Institute, Jodhpur, India.

ABSTRACT

Tissue culture methods have been exploited for various activities related to the improvement of moth bean crop for capturing genetic gains related to its productivity. *In vitro* selection, conservation, inducing genetic variability, multiplication, pathogen free, disease resistance and gene transfer approaches have been applied for achieving various targets of crop development. Moth bean plant has high degree of capabilities for the development of organs and embryos. Such genetic ability can be used for various biotechnological approaches. Soma-clonal variation could be a quick method for development new moth bean varieties. Thus cell/callus/protoplast cultures could be integrated with conventi-onal as well as molecular marker based breeding programmes of *Vigna aconitifolia*.

Key Words: *In Vitro*, Tissue Culture, *Vigna*, moth bean and *Vigna aconitifolia*

Introduction

Moth bean (*Vigna aconitifolia* (Jacq.) Marechal) crop is a native to hot and dry land of Northern-Western parts of India (1). In Rajasthan, districts of Barmer, Bikaner, Churu, Jodhpur and Nagaur are major growing areas of moth bean production. Dana (2) reported that its wild forms are also distributed from Sonaram desert of Mexico to Guatemala in South America. The crop is one of the important sources of protein, calories, minerals and vitamins in arid and semi arid regions. Plants are also used for checking soil erosion, green manure and pasture. Large number of collections from different agro-ecological zones and farming systems have been done by the various workers. Variability with regard to erectness, early maturity of seeds, size and colour of seed and pod, resistance to YMV and shattering quality has been reported (3).

Crop is endowed with unique morphological and physiological features for quickly adjusting to fast changing moisture during kharif season. In low rainfall and soil moisture deficiency, moth bean is considered as most economical and sustainable grain legume crop due to having high adoption ability of genome in low moisture conditions. These genetic buffering capabilities placed moth bean crop as an important component of integrated arid farming system. Crop is grown in more than13 lakh ha of area and produces around 3 lakh ton annually with a productivity level of 215 kg ha^{-1} (4). The crop occupies nearly 80% area and contributing 73% production but with low productivity of 198 kg ha-1. Looking to its genetic capabilities, moth bean crop needs special attention of plant biotechnologist for its genetic improvement.

Earlier, major thrust for the improvement of moth bean was given to the selection of germplasm, which produced genotypes like Jadia and Jwala. But availability of limited genetic variability in germplasm hampered hybridization programme of moth bean to a great extent. Existing genotypes of moth bean are thermo and photo sensitiveness, indeterminate habit, non-synchronous maturity, spreading habit and susceptibility to YMV disease (5,6). Some of the improved varieties have determinate habit, semi-erect plant geometry, near synchronous maturity, photo insensitiveness and resistance to the viral diseases (5-8). *V. aconitifolia* contains antimicrobial compounds such as quercetin and kaempferol, which have been isolated and identified for their antimicrobial activity against gram positive, gram negative and fungal pathogens. These compounds were found in seeds, leaves and callus cultures of *V. aconitifolia* (9).

Promising germplasm of moth bean could be genetically improved by integrating conventional breeding methods with integration of *in vitro* biotechnological manipulations, as already successes have been achieved many annual crops. In many *Vigna* species a range of genotypes are amenable to a specific tissue culture type, whereas in others, only a specific genotype or wild relative can be used for tissue culture initiation. Expanding the range of genotypes within a species that undergo the requisite tissue culture process would provide a major contribution to improving the *in vitro* system. This can be accomplished by the conventional empirical approach of manipulating culture media composition, phytohormones, explant source, and tissue culture environment. Fortunately, moth bean plant has high morphogenetic potential under *in vitro* organogenesis and embryogenesis system of regeneration. Thus there is urgent need to utilize conventional as well as modern biotechnological tools for the improvement of moth bean.

MATERIALS AND METHODS

Various *in vitro* methods such as cell, callus, embryo, and protoplast cultures individually or combined with gene transfer techniques via *Agrobacterium* have been used. Improved germplasm of *V. aconitifolia* were cultured for the *in vitro* growth and differentiation by manipulation of medium constituents and cultural conditions for developing organogenesis and somatic embryogenesis. Tissue culture media such as MS, BS, B_5 and L-6 were employed for regeneration and embryo inductions. Standard tissue culture procedures were used for growth and differentiation of callus and protoplasts. Seeds from healthy and YMV infected plants were undertaken for *in vitro* culture and understanding behaviour of virus in infected seeds and seedlings.

RESULTS AND DISCUSSIONS

In vitro methods have various applications in crop improvement programs related to the selection, conservation, inducing genetic variability, multiplication, disease-free and resistance and gene transfer technology. Tissue culture is the first prerequisite for many plant biotechnology activities. Somatic embryogenic and organogenic tissue cultures are the primary sources of totipotent target cells/tissues used in grain legume. This genetic ability of *V. aconitifolia* can be exploited for its improvement. Production of somaclones may be of immediate utility by exploiting callus, somatic embryos and protoplast culture. Larkin and Scowcroft (10) has suggested usefulness of somaclonal variation for development new varieties having beneficial traits in many annual crops. Similarly, somaclones of moth bean could be a promising source for generating variability in recalcitrant germplasm. Since, single gene mutations and organelle gene mutations have been produced in somaclones, the obvious strategy is to include promising varieties into cell/callus/protoplast cultures to select for incremental improvements for various traits. Stabilized variations in moth bean could be further utilized easily due to extreme self-pollination behaviour and due to tight genetic linkages.

1. Callus growth and differentiation

In moth bean *in vitro* system, the morphogenesis is influenced by various biotic and abiotic factors viz., medium constituents, electrical impulses, hormones (concentrations, types and ratios), age of explant, stage of donor plant and genotype. Several workers have reported role of genomic constituent of moth bean in regeneration (11-13). Electrical impulses also showed marked increase in stimulation of shoot bud formation in the hypocotyls of moth bean (11). Saini (12) and Bhargava and Chandra, (13) have worked out the genotype x cultural environment interactions for the selection of regenerative genotypes.

They found that genotype limits the ability of regeneration but the frequency of regeneration could be achieved by altering the cultural environment. Thus many of *V. aconitifolia* germplasm have performed high regeneration on BAP/KN and IAA containing medium (Fig. 1 and 2).

Fig. 1.

Fig. 2.

Moth bean tissue culture work was initiated particularly for the establishment of callus cultures, regeneration and multiplication aspects. Tissue cultures have been raised and plantlets have been regenerated from various types of explants such as roots (14), hypocotyls (11,15,16), cotyledons (15,17,18), leaf (13,18). Plantlets have also been developed from callus cultures (5,17,19). Bhargava and Chandra (13) reported mass multiplication of *V. aconitifolia* variety Jwala through shoot tip cultures. Cultures were raised mostly on MS medium, which have been modified depending on type of explant and objective of culture. Interactions between genotypes and cytokinin hormones were evident as variation in differentiation in various moth bean genotypes. Technique can be used *in vitro* evaluation of germplasm. Godbole et al. (17) and Eapen et al. (18), however, regenerated plantlets from callus cultures on hormone free MS media via organogenesis path way. Leaf and seed explants were found best source for obtaining prolific organogenesis in moth bean.

Growth regulators such as triazoles negatively alter the growth of callus and regeneration capacity of moth bean. Sankhla et al (20) have reported that triazole plant growth regulators such as paclobutrazol and uniconazole reduced *in vitro* growth of moth bean callus by 20–25% when added to the culture medium at low concentration (1 mg/l). However, gibberellic acid (GA_3) counter acted the action of triazoles. This indicates that exposure of moth bean callus tissue to micromolar concentrations of triazoles or GA_3 can significantly alter *in vitro* growth and differentiation. Similarly, paclobutrazol has been found to reduce the differentiation of roots and shoots *in vitro* growth of *V. aconitifolia* (21).

2. Somatic embryogenesis

Embryo-like structures in the suspension cultures of moth bean was reported by Bhargava and Chandra (13). Plants were also regenerated from the suspension cultures originated from three-week-old leaf callus of VA-686 germplasm. Kumar et al (22) have used several media including MS, B_5 and L-6 for their effects on cell growth. Maximum growth was observed in L-6 medium containing 2,4-D. Cultures were maintained on L-6 medium with low 2,4-D and zeatin. Transferring the suspension cells to the medium with zeatin induced somatic embryos. The somatic embryos were also developed from the cell clusters on medium having BAP. *V. aconitifolia, V. mungo* and *V. radiata* were used for callus culture from immature cotyledons of on MS medium supplemented with NAA, picloram or 2,4-D (23). They also reported induction of somatic embryos in all three major *Vigna* sp. on L-6 medium supplemented with low concentrations of picloram, 2,4-D, GA_3 and cytokinins. The conversion of these embryos to plantlets was very low (1%). Embryogenic cells were spherical or filamentous proembryos. Somatic embryos of *Vigna* species have unique feature in somatic embryogenesis. In embryogenic cultures, filamentous proembryo developed embryos either at single or multiple sites. Development of somatic embryos from multiple sites resulted in several embryos connected by a common suspensor at the radicle end. Continued divisions of the proembryos led to globular, heart shaped and dicotyledonary stages of somatic embryogenesis. The somatic embryos of *V. mungo* and *V. aconitifolia* differentiated into tiny plantlets at low frequency in liquid suspension cultures. Plantlets were also obtained from the embryos with IBA containing medium. The regenerated plants were grown to maturity in the greenhouse (15). Similarly, plants have been regenerated via somatic embryogenesis in mung bean (*V. radiata*). Mature cotyledons, hypocotyls, nodal segment, and leaf explants were found best source for somatic embryogenesis on MS medium supplemented 2,4-D with BAP. A higher percentage of regeneration was observed in the cultures initiated from leaf explants of mung bean (24). Efforts are needed for understanding complete somatic embryogenesis process for development of plantlets and synchronous development of embryos.

Such highly regenerative embryogenic cell lines, retaining long term and high morphogenic potential, are convenient system for screening and selection of variants with desirable characteristics. A salt tolerant cell line, VA 986 was obtained at 2% NaCl. Developed plantlets were salt tolerant. Bhargava and Chandra (13) have reported NaCl tolerant cell line having halophytic ability but lost the regenerative capacity.

3. Protoplast Culture

Protoplasts are the good source for inducing variation in *Vigna* species. Arya et al (25) have reported regeneration of protoplasts and calli. Further, plants were also developed from the protoplasts and transferred to field conditions. Derived plants showed significant variation in characteristics of seed, pod and plant. Shekhawat and Galston (26) were first to isolate, culture and regenerate plants from mesophyll cells of moth bean. High division rates and plating efficiency of protoplasts were obtained by adding glutamine and asparagines to the culture medium. Protoplast derived calli regenerated through embryogenesis on an auxin free medium. Somatic embryos developed into complete plants. Later, Krishnamurthy *et al.* (19) and Eapen and Gill (16) reported simpler protocols of moth bean protoplast regeneration. However, Krishnamurthy (19) has reported induction of multiple shoot bud formation in the protoplast-derived callus on MS basal medium without hormones. Arya et al. (25) also analyzed protoplast-derived plantlets of moth bean genotype (Jadia) up to maturity level. Variation was observed in seed germination, pod length, pod and seed plant height, seed/pod and seed weight/pod. Protoplast can be used for gene delivery system of important traits.

4. *Agrobacterium* mediated transfer

Genetic transformation is emerging as an important crop improvement tool. Transformation theoretically expands the sources of genes for plant improvement to all organisms, far beyond the gene pool accessible via sexual hybridization. Transformation also offers strategies for over expressing or suppressing endogenous genes. Thus, introducing new genes or manipulating endogenous gene expression via transformation generates new phenotypic variation useful for investigating gene function and for *Vigna* improvement. Although there is a report of successful production of transgenic plants, further evidence of transmission of the transgene genotype to progeny has not been reported.

Moth bean has well-established transformation system by using cell, tissue, organ and protoplast delivery system. Plant transformation system has been reported in various *Vigna* species Table. 1. Since the success of *Agrobacterium* mediated transformation is largely dependant up on genotype and cultural conditions. Direct gene transfer is attractive system as so many complications can be avoided. Increased organogenesis efficiency of *Agrobacterium* mediated co-cultivation strategy could be used for the transfer of useful genes from wild/sister species of *Vigna* into moth bean. Kohler *et al.* (27) transformed moth bean protoplast using PEG with NPT II. Different expression signals of cauliflower mosaic virus (Plasmid pABDI) or from nopaline synthase gene of *A. tumefaciens* (plasmid pLGV neo 2103) were used for obtaining transformed colonies. Bhargava and Smigocki (28) also incorporated genes into three *Vigna* species. Marker genes were bombarded into ungerminated mature embryos of RMO-40. Transformed plants were multiplied through axillary bud enhancement with low levels of kanamycin (40 mg/ml). By this way direct gene transfer could be achieved in moth bean. But molecular gene integration was not shown. Similarly, protoplast regeneration system developed by Eapen *et al.* (29) can be used for the gene transfer in moth bean. By co-cultivation of protoplasts with *A. tumefaciens* containing the Ti plasmid derivative pGV 38501103 neo kanamycin-resistant colonies were obtained; 23% of the transformed lines showed expression of the non-selected co-transferred nopaline synthase gene. Transformation can be confirmed by Southern blot analysis using a non-radioactive detection system. In these cases protoplasts were isolated from mesophyll tissues of young leaves of of 60% was obtained in mesophyll protoplasts from 10-day-old seedlings having high plating efficiency (29).They have suggested that plant type is important in determination of transformation rate, as one of the cultivars has responded 85-fold higher than the other. A protocol for efficient direct gene transfer by using particle gun bombardment was also developed for *V. aconitifolia*. Hypocotyl explants were transformed with 3 plasmids: pBI121, pHS101, and pHS102. Stable transformants were regenerated on MS medium supplemented with BAP and NAA and kanamycin. The helium pressure, plasmid type and cultivar factors determined the stable transformation frequency. Complete plants were regenerated and transferred to soil. The integration of the stable transgenes and reporter genes in plant genomes was shown by means of PCR amplification of these genes from plant genomic DNA and Southern blot hybridization with gene-specific probes. Such methods allow high-efficiency production of transgenic plants in moth bean. Therefore, availability of direct gene transfer protocols for moth bean open up new avenues for its genetic amelioration through transfer of gene(s) of interest from sister species or even from unrelated sources.

Table: 1. Genetic Transformation in *Vigna* species

Vigna Species	Type of Genes	Method of Transfer
V. unguiculata	Hpt (hygromycin phosphotransferase) gene[a]	Agrobacterium
	uidaA (ß-glucuronidase, GUS), genes[b]	Electroporation
	uidaA and nptII genes[c]	*Agrobacterium*
	35S CaMV promoter[m]	Particle bombardment
V. radiata	uidaA and nptII genes[d]	*Agrobacterium*
	uidaA and nptII genes (hygromycin phosphotransferase)[e]	Particle bombardment
	uidaA, nptII and hpt genes[f]	
V. angularis	γ-AI (γ-amylase inhibitor) and nptII genes[g]	*Agrobacterium*
	sGFP (S65T) (modified green fluorescent protein),	*Agrobacterium*
	uidaA and nptII genes[h]	*Agrobacterium*
V. aconitifolia	nptII genes[i]	Electroporation
	nptII genes[j]	Agrobacterium
	uidaA and nptII genes[e]	Particle bombardment
V. mungo	uidaA and nptII genes[e]	Particle bombardment
	pCAMBIA2301 (nptII) and (GUS) gene (gusA)[k,l]	Agrobacterium
V. sesquipedalis	phosphinothricin-N-acetyltransferase gene and neomycin-3-phosphotransferase-II gene[n]	Agrobacterium

a-Muthukumar et al (30); b-Ignacimuthu (31); c-Chowrira *et al.* (32); d-Pal et al. (33); e-Bhargava and Smigocki (28); f-Jaiwal et al. (34); g-Ishimoto *et al.* (35); h-Yamada et al. (36); i-Kohler et al. (27); j-Eapen *et al.* (29); k-Saini and Jaiwal (12); l-Saini and Jaiwal (12); m-Ikea, et al., (37); n-Ignacimuthu (31).

Techniques such as RAPD and RFLP can provide genetic variability in somaclones, GMO and plants developed from protoplast cultures of moth bean during *in vitro* culture. Germplasm collected from various ecological zones may help in tagging important genes. Genetic analysis of *in vitro* regenerated plants indicated that molecular profiles of various genotypes have very less variability within varieties but more variability was present between the groups and *in vitro* regenerants.

6. Embryo rescue

Another very important application of *in vitro* techniques lies in the transfer of desirable traits from different *Vigna* species to moth bean. This may be achieved either through somatic hybridization between these two species or even through cross pollination coupled with embryo rescuing. Studies in this direction have revealed that embryo rescuing may be attempted following crossing between moth bean, as a female parent, and mung bean, as a male parent. Although fertilization takes place at a low frequency and embryos start aborting 8th day after fertilization, nevertheless, embryo rescuing may bring fruitful results.

7. Seed borne nature of yellow mosaic virus (YMV)

YMV of moth bean is serious threat for cultivation and production of moth bean in arid environment. Most of varieties (local as well as improved) are moderate to highly susceptible to viral diseases causing serious losses (70%). In some varieties the disease is so dominant that it becomes difficult to get disease-free plant under epiphytotic conditions (38). Initial symptoms mostly converted in to yellow patches (39). Diseased plant produce small, shrived pods as well as seeds. Primary source of viral infection is a major factor for spread and destruction of crop. Various workers have reported the insect (*Bemisia tabaci*) borne nature of YMV of moth bean. Insect vectors quickly spread the disease from one plant to another. The role of seeds in transmission of YMV has been reported (7). Virus transmission through seeds has been reported in many crops. Pollen and seed transmission are closely interrelated ways of virus transmission in many legume crops. Many factors are responsible for movement of viruses from mother tissues to seed parts and then embryo to seedlings. Such viruses mostly belong to group potyvirus, carmovirus, comovirus, sobemovirus and geminivirus having ssRNA and ssDNA (38).

The transmission of YMV of moth bean through seeds has been studied (7&8). *In vitro* methods were used to ascertain the seed borne nature of the YMV. Initially, seedlings remained healthy up to 2-3 weeks but first symptom was observed on cotyledonary leaves. The symptoms varied from complete yellowing to mosaic forms. Maximum symptoms were noted in GMO 9003 followed by RMO 40 (Fig. 3). Later on apical tips and inter nodal buds were excised and cultured on MS medium supplemented with BAP and KN. Most of cultures exhibited yellowing symptoms in newly emerged multiple shoots with 4-6 weeks (Fig. 4). Thus *in vitro* technique has opened new window of research in understanding the seed borne viruses of plants through tissue culture methods.

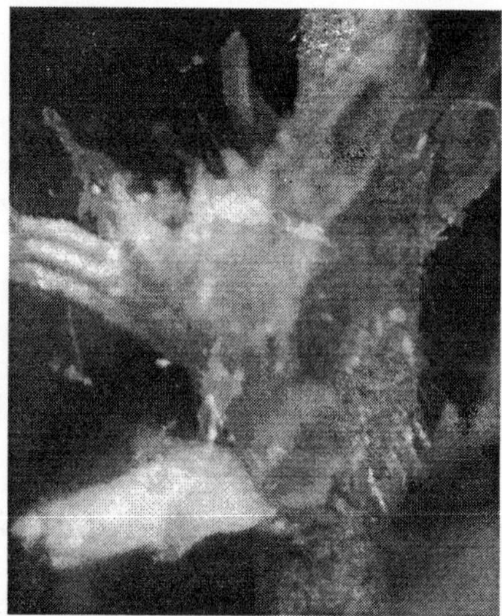

Fig. 3.

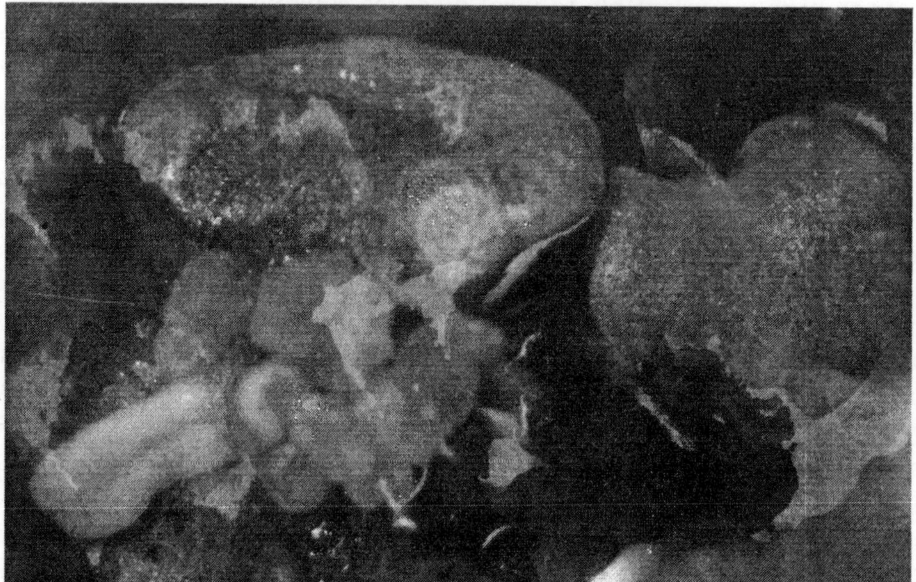

Fig. 4.

Raj Bhansali (7) has reported seed borne nature of viruses of some legume crops. Similarly broad-bean mosaic virus has been found to be seed borne in nature (40&41). Whether seed-transmitted-viruses are carried with in embryo to seedling, need more verification. Infection in seeds may be through pollen/ovule. Infected seeds may act as primary source of infection, which is then quickly transmitted by the insect vectors.

Conclusion

Legumes are a large, diverse family ranging from herbaceous annuals to woody perennials that, because of their capacity to fix nitrogen. In accordance, arid legumes are being rapidly developed as experimental systems to pursue a number of important biological questions unique to the *Vigna* species using molecular tools including genomics. A key component of most functional genomics approaches are a high-throughout transformation system useful for developing various gene identification strategies. Being moth bean crop is drought hardy having limited genetic variability and difficulty in manual hybridization can be improved by the utilization of *in vitro* techniques such as cell, tissue and suspension culture, somaclonal variations, genetic transformation, embryo rescue and protoplast culture. These methods are highly reproducible in terms of regeneration.

Many adopted moth bean local and improved germplasm to harsh arid environment could be excellent sources for further improvement for increasing production. In this process there has been a great accumulation of beneficial genes in many promising germplasm. Important traits such as disease resistant and yield traits should be characterized for development of improved varieties. Sometimes, the selection pressure under subsistence agriculture is responsible for erosion or dilution of high yielding genes. Thus *in vitro* improvement is required through restructuring leading varieties, which are early to medium maturity, high yielding, responsive to improved cultural practices, well adapted to drought conditions and resistant to pest and diseases. Moreover, variability generated through *in vitro* techniques should be conserved and utilized for the development of new varieties. The germplasm of moth bean requires modification of plant type from its primitive nature having better partitioning ability in promising genomes in boosting grain yield and induction of YMV resistance genes. Such strategies will help in removing various bottlenecks as well as in obtaining quick genetic gains. Further research is needed in developing new sources of totipotent regenerative cells as targets for transformation, especially in less dedifferentiated tissues, for selection or identification of transgenic plants. Substantial progress has been achieved in many legumes. Recent progress in legume transformation suggests that some systems will achieve the transformation efficiencies required for functional genomics applications in the near future.

REFERENCES

1. Piper, C.V. and Morse, W.J. 1974. Five oriental species of beans. US Dept. Agric. Bull. 119 pp.
2. Dana, S. 1976. Origin, evaluation and distribution of some grain legumes. Indian J. Genet. 36: 125-131.
3. Dwivedi, N.K. and Bhatnagar, N. 2002. Moth bean biotechnology. In: Moth bean in India, Eds: D. Kumar and N.B. Singh, Scientific Publishers (India), Jodhpur. pp. 49-56.
4. Kumar, D. and Singh, N.B. 2002. Moth bean in India. Scientific Publishers (India), Jodhpur.
5. Kumar, D. and Henry, A. 2008. An arid legume: An introduction. In: Souvenir, 3rd National Symposium on enhancing productivity, nutritional security and export potential through arid legumes, June 28-30, 2008.
6. Kumar, D., Solanki, K. and Solanki, Z.S. 2008. Moth bean. In: Souvenir, 3rd National Symposium on enhancing productivity, nutritional security and export potential through arid legumes. Eds. Kumar, D. and A. Henry, June 28-30, 2008, 87-99 pp.
7. Raj Bhansali, R. 2005. In quest of seed borne nature of yellow mosaic virus (YMV) in moth bean by using tissue culture technique, J. Arid Legume, 2: 354-356.
8. Raj Bhansali, R., Kumar, A. and Vyas, N.L. 2000. Is yellow mosaic virus (YMV) of moth bean is seed borne? Symposium on Biotechnology of Plant Protection, Application and Technology Development, Feb. 25-27, 2000, CAS, BHU, Varanasi, 98 pp.
9. Tyagi, S. and Nag T.N. 2004. Flavonoids of moth bean cultivars as antimicrobial agent, Indian Journal of Plant Physiology, 9: 1.
10. Larkin, P.J. and Scowcroft, W.R. 1981. Somaclonal variation a novel source of variability from cell cultures for plant improvement. Theor. Appl. Genet. 60: 197-214.
11. Gill, R., Mishra, K.P. and Rao, P.S. 1987. Stimulation of shoot regeneration of *Vigna aconitifolia* by electrical control Annals of Botany 60: 399-403.
12. Saini, R. and Jaiwal, P.K. 2005. Transformation of a recalcitrant grain legume, *Vigna mungo* L. Hepper, using *Agrobacterium tumefacjens*, Plant Cell Reports, 24: 164-171.
13. Bhargava, S. and Chandra, N. 1989. Factors affecting regeneration from leaf explants of moth bean *Vigna aconitifolia* (Jaca) Marechal. Indian Journal Experimental Biology, 27: 55-57.

14. Eapen, S. and Gill, R. 1986. Regeneration of Plants from cultured root explants of moth bean (*Vigna aconitifolia* L. Jacq. Marechal). Theoretical Applied Genetics, 72: 384-87.

15. Bhargava, S., Upadhyaya, S., Garg, K. and Chandra, N. 1983. Differentiation of shoot buds in hypocotyls explants and callus cultures of some legumes. In: Sen, S.K. & Giles, K.L. (eds.) Plant Cell Cultures in Crop Improvement, Plenum Press, New York, 431-33 pp.

16. Eapen, S. and R. Gill 1986. Regeneration of Plants from cultured root explants of moth bean (*Vigna aconitifolia* L. Jacq. Marechal). Theoretical Applied Genetics, 72: 384-87.

17. Godbole, G., Kunachgi, M.N., Potdar, U.A., Krishnamurthy, K.V. and Mascarenhas, A.F. 1984. Studies on a drought resistant legume: The moth bean, *Vigna aconitifolia* (Jaca)Marechal II Morphogenetic studies Plant Cell Report, 3: 75-78.

18. Eapen, S., Gill, R. and Rao, P.S. 1986. Tissue culture studies in moth bean. Current Science, 55: 07-9.

19. Krishnamurthy, K.V., Godbole, D.A. and Mascarenhas, A.F. 1984. Studies on a drought resistant legume: The moth bean, *Vigna aconitifolia* (Jacq) Marechal. I. Protoplast culture and organogenesis Plant Cell Reports, 3: 30-32.

20. Sankhla, N., Davis, T.D., Gehlot, H.S., Upadhyaya, A., Sankhla A. and Sankhla, D. 1991. Growth and organogenesis in moth bean callus cultures as influenced by triazole growth regulators and gibberellic acid, Journal Plant Growth Regulation, 10: 41-45.

21. Gehlot, H.S., Davis, T.D., Sankhla, A., Sankhla, D., Upadhyaya, A. and Sankhla, N. 1987. Influence of triazole growth regulators and GA_3 on *in vitro* differentiation from moth bean callus. Proc. Plant Growth Reg. Soc. Amer. 14:103-107.

22. Kumar, A.S., Gamborg, O.L. and Nabors, M.W. 1988. Plant regeneration from cell suspension cultures of *Vigna aconitifolia*. Plant cell reports, 4: 138-141.

23. Eapen, S. and George, L. 1990. Ontogeny of Somatic Embryos of *Vigna aconitifolia, Vigna mungo* and *Vigna radiata, Annals of Botany* 66: 219-226.

24. Radha, P, D., Syamala, P.D., Seethamahalakshmi, L. and Kumar, M.V. (2004). Plant regeneration via somatic embryogenesis in mung bean [*Vigna radiata* (L) Wilczek]. Scientia Horticulturae (Netherlands) 99: 1-8.

25. Arya, I.D., Arya, S., Rao D.V. and Shekhawat, N.S. 1990. Variation amongst protoplast-derived moth bean *Vigna aconitifolia* plants. Euphytica, 47: 333-38.

26. Shekhawat, N.S. and Galston, A.W. 1983. Isolation, culture, and regeneration of moth bean *Vigna aconitifolia* leaf protoplasts. Plant Science Letters, 32: 43-51.

27. Köhler, F., Golz, C., Eapen, S., Kohn, H. and Schieder, O. 1987. Stable transformation of moth bean *Vigna aconitifolia* via direct gene transfer, Plant Cell Reports, 6: 313-317.

28. Bhargava, S. and A. Smigocki 1994. Transformation of tropical grain legumes using particle bombardment. Current Science. 66: 439-42.

29. Eapen, S., Köhler, F., Gerdemann, M. and Schieder, O. 1987. Cultivar dependence of transformation rates in moth bean after co-cultivation of protoplasts with *Agrobacterium tumefaciens,* Theoretical and Applied Genetics, 75: 207-210.
30. Muthukumar, B., M. Mariamma, K. Veluthambi and A. Gnanam. 1996. Genetic transformation of cotyledon explants of cowpea (*Vigna unguiculata* L. Walp) using *Agrobacterium tumefaciens*. Plant Cell Reports 15: 980-985.
31. Ignacimuthu S. 2000. *Agrobacterium* mediated transformation of *Vigna sesquipedalis* Koern (Asparagus bean), Indian J. Exp. Biol., 38: 493-498.
32. Chowrira, G.M., Akella, V., Fuerst, P.E., and Lurquin, P.F. 1996. Transgenic grain legumes obtained by *in planta* electroporation-mediated gene transfer. Mol. Biotech. 5: 85-95.
33. Pal, M., Ghosh, U., Chandra, M., Pal, A. and Biswas, B.B. 1991.Transformation and regeneration of mung bean (*Vigna radiata*). *Ind. J. Biochem. & Biophys.* 28: 449- 455.
34. Jaiwal, P.K., Kumari, R., Ignacimuthu, S., Potrykus, I. And Sautter, C. 2001. *Agrobacterium tumefaciens*-mediated genetic transformation of mung bean (*Vigana radiata* L. Wilczek)-a recalcitrant grain legume. Plant Science, 161: 239-247.
35. Ishimoto, M., Sato, T., Chrispeels, M.J. and Kitamura, K. 1996. Bruchid resistance of transgenic azuki bean expressing seed alpha-amylase inhibitor of common gene. Entomol. Exp. Appl. 79: 309-315.
36. Yamada, T., Teraishi, M., Hattori, K., and Ishimoto, M. 2001. Transformation of azuki bean by *Agrobacterium tumefaciens*. Plant Cell Tissue Organ Cult 64: 47–54.
37. Ikea, J., Ingelbrecht, I., Uwaifo, A. and Thottappilly, G. 2003. Stable gene transformation in cowpea (*Vigna unguiculata* L. walp.) using particle gun method, African Journal of Biotechnology, 2: 211-218.
38. Raj Bhansali, R. 2004. Viral diseases of arid and semi-arid legume crops. In: Management of planrt diseases, Eds. Kapoor, B.B.S. and Khatri, N.K., Madhu Publications, Bikaner, 85-100 pp.
39. Raj Bhansali, R. 2008. Variations in symptoms of yellow mosaic virus disease of moth bean, J. Arid Legumes, 5: 21-24.
40. Kadian, A.O.P. and Varma, J.P. 1981. Some properties of virus causing leaf crinkle disease of urd bean and mung bean in India. 5th International Conference on Virology, Abstr. 22/06.
41. Baker, K.F. and Smith, S.H. 1966. Dynamics of seed transmission of plant pathogens. Annual Review Phytopathology, 4: 311-334.

Performance of moth bean + pearl millet intercropping systems as influenced by delayed sowing in extreme arid environment

B.K. Kandpal[1], R.S. Mertia, R.N. Kumawat and S.S. Mahajan

Central Arid Zone Research Institute, Regional Research Station, Jaisalmer-345001, Rajasthan

[1.]Corresponding author (Email: basantkandpal@yahoo.com)

ABSTRACT

An intercropping study was conducted during *Kharif* 2006 to find out suitable intercropping combination of moth bean (var. CZM-3) with pearl millet (CZP 9802) at Chandan Farm, Jaisalmer. Five intercropping combinations of moth bean and pearl millet (2:1, 1:2, 3:1, 1:3 and 2:2) along with sole crop of moth bean and pearl millet were sown in randomized block design with three replications in the first week of August after a good shower of rain. Successive increase in the rows in per unit area increased the respective yield of moth bean and pearl millet. However, increase in rows of pearl millet could not make up for the reduction in moth bean yield. It may be due to delayed sowing of pearl millet results into poor growth, tillering and grain filling. Over all sole crop of moth bean produced significantly maximum moth bean equivalent grain yield (484 kg/ha) than rest of the combinations. It was followed by moth bean: pearl millet intercropping at 2:1 ratio (344 kg/ha) and 3:1 ratio (340 kg/ha). Moth bean equivalent straw yield was statistically at par among the treatments. Overall maximum gross return of Rs.14475/- was recorded in sole moth bean crop while minimum gross return of Rs. 7732/- was with sole pearl millet crop. Thus it could be concluded that at extreme arid environment with delayed monsoon sole crop of moth bean is advantageous than any moth bean and pearl millet intercropping combinations.

Key words: Moth bean, pearl millet, intercropping, performance, yield

INTRODUCTION

The extreme arid thar desert is characterized by extremes of temperature, low and erratic rainfall, high wind speed and thus, very high evapo-transpiration. The combined effect of climatic factors results in frequent crop failure and agricultural economy in the region (1). The farmers of the region thus believe in low input low risk low yield concept (2). Rainfed farmers traditionally grow a mixture of crops viz. pearl millet, guar seed, moth bean, seasamum, sonamukhi etc. depending upon the onset of monsoon. Mixed cropping provides an insurance against failure of some crops due to vagaries of nature, particularly drought but plants do not get the opportunity to properly utilize available resources like land, nutrients, moisture etc. Moreover, mixed cropping creates the hindrance in many cultural operations like weeding, hoeing, pest control and harvesting. Thus, intercropping, which involves component crops of different growth pattern for extended use of resources, is viable options for arid conditions. Under extreme arid conditions, generally arid legumes (moth bean and guar seed) are selected as one of the components crops with pearl millet in mixed cropping. A study carried out by Singh and Joshi (3) revealed that pearl millet and guar seed grown in alternate rows (1:1) gave mean combined dry matter yield at par with sole cropping of pearl millet, but recorded significantly higher yield over mixed cropping systems. Singh and Singh (4) found that total productivity of intercropping system of cow pea and moth bean in sunflower was 74% higher over pure cropping of sunflower. Similarly, Yadav and Beniwal (5) found highest yield advantage of moth bean (1:1) and cluster bean (2:2) intercropping in pearl millet.

During recent decades frequent delayed onset of monsoon was experienced at Jaisalmer. Short duration moth bean is a viable option under such conditions. But farmers still followed mixed pearl millet with moth bean to fulfill fodder/grain requirement of the household, although delayed sowing causes failure of seed maturity in pearl millet. But the information on the over all productivity of moth bean and pearl millet intercropping systems under delayed monsoon situation is not available. Hence present investigation was aimed to study the compatibility of pearl millet as an intercrop with moth bean under late rainfed sowing in extreme arid conditions of Jaisalmer.

MATERIALS AND METHODS

The geographical setting

The experiment was conducted during *Kharif,* 2006 at the experimental Chandan Farm (26°59'42"N Latitude, 71°20'08"E Longitude) of Central Arid Zone Research Institute, Regional Research Station, Jaisalmer. It is situated at the heart of thar desert in western most part of the country.

The climate

The agro-climatic situation of Jaisalmer district is characterized by low and erratic rainfall, high wind speed, extreme of temperatures, high evapotranspiration rate, severe wind erosion and thus, frequent droughts and famines. The overall climate is categorized under *Marusthali,* hot, hyper arid eco-sub region (6). Long term normal rainfall in the area is 164.0 mm. Generally the monsoon sets in during the second week of July and starts receding by the end of August or in the first week of September (7).

During 2006, first rain spell was on 7 August and total 221.0 mm rainfall was received in 11 rainy days at regular intervals up to 23August. Thereafter, there was almost a complete dry spell for the remaining crop season. The mean daily temperature was above 30°C during August and September (30.7 to 32.3°C). Thereafter it gradually declined around 24°C at the end of October. The average wind speed was 12.45, 7.16 and 5.11 km/hr during August, September and October months, respectively. The potential evapotranspiration ranged between 4.9 to 5.2 mm/day during crop season.

The soils

The experiment was laid out in almost level (0-1% slope) aeolian plain. The soils of the sites were classified as deep, mixed, hyperthermic, calcareous family of Typic Torripsamments (Table 1). The soils were low in organic carbon (0.12%) and available N content (207 kg $KMnO_4$-N/ha), medium in P (14 kg P/ha) and K content (138 kg K/ha). These almost level, well drained, very deep sandy loam soils with 4.0 to 6.2% free $CaCO_3$ content provide sufficient soil volume and condition for in situ rain water conservation and penetration of crop roots deeper into sub surface soils. However, alkaline condition of the soils (pH> 8.0) imposed slight limitation for successful rainfed cultivation.

Table 1. Physio-chemical properties of the soils of the experimental site

Horizon	Depth (cm)	Particle Size Distribution			$CaCO_3$ (%)	pHe	ECe dSm^{-1}	Organic Carbon (%)	CEC (Cmol (+p)/kg)
		Sand (%)	Clay (%)	Texture					
Mixed, hyperthermic, calcareous family of Typic Torripsamments									
Ap	0-20	77.4	6.1	ls	5.4	8.7	0.8	0.12	3.5
C1	20-55	83.2	6.8	ls	4.0	8.6	0.5	0.08	2.8
C2	55-108	86.1	6.5	ls	5.2	8.6	0.3	0.06	2.4
C2	108-137	86.7	6.7	ls	6.2	8.7	0.4	0.06	2.4

ls= loamy sand

The experimental details

An intercropping experiment on moth bean (var. CZM-1) and pearl millet (var. CZP 9802) was laid out in randomized block design with three replications.

The seven treatments consisted of moth bean (sole), pearl miller (sole) and moth bean + pearl millet intercrops in the ratio of 2:1, 1:2, 3:1, 1:3 and 2:2, respectively. The crops were sown in second week of August during 2006, immediately after the onset of monsoon. One hand weeding and thinning was carried out at 20 days after sowing (DAS). Uniform basal application of 30 kg N + 8.75 kg P/ha was applied at the time of field preparation. The crops were harvested at maturity. Moth bean equivalent yield (MEY) was worked out on the basis of existing local market price of each crop.

For economic evaluation of the intercropping systems, prevailing market prices were used for different inputs and outputs. The prices of different produce per kilogram used for calculation were Rs. 26.25 for moth bean grains and Rs. 3.00 for moth bean straw, Rs. 8.00 for pearl millet grains and Rs. 2.00 for pearl millet straw, respectively. The input costs used for calculation of net returns were based on the reports of the Commission for Agricultural Costs and Prices (8). The intercropping indices, viz. relative crowding coefficient (RCC), competitive ratio (CR), aggresivity (A) and land equivalent ratio (LER) were worked out to evaluate the treatment effects reciprocity function for intercropping. Grain yield equivalent to straw yield based on prevailing local market prices was added to grain yield of both crops, since fodder is equally important in predominantly livestock based farming region. Pooled analysis of variance technique was used to compare years (environments) and means of different growth and yield attributes to draw logical conclusions (9).

RESULTS AND DISCUSSION

Crop growth

The moth bean-pearl millet intercropping systems significantly influenced growth of both moth bean and pearl millet plants under extreme arid rainfed conditions (Table 2). Growth attributes shows advantage of sole moth bean crop over any intercropping combination with pearl millet. Total biomass/ plant of moth bean was highest under sole cropping. Replacing moth bean rows with pearl millet up to 25% (3:1 intercropping ratio) did not decrease biomass/ plant significantly. However, beyond this ratio it decreased significantly. Almost similar trend was found in plant height, pods/plant and seed weight/plant. Under rainfed condition moisture is the most limiting factor (10, 11). The fibrous root system of pearl millet shares large rhizosphere than tap rooted moth bean and this might have significantly reduced the availability of moisture beyond a threshold level of intercropping (2:1). Contrary to it, sole pearl millet crop produced lowest biomass/plant. It may be due to inter and intra row competition among pearl millet plants for limited moisture. Intercropping pearl millet with moth bean successively improved the biomass/plant in proportion to its population per unit area and significantly higher biomass/plant was recorded at intercropping treatments of 2:2, 3:1 and 2:1. Intercropping pearl millet with moth bean at low proportion (3:1, 2:1 and 2:2) might have provided comparative larger rhizosphere to its prolific fibrous root system and thus, more moisture availability and significantly higher biomass/plant. Secondly, being leguminous crop moth bean nodulation also contributes some nitrogen for pearl millet protoplasm synthesis. Several workers (12, 13 and 14) also reported beneficial effect of arid legumes on pearl millet growth and yield.

Table 2. Effect of intercropping treatments on growth and yield attributes of moth bean and pearl millet.

Treatment	Moth bean				Pearl millet			
	Plant Height (cm)	Pods/ plant	Seed weight/ plant (g)	Total biomass/ plant (g)	Plant Height (cm)	Cobs/ plant	Seed weight/ plant (g)	Total biomass/ plant (g)
Intercropping System (moth bean: pearl millet)								
2:1	21.3	24.5	2.52	6.05	109.7	2.0	8.16	16.55
1:2	19.7	17.5	2.09	5.78	104.3	1.9	6.37	13.37
3:1	22.1	24.8	2.57	8.18	118.3	2.7	9.11	16.17
1:3	19.5	14.5	1.81	5.78	102.7	1.6	5.93	12.64
2:2	20.0	20.2	2.41	5.15	104.7	1.9	6.49	14.26
Sole Moth bean	22.5	31.5	3.10	8.30	-	-	-	-
Sole pearl millet	-	-	-	-	102.0	1.6	5.44	12.16
SEm±	1.5	1.2	0.18	0.38	2.7	0.2	0.46	0.48
CD(p=0.05)	3.9	3.7	0.57	1.19	7.7	0.6	1.44	1.38

Crop productivity

The intercropping systems significantly influenced crop productivity of moth bean and pearl millet (Table 3). Under late monsoon situations of Jaisalmer sole moth bean crop produced significantly higher grain and straw yield followed by 3:1 and 2:1 moth bean-pearl millet intercropping. The harvest index of pearl millet crop varied non-significantly among the treatments. The proportional successive reduction in yield of moth bean and pearl millet reveals lower intra-row competition among the crop at lower ratios. It may be due to more vertical than horizontal development of plant rhizosphere in search of water under extreme arid situations (2). Overall non-significant harvest index may be due to uniform insolation, fertility and moisture availability conditions in all the intercropping treatments. However, comparative low harvest index of pearl millet may be due to low temperature at grain filling stage. At Jaisalmer, the mean temperature drastically reduce below 28°C beyond mid October. Low temperature at late reproductive stage of pearl millet caused poor grain filling and hence lower harvest index (15).

Table 3. Effect of intercropping treatments on yield of moth bean, pearl millet and moth bean equivalent yield (MEY).

Treatment	Moth bean			Pearl millet			MEY	
	Grain Yield (kg/ha)	Straw Yield (kg/ha)	HI (%)	Grain Yield (kg/ha)	Straw Yield (kg/ha)	HI (%)	Grain Yield (kg/ha)	Straw Yield (kg/ha)
Intercropping System (moth bean: pearl millet)								
2:1	279	510	35.7	213	568	27.6	344	889
1:2	122	275	31.1	433	996	27.4	254	938
3:1	295	597	33.0	148	444	25.0	340	893
1:3	94	195	33.0	466	1139	29.0	236	954
2:2	204	423	32.2	311	816	25.3	298	967
Sole Moth bean	484	885	35.6	-	-	-	484	885
Sole pearl millet				665	1811	26.9	203	1207
SEm±	21	44	2.1	11	82	1.1	22	74
CD(p=0.05)	55	114	NS	28	211	NS	55	NS

Table 4. Economics of moth bean and pearl millet intercropping systems.

Treatment	Gross Return (Rs./ha)	Net Return (Rs./ha)	B:C ratio
Intercropping System (moth bean: pearl millet)			
2:1	10814	5734	2.13
1:2	8550	4061	1.91
3:1	10406	5825	2.21
1:3	8101	3764	1.87
2:2	9772	5027	2.06
Sole Moth bean	14475	8136	2.28
Sole pearl millet	7732	3299	1.75
SEm±	648	373	0.03
CD(p=0.05)	1633	941	0.07

Sole moth bean crop produced highest grain MEY. It was followed by 2:1 and 3:1 moth bean and pearl millet intercropping. Sole pearl millet crop produced highest straw MEY followed by moth bean and pearl millet intercropping ratios 2:2, 1:3 and 1:2 but straw MEY was statistically at par among all the treatments.

Economics

Among the intercropping treatments sole moth bean crop produced significantly maximum gross return followed by 3:1 and 2:1 moth bean-pearl millet intercropping ratios. The lowest gross return was recorded in sole pearl millet followed by 1:3 and 1:2 intercropping ratios. Almost similar trend was recorded for net return and B:C ratios as well. Low grain yield and lower harvest index of pearl millet under delayed sowing conditions caused significantly lowest returns from pearl millet and hence influenced the comparative economics of pearl millet production either as a sole or intercrop.

Competition functions

Relative crowding coefficient (RCC) of moth bean, pearl millet and system as well was found less than unity (Table 5), and thereby confirming disadvantage of moth bean-pearl millet intercropping in extreme arid region where onset of monsoon is usually late.

Table 5. Effect of moth bean and pearl millet intercropping systems on relative crowding coefficient (RCC), aggresivity (A), competitive ratio (CR) and land equivalent ration (LER) (mean of 3 years).

Treatment	RCC			CR		A	LER
	Moth bean	Pearl millet	System	Moth bean	Pearl millet		
Intercropping System (moth bean: pearl millet)							
2:1	0.695	0.932	0.804	0.913	1.097	-0.084	0.897
1:2	0.729	0.791	0.737	0.905	1.201	-0.111	0.866
3:1	0.542	0.917	0.700	0.867	1.166	-0.124	0.843
1:3	0.757	0.608	0.676	0.905	1.131	-0.084	0.864
2:2	0.743	0.863	0.791	0.930	1.109	-0.076	0.882

The results are further confirmed by aggressivity value (A) and land equivalent ratio (LER). Aggressivity value near zero and LER values less than one suggest superiority of sole crop over intercropping. Competitive ratio (CR) of almost unity for moth bean and pearl millet in all intercropping treatments reveals minimal intra-row competition for limited water as roots of both crops become more vertically pronounced in search of water. Further, equal CR values of moth bean and pearl millet shows no advantage of intercropping over sole cropping.

CONCLUSION

Thus, it was concluded that sole cropping of moth bean is advantageous than any moth bean and pearl millet intercropping combinations in extreme arid rainfed areas where onset of monsoon is usually late.

REFERENCES

1. Rao, A.S. and Singh, R.S. (1998). Climatic features and crop productivity. In: Fifty years of Arid Zone Research in India, pp 17-38. Faroda A.S. and Singh, Manjeet (Eds.), Central Arid Zone Research Institute, Jodhpur.
2. Kandpal, B.K.; Kumawat, R.N. and Mertia, R.S. (2006). Agronomic evaluation of promising genotypes of moth bean under hyper arid conditions of Jaisalmer. *Journal of Arid Legumes* 3(1):1-10.
3. Singh, M. and Joshi, N.L. (1994). Effects of mixed and intercropping systems on dry matter and grain yields of component crops in arid environment. *Annals of Arid Zone* 33:125-128.
4. Singh K.C. and Singh, R.P. (1977). Intercropping of annual grain legumes with sunflower. *Indian Journal of Agricultural Sciences* 47:563-567.
5. Yadava, N.D. and Beniwal, R.K. (2003). Competitive relationship of arid legume and pearl millet intercropping system under rainfed condition of western Rajasthan. In: Advances in Arid Legume Research, pp 272-274. Henery, A.; Kumar D. and Singh, N.B. (Eds.). Indian Arid Legume Society, Jodhpur. Scientific Publishers (India). Jodhpur.
6. Velayutham, M.; Mandal, D.K.; Mandal, C. and Sehgal, J. (1999). Agro-ecological sub regions of India for planning and development. NBSS & LUP Pub. 35, 372p. National Bureau of Soil Survey and Land Use Planning, Nagpur, India.
7. CAZRI (1992). Integrated natural and human resource appraisal of Jaisalmer district. CAZRI Pub. 39, 156p. Chatterji, P.C. and Kar, A. (Eds.). Central Arid Zone Research Institute, Jodhpur.
8. FAI (2007). Fertilizer Statistics 2006-07. The Fertilizer Association of India, New Delhi.
9. Gomez, K.A. and Gomez, A.A. (1984). Statistical procedures for agricultural research (2nd edition), 680p. John Wiley & sons, Inc.
10. Kandpal, B.K.; Kumawat, R.N.; Mahajan, S.S. and Mertia, R.S. (2008). Moisture stress management behaviour of moth bean (*Vigna aconitifolia*) as influenced by planting pattern and urea foliar application. *Journal of Arid Legumes* 5(1):33-37.

11. Singh, Y.V. and Saxena, Anurag. (1998). Management of limited water resources for crop production. In: Fifty years of Arid Zone Research in India, pp 189-213. Faroda A.S. and Singh, Manjeet (Eds.), Central Arid Zone Research Institute, Jodhpur.

12. Panjab-Singh and Joshi, N.L. (1980). Intercropping of pearl millet in arid areas. *Indian Journal of Agricultural Sciences* 50:338-341.

13. Singh, R.P., Singh, K.C. and Ramkrishna, Y.S. (1978). Effect of systems of planting pearl millet on yield, total productivity, moisture use and monetary returns. *Indian Journal of Agricultural Sciences* 48:138-142.

14. Singh, M. and Joshi, N.L. (1997). Pearl millet based cropping systems in relation to plant population and moisture availability. *Annals of Arid Zone* 36:37-42.

15. Sys, C.; Van Ranst, E.; Debaveye, J. and Beernaert, F. (1993). Land Evaluation-III: Crop requirements, 199p. International Training Centre for Post Graduate Soil Scientist, Ghent, Belgium.

Growth and yield of fodder cowpea (*Vigna unguiculata* L. Walp) grown in association with pasture grasses in arid zone

M. Patidar, B.K. Mathur and M.P. Rajora

Central Arid Zone Research Institute, Jodhpur-342 003

ABSTRACT

An experiment was conducted at CAZRI, Jodhpur during Kharif 2007 to assess the productivity and suitability of cowpea in strip cropping with two pasture grasses viz. *Cenchrus ciliaris* and *Lasiurus sindicus* with and without fertilization. Plant height and dry matter accumulation of cowpea on per plant basis was not much affected by strip cropping of grasses except plant growth of cowpea in adjacent row was slightly suppressed by presence of grasses. Per ha green and dry fodder yields of cowpea were reduced in strip cropping as compared to sole cowpea as the area under strip cropping was decreased to half. However, the total forage productivity of strip cropping (51-52 q/ha green fodder yield and 11-12 q/ha dry fodder yield) was higher as compared to sole cowpea (40.3 q/ha green fodder yield and 6.6 q/ha dry fodder yield). Application of 40 kg N/ha enhanced the fodder yield of cowpea and grasses by 10-15% over control. Further the forage yields of sole grasses were higher than under grass-legume strip cropping. Thus land equivalent ratio (LER) indicate no real yield advantage of strip cropping but the nutritive value in terms of protein content was increased in strip cropping without affecting the growth of cowpea by grasses.

Key words: Cowpea, grasses, Growth, Fodder yield, arid regions, farming system.

INTRODUCTION

In the semi-arid tropics of India, large section of people rely on agriculture for a significant part of the livelihood. Livestock i.e. important aspects of this group and play a critical role in both sustainability and intensification of agricultural productivity in most of the farming systems. Increasing human population and changing dietary habits associated with urbanization and higher income are causing increased demand of animal origin food like milk, egg, chicken, meat etc. This in turn will gradually decrease cereal consumption with pressing demand for livestock products (1). Therefore, to increase the productivity of livestock, producers have to feed animal adequately throughout the year but the availability of fodder including crop residues, grasses and cultivated forages are limited in arid zone due to harsh climatic condition and decreasing availability of grazing land. Further pasturelands are degrading and infested by unpalatable and less nutritious plant species. The net sown area in western Rajasthan has increased from 35.5% in 1956-57 to 53.4% in 1997-98 and nearly 1/3rd area of arid zone of Rajasthan is wasteland (2). The productivity of available grazing lands is hardly 300-400 kg/ha (3). Due to this reason the fodder availability is not sufficient to feed about 23.62 million of livestock (4) in arid zone of Rajasthan.

The deficit in fodder availability could be minimized to some extent through improvement of pastureland, cropland and current fallows by adopting improved technologies for forage production. Generally grasses and cereal fodder have low protein content, which adversely affects the animal performance (4). It has been found that by mixed cropping of grasses/cereals and legume, the nutritive value of mixed herbage can be greatly improved without affecting the forage yield. Mixed cropping or intercropping can improve the productivity of land and ensure better utilization of natural resources such as plant nutrients and water from different soil layers enriching herbage in term of nutritive value since legume component contain higher amount of protein, calcium and phosphorus. Legume is also palatable to most of the livestock. Cowpea (*Vigna unguiculata* L. Walp) is one of the important food legumes and a valuable component of the traditional cropping system of the semi-arid tropics. It is a deep rooted legume and well suitable under harsh agro-climatic condition of arid zone. It gives a valuable contribution towards human food and livestock fodder and its dual purpose character makes it a very attractive crop where resources are becoming scarce. Cowpea has the unique ability to fix atmosphere nitrogen through its nodules and grow well even in poor soils with more then 85% sand, less then 0.2% organic matter and low levels of phosphorus (5,6). Cowpea is shade tolerant and therefore compatible as an inter crop with maize, millet, sorghum, sugarcane, cotton (7) coupled with these attributes its quick growth and rapid ground cover checks soil erosion and root decay in situ produce nitrogen rich residue that improves soil fertility and structure. Cowpea is equally important as a nutritious fodder for livestock (8). Increased availability and incorporation of the leguminous straws in the diet of the animals would ensure a balanced ration to the animals (9). The nutritive value of cowpea grain, leaves and haulms is very high. The crude protein content ranges from 22 to 30% in seed and leaves on dry weight basis (10, 11), and from 13 to 17% in the haulms with high digestibility and low fibre level (12).

The statistics of area under fodder cowpea is not available but it is grown in about 1.26 lakh ha in Rajasthan and about 88% of this area lies in arid zone of Rajasthan comprising 11 districts. Lack of adequate soil moisture during the crop growth stages coupled with poor soil fertility condition restricts plants to express their full growth and yield potential resulting low productivity of land in arid zone. Application of nitrogen enhances early crop growth and root development. Growing cowpea in association with grasses helps in increasing nutritive value of fodder in arid zone. Thus, growing legume in strip cropping and applying fertilizer assumed to increase the productivity of land. Keeping this in the view a field experiment was conducted to assess the productivity and suitability of cowpea in strip cropping along with pasture grasses and fertilizer application under arid zone conditions.

MATERIALS AND METHODS

A field experiment was conducted at the Central Arid Zone Research Institute, Jodhpur during kharif season of 2007. The treatments comprised of five cropping systems, ie *Cenchrus ciliaris, Lasiurus sindicus* Cowpea, *Cenchrus ciliaris* + cowpea, and *Lasiurus sindicus* + cowpea and two fertility levels, ie 0 and 40 kg N/ha. These treatments were evaluated in RBD with three replications. The soil of experiment field was low in available nitrogen (140 kg/ha), medium in available phosphorus (12.32 kg/ha) and potassium (191 kg/ha^{-1}) having pH 8.3. The moisture content at field capacity and permanent wilting point (-15 bar tension) was 10.5 and 3.4% (w/w) respectively. The region is characterized by a monsoon climate with the wet season occurring from the end of June to September. Total rainfall received during the crop growth period was 256.6 mm. Cowpea var. Bundel lobia-1 was grown with and without nitrogen fertilizer application (40 kg N/ha) in association with *Cenchrus ciliaris and Lasiurus sindicus* Cowpea was sown on July 7, 2007. The row spacing for grasses and cowpea was maintained at 60 cm. There were four strips of 3.6 m width in each plot. Grasses and legume were planted in alternate strips in grass-legume system. Soil water content to a depth of 60 cm was determined with gravimetric method at 0.2m depth interval in each treatment. Soil moisture measurements were taken prior to sowing and thereafter at weekly interval through at in growing season. Total dry matter yield included only the above grown parts of the crop was recorded. The data were analyzed statistically as per standard procedure (13). Water use and water use efficiency of the cowpea was calculated as under:

mm water use in 0.20 m depth = % water content x bulk density x 0.20

Water use efficiency = Dry matter production (kg/ha)/Water use (mm)

RESULTS AND DISCUSSION

Plant height

Plant height of cowpea was not influenced by strip cropping of grasses ie. *Cenchrus ciliaris and Lasiurus sindicus* but the growth of cowpea in the row near to grass row was suppressed while grass growth was better in adjacent row. This was due to the beneficial effect of legume on grass growth. On contrary grasses suppressed the legume growth due to hardy nature and fast growing habits of grasses resulting in more uptake of water at early stage of crop growth. Application of nitrogen significantly enhanced the plant height of cowpea by 10% over conrtrol (Table 1). The improvement in plant height with nitrogen application was due to its role in protein synthesis and translocation of growth hormones which generally act as stimuli for certain steps of growth and development (14). Further better interception and utilization of radiant energy leading towards higher rate of photosynthesis and finally more accumulation of dry matter of individual.

Table 1. Plant height of cowpea and grasses as affected by strip cropping and nitrogen levels.

Treatment	Plant height (cm)		
Cropping system	*Cenchrus ciliaris*	*Lasiurus sindicus*	Cowpea
Cenchrus ciliaris (CC)	98.8		
Lasiurus sindicus (LS)		95.5	
Cowpea			41:05
CC+Cowpea	103.0		40.33
LS+Cowpea		95.8	31.34
Nitrogen			
Control	97.8	93.9	35.9
40 kg N/ha	104.0	97.4	39.5

Forage and protein yield

Green and dry fodder yields of grasses and legume were significantly influenced by strip cropping (Table 2). Green and dry fodder yields of cowpea were lower in strip cropping with grasses as compared to sole cowpea as the area under cowpea was decreased to half. The productivity of cowpea on per hectare basis was not much influenced by strip cropping of *Cenchrus ciliaris*

(Fig. 1). However, it was reduced significantly when cowpea grown in association with *Lasiurus sindicus*. This shows that *Lasiurus sindicus* exert more competition than *Cenchrus ciliaris*. The total forage productivity of strip cropping (52-54 q/ha green fodder yield and 12-13 q/ha dry fodder yield was higher as compared to sole cowpea (40.3 q/ha green fodder yield and 6.6 q/ha dry fodder yield) and it was lower as compared to sole grasses. Further, application of 40 kg N/ha significantly increased fodder yield of cowpea and grasses by 10-15% over control. The better interception and utilization of radiant energy leading towards higher photosynthesis and finally more accumulation of dry matter of individual plant due to nitrogen fertilization. Singh *et al.* (15) observed a positive response of fodder cowpea to nitrogen fertilizer up to 80 kg/ha and (16) also observed the response of cowpea with application of 40kg N/ha. Protein yield under different cropping system was at par. However, maximum protein yield ie. 122.7 kg/ha was recorded with *Lasiurus sindicus* + cowpea and lowest in sole *Cenchrus ciliaris*. Growing legume in association with *Cenchrus ciliaris* also found to enhance the protein yield (17). Cowpea grown in association with *Lasiurus sindicus* did not increase protein yield as compared to sole *Lasiurus sindicus*. This might be due to poor growth of cowpea in the system. Further LER of strip cropping was not much improved but it was slightly higher in *Cenchrus ciliaris* + Cowpea as compared to *Lasiurus sindicus* + Cowpea. This indicates that no real advantage of strip cropping but nutritive value in term of protein yield was increased in strip cropping of *Cenchrus ciliaris* + cowpea without affecting the growth.

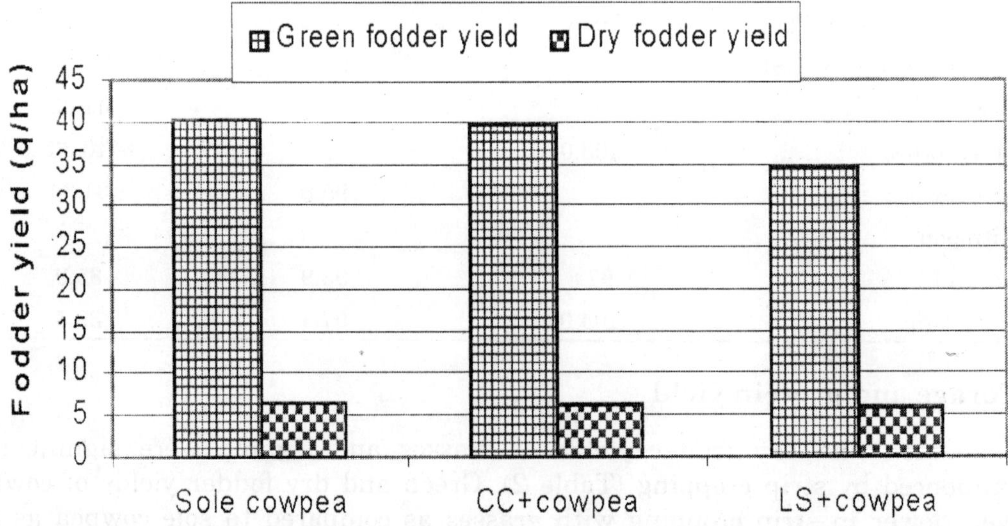

Fig. 1. Productivity of cowpea in different cropping system

Table 2. Fodder and protein yield of cowpea and grasses as affected by strip cropping and nitrogen levels.

Cropping system	Green Fdder (q/ha)	Dry matter yield (q/ha)	Protein yield (kg/ha)	LER
Cenchrus ciliaris (CC)	62.55	16.13	110.17	1
Lasiurus sindicus (LS)	65.89	17.66	122.34	1
Cowpea	40.28	6.56	119.55	1
CC+Cowpea	53.34	12.16	120.18	1.07
LS+Cowpea	52.36	12.27	122.69	0.99
SEm	1.88	0.79	5.91	
CD (P=0.05)	5.64	2.37	NS	
Nitrogen (kg N/ha)				
0	51.33	12.28	112.52	-
40	58.44	13.62	125.46	-
SEm±	1.19	0.50	3.74	-
CD (P=0.05)	3.57	1.5	11.22	

Soil moisture storage, water use and water use efficiency

Soil moisture content at the time of harvest was less in cowpea as compared to grasses due more extraction of moisture by cowpea from different layer (Fig 2). Further soil moisture content was also less with 40 kg N application as compared to control in both grasses and legume this showed that soil moisture extraction was more in the plots with fertilizer application Water use and water use efficiency of cowpea was improved with nitrogen application (Table 3). Maximum water use efficiency of cowpea (3.01 kg dry matter/ha/mm) was recorded with application of 40 kg N/ha. This improvement was attributed to better plant growth and higher yield under fertilizer application. Similar results were also reported by Patidar *et al.* (16). This increase may be attributed to adequate fertility stimulating early growth, therefore increasing leaf surface for photosynthesis, and increasing root development, often into deeper soil zones leading to increased supply of water to plants. Water use of cowpea was higher as compared to grasses. However, water use efficiency of cowpea decreased as compared to grasses. Maximum water use efficiency was recorded in *Lasiurus sindicus* followed by *Cenchrus ciliaris* and cowpea. The higher water use efficiency in grasses was due to more dry matter accumulation (Table 3).

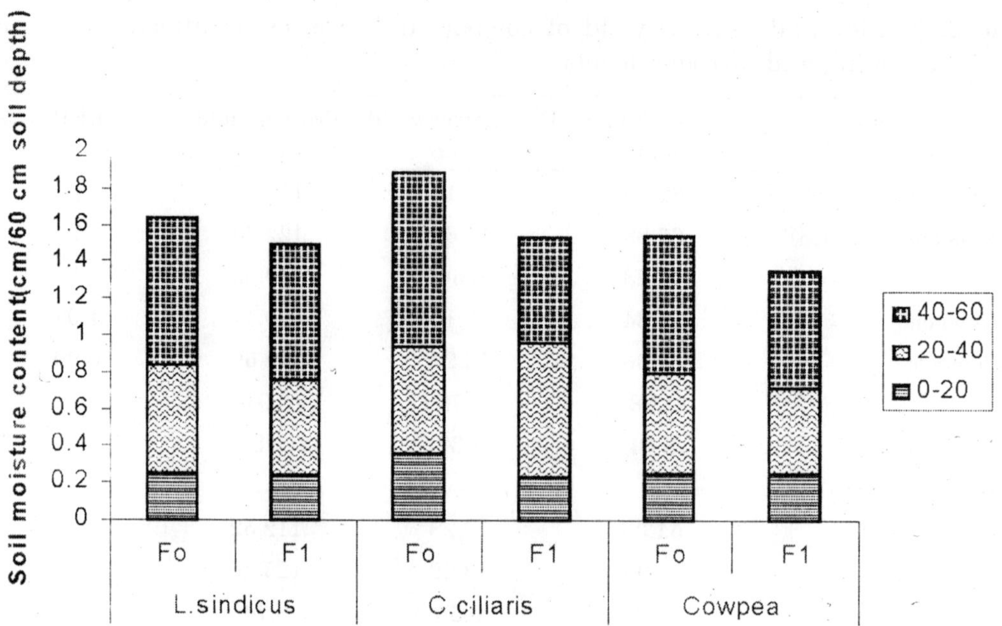

Fig. 2. Soil moisture content at harvest of grasses and legume under strip cropping system

Table 3. Water use and water use efficiency of cowpea and grasses under different fertility levels

Treatment	Water use(mm)			Water use efficiency (kg/ha/mm)		
	Control	40 kg N/ha	Mean	Control	40 kg N/ha	Mean
Cropping system						
Cenchrus ciliaris (CC)	244.6	226.9	225.8	6.73	7.56	7.15
Lasiurus sindicus (LS)	225.8	227.1	226.4	7.00	8.59	7.80
Cowpea	225.4	228.4	226.9	2.78	3.01	2.90
CC+Cowpea	224.8	227.6	226.2	5.22	5.53	5.38
LS+Cowpea	225.8	227.7	226.8	5.55	5.27	5.41

REFERENCES

1. Mathur, B. K. 2004. Drought proofing through utilization of low – grade roughage and non-conventional feed resources for small ruminants. *In National Seminar on Opportunities and Challenges in Nutrition and Feeding Management of Sheep, Goat and Rabbit for sustainable production* pp. 191-201. *Indian Society of Sheep and Goat Production and Central Sheep and Wool Research Institute,* Avikanagar-304501, Rajasthan February, 10-12.

2. Mathur B.K. and Sharma Tribhuwan 2005. Nutritional Technologies for Sustainable Animal Production in Arid Region. *Veterinary Practitioner.* 6, 1:58-61

3. Balak Ram. 2003. Impact of human activities on land use changes in arid Rajasthan: Retrospect and prospects. *(In) Human Impact on Desert environment*, pp 44-59. Pratap Narain, Kathju S, Amal Kar, Singh M P and Praveen Kumar (Eds). Scientific Publishers, Jodhpur, India.

4. Pratap Narain and Rajora, M.P. 2005. Managing drought stress for augmenting forage productivity in arid western Rajasthan. *(in) Livestock Feeding Strategies for Dry Regions*, pp 41-56. Pathak P S and Kundu S S (Eds). International Book Distributing Co., Lucknow, India.

5. Kolawale, G.O., Tian, G. and Singh, B.B. 2000. Differential response of cowpea varieties to aluminum and phosphorus application. *Journal of Plant Nutrition* 23: 731-740.

6. Sanginga, N., Lyasse, O. and Singh, B.B. 2000. Phosphorus use efficiency and nitrogen balance of cowpea breeding lines in a low P soil of the derived savanna zone in West Africa. *Plant Soil* 220 : 119-128.

7. Singh, B.B. and Emechebe, A.M. 1998. Increasing productivity of millet-cowpea intercropping systems. In Pearl Millet in Nigeria Agriculture: Production, Utilization and Research Priorities, pp. 88-75 Emechebe, A.M., Ikwelle, M.C., Ajayi, O., Aminu Kano, M., Anaso, A.B. (Eds.), Proceedings of the Pre-season Planning Meeting for the Nationally Coordinated Research Programme for Pearl Millet Maiduguri,: Lake Chad Research Institute, Maiduguri, Nigeria, April 21-24, 1997.

8. Singh, B.B., Tarawali, S.A., 1997. Cowpea and its improvement: key to sustainable mixed crop/livestock farming systems in West Africa. In: *Crop Residues in Sustainable Mixed Crop Livestock Farming Systems*. Renard, C. (Ed.). pp. 79-100. CAB International in Association with ICRISAT and ILRI, Wallingford, UK,

9. Mathur,B.K., Patel A.K, Bohra H.C., Mathur A.C and Kaushish S.K. 2005. Acceptability, palatability and utilization of clusterbean and dewbean fodder v/s chaffed buffel grass in sheep. *Journal of Arid Legumes* 2(1):39-42

10. Bressani, R. 1985. Nutritive value of cowpea. In: *Cowpea Research, Production and Utilization*. pp. 353-359. Singh. S.R., Rachie, K.O. (Eds.), Wiley. Winchester, UK,.

11. Nielsen, S.S., Ohler, T.A., Mitchell, C.A. 1997. Cowpea leaves in for human consumption: production, utilization and nutrient composition. In: *Advance in Cowpea Research* pp. 326-332. Singh, B.B. Mohan Raj, D.R., Dashiell, K.E., Jackai L.E.N. (Eds) Co-Publication of International Institute of Tropical Agricultural and Japan International Research for Agricultural Science IITA Ibadan, Nigeria.

12. Tarawali, S.A., Singh, B.B., Pelers M. and Bladc, S.F. 1997. Cowpea haulms as fodder. In: *Advances in Cowpea Research*. Singh B.B., Mohan Raj, D.R. Dashiell, K., Jackai, L.E.N. (Eds.). pp. 313-325 Co-publication of International Institute of Tropical Agricultural Sciences and the JIRCAS, IITA, Ibadan, Nigeria.

13. Gomez, K.A. and Gomez, A.A 1984. Statistical procedures for Agricultural research. John Wily and Sons, New Yark.

14. Beringer, H. 1980. Nutritional and environmental effect on yield formation physiological aspects of crop productivity. *International Potash Research Institute,* Wageningen (Netherlands).

15. Singh, S., Singh, H.N., Singhania, R.A and Singh, K. 1999. Response of soil and applied N, P and K on fodder cowpea (*Vigna unguiculata* L. Walp.) in mollisol of U.P.Tarai. *Forage Research* 25 (1): 69

16. Patidar, M., Saxena, A. and Siyak, S.R. 2003. Effect of nitrogen fertilization on fodder cowpea grown in association with kheri in arid zone. *Forage Research.* **28** (4): 201-203.

17. Patidar, M., Mathur, B.K., Rajora, M..P. and Mathur, D. 2008. Effect of grass – legume strip cropping and fertility levels on yield and quality of fodder in silvipastoral system under hot arid condition. *Indian Journal of Agricultural Sciences*.78 (5):394-398.

Legumes in rainfed organic farming: Profitable success with nutritional security

Arun K. Sharma

Central Arid Zone Research Institute, Jodhpur-342 003

E-mail:arun.k_Sharma@yahoo.co.in

ABSTRACT

Organic farming is a system of production in which optimum use of local resources is ensured for sustainable production .In rainfed areas due to climatic uncertainties, use of external inputs is highly risky and in such situation organic farming can be a viable option. Legumes and organic farming both are not only complementary to each other but also their integration can make rainfed farming profitable. Possibilities on all these aspects of sustainable as well as profitable rainfed farming were discussed in this paper. Organic farming is considered to be one of the most feasible and sustainable farming system for rainfed areas because it utilizes efficiently, the limited resources of these areas. Additionally, in organic farming use of synthetic fertilizers and pesticides is prohibited, therefore, environment remains safe for living beings in these comparatively less polluted regions.

Key words: Organic farming, rainfed areas, legumes.

Organic farming is the most widely recognized sustainable farming system. Modern organic farming evolved as an alternative to chemical agriculture in the 1940s, largely in response to the publications of J.1. Rodale in the U.S., Lady Eve Ballfour in England, and Sir Albert Howard in India.

In 1980, U.S.D.A. released a landmark report on organic farming. The report defined organic farming as:

Organic farming is a production system, which avoids or largely excludes the use of synthetically compounded fertilizers, pesticides, growth regulators, and livestock feed additives. To the maximum extent feasible, organic farming systems rotations, crop residues, animal manures, legumes, green manures, off-farm organic wastes, mechanical cultivation, mineral-bearing rocks. and aspects of biological pest control to maintain soil productivity and tilth, to supply plant nutrients, and to control insects, weeds, and other pests".

In simple words organic farming is system developed with integrated application of eco-friendly technologies to maintain soil fertility and economic level of farmer on a sustainable basis. The use of synthetic chemicals like fertilizers and pesticides - is prohibited in this system.

Possibilities of organic Farming in rainfed areas

During last 4-5 decades, the traditional organic systems in rainfed areas has been adulterated with the use of synthetic inputs, mainly in the cash crops like species, vegetables, fruits etc. Although, this has increased production for 10-15 year but later on not only productivity was going down but also these produce of chemical farming is not preferred by the consumers. Therefore to sustain the production and demand, it is necessary to restore and improve and traditional organic farming, that have immense possibilities in rainfed areas. Some of these exploitable possibilities are-

1. **Efficient use of limited water:** Water is the most scarce resource of these regions. Use of synthetic fertilizers not only increases water demand of crop but also reduces water holding capacity of already light soils of these areas. In these regions at many places ground water combined with fertilizers is being used lavishly for production of rice, wheat, cotton and vegetables. This has resulted in severe depletion of ground water and barren soil. There is strong possibility of desertification of these areas. This type of desertification has already been started is southern Punjab (Bhatinda, Sangroor etc.) and Vidharbh region of Maharashtra. Contrary to this chemical intensive farming, it has been found by experiments and experience that use of organic manure increases soil water holding capacity and crop water use efficiency, resulted in decrease in number of irrigation by 2-4 times in food crops. This water use is further economize by growing low water demanding crops like pulses spices and medicinal plants. By this way there is great possibilities of saving precious water without decreasing yield and income.
2. **Low fertilizer use-early conversion** : 89% of the agricultural area comes under rainfed and the rainfall pattern is erratic. Due to this rate of fertilizer application is very low (36.4kg/ha) as compared to national average of 76.8 kg/ha (1). According to the priority areas of National

organic farming policy about 90% part of the rainfed areas comes under the priority I and II (2). Therefore, early conversion of these areas into organic in possible.

3. **Diversified farming system** : Farming systems in the region are highly diversified in nature with crops, trees, animals, grasses etc (3). This system is scientifically efficient in nutrient recycling and restoration of soil fertility. In these systems 10-30 trees/ha are available and 2-5 animals are reared by a farm family. This integrated farming system minimizes pest incidence as well as favours organic farming.
4. **Rich traditional wisdom** : Rich traditional wisdom in these areas for restoration of soil fertility and for pest control, further strengthen and provide strong infrastructure for organic system(2).
5. **Natural Availability of inputs**: Plants like neem, pongamia, calotropis etc. are the best sources of biopesticides, are abundantly available in these areas. Minerals like rock phosphate, gypsum and lime are available in large quantity. These are good soil ameliorator as well as good nutrients supplier. Further the farming systems are dominated by animals. Waste and product of huge animal population can be a best source of balanced nutrient supply..
6. **Employment opportunities** : High density as well as high growth of human resource remains unutilized through out the year due to erratic rainfall and limited irrigation facilities. Migration of human resources during drought imbalances the development of the region. Since the organic farming is labour intensive and input supply made at local level, there is ample opportunity for employment and proper utilization of human resource.
7. **Soil improvement** Soils are poor in water holding capacity and deficient in most of the essential nutrients. Addition of organic matter not only improves the water holding capacity but also make the soil to supply nutrients in balanced manner.

Therefore there are ample opportunities in rainfed areas for promotion of organic farming.

PROFITABLE SUCCESS WITH ORGANIC FARMING

With the use of external inputs like fertilizers, pesticides etc. crop production can be taken profitably for some time but afterwards due to decrease in soil fertility, increasing cost of production due to use of higher amount of external inputs year after year, the crop production does not remains profitable.

However, in Organic farming as system is developed by recycling and efficient utilization of local resources, the system remains profitable forever.

Efficient recycling of resources is possible with

1. Harvesting and efficient utilization of rainwater
2. Use of locally improved seed
3. Maintaining diversity by crop rotation, crop combination etc.
4. Recycling of nutrients by composting, mulching etc.

Legumes help a lot in resource recycling by adding nitrogen and organic matter in the soil. Legumes also support to a system that can be developed without use of fertilizers, pesticides and weedicides by-

(A) Nitrogen fixation as leguminous crop

(B) Weed suppression as cover crop

(C) Pest repellence as barrier crop

Cluster bean : Best fit in rainfed organic farming

Cluster bean is an appropriate example to show multifarious benefits of legume to rainfed farming by the following ways:

Soil fertility improvement

- Nitrogen fixation (5)
- Phosphorus mobilization (6)
- Roots encourage microbial activity (5)
- Can be used as green manure
- Leaf fall as mulch
- Good effect on companion/succeeding crops

Pest control

- Pest repellent for companion crop

Feasibility for arid zone

- Utilize erratic rainfall (7)
- Uniform maturity period/least seed shattering
- No damage by birds
- Least liking by blue bull

- Crop and seed residues (after gum extraction) are good feed for animals

Economics

- Last external inputs required
- Good export value of gum
- Every part of plant having economic value

Social acceptance

- Very well adopted in traditional farming and food system

LEGUMES AND ORGANIC FARMING

Legumes and organic farming are complimentary to each other, two types of complimentarity exist between two, that may also be better for promoting both legumes and organic farming. These are:

(A) Legume supplies nitrogen to organic farming system. Since nitrogen supply is the major concern in organic farming and legumes either as crop or green manure provide nitrogen in sufficient quantity that makes success of organic farming. Therefore, legumes need to be integral part of organic farming. On the other side organic manure is the major input in organic farming and this manure can supply most of the nutrients and conditions for symbiotic nitrogen fixation. by this way a complimentary relation developed.

(B) Another type of complementarity between legumes and organic farming can be possible in term of resource arrangement to each other. On one side leaf fall in legumes, first act as mulch and later become manure. While on the other side better moisture retention in soil with manure application leads to successful legume production under rainfed conditions (7).

With the above two types of complimentary relations between legumes and organic farming it becomes clear that promotion of legumes and organic farming simultaneously, is having immense possibility to increase sustainability and profitability

IMPROVEMENT WITH ORGANIC INPUT IN ARID ZONE FARMING

Under organic system, besides, use of improved manure, several other eco-technologies like water harvesting, mulching, improved local crop species, use of beneficial microorganisms were used and studied for improvement in the system. Planting of local tree species was done in rows to facilitate crop operations. Neem cake was applied in soil for termite (the major soil born pest) control.

After two year of experiments the results are very encouraging. Application of manure significantly improved the moisture retention capacity of soil which not only made possible to survive crop during 50 days long dry spell but also' given yield, while crop without manure was failed. Application of organic manure 5 t/ha resulted in 48% increase in clusterbean seed yield at medium depth soil as compared to without manuring, Spray of biopesticide made with cow urine+ neem leaves was found effective for controlling pest. Similarly use of neem cake for termite control in crop was found effective. 'Therefore with these results it can be inferred that integrated use of modern eco technologies can enhance the productivity of traditional default organic system.

Suggestions for promotion of arid legumes and organic farming in rainfed areas

Research

Multiple role of legumes need to be integratedly studied e.g. nitrogen availability + soil life + pest and weed control + economize water use to generate results having practical use.

Policy

- Arid legumes as green vegetable (cluster bean, cow pea etc.) may be included in all organic farming promoting programs e.g. National Horticulture Mission.
- Cluster village of watershed may be most appropriate for promoting organic farming and arid legumes and therefore, it may be made amendatory practice in all the watershed area.
- Two very useful arid legume tree species i.e. *Prosopis cineraria* (Khejri) and *Acacia senegal* (Kumat) may be included in arid legume Program to make a holistic approach for arid farming system.
- Integrated promotion of all proven eco-technologies e.g. legumes in rotation, use of compost, green manuring, use of bio-pesticides, man-ual weeding etc. may be done. This will make platform/ infrastructure for smooth shifting to organic farming.

REFERENCES

1. Fertiliser Association of India.1998. Fertilizer statistics .1997-98.New Delhi. pp11.

2. Gauri, P.V.S.M. 2005. National Programme for organic production. In : Proceedings of national seminar on national policy on promoting organic farming. NCOF, Gaziabad. pp 44-48.

3. Shankarnarayan K.A., Harsh L.H and Kathju, S. 1987. Agroforestry in the arid zones of India. *Agroforestry System*, 5; 69-88.

4. Sharma,A.K. and Goyal R.K.2000. Additional in tradition on agroforestry in arid zone. *LEISA- INDIA*,2(3);19-20.

5. Rao, A.V. 2002. Microbiology: In: Guar in India (Eds. D. Kumar and N.B. Singh). Scientific Publishers, Jodhpur. pp 105-125.

6. Tarafdar, J.C., Yadav, R.S. and Sharma, B.M. 2005. Role of arid legumes in P obilization for plant unable sources. J. of Arid Legumes, 2(1); 20-24.

7. Joshi, N.L. 2005. Sustaining productivity of arid legumes through agronomic approaches in rainfed situations. *J.Arid Legumes*, 2(1); 162-170.

Improving productivity and profitability of clusterbean (*Cyamopsis tetragonoloba* (L.) Taub) + sesame (*Sesamum indicum* L.) intercropping system with optimum row ratio & balanced fertilization under arid region of Gujarat

Samrath Lal Meena*, Shamsudheen, M. and Devi Dayal

Central Arid Zone Research Institute, Regional Research Station,
Kukma, Bhuj (Gujarat) – 370 105
E-mail: dr_slmeena@yahoo.co.in (Corresponding author*)

ABSTRACT

A three year field study was conducted during rainy (*kharif*) season of 2005, 06 & 07 at the research farm, central Arid Zone Research Institute, Regional Research Station, Kukma-Bhuj to find out possibility of increasing production of pulses and oilseeds through intercropping system in relation to spatial arrangement of crops and integrated nutrient supply. The soil of the experimental site was gravelly-sandy loam with shallow depth, EC from 2 to 6.38 dS/m and pH from 8.4 to 9.2 with 0.25% organic carbon, 7.35 kg P_2O_5 and 215 kg K_2O ha^{-1}. Treatments consisting of 15 combinations of cropping systems, viz. sole clusterbean, sole sesame, clusterbean + sesame in 1:2, 1:1 and 2:1 row proportion and nutrient management i.e. control, 40 kg N/ha and 20 kg N + 5 t FYM (farmyard manure)/ha applied to the crops were tested in randomized block design and replicated thrice. The grain yield of clusterbean was significantly higher under sole cropping (710 kg/ha) compared to that of intercropping (495.1 kg/ha). However, clusterbean-equivalent yield (919 kg/ha), net returns (Rs 7440/- per ha) and benefit: cost ratio (1.80) were higher with clusterbean + sesame (2:1) intercropping system over sole crop of clusterbean (710 kg/ ha, Rs 5945/- per ha and 1.68, respectively). Irrespective of cropping systems, application of 20 kg N/ha + 5 tonne FYM/ha recorded significantly higher clusterbean-equivalent yield (1036 kg/ha), net return (Rs. 11171/- per ha,) & benefit: cost ratio (2.14) over 40 kg N/ha alone and absolute control. Addition of 5 t FYM/ha along with 20 kg N/ha resulted in 8.5 and 9.8% higher uptake of N over 40 kg N/ha and control, respectively. Intercropping of clusterbean with sesame in the row ratio of 2:1 fertilized with 20 kg N + 5 tonne FYM/ha may be recommended to get more economic benefit and sustained yield in arid region of Gujarat.

Key words: Clusterbean and sesame inter cropping, row rates, fertilizer.

Introduction

Clusterbean is a major rainfed crop of arid zone, mostly grown as a mixed crop with pearl millet, mothbean and sesame but the productivity is low (1). Plant population and spatial arrangement in intercropping have important effects on the balance of competition between component crops and their productivity. Intercropping of oilseed and pulse crops is one of the ways to increase their production because intercropping is more advantageous than sole cropping of either oilseeds or pulses (2).

The greatest limitation of increasing in productivity of these crops is inadequate supply of nutrients since the soils of arid region are poor in native fertility (3). The continuous application of inorganic fertilizers even in balanced form may not sustain soil fertility and productivity. However, judicious use of chemical fertilizers in combination with organic manures is required to improve the soil health as well as to achieve sustainable crop production (4). Farmyard manure (FYM) is the proven source of nutrition to agricultural crops. Thus, balance fertilization along with sound crop husbandry offers a great scope for increasing productivity. A lot of information is available on cereal + legume intercropping but it lacks for legume + oilseed system. The information on comparative performance, nutrient mana-gement, competition relations and sustainability of the system in arid region of Gujarat is lacking. Keeping this in view, an experiment was undertaken to assess yield performance of the crops under different row ratio and nutrient management and evaluate economics of the system for Kachchh region of arid Gujarat.

Materials and Methods

A field experiment was laid out at the research farm, Central Arid Zone Research Institute, Regional Research Station, Kukma, Bhuj, Kachchh, Gujarat during rainy (*kharif*) season of 2005 to 2007, that is located between 23°12' to 23° 13' N latitude and 69°47' to 69°48' E longitude. The soil of the experimental site was gravelly sand to loamy sand in texture with shallow depth (21 cm). The soil was low in organic C (0.25%), available N (121 kg/ha^{-1}), available P (7.35 P_2O_5 mg/kg soil), and medium in available K (215 kg K_2/ha^{-1}) with pH 8.6. The region is characterized by low and erratic rainfall, high temperature, high wind velocity and high potential evapotranspiration. A total rainfall received during the cropping period (June to October) was 238.2, 689.4 and 701.6 mm in 2005,

2006 and 2007, respectively, compared with the average annual rainfall of 340 mm. In the first year (2005) of experimentation the yield of both the crops were adversely affected due to low and erratic rainfall. Cropping season of 2006 and 2007 was quite normal for the growth of both the crops and comparatively higher yields were obtained owing to adequate and well distribution of rainfall. However, the yield of both the crops was slightly low in 2007 as compared with 2006 because of continuous rainfall during maturity stage. Treatments consisting of 15 combinations of cropping systems, viz. sole clusterbean, sole sesame, clusterbean + sesame in 1:2, 1:1 and 2:1 row proportions and nutrient management i.e. control, 40 kg N/ha and 20 kg N + 5 t FYM/ha applied to the crops, were tested in randomized block design and replicated thrice. The net plot size of 4.0 x 4.5 m and spacing of 45 x 15 cm were adopted for both the crops and were sown in the last week of June in all the years of experimentation. A common recommended dose of phosphorus and potash was applied to the crops at the time of sowing. The cultivars used in the study were 'RGC–936' (clusterbean) and 'GUJ–1' (sesame). The remaining agronomic practices were followed as per recommendations for the region. The economics was calculated on the basis of prevailing market prices of different inputs and outputs. The clusterbean equivalent yield was calculated by converting the seed yield of sesame into clusterbean yield on the basis of existing market price of the crops. The yields were further used for computation of land-equivalent ratio (LER), as suggested by Willey (5). Sustainable yield index (SYI) and sustainable value index (SVI) were calculated as per the procedure described by Singh *et al.* (6). Chemical analysis for estimation of N was done by micro Kjeldahl method.

Results and Discussion

Growth and yield

Clusterbean

Sole clusterbean recorded significantly lower plant height and yield attributing characters than that recorded in its intercropping with sesame. Plant height and yield attributing parameters of clusterbean i.e. pods/plant, seeds/pod, 1000-seed weight and harvest index were significantly higher under intercropping system than sole cropping (Table 1). Among the intercropping treatments clusterbean with sesame under 2:1 row ratio recorded higher plant height, yield attributes and seed yield (Table 2) of clusterbean than other row ratios of 1:1 and 1:2 in all three 3 years of experiment and could recover maximum (78%, on mean basis) of its sole crop seed yield, owing to higher plant density of main crop and complementary effect of intercrop. Similar trend of results, on chickpea based intercropping systems, was observed by Ahlawat *et al.* (7). Application of 20 kg N/ha along with FYM 5 t/ha recorded signifi-cantly higher plant height, number of pods/plant, seeds/pod, 1000-seeds weight, harvest index (Table 1) and seed yield (589) of clusterbean and was 33% more over control (Table 2). This might be due to increased availability of nutrients to the crop plant through combined use of organic and inorganic sources (8).

Sesame

Plant height and yield attributing parameters of sesame were better under intercropping system than sole stand but sesame crop produced higher seed yield of 215 kg/ha (mean basis) in sole stand. The lower seed yield in inter-cropping systems might be due to lesser number of plants per unit area. Among intercropping systems, plant height, capsules/plant, seeds/capsule, 1000-seed weight and harvest index (Table 1), and seed yield (Table 2) of sesame were significantly higher in 1:2 row ratio than other ratios. This could be attributed to the variations in the magnitude of competition among component crops grown in various proportions. The results are conformity with the findings of Sarkar *et al.* (9). The integrated nutrient supply system of farmyard manure at 5 tonne with 20 kg N/ha favourably improved plant height, yield attributes, harvest index and seed yield of sesame and produced the highest seed yield which was 16 and 24% higher over application of 40 kg N/ha and control, respectively. This may be ascribed to the better macro and micronutrient availability as well as physical condition of the soil. These findings support the results of Singh (10) and Imayavaramban *et al.* (11).

Clusterbean equivalent yield

All the intercropping systems, irrespective of the row ratios had more system productivity in terms of clusterbean equivalent yield compared to sole clusterbean (Table 2). Higher clusterbean equivalent yield under various row patterns was due to additional yield of sesame and better utilization of natural resources, viz., light, space, nutrients, etc. than sole stand. Among different intercropping patterns, 2:1 row ratio gave maximum clusterbean equivalent yield of 919 kg/ha which was 29% higher over sole cropping. The differential behavior in clusterbean equivalent yield was on account of productivity of crops in intercropping systems and their relative market prices (12). This revealed a greater degree of efficiency and compatibility of sesame in clusterbean. The fertilizer application (20 kg N + 5 t FYM/ha) to component crops improved the equivalent yields of system due to proper nourishment and less competition for nutrients.

Table 1. Effect of row ratio and nutrient management on plant height, yield attributes and harvest index of clusterbean and sesame (mean of 3 years)

Treatment	Plant height (cm)		Pods or capsules/ plant		Seeds/pod or capsule		1000-seed weight (g)		Harvest index (%)	
	Cluster bean	Sesame	Cluster bean	Sesame	Cluster bean	Sesame	Cluster bean	Sesame	Cluster bean	Sesame
Cropping system										
Sole cropping	54.4	35.3	31.1	10.7	5.5	31.2	27.4	2.0	24.29	17.38
Clusterbean+sesame (1:2)	60.3	45.9	38.5	17.8	6.0	34.4	29.2	2.1	26.13	19.00
Clusterbean+sesame (1:1)	63.5	43.3	42.1	16.0	6.4	35.5	29.1	2.2	26.67	18.27
Clusterbean+sesame (2:1)	67.3	42.2	47.6	15.2	6.6	35.8	30.7	2.2	27.12	18.03
SEm ±	1.2	0.5	1.4	0.5	0.2	1.4	0.4	0.1	0.13	0.21
CD (P= 0.05)	3.5	1.9	4.7	1.3	0.7	4.1	1.2	NS	0.38	0.67
Nutrient management										
Control	52.4	38.3	32.6	12.8	6.2	33.3	28.5	2.0	24.65	17.96
40 kg N/ha	69.3	49.0	55.4	22.8	6.7	39.6	31.3	2.2	28.31	19.48
20 kg N + 5 t FYM/ha	75.5	52.3	64.0	26.3	7.3	45.4	33.4	2.3	28.47	20.14
SEm ±	1.5	1.0	2.1	0.8	0.3	1.6	0.6	0.9	0.06	0.18
CD (P= 0.05)	4.6	2.8	6.4	2.7	0.9	4.7	1.9	NS	0.19	0.54

Table 2. Effect of row ratio and nutrient management on seed yield, clusterbean equivalent yield, land equivalent ratio, N uptake, economics and sustainability of clusterbean and sesame intercropping (mean of 3 years)

Treatment	Seed yield (kg/ha)		CEY (kg/ ha)	LER	Total N uptake (kg/ha)			Net returns (Rs/ha)	B: C ratio	Sustainability	
	Cluster-bean	Ses-ame			Cluster-bean	Ses-ame	Comb-ined			SYI	SVI
Cropping system											
Sole cropping	710	215	710	1.00	55.1	22.2		5945	1.68	0.51	0.52
Clusterbean + sesame (1:2)	433	156	867	1.32	32.6	14.8	47.0	7172	1.77	0.73	0.73
Clusterbean + sesame (1:1)	493	126	879	1.33	36.0	13.6	49.5	6918	1.75	0.72	0.72
Clusterbean + sesame (2:1)	589	102	919	1.38	39.7	13.0	52.7	7440	1.80	0.74	0.76
SEm ±	26	7	11		1.1	0.4	0.9	143	0.01		

CD (P= 0.05)	78	21	33		3.7	1.1	3.0	431	0.04		
Nutrient management											
Control	445	124	799	1.21	34.0	12.7	46.7	6479	1.74	0.75	0.75
40 kg N/ha	579	153	976	1.48	39.4	13.9	53.3	9131	1.93	0.80	0.81
20 kg N + 5t FYM/ha	644	173	1036	1.56	41.0	14.8	55.7	11171	2.14	0.81	0.82
SEm ±	18	6	19		0.4	0.3	0.7	315	0.02		
CD (P= 0.05)	56	18	57		1.4	0.8	2.1	948	0.07		

Land equivalent ratio

All the intercropping systems had land-equivalent ratio greater than unity (Table 2). The maximum land-equivalent ratio was recorded under 2:1 row ratio followed by 1:1 and 1:2 ratios. Application of 20 kg N + 5 t FYM/ha also recorded the highest land-equivalent ratio (1.56) among the nutrient management practices.

Total N uptake

Sole clusterbean crop removed maximum quantity of N (55.0 kg/ha) which was at par with intercropping in 2:1 ratio (Table 2). Among the intercropping systems, 2:1 ratio removed significantly highest quantity of N than other ratios. Addition of 5 t FYM/ha along with 20 kg N/ha resulted in 8.5 and 9.8% higher uptake of N over 40 kg N/ha and control, respectively. This might be due to additional amount of nutrients supplied by farmyard manure as well as the beneficial effects of organic matter addition, which were derived in connection with the physical and chemical properties of the soil (11).

Economics

All the intercropping systems were more remunerative than the sole clusterbean with higher net monetary returns and benefit: cost ratio. Among the intercropping cropping systems, planting of clusterbean + sesame (2:1) row proportion appeared more net return of Rs. 7440 per ha over sole cropping of clusterbean (Table 2) with B: C ratio of 1.80. The highest net return (Rs. 11171 per ha) and B: C ratio (2.14) over 40 kg N/ha alone (Rs. 9131 per ha, 1.93) and absolute control (Rs. 6479 per ha, 1.74), respectively, were accrued with the application of 20 kg N + 5 t FYM/ha. This is close conformity with the findings of Kumar (12) and Singh (10) under different intercropping systems with nutrient management. This system was economically viable intercropping with the higher net returns.

Sustainability

The sustainable yield index (SYI) and sustainable value index (SVI) were higher with clusterbean + sesame under 2:1 (0.74, 0.76) row ratio (Table 2). This was followed by 1:2 (0.73, 0.73) and 1:1 (0.72, 0.72), indicating that the intercropping system of 2:1 was more stable than the others. Bastia *et al.* (2008) also reported higher sustainability of rice based cropping systems with different oilseeds and pulses. The highest SYI (0.81) and SVI (0.82) indices were observed under the application of 20 kg N/ha with 5 t FYM/ha followed by 40 kg N/ha alone (0.80, 0.81) and absolute control (0.75, 0.75), respectively.

Thus it was concluded that an intercropping of clusterbean with sesame in the row ratio of 2:1 may be followed for fetching higher yield and monetary advantages and better utilization of resources in arid region of Gujarat. Application of 20 kg N with 5 tonne farmyard manure/ha may be recommended for the system as an alternative of 40 kg N/ha dose to reduce the chemical fertilizer as well as to get more economic benefit and sustained yield.

References

1. Faroda, A.S., Joshi, N.L, Singh, Raj and Sexena, Anurag. 2007. Resource management for sustainable crop production in arid zone – A review. *Indian Journal of Agronomy* **52**(3): 181-193.
2. Padhi, A.K. and Panigrahi, R.K. 2006. Effect of intercropping and crop geometry on productivity, economics, energetic and soil-fertility status of maize (*Zea mays*)-based intercropping systems. *Indian Journal of Agronomy* **51**(3): 174-177.
3. Singh, Raj and Khan, M.A. 2003. Response of clusterbean varieties to fertility levels and cropping systems under arid conditions. In: *Advances in Arid Legume Research*, pp. 225-228. Henry, A., Kumar, D. and Singh, N.B. (Eds.). Scientific Publishers and Indian Society of Arid Legumes, Jodhpur.
4. Mankotia, B.S. 2007. Effect of fertilizer application with farmyard manure and *in-situ* green manures in standing rice (*Oryza sativa*) – wheat (*Triticum aestivum*) cropping system. *Indian Journal of Agricultural Sciences* **77**(8): 512-515.
5. Willey, R.W. 1979. Intercropping – its importance and research needs. Part 1. Competition and yield advantages. *Field Crops Abstracts* **32**(1): 1-10.
6. Singh, R.P., Das, S.K., Bhaskar Rao, V.M. and Narayan Reddy, M. 1990. *Towards Sustainable Dryland Agricultural Practices,* pp.5-8. Central Research Institute for Dryland Agriculture, Hyderabad.
7. Ahlawat, I.P.S., Gangaiah, B. and Singh, Ompal. 2005. Production potential of chickpea (*Cicer arietinum*)-based intercropping systems under irrigated conditions. *Indian Journal of Agronomy* **50**(1): 27-30.
8. Sarkar, R.K., Goswami, S. and Pal, P.K. 2003. Production potential and economic feasibility sunflower (*Helianthus annuus*) and pigeonpea (*Cajanus cajan*)

intercropping system on rainfed upland condition. *Indian Journal of Agronomy* **40**(4): 263-266

9. Nambiar, K.K.M. 1994. Soil fertility and crop productivity under long-term fertilizer use in India. ICAR. New Delhi.
10. Singh, R. 2002. Effect of integrated nutrient management in cluster bean-wheat cropping system under western Rajasthan condition. *Indian Journal of Agronomy* **47**(1): 41-45.
11. Imayavaramban, V., Thanunathan, K., Singhravel, R. and Manikam, G. 2002. Studies on the influence of integrated nutrient management on growth, yield parameters and seed yield of sesame. *Crop-Research Hisar* **24**(2): 309-313.
12. Kumar, S. 2002. Effect of planting pattern and fertilizer management on castor-based intercropping system. *Indian Journal of Agronomy* **47**(3): 355-360.

Response of moth bean [*Vigna aconitifolia* (Jacq.) Marechal] to *Chaetomium globosum*

B. K. Yadav, Ram Niwas and J. C. Tarafdar

Central Arid Zone Research Institute, Jodhpur-342 003 (Rajasthan)

ABSTRACT

Chaetomium globosum was found to be an efficient phosphatase and phytase releasing organisms under arid environment to exploit native phosphorus from soils. The present paper describes the effect of *Chaetomium globosum* to moth bean in a sandy soil under higher organic matter level. To determine the effect of *Chaetomium globosum* on higher organic matter condition, we applied vermi-compost to a sandy soil to maintain 3% organic matter level under green house condition. There were three treatments viz. control, organic matter (3%) with or without *Chaetomium globosum*. The sampling was done at 28, 42, 56 and 63 (crop harvest) days after germination.

The pH of the soil decreases up to 4.0% and 5.8% with organic matter and *Chaetomium globosum* treatment respectively, as compared to control. The electrical conductivity was significantly increased up to 8th weeks with both treatments. Significantly increase in acid phosphatase ($r^2 = 0.73$, n = 36, p = 0.01), alkaline phosphatase ($r^2 = 0.93$, n = 36, p = 0.01) and phytase ($r^2 = 0.99$, n = 36, p = 0.01) activities were observed due to combined effect of organic matter and *Chaetomium globosum*, which shows organic matter and *Chaetomium globosum* helps to increase in the level of soil enzyme activities due to favorable condition in plant rhizosphere. The nutrient (N, P and K) content of plants increased (13% N, 115% P and 11% K) in organic matter treatment, which was further enhanced due to *Chaetomium globosum* up to 5.7% for N, 39% for P and 3.9% for K. The straw and seed yield of moth bean also increased by 10.7% and 14% and respectively, due to *Chaetomium globosum* as compared to organic matter. The results clearly showed that efficiency of *Chaetomium globosum* has been enhanced in presence of organic matter, which ultimately helped in greater nutrient mobilization from soil and increased in nutrient concentration as well as moth bean yield.

Key words: Organic manure, *Chaetomium globosum*, acid and alkaline phosphatase, phytase, nutrient content, moth bean.

INTRODUCTION

Moth bean [*Vigna aconitifolia* (Jacq.) Marechal] is one of the important Indian legume grown mostly in arid and semi-arid areas. It covers about 12 lakhs hectare area in Western Rajasthan. The crop is highly drought resistant and well adopted to sandy soil. The grain yield of the crop in arid soils is low because of low and erratic rainfall distribution and poor soil conditions besides the low availability of various nutrients especially phosphorus. Application of organic matter, for improving the availability of nutrients for plants and physical properties of soil, is a conventional practice in Indian agriculture. A large proportion of P that is applied to soil as fertilizer rapidly becomes unavailable to plants, accumulating in inorganic P fractions that are fixed by chemical adsorption and precipitation, and organic P fractions that are immobilized in soil organic matter [1]. About 20- 30% of total phosphorus in the arid soils of Western Rajasthan of India is organic in nature [2 and 3] and 68% organic phosphorus in the soil was present as phytin [3 and 4]. In order to become available to plants, these P compounds must be hydrolyzed by phosphatases, which are of plant and microbial origin. The ability of soil fungi to solubilize various forms of insoluble P is well known [3 and 4]. *Chaetomium globosum* was found to be an efficient phosphatase and phytase releasing organi-sms under arid environment to exploit native phosphorus from soils [4]. The present paper describes the effect of *Chaetomium globosum* to moth bean in a sandy soil under native as well as higher organic matter level.

MATERIAL AND METHODS

The soil used under the study was loamy sand (hypothermic typic haplocambids) collected from Research Farm of Central Arid Zone Research Institute, Jodhpur (Latitude 26°18'N, Longitude 73°01′ E). The characteristics of the experimental soil are: sand-85.5%, silt-9.5%, clay-4.5%, pH-8.2, EC-0.18 dSm^{-1}, organic matter 0.18%, total P- 1261 mg kg^{-1}, organic P- 367 mg kg^{-1}, available P- 8.3 mg kg^{-1} . The experiment was carried out during *kharif* 2006 with 3 treatments: (T_1) – control (0.2% organic matter); (T_2) - 3% organic matter; (T_3) -3% organic matter + *Chaetomium globosum* . The organic matter gradient was established by application of vermicompost (contain 14% organic matter) @ 1.200 kg pot^{-1} for 3% organic matter gradient. The organic matter build up in pots were checked once in 15 days and the experiment was started when it's reached to a constant value. Twelve pots (5 kg capacity) were used for each treatment with three replicates. Moth bean [*Vigna aconitifolia* (jacq.) Marechal cv, RMO-40) was inoculated by the method as described by Yadav and Tarafdar [3].Ten seed were shown in each pot; four plants were maintained after germination. The water was added at 50% water holding capacity (WHC) of the soil and watered every second day. The pots were completely randomized and re-positioned weekly to minimize any effect of uneven environmental factors. The sampling was done at 28, 42, 56 and 63 (crop harvest) days after germination.

Total, mineral and organic P was estimated as described by Seeling and Jungk [5]. The phytin P was estimated by extraction of phytate with 15% CCl_3-COOH (trichloro-acetic acid) as described by Mega [6]. Following the standard methods [7] estimations of organic carbon, pH (1:2), EC (1:2), partical size distribution and available P (Olsen's) was determined.

Acid and Alkaline phosphatase was assayed by adopting the standard method of Tabatabai and Bremnar [8] using acetate buffer (pH 5.4) and sodium tetra borate-NaOH buffer (pH 9.4), respectively. Phytase activity was assayed by measuring the amount of inorganic phosphate (Pi) released by hydrolysis of sodium phytate (1mM) in 100M sodium acetate buffer (pH 4.5) and incubated at 37°C for 1h [9].

After each harvest plants were thoroughly washed free of soil, were dried at 60°C and weight of dry matter was recorded. Plant N, P and K contents were analyzed by standard procedures [7].

Microsoft Excel 2000 was used in statistical processing of the data (standard error and correlation analysis).

RESULTS AND DISCUSSION

The pH of the soil decreases with organic matter and crop age which varies between 3.8 to 4.7% with 3% organic matter (Table 1). *Chaetomium globosum* enhanced decrease in soil reaction which was 5.4 to 7.0% under 3% organic matter respectively as compared to control soil. In general 2.0% more decline in pH was observed due to inoculation of *Chaetomium globosum*. The initial decrease in pH in presence of organic matter may be due to higher humic and fulvic acid in vermicompost [10]. *Chaetomium globosum* further reduced soil pH due to the release of different organic acids, as the production of organic acid such as malate, citrate and oxalate by different fungi are well known [11, 12, 13, 14 and 15]

Application of organic matter positively influenced soil EC (Table 1) which increased between 7% to 47% due to organic matter and 6% to 53% when *Chaetomium globosum* was inoculated in higher organic matter soil. This may be due to the production of CO_2 during decomposition of organic matter, which lowers the pH and dissolution of minerals and release of ions such as Ca^{++}, Mg^{++} and Na^{+} in soil solution [16]. *Chaetomium globosum* further enhanced EC due to release of organic acids which improves decrease in pH and release of ions in soil solution.

Table 1. Effect of organic matter and *Chaetomium globosum* on soil reaction and electrical conductivity (± represent standard errors of the mean)

Treatments	pH				EC (dSm^{-1})			
	Days after germination				Days after germination			
	28	42	56	63	28	42	56	63
Control (0.2% organic matter)	8.05 ± .39	7.99 ± .40	8.04 ± .38	8.03 ± .41	0.14±.002	0.15±.004	0.15±.006	0.15 ± .004
Organic matter (3%)	7.73 ± .37	7.68 ± .37	7.67 ± .36	7.65 ± .38	0.16±.003	0.19±.005	0.22±.006	0.16 ± .004
Organic matter (3%) + *Chaetomium globosum*	7.58 ± .35	7.56 ± .36	7.48 ± .34	7.48 ± .34	0.17±.004	0.21±.007	0.23±.004	0.16 ± .006

Significant ($p<0.05$, n= 48) increase in acid phosphatase was observed with increase in organic matter content of arid soil, which was progressively increased with crop age up to 42 days (Table 2). The net increase in activity was observed about 1.2 to 2.3 times with 3% organic matter as compared to control. *Chaetomium globosum* further increased the activity by 1.3 to 2.4 times. In general the increase in acid phosphatase activity was observed 45% with organic matter alone and additional 5% after inoculation of *Chaetomium globosum* as compared to control soil. Alkaline phosphatase activity between 28 to 63 days period in the rhizosphere of moth bean (Table 2) resulted a significantly ($p<0.05$, n= 48) higher activity during crop age as compared to control. The increase in alkaline phosphatase activity was more (7 %) due to inoculation of *Chaetomium globosum* as compared to application of only organic matter. The phytase activity was significantly ($p<0.05$, n= 48) increased with increase in level of organic matter and crop age up to 42 days (Table 2), which ranges between 1.7 to 3.0 times with 3% organic matter. In general increase in phytase activity was observed 113% more in organic matter with *Chaetomium globosum*, whereas only 105% higher activity was observed only organic matter application as compared to control.

Soil enzyme activities increased with increase in organic matter content [10 and 17]. Addition of organic matter stimulated the root growth and activity of microorganisms [18 and 19]. Phosphatase and phytase may release by both plant roots as well as microorganisms [20, 21 and 22). Inoculation of *Chaetomium globosum* significantly increased the enzyme activities because organic matter helps to increase in microbial population and their activity [23, 24 and 25].

Table 2. Different soil enzyme activities of moth bean rhizosphere during growth period

Treatments	Acid phosphatase (EU × 10^{-5})				Alkaline phosphatase (EU × 10^{-5})				Phytase (EU × 10^{-5})			
	Days after germination				Days after germination				Days after germination			
	28	42	56	63	28	42	56	63	28	42	56	63
Control (0.2% organic matter)	12.14	13.89	13.24	5.26	14.81	14.80	13.91	9.70	147.89	190.22	171.03	76.86
Organic matter (3%)	18.90	17.99	16.16	12.49	15.81	15.42	14.27	11.22	351.10	321.93	295.88	235.76
Organic matter (3%) + *Chaetomium globosum*	18.97	19.41	17.79	12.73	17.51	16.53	14.89	11.67	373.77	338.59	298.04	240.66
LSD (p=0.05)	0.05	1.02	0.85	0.04	0.06	0.75	0.48	0.51	21.85	15.40	2.74	4.75

Nitrogen content was increased 12.8% with organic matter alone as compared to control however it was observed 18.9 % more with organic matter when *Chaetomium globosum* was introduced (Table 3), which shows that *Chaetomium globosum* helps more (6%) accumulation of nitrogen in plants. The plant P content was observed 2.2 times and 3.0 times higher with organic matter alone and *Chaetomium globosum* in presence of organic matter respectively as compared to control (Table 3). In general 37% more P accumulation was observed with *Chaetomium globosum* combined with organic matter as compared to organic matter alone. The K accumulation was 1.1 times higher in organic matter and 1.2 times higher in organic matter with *Chaetomium globosum* than control (Table 3). In general, an overall improvement (4.0%) in K content due to *Chaetomium globosum* was observed.

Table 3. Nutrient content of moth bean at different plant growth stage

Treatments	Nitrogen concentration (%)				Phosphorus concentration (%)				Potassium concentration (%)			
	Days after germination				Days after germination				Days after germination			
	28	42	56	63	28	42	56	63	28	42	56	63
Control (0.2% organic matter)	0.87	1.00	0.97	0.91	0.11	0.12	0.14	0.15	0.015	0.016	0.016	0.013
Organic matter (3%)	0.95	1.10	1.12	1.06	0.26	0.28	0.35	0.20	0.018	0.018	0.017	0.015
Organic matter (3%) + *Chaetomium globosum*	1.04	1.15	1.19	1.08	0.38	0.41	0.43	0.36	0.019	0.019	0.017	0.016
LSD (p=0.05)	0.04	0.06	0.07	0.05	0.05	0.07	0.08	0.05	0.001	0.001	0.001	0.001

Increased N content with organic matter could be assigned to supplementation of soil reservoir on mineralization of organic nitrogen of organic matter as well as enhanced microbial activity of ammonifiers and nitrifiers in particular, due to available organic carbon [18]. Content of P is synergistic to content of N and higher content in clusterbean could be due to solubilization effect of organic acids produced during decomposition of organic matter, improved aeration and better root proliferation [19]. Application of organic matter also supplied K and solubilized K from K-bearing minerals by the organic acids released and hence increased its availability in soil and content in plants. The beneficial effects of organic matter addition are also related to improvement in soil physical properties [26], which improves the microbial population and root growth of plants after inoculation of *Chaetomium globosum*, may helped in higher N, P and K mobilization from soil and uptake by clusterbean plant.

The seed yield of moth bean was increased significantly ($p<0.05$, $n= 15$) by 8.6% and 24% respectively due to organic matter alone and inoculation of *Chaetomium globosum* in presence of organic matter (Table 4). A significant increase in straw yield ($p<0.05$, $n= 15$) was observed (Table 4) due to organic matter alone and inoculation of *Chaetomium globosum* as compared to control. Increase in dry matter was observed 10% with 3% organic matter, while 22% increase in dry matter was observed after inoculation. In general, *Chaetomium globosum* with organic matter enhanced seed yield by 14% and straw yield by 10.7% as compared to organic matter alone.

Table 4. Grain and straw yield of moth bean

Treatments	Yield g pot^{-1}	
	Grain	Straw
Control (0.2% organic matter)	4.63	11.70
Organic manure (3%)	5.03	12.88
Organic manure (3%) + *Chaetomium globosum*	5.74	14.26
LSD (p=0.05)	0.03	0..89

The increase in yield due to organic matter has associated with the release of macro and micronutrients during the course of microbial decomposition. Organic matter also functions as source of energy for soil microflora which brings about the transformation of inorganic nutrients held in soil that is readily utilized by growing plants. The beneficial response of organic matter to yield might also be attributed to the availability of sufficient amounts of plant nutrients throughout the growth period and especially at critical growth periods of crops resulting in better uptake, plant vigour and superior yield attributes. *Chaetomium globosum* with organic matter would provide all the essential nutrients in proper ratio to plants and also reduce the probabilities of emergence of multiple micronutrient deficiencies in soil and increase yield of clusterbean.

ACKNOWLEDGEMENTS

Indian Council of Agricultural Research under National Fellow Scheme financially supported this research work. Authors are grateful to the Director, Central Arid Zone Research Institute, Jodhpur, for providing necessary facilities for the study. The help rendered by Shri B.N. Sharma for technical support is thankfully acknowledged.

REFERENCES

1. Sanyal, S. K. and De Datta, S. K. 1991. Chemistry of phosphorus transformation in soil. *Advances in Soil Science* 16: 1-120.
2. Tarafdar, J.C., Bala, K. and Rao, A.V. 1989. Phosphatase activity and distribution of phosphorus in arid soil profiles under different land use patterns. *Journal of Arid Environments* 16: 29-34.
3. Yadav, B.K. and Tarafdar, J.C. 2007. Ability of *Emericella rugulosa* to mobilize unavailable P compounds during Pearl millet (*Pennisetum glaucum* L.) crop under arid condition. *Indian Journal of Microbiology* 41: 57-63.
4. Tarafdar, J.C. and Gharu, A. 2005. Mobilization of organic and poorly soluble phosphate by *Chaetomium globosum*. *Applied Soil Ecology* 32: 273-283.
5. Seeling, B. and Jungk, A. 1996. Utilization of organic phosphorus in calcium chloride extracts of soil by barley plants and hydrolysis of acid and alkaline phosphatases. *Plant and Soil* 178: 179-184.
6. Mega, J.A. 1982. Phytate: Its Chemistry, Occurrence, Food interactions nutritional significance and methods of analysis. *Journal of Agriculture and Food Chemistry* 30: 1-9.
7. Jackson, M.L. 1967. Soil Chemical Analysis. Prentice-Hall of India, Delhi, 498 p.
8. Tabatabai, M.A. and Bremner, J.M. 1969. Use of p-nitrophenyl phosphate for assay of soil phosphatase activity. *Soil Biology and Biochemistry* 1: 301-307.
9. Ames, B.N. 1966. Assay of inorganic phosphate, total phosphate and phosphatases. *Method in Enzymology* 8:115-118.
10. Zachariah, A.S. and Chhonkar, P.K. 2004. Biological properties of compost as influenced by earthworm and feed materials. . *Journal of the Indian Society of Soil Science* 52:155-159.
11. Cunningham, J.E. and Kuiack, C. 1992. Production of citric and oxalic acids and solubilization of calcium phosphate by *Penicillium billai*. *Applied and Environmental Microbiology* 52: 1451-1458.

12. Illmer, P. and Schinner, F. 1995. Solubilization of inorganic calciumphosphates-solubilization mechanisms. *Soil Biology and Biochemistry* 27: 257-263.

13. Jones, D.L. 1998. Organic acids in the rhizosphere – a critical review. *Plant and Soil* 205: 25-44.

14. Reyes, I., Bernier, L., Simard, R.R. and Antoun, H. 1999. Effect of nitrogen source on the solubilization of different inorganic phosphates by an isolate of Penicillium rugulosum and two UV-induced mutants. *FEMS Microbial Ecology* 28: 291-295.

15. Tarafdar, J.C., Bareja, M. and Panwar, J. 2003. Efficiency of some phosphatase producing soil-fungi. *Indian Journal of Microbiology* 43: 27-32.

16. Kumar, C., Swarup, A. and Sharma, D.R. 2006. Release of ions from calcium carbonate treated alkaline soil as influenced by organic matter and moisture contents. *Journal of the Indian Society of Soil Science* 54: 179-184.

17. Tiwari, S.C., Tiwari, B.K. and Mishra, A. 1989. Microbial community, enzyme activity and CO_2 evolution in pineapple orchard soil. *Tropical Ecology* 30: 265-273.

18. Shuman, M.L. and Hargrover, W.L. 1985. Effect of tillage on the distribution of manganese, copper,Iron and zinc in soil fractions. *Soil Science Society of Amercian Journal* 49: 1117-1121.

19. Gopal Reddy, B. and Reddy, Suryanarayana 1998. Effect of organic manures and Nitrogen levels in soil available nutrients status in maize- soybean cropping system. *Journal of the Indian Society of Soil Science* 46: 474-476

20. Richardson, A.E., Hadobas, P.A., Hayes, J.E., O' Hara, C.P. and Simpson, R. J. 2001. Utilization of phosphorus and pasture plants supplied with *myo*-inositol hexaphosphate is enhanced by the presence of soil micro-organisms. *Plant and Soil* 229: 47-56.

21. Tarafdar, J.C., Yadav, R.S. and Niwas, R. 2002. Relative efficiency of fungal intra- and extracellular phosphatases and phytase. *Journal of Plant Nutrition and Soil Science* 165: 17-20.

22. Yadav, B.K. and Tarafdar, J.C. 2004. Phytase activity in the rhizosphere of crops, trees and grasses under arid environment. *Journal of Arid Environments* 56: 285-293.

23. Kiss, S., Dragon Bularda, M. and Rudubeson, R. 1975. Biological significance of enzymes accumulated in soil. *Advances in Agronomy* 27: 25-87.

24. Goyel, S., Mishra, M. M., Dhankar, S.S. and Chander, K. 1994. Relationship of Soil microbial biomass and mineralization of Nitrogen following Incorporation of organic materials to soil. *Journal of the Indian Society of Soil Science* 42: 474-476.

25. Ramesh, P.R., Kotur, S.C. and Sudhir, K. 2003. Releationship of activities of urease and dehydrogenase with physico-chemical properties of some coffee- growing soils of Karnataka. *Journal of the Indian Society of Soil Science* 51: 78-80.

26. Nambiar, K.K.M. and Abrol, I.P. 1989. Long- terms fertilizer experiments in India – An overview. *Fertilizer News* 34: 11-20.

Studies on sowing time, row spacing and seed rate in mothbean (*Vigna aconitifolia* (Jacq.) Marechal]

B.D. Yadav, P. P. Gupta, U.N. Joshi and J.S. Yadav*

Forage Section, CCS Haryana Agricultural University, Hisar-125 004 (Haryana), India
*Regional Research Station, Bawal-123 501 (Haryana)

ABSTRACT

A field experiment was conducted during *kharif* season of 2003, 2005, 2006 and 2007 to work out the optimum sowing time, row spacing and seed rate of mothbean. Highest grain yield was obtained when sowing was done in the first (8.09 q/ha) and second week of July (7.42 q/ha). Delay in sowing time lead to decrease in grain yield when sowing was done beyond July (3.73 q/ha). Crude protein content was significantly higher in the grain when sowing was done in the second week of July. Disease incidence increased with the enhancing of sowing time and maximum incidence was observed in early August sowing. Higher grain yield and crude protein content was recorded with wider row spacing of 45 cm. On the other hand there was higher disease intensity and low protein content with closer row spacing of 30 cm. Use of 10 kg seed/ha gave higher grain yield and crude protein content with low disease incidence.

Key words: Mothbean, seed rate, sowing time, disease incidence.

INTRODUCTION

Mothbean (*Vigna aconitifolia* Jacq.) is an important crop of arid and semi-arid regions of North-Western States of India. It is a source of food, fodder, green manuring and used as pasture, hence, serves as a multipurpose crop. Being a legume crop, it has the capacity to six atmospheric nitrogen and increases soil fertility. Mothbean has not been rated as a national pulse as its contribution in total pulse production is hardly 5.9 per cent in area and 1.6 per cent in production, however, it appears to be a major pulse as far as the arid and semi-arid regions of India are concerned [1]. Mothbean is a rainfed crop and distribution of rainfall is erratic; however, it is sown with the onset of monsoon.

Thus a short period is available for sowing of this crop. The crop suffers heavy damages due to various kinds of foliar diseases. Among these, mungbean yellow mosaic virus (MYMV) disease is the most important which causes a yield loss of 75 to 90 per cent [2]. Severity of this disease depends on time of sowing, row spacing and climatic conditions prevailing during crop growth and nature of cultivar used. Optimum plant population plays a paramount role in realizing the yield potential of a crop. To harvest higher grain yield from a particular genotype in addition to sowing time it is imperative to study row spacing and seed rate requirement of mothbean.

MATERIALS AND METHODS

The field experiment was carried out in Dryland Research Area at CCS Haryana Agricultural University, Hisar during *kharif* seasons of 2003, 2005, 2006 and 2007. The soil of the experimental site was sandy loam in texture, low in organic carbon and available nitrogen, medium in available phosphorus and slightly alkaline in reaction. The experiment was laid out in randomized block design with three sowing time (D_1-First soaking rain; D_2-At subsequent rains after a gap of minimum 10 days of first sowing; D_3-At subsequent rains after a gap of minimum 10 days of second sowing) alongwith two row spacing (30 and 45 cm) and two seed rates (10 and 15 kg/ha) replicated thrice. A uniform dose of 20 kg N and 40 kg P_2O_5/ha was applied at the time of sowing. The crop was raised completely rainfed during all the years. During 2003 and 2005 variety RMO 257 and during 2006 and 2007 genotypes MH-61 was taken. A total rainfall of 417.2, 371.8, 168.8 and 400.8 mm was received during 2003, 2005, 2006 and 2007, respectively. Recommended package of practices were fallowed for raising the crop.

RESULTS AND DISCUSSION

The results revealed that sowing done after first soaking rains (first week of July) and second soaking rains (Second week of July) produced significantly higher grain yield (8.09 and 7.42 q/ha) over third soaking rains received in early August (3.73 q/ha) recording 116.9 and 98.9 per cent increase in grain yield, respectively. The data also indicated that the disease incidence was low in early dates of sowing and it was higher when the sowing was done beyond July (Table 3). Higher grain yield obtained in first and second soaking rains was mainly due to less disease incidence and favourable weather conditions experienced by the crop during the crop growth period. These results are in close conformity of Bhati [3] who obtained higher grain yield in second week of July sowing in Jobner. Similarly, study conducted at Fatehpur Shekhawati showed that sowing of mothbean beyond last week of July decreased the grain yield [4]. Higher grain yield was also obtained by Shekhawat *et al.* [5] at Mandore when sowing of mothbean was done at first soaking rain. Lowest grain yield was obtained in third soaking rains (early August sowing) which may be due to reduced photosynthetic activity and translocation of assimilates as a result of abnormal weather conditions and higher disease incidence. Significantly higher crude protein content was recorded in the grain when sowing was done in second week of July (Table 1). Disease incidence also increased with the delay in sowing beyond July (Table 3). Similar findings have also been reported by Gupta *et al.* [2] at Hisar. Conspicuous reduction in yield and higher disease incidence beyond second rain and higher grain yield was recorded with second rain after onset of monsoon [6].

Table 1 : Effect of sowing time, row spacing and seed rate on yield and grain quality of mothbean

Treatments	Grain yield (q/ha)					Crude protein (%)				
	2003	2005	2006	2007	Mean	2003	2005	2006	2007	Mean
Sowing time										
D_1–Ist soaking rains 5 July (2003, 2005, 2006) and 28 June (2007)	7.32	9.22	10.02	5.83	8.09	22.62	21.36	21.80	21.26	21.76
D_2–At subsequent rains after a gap of min. 10 days of 1st sowing 26 July (2003) 16 July (2005), 15 July (2006), 12 July (2007)	4.23	8.65	9.70	7.10	7.42	22.42	22.44	23.37	21.84	22.51
D_3–At subsequent rains after a gap of mini. 10 days of 2nd sowing 14 August (2003) 1st August (2005), 29 July (2006), 9 August (2007)	2.54	3.57	5.11	3.72	3.73	21.74	20.86	22.34	21.12	21.51
SEm±	0.19	0.23	0.28	0.20	0.21	0.07	0.03	0.07	0.08	0.05
C. D. at 5%	0.55	0.68	0.82	0.57	0.61	0.20	0.09	0.21	0.23	0.14
Row Spacing (cm)										
30	4.63	6.97	7.96	5.33	6.22	21.99	21.40	22.21	21.34	21.73
45	4.77	7.32	8.59	5.78	6.61	22.53	21.71	22.80	21.48	22.13
SEm±	0.15	0.19	0.23	0.16	0.17	0.06	0.02	0.05	0.06	0.04
C.D. at 5%	NS	NS	NS	NS	NS	0.16	0.07	0.17	NS	0.12
Seed rate (kg/ha)										
10	4.90	7.41	8.38	5.70	6.59	22.54	21.73	23.03	21.61	22.22

15	4.49	6.88	8.17	5.41	6.24	21.98	21.38	21.98	21.21	21.63
SEm±	0.15	0.19	0.23	0.16	0.17	0.06	0.02	0.05	0.06	0.04
C.D. at 5%	NS	NS	NS	NS	NS	0.16	0.07	0.17	0.19	0.12

A row spacing of 45 cm was found appropriate. Wider row spacing of 45 cm possessed more crude protein content whereas closer row spacing showed less protein content followed by higher disease incidence. Higher disease intensity with closer row spacing have also been reported [2]. Higher grain yield was also obtained at 45 cm row spacing [7]. Whereas, in a similar study at Mandore in erect type variety RMO 40 row spacing of 30 cm was found optimum [5].

Table 2: Interaction effect of sowing time x seed rate on grain yield (q/ha) of mothbean

Sowing time	Seed rate (kg/ha)							
	2003		2005		2006		2007	
	10	15	10	15	10	15	10	15
D1	7.87	6.77	9.95	8.50	10.57	9.46	6.45	5.21
D2	4.72	3.75	9.24	8.06	10.17	9.24	7.62	6.58
D3	2.12	2.97	3.04	4.09	4.39	5.83	3.03	4.42
SEm±	0.26		0.33		0.31		0.28	
C.D. at 5%	0.78		0.97		0.91		0.82	

Table 2a: Interaction effect of sowing time x row spacing on grain yield (q/ha) of mothbean

Sowing time	Row spacing (cm)					
	2005		2006		2007	
	30	45	30	45	30	45
D1	8.72	9.73	9.25	10.79	5.26	6.39
D2	8.21	9.10	8.98	10.43	6.42	7.78
D3	4.00	3.14	5.66	4.56	4.30	3.15
SEm±	0.33		0.31		0.28	
C.D. at 5%	0.97		0.91		0.82	

Amongst varying seed rate, 10 kg/ha was found optimum. Use of 10 kg/ha showed significantly higher crude protein value and less disease incidence as compared to 15 kg/ha (Table 1 and 3). A seed rate of 10 kg/ha was found to be optimum for spreading types of varieties whereas erect type varieties responded significantly upto 15 kg seed/ha [5]. There was significant interactions between sowing time x row spacing and sowing time x seed rate. From the Table 2 it is clear that 45 cm row spacing was more appropriate for first soaking rain (first week of July) and second soaking rains (second week of July) while 10 kg seed/ha was found sufficient. Higher seed rate of 15 kg/ha required when sowing is delayed to early August. Therefore, mothbean crop should be sown in the first fortnight of July at 45 cm row spacing using 10 kg seed/ha to get higher yield and better grain quality.

Table 3. Effect of sowing time, row spacing and seed rate on diseases intensity of mothbean

Treatments	Mothbean mosaic (%DI)			Myrothecium (%DI)			Leaf curl (%DI)		
	2003	2005	2006	2003	2005	2006	2003	2005	2006
D_1– Ist soaking rains 5 July (2003, 2005, 2006) and 28 June (2007)	18.4	19.30	13.55	10.1	15.50	12.92	8.90	7.10	6.10
D_2–At subsequent rains after a gap of min. 10 days of 1st sowing 26 July (2003) 16 July (2005), 15 July (2006), 12 July (2007)	22.7	24.67	17.32	12.6	20.72	16.32	13.53	12.42	11.77
D_3–At subsequent rains after a gap of mini. 10 days of 2nd sowing 14 August (2003) 1st August (2005), 29 July (2006), 9 August (2007)	22.8	31.85	36.47	17.4	24.00	22.95	16.42	18.12	20.87
SEm±	0.71	0.55	0.56	0.60	0.60	0.45	0.55	0.30	0.34
C. D. at 5%	2.08	1.61	1.64	1.77	1.75	1.33	1.62	0.88	1.00
Row Spacing (cm)									
30	27.6	30.30	25.70	15.9	23.31	20.60	15.03	14.76	15.41
45	18.9	20.25	19.20	10.8	16.83	14.20	10.86	10.33	10.42
SEm±	0.58	0.45	0.46	0.49	0.49	0.37	0.45	0.24	0.28
C.D. at 5%	1.70	1.32	1.34	1.45	1.43	1.08	1.32	0.72	0.82
Seed rate (kg/ha)									
10	20.3	21.86	19.48	11.9	18.03	15.50	11.80	10.96	11.15
15	26.2	28.68	25.42	14.8	22.12	19.30	14.09	14.13	14.68
SEm±	0.58	0.45	0.46	0.49	0.49	0.37	0.45	0.24	0.28
C.D. at 5%	1.70	1.32	1.34	1.45	1.43	1.08	1.32	0.27	0.82

REFERENCES

1. Kumar, D. and Singh, N.B. 2002. Mothbean in India. Scientific Publishers (India), Jodhpur.
2. Gupta, P.P., Kumar, S. and Yadav, B.D. 2007. Effect of date of sowing and row spacing on progression of MYMV disease incidence and whitefly population in mothbean. In: National Symposium on Innovative approaches for disease management in arid crops for enhancing export potentials held at ARS Mandore, Jodhpur from Nov. 2-3, 2007 at P-53.
3. Bhati, D.S. 1989. Effect of sowing date on mothbean. Farm. Syst. 5 (1-2) : 52-54.
4. Anonymous 1987. Ann. Progress Rep., RAU, Fatehpur Shekhawati. Pp. 27-40.
5. Shekhawat, B.S., Sharma, R.C. and Shekhawat, R.P.S. 2002. Mothbean Agronomy. In : Mothbean in India, Kumar D. and Singh, N.B. (eds.) Scientific Publishers (India), Jodhpur, pp. 63-85.
6. Shekhawat, B.S. 1992. Effect of sowing dates and row spacing on mothbean varieties. *Indian J. Pulses Res.* 5 (2) : 160-163.
7. Babu Ram, Yadav, B.D, Nanwal, R.K. and Satyajeet 2007. Growth and yield of mothbean (*Vigna aconitifolia* Jacq) as influenced by varieties, row spacing and seed rate under rainfed conditions. *Forage Res.* 33(2) : 95-97.

Role of legumes in improvement on soil health under arid environment

J.C. Tarafdar
Central Arid Zone Research Institute, Jodhpur 342 003

ABSTRACT

Experiments were conducted to find out the role of important leguminous crops grown under arid environment to the improvement in soil health. The results reveal that per centimeter root of arid legumes, at critical growth period, released 40 to 73% more acid phosphatase and 48 to 78% higher phytase than cereals grown under the same environment. The beneficial P mobilizer's population (P solubilizing bacteria, phosphatases producing fungi) in the legume rhizosphere was also increased by 27%, which leads to 14% more alkaline phosphatase activity. The dehydrogenase (representative of bacteria and actinomycetes population) and esterase (representative of fungi, bacteria and actinomycetes population) activities of the rhizosphere soils has been improved between 29 and 32% than cereal rhizosphere. In general, 22% more microbial biomass was noticed under legume rhizosphere as compared to pearl millet. Due to more P mobilizing enzyme activities (acid phosphatase, alkaline phosphatase, phytase), in the legume rhizosphere the P inflow increased by 3.8 to 8.6 times than cereals under arid condition. Organic C and water holding capacity was improved by 7 and 5%, respectively, under legumes. Legumes can exploit 35% higher phosphorus from organic sources and 6% more from unavailable inorganic sources. No definite trend on free living nitrogen fixing and mycorrhizal population under legumes was noticed but *rhizobium* population was significantly higher under legume field condition. The results clearly demonstrated that for improvement in soil health and its maintenance under arid condition legumes cultivation should get highest priority.

Key words: Legumes, soil health, improvement, arid regions.

INTRODUCTION

In the arid environment of India the rainfall is low and their distribution is erratic. Moreover, the soil conditions are poor besides the low availability of various nutrients [1]. In general, the crop yield could be improved in these regions by improvement of soil health. Many drought-tolerant legume crops are grown in these areas. In many places legumes are replacing by pearl millet, which is the important cereal grown under arid environment. Due to low microbial activity in the area, the native nutrient could not be recycled. For this purpose, the root must be able to absorb nutrients effectively under the conditions prevailing in the soil. Plant productivity under any agro-ecological situation primarily depend on the nutritional uptake capabilities of the roots. Differences in the uptake behaviour are of great significance in dry land situation, which are often nutritionally poor and periodically subjected to wind erosion [2]. Therefore, improvement in soil health is a must to choose any crop under these situations.

Hence, an attempt has been made to examine whether legumes are best fitted under arid environments than cereals.

MATERIALS AND METHODS

Field experiments were conducted with five legumes (clusterbean cv HG 75, cowpea cv RC19, horse gram cv PGH 9, moth bean cv Maru moth 1, mung bean cv K 851) and two cereals (pearl millet cv HHB 67, sorghum cv SSG 1000) under randomized block design with a plot size of 5 m × 4 m in triplicate under arid environment. The crops were harvested weekly, to know the nutrient inflow at different growth stages, and at maturity.

Microbial biomass was measured using the fumigation- extraction method of Vance *et al.* [3]. Fifty-gram soil was split into two parts of 25 g each, and formed fumigated and non-fumigated treatments. Each part was extracted with 100 mL 0.5 M K_2SO_4 by horizontal shaking (200 rev min^{-1}) for 30 min at a soil-to-solution ratio 1:2. The organic C and water holding capacity was estimated following the standard methods [4]. Dehydrogenase activity was assayed as described by Tabatabai [5]. Esterase activity was measured as described by Schnürer and Rosswall [6]. Acid and alkaline phosphatase activity was assayed by adopting the standard method [7] using acetate buffer (pH 5.4) and sodium tetra borate-NaOH buffer (pH 9.4), respectively. Phytase activity was assayed by measuring the amount of inorganic phosphate (Pi) released by hydrolysis of sodium phytate (1 mM) after following in method of Ames [8].

Shoot and roots at each harvest were dried at 60°C to a constant weight, ground to fine powder and subjected to a triacid digestion. The root length was measured, using a modified line intercept method [9]. The influx of P was calculated by using the formula after Roemer and Schenk [10] as follows:

$$P\,influx = \frac{\Delta U}{\Delta L}\left(\ln \frac{1_2/1_1}{\Delta t}\right)$$

where, ΔU is the change in P uptake (mol plant^{-1}), ΔL the change in total root length at t_1 and t_2 (cm plant^{-1}) and Δt the change in time (seconds). The P influx was expressed as (mol cm^{-1} root s^{-1}).

VAM spores were estimated by the wet sieving and decanting technique [11]. *Azotobacter*, *Azospirillum* and *Rhizobium* population was enumerated using standard method [12]. Standard errors of means were calculated and, when appropriate, analysis of variance and least significant difference (LSD) were carried out by means of standard procedure [13].

RESULTS AND DISCUSSION

After six years of continuous cropping, it was found that under legume field, there was 25% improvement in water holding capacity of the soil than the bulk soil, which was 5% higher than the fields where cereals were cultivated (Table 1). In general, 18% improvement of organic matter build up over bulk soil was noticed under legumes, which was 7% higher than the cereals. The microbial biomass C under legumes was enhanced by 132% than the bulk soil, which was 22% higher than the cereals. The results clearly demonstrated that legume cultivation under arid environment definitely improved the soil environment for microbial build up. The microbial soil biomass is part of the active soil organic pool, and has been proposed as an indicator of the state and change of total soil organic matter. It control key functions in soil such as decomposition and accumulation of soil organic matter or mineral nutrient transformations.

Table 1. Build up of microbial biomass, organic carbon and water holding capacity under cereals and legumes (± indicate the standard errors of mean)

Parameters	Microbial biomass (mg kg^{-1})	Organic C (%)	Water holding capacity (%)
Bulk soil	40±6.2	0.28±0.02	23.8±0.6
Cereals	76±9.8	0.31±0.03	28.4±0.9
Legumes	93±8.5	0.33±0.03	29.8±1.4

Changes in microbial activities (dehydrogenase and esterase) at critical growth stages (between 4 and 6 weeks) of different crops were estimated and the average values are presented (Table 2). The results clearly showed that legumes had 29% higher dehydrogenase and 32% more esterase activities than the cereal crops grown in the region. The significant improvement in dehydrogenase and esterase activities under legumes clearly indicates their higher build up than the soils where cereals were grown. Aseri and Tarafdar [14] demonstrated that both dehydrogenase and esterase is biological indicator of arid soils and esterase is a better indicator than dehydrogenase activity as it represents almost entire fungi, bacteria and actinomycetes population.

Table 2. Dehydrogenase and esterase activities under cereals and legumes at critical growth stages (between 4 and 6 weeks)

Crop type	Dehydrogenase activity (p kat g^{-1})	Esterase activity (p kat g^{-1})
Cereals	2.58	4.01
Legumes	3.33	5.29
LSD (p = 0.05)	0.29	0.38

The P solubilizing bacteria population and phosphatase producing fungal population was estimated under different crops and presented as Table 3. In general, 27% higher both bacterial and fungal population was noticed under legumes than cereals. The clusterbean rhizosphere showed more population build up as compared to the other crops tested. The higher population indicates more P mobilization from the native unavailable sources.

Table 3. P solubilizing bacteria and phosphatase producing fungal population in the rhizosphere of cereals and legumes under arid environment

Crops	Cultivars	P solubilizing bacteria ($\times 10^4$)	Phosphatase producing fungi ($\times 10^3$)
Cereals			
Pearl millet	HHB 67	7.2±0.5	10.2±0.8
Sorghum	SSG 1000	7.6±0.3	9.6±0.4
Legumes			
Clusterbean	HG 75	10.1±0.8	14.1±1.2
Cowpea	R 19	9.8±0.5	12.7±0.9
Horse gram	PHG 9	8.7±0.4	11.6±0.7
Moth bean	Maru moth 1	9.1±0.7	12.4±0.5
Mung bean	K 851	9.3±0.9	12.2±0.8

Acid phosphatase and phytase secretion by roots of different plants during the critical growth period (between 4 and 6 weeks of crop age) was estimated (Table 4). The results showed significant higher acid phosphatase and phytase secretion by legume roots. In general 38% more acid phosphatase secretion was observed under legumes than cereals. The phytase secretion also showed 27%

more under legumes. The secretion per centimeter of root was highest in clusterbean and least under pearl millet.

Table 4. A comparison between acid phosphatase and phytase secretion per centimeter of fresh root of cereals and legumes at critical growth stages under arid environment

Crops	Cultivars	Acid phosphatase activity (EU × 10^{-7})	Phytase activity (EU × 10^{-7})
Cereals			
Pearl millet	HHB 67	0.48±0.07	14.0±1.92
Sorghum	SSG 1000	0.55±0.04	18.4±1.85
Legumes			
Clusterbean	HG 75	0.83±0.05	24.9±1.45
Cowpea	R 19	0.69±0.01	20.8±2.21
Horse gram	PHG 9	0.68±0.05	20.7±2.15
Moth bean	Maru moth 1	0.74±0.03	23.7±1.68
Mung bean	K 851	0.67±0.02	20.9±1.73

Alkaline phosphatase activity in the soil was found 14% higher under legumes than cereals (Table 5). The more alkaline phosphatase activity was observed under clusterbean closely followed by moth bean. Pearl millet rhizosphere produced least alkaline phosphatase build up. As alkaline phosphatase is purely microbial in origin [15] more activity indicates higher microbial build up in the soil.

Table 5. Alkaline phosphatase activity in the rhizosphere of cereals and legumes under arid environment

Crops	Cultivars	Alkaline phosphatase activity (EU × 10^{-4})
Cereals		
Pearl millet	HHB 67	2.12±0.08
Sorghum	SSG 1000	2.24±0.15
Legumes		
Clusterbean	HG 75	2.72±0.12
Cowpea	R 19	2.37±0.14
Horse gram	PHG 9	2.42±0.06
Moth bean	Maru moth 1	2.65±0.18
Mung bean	K 851	2.29±0.04

Both *Rhizobium* and free living nitrogen fixing population was much higher under legumes than cereals (Table 6). In general 10.2% more *Azotobacter* and 16.7% higher *Azospirillum* population was noticed under legumes. No *Rhizobium* population was detected under cereals. The results clearly indicated that not only symbiotic organism's population was higher under legumes but free living nitrogen fixers (*Azotobacter*) and associative nitrogen fixers (*Azospirillum*) population was also much higher under legumes than cereals.

Table 6. *Rhizobium* and free living nitrogen fixing population under cereal and legumes (± indicate standard errors of mean)

Organisms	Cereals	Legumes	% improvement
Azotobacter ($\times 10^2$)	39±10.2	43±13.5	10.2
Azospirillum ($\times 10^2$)	18±3.5	21±4.9	16.7
Rhizobium ($\times 10^3$)	n.d.	38±3.9	100

n.d. not detected

The inflow of P (Fig. 1) clearly demonstrated higher inflow in legumes than cereals throughout the crop grown period. In general, 3.8 to 8.6 times higher inflow was noticed under legumes than cereals. The maximum P inflow was noticed between 4 and 5 weeks in legumes and 5 and 6 weeks in cereals.

The supply to plants of mineral nutrients is the results of interactions between two complex phenomena: availability of the nutrients in soil and ability of plants to acquire nutrient. Differences in P influx could be related to rhizosphere chemical mobilization, root exudation or root hair density. Foehse *et al.* [15] found that a higher concentration of P at the root hair surface caused a higher influx. If the influx per unit root is low, a large root length is required in order to take up the nutrients and support growth.

The increased enzyme activities indicate an increase in C and nutrient leakage in soil [17]. Microbial activity results in quantitative and qualitative alteration of root exudates composition due to degradation of exudates compounds and the release of microbial metabolites [18]. Exo-enzymes represent a permanent activity potential of biological origin, which is higher in soils where legumes are cultivated under arid environment.

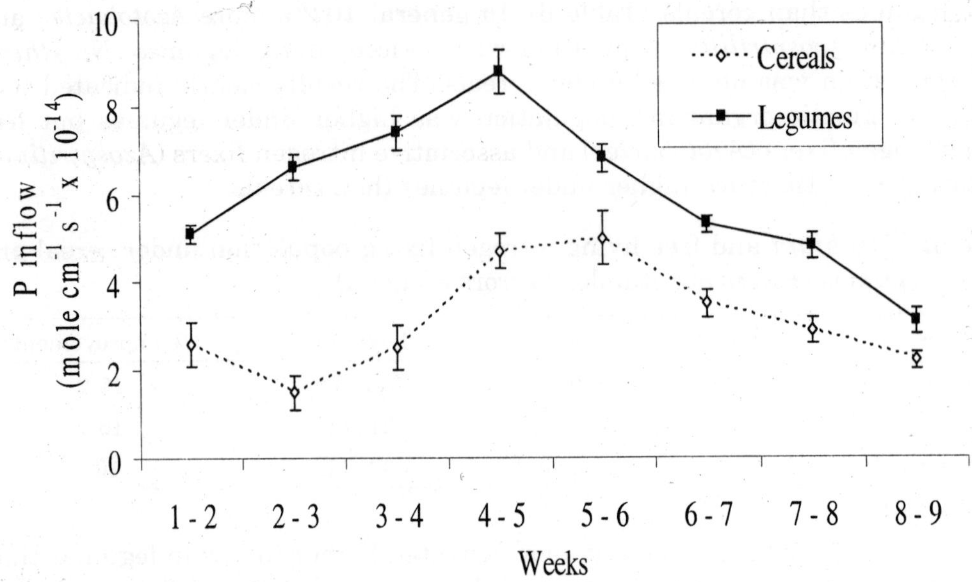

Figure 1. P inflow in legumes and cereals under arid environment

ACKNOWLEDGEMENT

The authors are grateful to ICAR for funding the National Fellow Scheme under which the present work has been carried out.

REFERENCES

1. Tarafdar, J.C., Rao, A.V. and Praveen-Kumar 1995. Role of phosphatase-producing fungi on the growth and nutrition of clusterbean (*Cyamopsis tetragonoloba* (L.) Taub.). Journal of Arid Environments, 29: 331-337.
2. Tarafdar, J.C. and Kumar, D. 2003. Moth bean (*Vigna aconitifolia*) cultivars compared for root nutrient uptake efficiency in arid NW India. Journal of Arid Environments 55: 215-222.
3. Vance, E.D., Brookes, P.C., Jenkinson, D.S. 1987. An extraction method for measuring soil microbial biomass C. Soil Biology and Biochemistry 19: 703-707.
4. Jackson, M.L. 1967. Soil Chemical Analysis, Prentic-Hall of India, Delhi, pp. 498.
5. Tabatabai, M.A. 1982. Soil enzymes, in: Page A.L. (Ed.), Methods of Soil Analysis Part 2, Chemical and Microbiological Properties, American Society of Agronomy, Madison, WI, pp. 903-947.
6. Schnürer, J. and Rosswall, T. 1982. Fluorescien di-acetate hydrolysis as a measure of total microbial activity in soil and litter. Applied Environmental Microbiology 43: 1256-1261.

7. Tabatabai, M.A. and Bremner, J. M. 1969. Use of p-nitrophenyl phosphate for assay of soil phosphatase activity. Soil Biology and Biochemistry 1: 301-307.

8. Ames, B.N. 1966. Assay of inorganic phosphate, total phosphate and phosphatases. Methods of Enzymology 8: 115-118.

9. Tennant, D. 1975. A test of a modified line intersect method for estimating root length. Journal of Ecology 63: 331-337.

10. Roemer, W. and Schenk, H. 1998. Influence of genotype on phosphate uptake and utilization efficiencies in spring barley. European Journal of Agronomy 8: 215-224.

11. Gerdemann, J.W. and Nicolson, T. H. 1963. Spores of mycorrhizal *Endogone* species extracted from soil by wet sieving and decanting. Transactions of British Mycological Society 46: 235-244.

12. Subba Rao, N. S. 1977. Soil Microorganisms and Plant Growth. Oxford and I.B.H. Publishing Co., New Delhi, India.

13. Sokal, R.R. and Rohlf, F.J. 1981. Biometry- The Principles and Practices of Statistics in Biological Research, Second ed. Freeman, New York.

14. Aseri, G.K. and Tarafdar, J.C. 2006. Fluorescein diacetate: a potential biological indicator for arid soils. Arid Land Research and Management 20: 1-13.

15. Tarafdar, J.C. 1989. Use of electrofocussing technique for characterizing the phosphatase in the soil and root exudates. Journal of the Indian Society of Soil Science 37:, 393-395.

16. Foehse, D., Claassen, N. and Jungk, A. 1991. Phosphorus efficiency of plants. II. Significance of root radius, root hairs and cation anion balance for phosphorus influx in seven plant species. Plant and Soil 132: 261-272.

17. Naseby, D.C., Pascual, J.A. and Lynch, J.M. 1999. Carbon fraction in the rhizosphere of pea inoculated with 2, 4 diacetylphloroglucinol producing and non producing *Pseudomonas fluorescens* F.B. 13. Journal of Applied Microbiology 87: 173-181.

18. Neumann, G. and Römheld, V. 2000. The release of root exudates as affected by the plants physiological status, pp 41-93, in R. Pinton, Z. Varanini and P. Nannipieri, eds., The Rhizosphere-biochemistry and organic substances at the soil plant interface. Dekkor, New York.

Response of cluster bean (*Cyamopsis tetragonoloba* (L.) Taub.) to panchgavya and plant leaf extracts in arid western Rajasthan

R.N. Kumawat, S.S. Mahajan, B.K. Kandpal and R.S. Mertia

Regional Research Station, CAZRI, Jaisalmer - 345 001

ABSTRACT

A field experiment on Clusterbean (*Cyamopsis tetragonaloba* (L) Taub.) was conducted during *kharif* 2006 under irrigated condition at CAZRI, RRS, Jaisalmer, Rajasthan to explore nutrient supplying capacity of neem (*Azadiracta indica*), datura (*Datura metel*) and tumba (*Citrullus colocynthis*) plant leaf extracts. The experiment was laid out in Factorial Randomized Block Design with five levels of foliar sources (control, *panchgavya*, neem, datura and tumba) and three levels of stages of application (branching, flowering and both at branching + flowering). The filtered leaf extracts of plants and *Panchgavya* (1:1) was used for foliar application after dilution to 30 times. The results of the study revealed that foliar application of tumba leaf extract significantly increased per plant dry matter by 66.84 and 64.59 per cent at 45 and 65 days after sowing (DAS) than control. Similarly, biological and grain yield was increased 12.88 and 25.63 per cent with tumba leaf extract. Dual application of foliar sources at branching + flowering gave significantly higher dry matter per plant at 45 DAS, 65 DAS and maturity than single spray either at branching or flowering. Compared to spray at flowering, dual application of the plant leaf extracts increased biological and grain by 11.01 and 13.52 per cent.

Key words: Clusterbean, *panchgavya*, leaf extracts, yield.

INTRODUCTION

In India, arid zone covers about 12% of the country's geographical area (3,29,000 sq. km) of which Rajasthan accounts 62%. The soils of arid western Rajasthan are wind transformed ancient aggraded alluvial plains covered either by wind blown sand or by stabilized and new moving sand. The climate of the region is arid with mean aridity index of 77% (1). Since ages, men lived in balance with the natural environment of this arid ecosystem. Over the time, people have developed several conservationist techniques helpful in mitigating the resource degradation. Thus, all kinds of life-human, animal and vegetation have been survived in this fragile ecosystem. However, with the advent of IGNP and exploration of under ground water, peoples of the region begun intensive cultivation of arable crops on desert soils with the use of modern agriculture inputs. Moreover, to meet the increasing needs of ever growing human and livestock population and temptation to get higher income, more and more marginal lands are becoming brought under cultivation. But prevailing climatic and edaphic constraints limits sustainable crop production in the region. Rapid hydraulic conductivity (> 0.25 m $hour^{-1}$), low soil fertility (less than 0.1% OC, 11 kg ha^{-1} available P_2O_5), low moisture retention capacity (3-5%), alkalinity (pH > 8.0, $CaCO_3$ > 7%), and low CEC (<5.1 meq 100 g^{-1} of soils) were common to all the desert soils (2). Because of high mean annual soil temperature (> 22.0°C), lesser weathering of minerals and low microbial mass, these soils are very low in soil organic carbon and other available plant nutrients. Efficiencies of applied nitrogenous and phosphatic fertilizers hardly reach to 20% due to meager presence of nitrosomonas and nitrobacter in the rhizosphere and fixation of applied phosphorus by Ca ions. Similarly, higher oxidation coupled with lower mineralization rate of applied organic inputs in the form of FYM, compost, etc. invariably creates deficiency of organic carbon. Thus economically efficient production of crops in these soils can only be possible through foliar feeding of crop plants. Hitherto, foliar application is generally restricted to urea and to some extent of Fe and Zn. But for proper development crops need all the essential nutrients in proper amount. Foliar application of FYM and compost is very cumbersome owing to long gestation periods and low nutrient content of these sources. Keeping these in view an experiment was conducted with leaf extracts of endemic desert plants in association with *panchgavya* on cluster bean to explore nutrient supplying capacity of these plants.

MATERIALS AND METHODS

The experiment was conducted at CAZRI, RRS, Jaisalmer, Rajasthan during *kharif* 2006 under irrigated condition. The sandy soils of the experimental field was shallow in depth (30 cm) having 0.08% organic carbon, 72.80 kg ha^{-1} available N, 6.45 kg ha^{-1} P, 215.78 kg ha^{-1} K, 6.92 kg ha^{-1} S, and 7.55% free $CaCO_3$ with pH 9.2. The experiment was laid out in factorial randomized block design with five levels of foliar sources (control, *panchgavya*, neem, datura and tumba) and three levels of stages of application (branching, flowering and both at branching + flowering). *Panchgavya* was prepared by thorough mixing and fermentation of fresh cow dung (7.0 kg), cow ghee (1.0 kg), fresh cow urine (10.0

liter), cow milk (3.0 liter) and cow milk curd (2.0 liter). The leaf extracts of neem (*Azadirachta indica)*, datura (*Datura metal*) and tumba (*Citrullus colocynthis*) were prepared by mixing grounded fresh leaves with cow urine in 1:1 ratio followed by fermentation. The filtered leaf extracts were mixed with the filtered *panchgavya* solution in the ratio 1:1 and used for foliar application after dilution to 30 times. Potential cluster bean variety, RGC 936, was sown on 22-07-2006 @ 15 kg seed ha^{-1} after applying pre-sowing irrigation. Foliar application of *panchgavya* and plant leaf extracts were done at branching, flowering and both branching + flowering as per the treatments. Five plants altogether were randomly taken from 2nd row from each sides of every plot at 45 days after sowing (DAS) (flowering), 65 DAS (podding) and maturity for per plant dry matter and yield attributes. Biological yield and grain yield was computed from the plants harvested from net plots in each treatment.

RESULTS AND DISCUSSION

Foliar application of plant leaf extracts and *panchgavya* have considerable effect on the accumulation of dry matter in all the plant parts at various growth stages of cluster bean (Table 1 & 2). Nevertheless, foliar application of leaf extracts of neem, datura and tumba in combination with *panchgavya* increased dry matter accumulation of plants and its distribution among leaf, stem, and pods significantly compared to alone application of *panchgavya* and control at 45 DAS, 65 DAS and maturity. Application of tumba leaf extracts recorded 55.27, 83.91 and 66.84 per cent higher dry matter in leaf, stem and plant, respectively as compared to control at 45 DAS (Table 1). Similarly, dry matter accumulation increased to the tune of 94.10, 106.79, 21.05 and 64.59 per cent in leaf, stem, pods and plant at 65 DAS with application of tumba leaf extract. At maturity, foliar application of neem and datura leaf extracts recorded significantly the highest plant dry matter accumulation in plant, straw (leaf + stem) and pods. Application of neem leaf extracts increased dry matter accumulation of straw, total pods and seeds per plant by 67.28, 48.32 and 25.91 per cent than control at maturity (Table 2).

Dual application of these sources at branching + flowering recoded significantly higher plant dry matter compared to single spray either at branching or flowering (Table 1 & 2). Dual application of all these sources both at branching + flowering increased total plant dry matter by 15.19 and 24.78 per cent at 45 and 65 DAS as compared to control. Similarly at maturity, dry matter accumulation in pods and seeds per plant increased to the tune of 20.39 and 7.31 per cent, respectively than control.

Table 1: Effect of foliar plant nutrient sources and stages of application on per plant dry matter of Cluster bean at different growth stages, g/plant

Treatments	45 DAS (Flowering)			65 DAS (Podding)			
	Leaf	Stem	Total	Leaf	Stem	Pod	Total
Foliar source							
Control	5.59	3.73	9.32	8.31	7.36	12.73	28.41
Panchgavya	6.66	4.47	11.13	11.66	11.47	13.53	36.65
Neem	7.68	5.76	13.44	14.24	13.41	16.68	44.34
Datura	8.28	5.94	14.22	14.78	13.95	13.83	42.56
Tumba	8.68	6.86	15.55	16.13	15.22	15.40	46.76
SEm±	*0.20*	*0.20*	*0.40*	*0.45*	*0.36*	*0.53*	*0.99*
CD at 5%	*0.58*	*0.58*	*1.16*	*1.31*	*1.05*	*1.55*	*2.87*
Stages							
Branching	7.91	5.50	13.40	12.69	11.91	14.54	39.14
Flowering	6.47	5.05	11.52	12.14	11.11	12.38	35.63
Bran+Flow	7.76	5.51	13.27	14.24	13.83	16.39	44.46
SEm±	*0.16*	*0.15*	*0.31*	*0.35*	*0.28*	*0.41*	*0.77*
CD at 5%	*0.45*	*0.45*	*0.90*	*1.02*	*0.81*	*1.20*	*2.22*

Table 2: Effect of foliar plant nutrient sources and stages of application on per plant dry matter and yield attributes of Clusterbean at harvesting, g/plant

Treatments	Straw	Pods (main branch)	Pods (Sec. branch)	Total pods	Total plant	Seed weight (Main branch)	Seed weight (Sec. branch)	Seed weight per plant
Foliar source								
Control	20.63	11.44	18.41	29.86	50.49	6.76	9.91	16.67
Panchgavya	25.04	14.71	19.55	34.26	59.29	6.45	9.94	16.38
Neem	34.51	15.74	28.55	44.29	78.79	6.96	14.02	20.99
Datura	33.78	13.27	26.44	39.72	73.49	7.14	13.97	21.11
Tumba	29.15	17.43	22.81	40.24	69.38	9.43	11.87	21.31
SEm±	*1.16*	*0.77*	*1.46*	*1.83*	*2.35*	*0.49*	*0.61*	*0.86*
CD at 5%	*3.37*	*2.24*	*4.22*	*5.29*	*6.81*	*1.43*	*1.76*	*2.49*

Stages								
Branching	27.62	12.89	23.50	36.39	64.01	7.40	12.28	19.68
Flowering	26.37	13.12	19.68	32.81	59.18	6.50	10.57	17.07
Bran+Flow	31.86	17.54	26.27	43.81	75.67	8.14	12.98	21.12
SEm±	*0.90*	*0.60*	*1.13*	*1.41*	*1.82*	*0.38*	*0.47*	*0.66*
CD at 5%	*2.61*	*1.73*	*3.27*	*4.10*	*5.28*	*1.11*	*1.36*	*1.93*

Biological yield and grain yield increased significantly with the foliar application of leaf extracts of neem followed by datura and tumba compared to control (Figure 1). Application of leaf extracts of tumba recorded 12.88 and 25.63 per cent higher biological and grain yield compared to control.

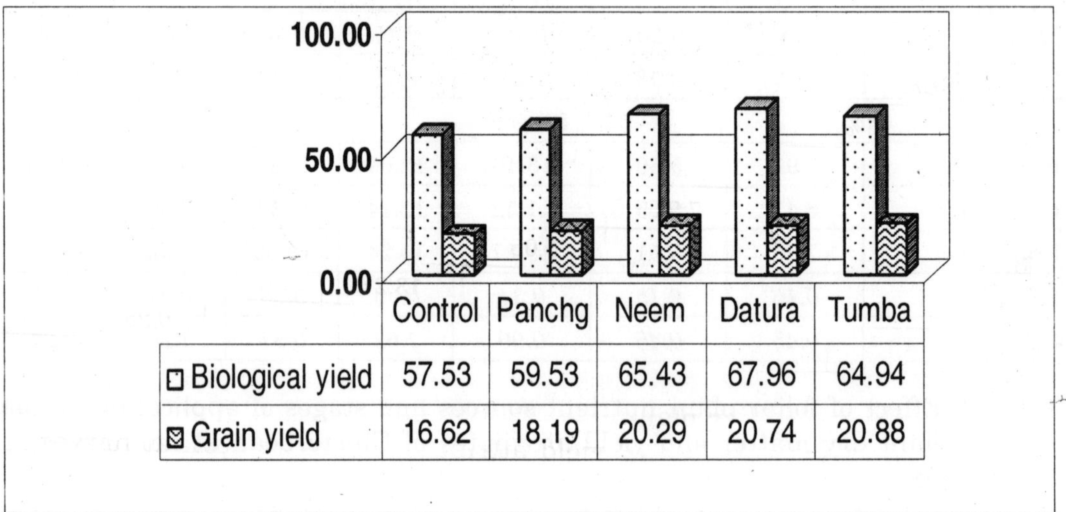

Figure 1: Effect of Sources of foliar application on biological and grain yield of Cluster bean, q/ha

Dual application of these plant sources at branching + flowering recorded significantly higher biological and grain yield than single spray either at branching and flowering (Figure 2). Compared to spray at flowering, dual application of the plant leaf extracts increased biological and grain by 11.01 and 13.52 per cent, respectively.

The increased dry matter and yield with the application of plant leaf extracts in the experiment might be due to increased supply of plant nutrients, increased chlorophyll content of the leaves and increased nodulation in the roots of the cluster bean, which in turn enhanced the processes of nitrogen fixation, chlorophyll synthesis, photosynthesis, assimilation of plant nutrients, synthesis of protein and carbohydrate and their translocation to the site of grain formation. The combined effects of all these biochemical reactions seems to be promoted vegetative and reproductive growth of cluster bean significantly which is reflected through increase in yield and dry matter production at all the stages of plant. The application of these plant nutrient sources either at branching or flowering may not supply nutrients in sufficient amount required for full development of the plants. Hence, higher plant dry matter per plant and biological and grain yields are observed in the experiment with dual application of these plants leaf extracts both at branching + flowering. Further, Selvaraj (3) observed higher plant growth and yield of crop plant due to hormonal effects and enhanced immunity of the plants against biotic stresses with the foliar application of *panchgavya* along with plant leaf extracts in many crops.

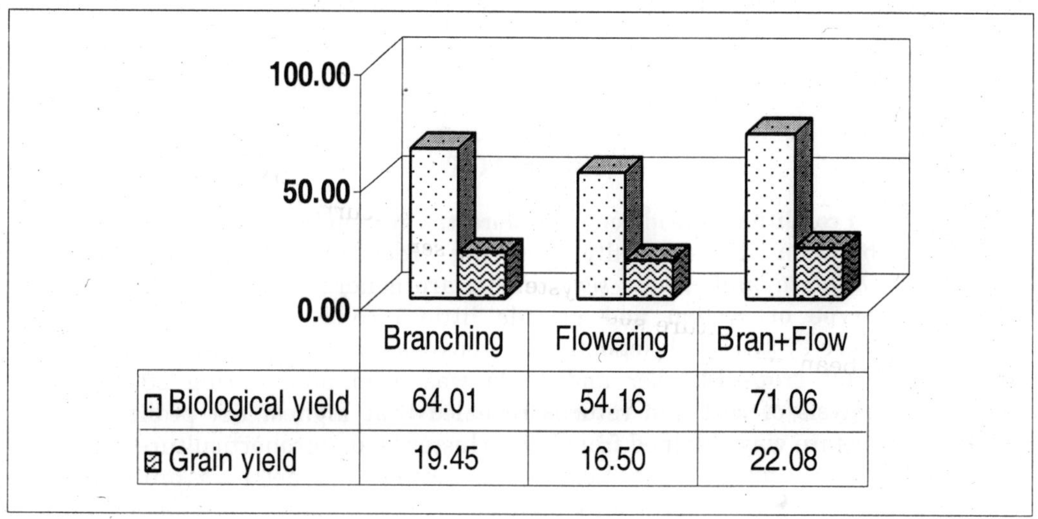

Figure 2: Effect of stages of foliar application on biological and grain yield of Cluster bean, q/ha

REFERENCES

1. Malhotra, S.P., Bharara, C.P. and Sen, M.L.A. 1976. Land utilization and agricultural situation in Bikaner. In: The Deccan Geographer, vol. XIV (1), p. 3.
2. Joshi, D.C. 1996. Soil constraints and sustainable land use. In: Land resources and their management for sustainability in arid regions (eds. Kolarkar, A.S., Joshi, D.C. and Kar, A.). Scientific Publisher (India), Jodhpur, p. 49-63.
3. Selvaraj, N. 2006. Dasagavya: Organic growth promoter for plants. In: The Hindu, p. 18.

Economic evaluation of different agrohorticulture systems under rainfed condition of arid zone

P.R. Meghwal, and A. Henry

Central Arid Zone Research Institute, Jodhpur

ABSTRACT

Under rainfed condition of arid zone, agroforestry system by growing cereals and pulses in between the naturally occurring trees like *Prosopis cineraria*, *Tecomella undulata* and *Acacia senegal* etc. are generally followed. This is a subsistence type of system and provide little income. Hence, an improve agrohorticulture system with leguminous crops such as mung bean and cluster bean with fruit crops like ber and aonla was attempted during 2006. The economic analysis of cost and returns revealed that highest net profit of Rs. 13,487 per hectare was obtained from ber + cluster bean agrohorticulture system followed by ber + mung bean. Least gross return and negative profits was obtained from aonla + cluster bean system,as the aonla plants were just in third year, however an additional income of Rs. 3720 was obtained by growing cluster bean which met out 79% of the expenditure on maintaining aonla plants and growing of cluster bean. The highest B:C ratio (2.15) was also achieved in ber+ cluster bean system followed by sole (2.10) ber and ber + mung bean system (2.02). On the basis of these results, it is advisable to grow leguminous crops like cluster bean and mung bean in association with ber as such system can give significantly higher returns than sole cropping of legumes under rain fed conditions.

Key words: Agro-horticulture systems, economic evaluation, rainfed, arid region.

INTRODUCTION

Arid agro ecosystem is characterized by environmental constraints (low, uncertain and erratic rainfall, high wind velocity, sandy soil intense radiation, sparse vegetative cover, saline irrigation water, heavy biotic pressure and poor socioeconomic conditions of inhabitants). Under such an environment, the ecosystem tends to be fragile since its equilibrium can easily be upset if it is managed in unsustainable manner. In arid zone more than 90% of the cultivable area is put to arable farming (1) which yield very low economic returns to the farmers and provide vegetative cover to the soil for a very brief period during the year. Introduction of perennial trees and shrubs in farming system is desirable for increasing farm income end enhancing system stability. Inclusion of fruit trees will further supplement the economic security of the system. There are limited options before the farmer for selecting the best combination of fruit tree-crops compatibility vis-à-vis economic output. Hence, an attempt has been made to have comparative evaluation of various agro horticulture systems through cost and return analysis in arid zone.

The most predominant existing system of arid land farming in practice under rain fed situation may be termed as agri-silvi-horticulture system with following annual and perennial components (Table 1)

Table : 1. Perennial and annual components of agri-silvi-horticulture system in Indian arid zone

Perennial components	Annual components
Khejri (*Prosopis cineraria*), Kumat (*Acacia senegal*) Kair (*Capparis decidua*) Ber (*Ziziphus mauritiana, Ziziphus rotundifoli Ziziphus nummularia*) Rohida (*Tecomella undulata*), Peelu (*Salvadora oleoides*)	Pearl millet, cluster bean, mung bean, moth bean, sesamum and various cucurbits (water melon, snap melon, round melon, kachra, tumba etc.

The important characteristic of the perennial component are that they grow naturally, multipurpose and well adopted in the area. The existing system of growing crops in between the naturally growing trees without uniform spacing ,though sustainable but offers very low economic benefit to the farmers.

In agro-horticulture system with jujube (*Ziziphus mauritiana*) at CAZRI, a spacing of 8x5m (200 plants/ha was found optimum for growing of dry land crops in the interspaces of ber plants. Among the dry land crops cluster bean showed better performance in drought years, whereas mung bean was the superior crop in good rainfall years and cowpea showed yield stability in most of the years (2)

Materials and Methods

Different agrohorticulture system involving ber and pomegranate as perennial fruit trees and mung bean and cluster bean as annual intercrops were tried. An old unproductive ber orchard was rejuvenated by top working during 2003 which developed fully in the year 2006. Aonla plants of variety N.A. 7 were planted in 2003. Both the species were planted at 8x5 m spacing (200 plants/ha) to facilitate growing of intercrops. The trial was conducted during kharif, 2006 under rainfed condition (208 mm). The details of varieties and treatments are as follows-

Crops/varieties	Treatments
• Ber, var. Gola • Aonla var. N.A.7 • Cluster bean,var. RGM-112 • Mung bean, var. RMG-62	• Sole ber • Sole clusterbean • Ber + Cluster bean • Ber + mung bean • Sole mung bean • Aonla + cluster bean

Results and Discussion

The data on yield of fruits, grain and other products are given in Table 2. There was only marginal reduction in fruit yield of ber due to intercrops while grain yield of both the intercrops was reduced to great extent in agrohoriculture system as compared to their respective sole crops. Leaf fodder and fuel wood production was at par in sole cropping and agrohorticulture system.

Table 2. Yield of fruits, grain, leaves and other products from different agrohorticulture system

system	Yield(kg/ha)				
	Fruits	Grain	Leaves	Pod straw	Fuel wood
Sole ber	2500	-	500	-	1000
Sole cluster bean	-	502	-	200	-
Ber+cluster bean	2400	280	500	115	942
Ber+mung bean	2400	180	500	100	1000
Sole mung bean	-	220	-	100	-
Aonla+cluster bean	-	350	-	200	-

The economic analysis of cost and returns (Table 3) revealed that highest net profit of Rs. 13,487 per hectare was obtained from ber + cluster bean agrohorticulture system followed by ber + mung bean (Rs. 11700). Ber, being a multipurpose species provided multiple incomes through sale of fruits, fodder (leaves) and fuel wood and interspaces were best utilized for growing of leguminous crops which also fixed atmospheric nitrogen, enriching the rhizosphere of ber plants. Least gross return and negative profits was obtained from aonla + cluster bean system, since aonla plants were just three years old and did not start fruiting yet, however an additional income of Rs. 3720/ha was obtained by growing cluster bean which met out 79% of the expenditure on maintaining aonla plants and growing of cluster bean. The highest B:C ratio (2.15) was also achieved in ber+ cluster bean system followed by sole ber (2.10) and ber + mung bean system (2.02). This was much higher than the B:C ratio obtained in growing either of sole crops of cluster bean or mung bean. The net profit is expected to increase further with increase in rainfall since it was only 208 mm during the experimental year, which was insufficient for ber. Bhati (1) earlier reported that among the dry land crops cluster bean showed better performance in drought years, whereas mung bean was the superior crop in good rainfall years and cowpea showed yield stability in most of the years.

Table 3. Cost and returns analysis of different agrohorticulture system under rain fed condition

System	Cost of cultivation (Rs.)	Gross returns (Rs.)	Net profit (Rs.)	B:C ratio
Sole ber	9500	20000	10500	2.1
Sole cluster bean	5500	10640	5140	1.93
Ber+cluster bean	11700	25187	13487	2.15
Ber+mung bean	11400	23100	11700	2.02
Sole mung bean	5200	7400	2200	1.42
Aonla+cluster bean	4700	3720	-980	0.79

References

1. Bhati, T.K. 1994. Agrihorticulture management system. In: Compedium of lecture of training on Horticultural crops for arid and semi arid areas, pp 82-90. January 24-31, CAZRI, Jodhpur.

2. Bhati, T.K. and Joshi N.L. 2008. Farming system for sustainable agriculture in Indian arid zone. In: Dryland Ecosystem, Indian Perspective (Eds. K.P.R. Vittal, R.L. Srivastava, N.L. Joshi, Amal Kar, V.P. Trivedi and S. Kathju, pp 35-52. Central Arid Zone Research Institute, Jodhpur and A.F.R.I. Jodhpur.

Effect of P concentration on release of P mobilizing enzymes by gram (*Cicer arietinum*)

Sushma Dave, Indira Rathore, Poonam Joshi and J.C. Tarafdar

Central Arid Zone Research Institute, Jodhpur 342 003

ABSTRACT

A hydroculture experiment was conducted under aseptic condition in growth chamber to find out the release of acid phosphatase and phytase by the gram (cv. RSG 44) roots between 7 and 21 days. The seedlings were raised in sterilized vermiculite (7, 14 and 21 days) and transferred to the sterilized nutrient solution having variable P concentrations between 0 and 5 mM applied as KH_2PO_4. The plants were kept in nutrient solution for 24 h at each harvest to study the release of enzymes and P uptake. The results clearly demonstrated that the release of acid phosphatase and phytase enzymes by the plant roots was optimum at 1.8 mM P concentration irrespective of the age of the plant. The activities of the secreted acid phosphatase and phytase increased with plant age till 14 days and declined thereafter. The P inflow to the plant was more between 14 and 21 days of plant age than 7 and 14 days of plant age and was higher at 1.0 mM P concentration. The P concentration of the plant was optimum at 1.8 mM P level, thereafter, no significant changes in P uptake with increase in P concentration in the solution was observed. Root length, shoot length and plant biomass was also more between 1.8 and 2.0 mM concentration during the entire growth stages. The present results clearly suggested that 1.8 mM P concentration is needed for optimum release of P mobilizing enzymes into the rhizosphere by gram roots, in the soil solution, at initial stages of plant growth.

Key words: Cicer ariethnum, phosphorus, enzymes, phosphorus mobilization.

Introduction

The Chick pea (*Cieer arietinum*) is an edible legume of family *fabaceae*. Chickpeas are high in protein and one of the earliest cultivated vegetables. They are very high in dietary fiber and a helpful source of zinc, iron and protein. Chickpeas are low in fat and most of this is polyunsaturated.

Phosphorus plays a pivotal structure and regulatory role at the nexus of photosynthesis, energy conservation, carbon metabolism, redox reactions, enzyme activation/inactivation, signaling and nucleic acid synthesis [1], but is one of the most immobile, inaccessible, and unavailable nutrients present in the soils. Although P is abundant in the lithosphere, limited Pi availability is a general phenomenon in agriculture.

Imbalanced and inadequate fertilizer application, organic and inorganic fixation, and fear of world's reserves of rock P [2] coupled with higher phosphate fertilizer prices, point to the need for alternate environmental friendly and economically feasible strategies to improve crop production in low-P soils. This might be achieved by developing precise methods to monitor crop 'P' status.

P mobilizing enzymes are important components involved in the dynamics of soil nutrient cycling. The mineralization and immobilization of phosphorus is brought about by the activities of phosphatase and phytases [3]. Plants can only release acid phosphatase and phytases, alkaline phosphatase are solely microbial in origin [4]. The purpose of this study is to identify the optimum concentrations of P to release of maximum P mobilizing enzymes by gram roots.

MATERIALS AND METHODS

Seeds of chick pea (cv. RGS 44) were collected from Agriculture Research Station, Mandore, Jodhpur. Healthy, viable seeds were surface sterilized in ethanol for five minutes followed by 0.03% H_2O_2 for two minutes. The sterilizing agents were drained off aseptically and the seeds were washed 10-12 times in sterile distilled water to remove all traces of the chemicals. The seeds were placed in a sterilized beaker containing sterilized vermiculture. They were supplied with nutrient solution without P on every third day of experimental schedule. The whole experiment was conducted in a aseptic growth chamber under 11 h day light and at 65 ± 5% relative humidity. The day temperature was 28±2°C and the night temperature was 16±2°C.

After 7, 14 and 21 days of plant growth two seedlings were placed in a sterilized flask containing 100 mL sterilized nutrient solution [5] to study the release of enzymes and P uptake. The nutrient solution was modified to maintain concentration of phosphorus from 0 to 5 mM using KH_2PO_4 as a P source. The treatments were replicated 3 times and the plants were kept under aseptic conditions. Plants of each harvest were immediately washed in distilled water and blotted dry using filter paper sheets. Root and shoots were separated and immediately put in craft paper bags and air dried in the laboratory for 12 h before oven drying. Dried roots and shoots were weighed and finely ground. Uniform samples of roots and shoots were then digested in a tri-acid mixture of perchloric acid: sulphuric acid:nitric acid in the ratio of 1:2:9 respectively [6].

Phosphorus concentrations in roots and shoots (mg P g^{-1} dry matter) were measured by the vanadate molybdate yellow color method [7]. The nutrient solution in which the plants were kept for 24 h was analysed for the enzyme activity.

Acid phosphatase activity was measured as described by Tabatabai and Bremnr [8] and phytase activity was analysed following the standard procedure [9]. Root length was measured by the method of Tennant [10]. The P-influx was calculated by using the formula given by Römer & Schenk [11].

Standard errors of means were calculated and, when appropriate, analysis of variance and least significant difference (LSD) were carried out by means of standard procedure [12].

RESULTS AND DISCUSSION

A wide variation in root length with changes of P concentration in soil solution was noticed at different crop growth stages (Table 1). In general, more root length was noticed in 1.6 mM concentration at 7 days and 5 mM concentration at 14 and 21 days. On an average there was no significant difference in root length was observed after 1.8 mM concentration in the entire growth stages.

Table 1. Variation in root length (cm) under different soil solution concentration

P concentration in solution (mM)	Crop age (days)		
	7	14	21
0	50.00	102.33	149.00
0.2	58.33	133.33	146.67
0.4	50.00	169.00	183.33
0.6	58.33	163.00	185.67
0.8	51.00	181.33	196.67
1	54.33	179.00	192.33
1.2	43.67	177.33	193.00
1.4	46.00	133.33	197.00
1.6	69.33	158.67	199.00
1.8	65.00	195.00	208.33
2	52.00	192.67	224.00
3	46.33	202.67	233.33
4	47.67	223.33	231.00
5	45.00	223.67	235.00
LSD (p = 0.05)	4.72	30.79	27.23

The maximum shoot length was observed between 1.8 and 2.0 mM concentrations. With increase in soil solution concentration further, there was no improvement in shoot length was noticed. The results clearly showed that optimum response to gram shoot was noticed at 1.8 mM of solution P concentration. The optimum biomass accumulation was also noticed at 1.8 mM concentration under different growth stages.

Table 2. Shoot length (cm) and plant biomass (g) under different soil solution concentration

P concentration in solution (mM)	Shoot length (cm)			Plant biomass (g)		
	Crop age (days)			Crop age (days)		
	7	14	21	7	14	21
0	13.50	17.17	18.50	0.34	0.46	0.46
0.2	14.67	15.67	18.00	0.40	0.60	0.38
0.4	14.33	17.67	17.67	0.41	0.46	0.46
0.6	14.50	17.00	17.33	0.37	0.44	0.48
0.8	14.67	17.33	19.50	0.42	0.46	0.44
1	14.50	17.67	18.17	0.40	0.40	0.52
1.2	14.50	16.50	19.17	0.37	0.50	0.44
1.4	14.67	17.00	17.67	0.36	0.53	0.42
1.6	13.83	17.33	18.00	0.35	0.50	0.46
1.8	14.00	18.17	18.67	0.37	0.45	0.38
2	16.17	19.00	18.00	0.39	0.45	0.41
3	16.17	17.33	18.67	0.34	0.52	0.44
4	15.33	16.33	18.33	0.39	0.49	0.36
5	15.67	16.67	18.00	0.32	0.47	0.47
LSD ($p = 0.05$)	0.72	0.57	0.68	0.02	0.07	0.04

The results (Table 3) clearly indicate that the release of acid phosphatase and phytase was optimum at 1.8 mM P concentration. The maximum acid phosphatase and phytase was released at 14 days of crop age. In general, the release of both the enzymes was increased with crop age till 14 days and declined thereafter.

Table 3. Acid phosphatase and phytase release by the plant roots at various solution P concentration

P concentration in solution (mM)	Acid phosphatase activity ($EU \times 10^{-3}$)			Phytase activity (EU)		
	Crop age (days)			Crop age (days)		
	7	14	21	7	14	21
0	4.40	267	113	95.00	77.11	57.37
0.2	20.70	270	93.7	103.89	50.49	51.53
0.4	57.30	295	63	104.15	58.62	54.39
0.6	86.70	299	148	101.57	83.25	61.81
0.8	116.80	318	145	93.42	79.20	57.81
1	149.50	317	210	88.23	84.21	57.82
1.2	171.20	320	233	99.48	103.02	67.90
1.4	215.10	340	258	103.30	107.21	70.64
1.6	228.30	345	325	99.96	115.45	72.33
1.8	300.40	376	342	115.62	118.11	78.51
2	199.10	399	350	101.63	122.10	75.24
3	275.70	320	352	112.93	122.47	79.47
4	289.70	328	365	103.04	132.72	79.92
5	295.50	328	363	101.26	134.14	80.13
LSD ($p = 0.05$)	11.25	24.12	9.35	3.05	4.21	2.16

The less enzyme release in the solution at higher P concentration, may be, at lower P concentration leads to increased cellular levels of phosphatase enzymes which may leak into the cell walls and into the surrounding medium.

The inflow of P was estimated between 7 and 14 days and again between 14 and 21 days. The maximum inflow was observed between 14 and 21 days except at 1.8 mM concentration (Fig. 1). In general, the more P inflow was noticed at 1.0 mM P concentration irrespective of crop age.

The P concentration of the plant was optimum at 1.8 mM P level (Table 4), thereafter, no significant changes in P concentration was noticed with increase in solution P concentration. Although there was little improvement in P concentrations till 5.0 mM solution P level.

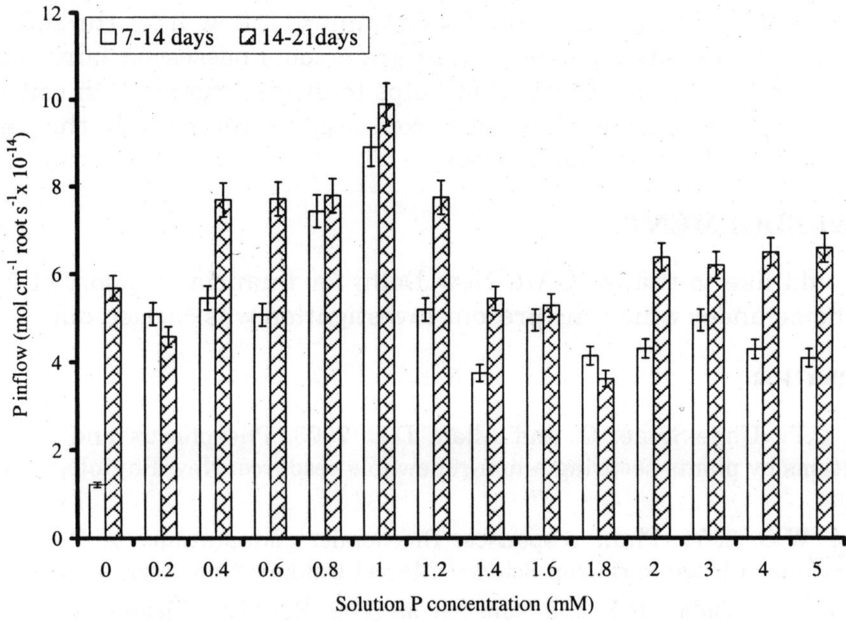

Figure 1. P inflow at various solution P concentrations

Table 4. Plant P concentration (μg g^{-1}) at different solution P concentration

P concentration in solution (mM)	Crop age (days)		
	7	14	21
0	15.39	18.24	17.06
0.2	20.19	12.56	18.38
0.4	20.03	14.73	19.81
0.6	21.16	17.49	20.76
0.8	22.59	15.51	22.17
1	23.82	17.28	24.55
1.2	24.18	14.15	25.71
1.4	28.46	18.60	26.14
1.6	23.31	20.82	27.17
1.8	28.35	26.52	32.01
2	24.49	26.11	31.08
3	36.42	26.61	33.12
4	36.14	28.47	36.28
5	36.05	30.70	37.48
LSD (p = 0.05)	8.65	4.31	5.48

Results of this study could be useful to know the exact solution P concentration needed by gram plants for optimum P inflow from the soil solution to the plants. The results also shows that gram roots possessed fairly high acid phosphatase and phytase activity, that helps to utilize organic P in soil solution efficiently by releasing phosphate ions from organic compounds through their hydrolysis by phosphatases and phytase.

ACKNOWLEDGEMENT

We would like to thank ICAR, New Delhi for financial support of National Fellow Scheme under which the present investigation was carried out.

REFERENCES

1. Vance, C.P., Uhde-Stone, C. and Allan, D.L. 2003. Phosphorus and use: Critical adaptations by plants securing a non-renewable resource. New Phytologist 157: 423-457.
2. Plaxton, W.C. 2004. Plant responses: Biochemical adaptations to P-deficiency in Encyclopedia of Plant and Crop Science. Marcel Dekkar, New York p. 976-980.
3. Tarafdar, J.C., Yadav, R.S. and Ram Niwas 2002. Relative efficiency of fungal intra and extra cellular phosphatase and phytase. Journal of Plant Nutrition and Soil Science 165: 17-20.
4. Tarafdar, J.C. 1989. Use of electrofocussing technique for characterizing the phosphatases in the soil and root exudates. Journal of the Indian Society of Soil Science 37, 393-395.
5. Tarafdar, J.C. and Claassen, N. 1988. Organic phosphorus compounds as a phosphorus source for higher plants through the activity of phosphatases produced by plant roots and microorganisms. Biology and Fertility of Soils 5: 308-312.
6. Jackson, M. L. 1967. Soil chemical Analysis. Prentice – Hall of India, Delhi pp. 498.
7. Kitson, R. E. and Mellon, M. G. 1944. Colorimetric determination of phosphorus molybdovanadophosphoric acid. Indian Eng. Chem. Anal. Ed., 16: 379-383.
8. Tabatabai, M. A. and Bremner, J. M. 1969. Use of p-nitrophenyl phosphate for assay of soil phosphatase activity. Soil Biology and Biochemistry 1: 301-307.
9. Ames, B.N. 1966. Assay of inorganic phosphate, total phosphate and phosphatases. Methods in Enzymology 8: 115-118.
10. Tennant, D. 1975. A test of a modified line intersect method of estimating root length. Journal of Ecology 63: 995-1001.
11. Römer, W. and Schenk, H. 1998. Influence of genotype on phosphate uptake and utilization efficiencies in spring barley. European Journal of Agronomy 8: 215-224.
12. Sokal, R.R. and Rohlf, F.J. 1981. Biometry- The Principles and Practices of Statistics in Biological Research, Second ed. Freeman, New York.

Phosphorus nutrition of arid legumes for higher drought tolerance and improved yields

B.K. Garg

Division of Soil-Water-Plant Relationships, Central Arid Zone Research Institute, Jodhpur 342 003

ABSTRACT

The seed yield of arid legumes is often low due to low and erratic rainfall and poor soils fertility conditions in arid regions. The role of phosphorus nutrition is very important for yield improvement of legumes as P supply is crucial for leaf growth, photosynthetic efficiency, root growth and its functioning and nitrate assimilation. Phosphorus nutrition is also known to impart drought tolerance in several crops under water limited conditions. Studies conducted on phosphorus nutrition under water stress and rainfed conditions have revealed the beneficial role of P application in yield enhancement of clusterbean and moth bean. P nutrition has been found to significantly increase leaf area, net photosynthetic rates, concentrations of chlorophyll, starch, soluble protein and activity of nitrate reductase under drought or water deficit conditions. P application has further been observed to enhance drought tolerance under different intensities of water stress in clusterbean and moth bean genotypes. Although the potential gains may be variable in different genotypes and under various rainfall situations, but the merit of P application for growth and yield improvement of arid legumes is well documented.

Key words: Arid legumes, phosphorus nutrition, leaf area, photosynthesis, nitrogen metabolism, drought tolerance, seed yield.

Introduction

Arid legumes comprising clusterbean, moth bean, cowpea and horse gram are the integrated component of dryland farming, being the mainstay of pulse and forage production in arid and semi-arid regions of the country. However, their yields are generally low due to poor soil fertility and erratic rainfall, leading to frequent droughts. In this regard the supply of phosphorus is particularly critical as declining water availability causes reduced P availability and absorption by different legumes under dryland conditions (1, 2). Improved soil fertility is reported to alleviate the adverse effects of drought on crop productivity in arid environments (3, 4).

For legume crops, the rolc of phosphorus nutrition is very crucial (5, 6). Phosphorus nutrition is known to affect plant responses to water deficits in a number of crops and increase drought tolerance of cotton (7), white clover (8), soybean (9) and wheat (10). Furthermore, the relative leaf growth rate is one of the most sensitive parameters to P deficiency and it affects the photosynthetic rate per unit area (11). Decline in leaf growth and photosynthetic rate was ascribed to a reduction in stomatal conductance and RuBP carboxylase regeneration capacity (12). Phosphorus nutrition is also known to enhance drought tolerance of several crops as a result of increased root growth and its more efficient functioning (13, 14).

For the last several years, research work has been carried out at CAZRI, Jodhpur on the phosphorus nutrition of various arid legumes with a view to improve their productivity under water limited conditions. Results have revealed that detrimental effects of water deficits can be significantly ameliorated on various physiological processes like leaf area development, net photosynthesis, carbohydrate and nitrogen metabolism resulting in improved growth and seed yield of various crops. A brief account of significant finding is presented here.

Phosphorus supply, plant growth and composition

In plants suffering form P deficiency, reduction in leaf expansion and leaf surface area (15) and also number of leaves (16) are the most striking effects. Leaf expansion is strongly related to extension of epidermal cells and this process might be particularly impaired in P deficient plants due to low P content of epidermal cells and decrease in root hydraulic conductivity (17, 18). Furthermore, the photosynthetic efficiency permit of chlorophyll is much lower in P deficient leaves (19). Sugarbeet plants grown with adequate amounts P displayed a higher CO_2 fixation compared to P-deficit plants and this has been attributed to high RuBP carboxylase activity (20). Increased cell expansion leading to a large leaf area, increased photosynthesis with P application (21).

In contrast to shoot growth, root growth is much less inhibited under P deficiency, leading to a typical decrease in shoot-root dry weight ratio. In bean (*P. vulgaris*) this ratio decreased from 5.0 in P sufficient to 1.9 in P deficient plants. As a rule, the decrease in shoot-root dry weight ratio in P deficient plants is correlated with an increase in partitioning of carbohydrates towards the roots, indicated by a steep increase, particularly of sucrose content of the roots of P deficient plants (22). The reported accumulation of starch indicates that photosynthates cannot be efficiently used for plant growth under P-limited condition (15). Phosphorus deficiency is also known to reduce the uptake rate of nitrate and its assimilation by the nitrate reductase (23). Several field studies have indicated, interaction between P and water availability on cereal yields in semi-arid regions (4, 10) but information is lacking on the response of arid legumes to P nutrition and drought.

Phosphorus-drought-interaction in clusterbean

Numerous reports are available on the phosphorus application for improvement of growth and yield of clusterbean (24, 25). Work carried out at CAZRI, Jodhpur showed that clusterbean responded upto 40 kg P_2O_5 ha^{-1} and increase in yield beyond this level was marginal (5, 26). As drought is the major constraint for crop production in arid regions, the interaction between P application and water stress assumes importance. Water stress at the pre-flowering and post-flowering stages in clusterbean significantly reduced leaf area, rates of net photosynthesis, contents of chlorophyll, soluble protein as well as nitrate reductase activity (27). However, P application (40 kg ha^{-1}) increased all of these parameters leading to significant improvement in plant growth and seed yield under water stress condition (Table 1).

Table 1. Influence of phosphorus on leaf metabolites, photosynthesis, NRA and seed yield of clusterbean under water stress (*Burman et al.*, 2004)

Parameters	P level (kg ha^{-1})	Control plants	Droughted plants
Leaf area (cm^2 $plant^{-1}$)	0	234.7	143.1
	40	264.8	167.7
Net photosynthesis	0	9.34	2.64
(μ mol m^{-2} s^{-1})	40	10.76	3.12
Chlorophyll	0	7.39	6.36
(mg g^{-1} dw)	40	7.90	7.15
Soluble protein	0	45.06	34.79
(mg g^{-1} dw)	40	58.82	45.47
Nitrate reductase	0	399.6	45.72
(μg NO_2 g^{-1} dw^{-1})	40	479.2	104.4
Seed yield	0	3.71	2.65
(g $plant^{-1}$)	40	4.38	3.15

Responses of clusterbean to phosphorus application (40 kg ha^{-1}) under different intensities of water stress at the critical pre-flowering stage revealed that increasing stress intensity progressively decreased plant water potential, relative water content, levels of different leaf metabolites, photosynthetic rate and nitrate reductase activity (Table 2). However, P-fertilized plants displayed higher photosynthetic efficiency, leaf metabolism and enzyme activity than unfertilized plants, though decrease in plant water status was more in P-fertilized plants (28). Recovery upon rewatering was also more under P application. Consequently, P nutrition enhanced dry matter production and seed yields at all stress intensities (Table 3). On an average of P-fertilized plants had 10.4% higher DMP and 17.0% higher seed yield over unfertilized plants. P application maintained both dry matter and seed yield at significantly higher level upto 6 days of water stress, compared to unfertilized control plants. These results indicate that alleviation of water stress effects by P application can be achieved, at least upto moderate stress levels.

Table 2. Influence of P nutrition on water status, photosynthesis and NRA of clusterbean under different intensities of water stress (28)

Days without water	Ψ plant (-MPa)		RWC (%)		Net photosynthesis (μ mol m^{-1} s^{-1})		NRA (μg NO_2g^{-1} dwh^{-1})	
	P_0	P_{40}	P_0	P_{40}	P_0	P_{40}	P_0	P_{40}
0	0.93	1.00	83.0	80.7	5.80	6.65	442.1	566.8
3	1.53	1.85	70.1	68.7	3.30	3.98	134.7	161.0
6	1.87	2.06	61.8	60.0	2.84	3.03	74.7	99.9
9	2.08	2.37	55.7	53.9	2.45	2.62	56.8	79.9
Mean	1.60	1.82	67.9	65.8	3.60	4.07	177.1	226.9

Table 3. Effect of phosphorus application on seed yield and shoot dry matter of clusterbean under different intensities of water stress (28)

Days without water	Seed yield (g $plant^{-1}$)			Shoot weight (g $plant^{-1}$)		
	P_0	P_{40}	Mean	P_0	P_{40}	Mean
0	5.11	5.92	5.52	16.3	17.8	17.0
3	4.53	5.32	4.92	15.1	16.8	15.9
6	3.68	4.36	4.02	11.6	12.9	12.3
9	3.11	3.64	3.37	10.5	11.6	11.0
Mean	4.11	4.81	-	13.3	14.8	-
LSD (0.05)	D	P	D x P	D	P	D x P
	0.53	0.37	NS	0.94	0.66	NS

Phosphorus nutrition under rainfed conditions

On soils low in phosphorus, clusterbean responds well to the application of phosphatic fertilizers. Application of 40 kg P_2O_5 ha^{-1} drilled with the seed alongwith a basal application of 20 kg N ha^{-1} provided economically high yields. Phosphate application to legumes enhanced the per cent N in nodules and dry weight of nodules (6). Kathju *et al.* (5) also found that P application to various legumes led to an increase in soil N, particularly of the hydrolysable organic N fraction. Phosphorus application was also reported to increase elongated root systems beside improving the growth and yield of clusterbean (29). On loamy sand soil at Jobner, application of 40 kg P_2O_5 ha^{-1} was found superior to 20 kg P with regard to increase in growth, yield attributes, seed yield and stover yield (30). Work at Bikaner revealed significantly higher response on yield of seed and straw with the application of 40 kg ha^{-1} P_2O_5 over 20 kg and control (25). Similarly, in a study on nutrient management in moth bean, optimum levels of N and P were found to be 10 and 40 kg ha^{-1}, respectively (31). In another study increasing levels of P upto 40 kg ha^{-1} increased seed yield of moth bean (32). There are several other reports indicating the positive effects of phosphorus application alongwith N on moth bean yield under rainfed conditions (2).

A two-year field study with clusterbean under rainfed conditions indicated that in dry lands, P application (0, 20, 40 kg ha^{-1}) alongwith an initial dose of N (20 kg ha^{-1}) may improve seed yield and dry matter production, even when the availability of soil moisture is limited due to variable and erratic distribution of rainfall (33). Application of 20 and 40 kg ha^{-1} P increased seed yield by 15.8% and 23.5% in 2001 and by 14.1% and 24.3% in 2003, respectively, relative to unfertilized plants (Table 4). Similarly N application (20 kg ha^{-1}) enhanced seed yield by 15.9% in 2001 and by 15.1% in 2003. However, their combined application (40 kg ha^{-1} P and 20 kg ha^{-1} N) enhanced seed yield by 55.1% in 2001 and by 44.0% in 2003 over controls, indicating synergistic effects of P with N fertilization. It has been found that P application either alone or in combination with nitrogen resulted in significantly higher net photosynthetic rates and concentrations of chlorophyll, starch, soluble protein as well as nitrate reductase activity compared to control plants (Table 5).

Interactive effects of phosphorus nutrition (0 and 40 kg ha^{-1}) and row spacings (30 cm and 45 cm) in clusterbean genotypes under rainfed conditions during 2004 and 2005 at Jodhpur revealed that P application increased seed yield of all the three genotypes (RGC-1003, RGC-197 and Maru guar) under both row spacings (Table 6). Phosphorus application increased net photosynthesis, levels of total chlorophyll, starch, reducing sugars, soluble protein and nitrate reductase activity in all the genotypes at both row spacings in both the years (34). It has been concluded that cultivation of genotypes of clusterbean having early to medium maturity with adequate P nutrition and row spacing of 45 cm may be most beneficial for obtaining sustainable yields under arid conditions.

Table 4. Interactive effects of P and N on seed yield and dry matter production of clusterbean in 2001 and 2003 under rainfed conditions (33)

P levels (kg ha^{-1})	Seed yield (kg ha^{-1})			Dry matter production (kg ha^{-1})		
	N_0	N_{20}	Mean	N_0	N_{20}	Mean
			2001			
0	296	385	341	1578	1795	1687
20	378	432	405	1810	2010	1910
40	397	459	428	1904	2072	1988
Mean	357	426	-	1764	1959	-
			2003			
0	518	652	585	2208	2692	2450
20	624	716	670	2576	3072	2824
40	712	746	729	3009	3124	3067
Mean	618	705	-	2598	2963	-
LSD (0.05)	P	N	P x N	P	N	P x N
2001	29	16	NS	87	61	98
2003	41	38	66	291	259	NS

Table 5. Effect of P and N on net photosynthesis and concentrations of chlorophyll, starch and soluble protein in clusterbean at the flowering stage in 2001 and 2003 (33)

Treatments	Net photosynthesis (μ mol m^{-1} s^{-1})		Total chlorophyll (mg g^{-1} dw)		Starch (mg g^{-1} dw^{-1})		Soluble protein (mg g^{-1} dw^{-1})	
	2001	2003	2001	2003	2001	2003	2001	2003
P (kg ha^{-1})								
0	17.78	12.06	5.26	4.83	214.0	261.2	66.1	41.3
20	19.22	12.85	6.13	5.15	236.5	275.6	72.3	47.2
40	20.60	13.36	6.51	5.36	251.9	309.4	79.0	50.7
LSD (p = 0.05)	0.95	0.77	0.40	0.22	12.6	14.5	6.9	3.1
N (kg ha^{-1})								
0	18.38	12.35	5.60	4.91	225.5	268.8	63.1	43.9
20	20.03	13.16	6.33	5.30	238.8	295.2	81.8	48.8
LSD (p = 0.05)	0.85	0.42	0.12	0.22	10.4	13.7	5.8	2.0

Table 6. Influence of P application and row spacing on seed yield of clusterbean genotypes during 2004 and 2005 under rainfed conditions (34).

Treatments	Seed yield (kg ha^{-1})	
	2004	2005
Phosphorus (kg ha^{-1})		
0	190	428
40	221	474
LSD (p=0.05)	20	18
Row spacing (cm)		
30	215	427
45	196	475
LSD (p=0.05)	NS	27
Genotypes		
RGC-1002	235	546
RGC-197	199	567
Maru guar	174	341
LSD (0.05)	25	23

Phosphorus nutrition and genotypic responses

Differential response patterns to P nutrition are likely to occur in genotypes differing in their flowering and maturity period. Therefore, several workers have studied the interactive effects of P nutrition and drought on genotypes of arid legumes (2, 35). The work carried out under all India coordinated project on arid legumes to find out suitable genotypes of moth bean under different agronomic management revealed that RMO-257 proved superior than RMO-40 and Jawala with application of 20 kg N and 40 kg P_2O_5 ha^{-1} under 30 cm row spacing. In case of clusterbean RGC-936 gave highest yield with application of 20 kg N + 40 kg P_2O_5 ha^{-1} under 30 cm spacing. In cowpea, the most suitable genotype was found to be GC-89105 with application of recommended dose of N + P_2O_5 under 45 row spacings.

The factors controlling the low P tolerance in 9 cultivars of moth bean (three for each early, medium and late maturity groups) were evaluated in moth bean (36). It was observed that early and medium maturing cultivars were more susceptible to low P conditions in terms of biomass production than late genotypes because of the larger reduction in photosynthesis, due to inhibition of leaf emergence and development.

Table 7. Effects of drought and phosphorus nutrition on certain metabolic parameters and seed yield of moth bean (37).

Parameters	Stress	P levels (kg ha^{-1})		Percent increase by Phosphorus
		P_0	P_{40}	
RWC (%)	C	86.3	83.7	-3.0
	D	69.3	73.0	+5.3
Chlorophyll (mg g^{-1} dw)	C	8.25	8.83	7.0
	D	7.16	7.90	10.3
Starch (mg g^{-1} dw)	C	241.9	275.0	13.8
	D	173.2	200.6	15.8
Soluble protein (mg g^{-1} dw)	C	36.8	39.7	7.9
	D	32.9	36.4	10.6
NRA (µg NO_2g^{-1}dwh^{-1})	C	119.42	125.7	5.2
	D	38.2	60.4	58.1
Seed yield (g plant^{-1})	C	2.57	2.90	12.8
	D	1.53	1.96	28.0

Data mean of four genotypes (RMO-257, CZM-18, CAZRI moth-1 and Jawala)

C – Control plants; D – Droughted plants

The influence of phosphorus nutrition on the physiological response of moth bean genotypes including two early (RMO-257 and CZM-18) and two late flowering (CAZRI moth-1 and Jawala) subjected to drought at the pre-flowering stage revealed that P application significantly ameliorated the detrimental effects of drought on seed yield and physiological parameters (RWC, net photosynthesis, chlorophyll, starch, soluble protein and nitrate reductase activity), particularly in the late genotypes (Table 7). It has been found that P application to moth bean may be justified even under conditions of low rainfall, particularly when late flowering genotypes are used (37). However, drought as such had less detrimental effects in early than late flowering genotypes of moth bean and these adverse effects could be significantly ameliorated by application of 40 kg P ha^{-1} in all the genotypes. The merit of P application to clusterbean genotypes under rainfed condition of arid zone has also been reported (34).

CONCLUSIONS

A number of studies carried out on the application of phosphorus in arid legumes either under rainfed conditions on under water stress has established the beneficial role of P nutrition for yield improvement. The magnitude of physiological and biochemical changes induced by P nutrition either alone or in combination with N may vary depending on the soil moisture conditions of arid regions. Thus the potential gains may be variable depending upon moisture supply and genotypes but the merit of P nutrition in yield enhancement is quite obvious.

REFERENCES

1. Garg, B.K. and Burman, U. 2002. Physiology (in) *Guar in India.* pp.63-83. D. Kumar and N.B. Singh (Eds.). Scientific Publishers, Jodhpur.
2. Garg, B.K. and Burman, U. 2002. Moth bean physiology (in) *Moth bean in India.* pp.87-102. D. Kumar and N.B. Singh (Eds.). Scientific Publishers Jodhpur.
3. Lahiri, A.N. 1980. Interaction of water stress and mineral nutrition on growth and yield (in) *Adaptations of Plants to Water and High Temperature Stress.* pp. 341-352. N.C. Turner and P.J. Kramer (Eds.). John Wiley, New York.
4. Payne, W.A., Drew, M.C., Hosner, L.R., Lascano, R.J., Onken, A.B. and Wendt, C.W. 1992. Soil phosphorus availability and pearl millet water use efficiency. *Crop Sci.* 31:1010-1015.
5. Kathju, S., Aggarwal, R.K. and Lahiri, A.N. 1987. Evaluation of diverse effects of phosphate application on legumes of arid areas. *Trop. Agric.* 64:91-96.
6. Aggarwal, R.K. and Praveen-Kumar 1993. Changes in soil nitrogen due to cultivation of green legumes fertilized with phosphate in an arid loamy sand soil. *Ann. Arid Zone.* 32:195-196.
7. Radin, J.W. 1981. Stomatal responses to water stress and to abscisic acid in phosphorus deficient cotton plants. *Plant Physiol.* 72: 392-394.
8. Singh, D.K., Sale, P.W.G. and Mckenzie, B.M. 1997. Water relations of white clover (*Trifolium repens* L.) in a drying soil, as function of phosphorus supply and defoliation frequency. *Aust. J. Agric. Res.* 48:675-681.
9. Gutierrez-Boem, F. and Thomas, G.W. 1998. Phosphorus nutrition and water deficits in field grown soybeans. *Plant Soil.* 207:87-96.
10. Gutierrez-Boem, F. and Thomas, G.W. 1999. Phosphorus nutrition affects wheat responses to water deficits. Agron. J. 90:166-171.
11. Kirschbaum, M.U.F. and Tompkins, D. 1990, D. 1990. Photosynthetic responses to phosphorus nutrition in *Eucalyptus grandis* seedlings. *Aust. J. Plant Physiol.* 17: 527 – 535.
12. Brooks, A. 1986. Effects of phosphorus nutrition on ribulose biophosphate carboxylase activation, photosynthetic quantum yield and amounts of some calvin cycle metabolites in spinach leaves. *Aust. J. Plant Physiol.* 13: 221-237.
13. Radin, J.W. and Eidenback, M.P. 1984. Hydraulic conductance as a factor limiting leaf expansion of phosphorus deficient cotton plants. *Plant Physiol.* 75:372-377.
14. Singh, D.K. and Sale, P.W.G. 1998. Phosphorus supply and growth of frequently defoliated white clover (*Trifolium repens* L.) in dry soil. *Plant soil.* 205:155-162.

15. Fredeen, A.L., Rao, I.M. and Terry, N. 1989. Influence of phosphorus nutrition on growth and carbon partioning in *glycine max*. L. *Plant Physiol.* 89:225-230.

16. Lynch, J., Lauchli, A. and Epstein, E. 1991. Vegetative growth of the common bean in response to phosphorus nutrition. *Crop Sci.* 31: 380-387.

17. Treeby, M.T., Vansteveninck, R.F.M. and de Vries, H.M. 1987. Quantitative estimates of phosphorus concentrations within *Lupinus luteus* leaflets by means of electron probe X-ray micro analysis. *Plant Physiol.* 85:331-334.

18. Radin, J.W., 1990. Responses of transpiration and hydraulic conductance to root temperature in nitrogen and phosphors-deficient cotton seedlings. *Plant Physiol.* 92:855-857.

19. Lauer, M.J., Pallardy, S.G., Blevins, D.G. and Randall, D.D. 1989. Whole leaf carbon exchange characteristics of phosphate deficient soybeans (*Glycine max* L.). *Plant Physiol.* 91: 848-854.

20. Rao, I.M. and Terry, N. 1995. Leaf phosphate status, photosynthesis and carbon partitioning in sugarbeet. *Plant Physiol.* 107:1313-1321.

21. Longnecker, N. 1994. Nutrient deficiency and vegetative growth (in): *Mechanism of Plant Growth and Improved Productivity*. pp.137-172. A.S. Basra, Marcel Dekker, Boca Raton, FL.

22. Khamis, S., Chaillou, S. and Lamaze, T. 1990. CO_2 assimilation and partioning of carbon in maize plants deprived of orthophosphate. *J. Exp. Bot.* 41:1619-1625.

23. Pilbeam, D.J., Cakmak, I., Marschnel, H. and Kirkly, E.A. 1993. Effect of withdrawl of phosphorus on nitrate assimilation and PEP carboxylase activity in tomato. *Plant Soil.* 154:111-117.

24. Singh, R.V. and Singh, R.R. 1989. Effect of nitrogen, phosphorus and seedling rate on growth, yield and quality of guar under rainfed conditions. *Indian J. Agron.* 34: 53-56.

25. Solanki, N.S., Shekhawat, R.P.S. and Shekhawat, B.S. 1998. Efficacy of phosphate culture with phosphorus levels on growth and yield of clusterbean. *Ann. Arid Zone.* 37:417-418.

26. Singh, R.P. 1998. Guar agronomy – a review article (in): *Guar – its Improvement and Management*. pp.41-48. R.S. Paroda and S.K. Arora (Eds.). *Forage Res.*4A.

27. Burman, U., Garg, B.K. and Kathju, S. 2004. Interactive effects of thiourea and phoshorus on clusterbean under water stress. *Biol. Plant.* 48:61-65.

28. Burman, U., Garg, B.K. and Kathju, S. 2009. Effect of phosphorus application on clusterbean under different intensities of water stress. *J. Plant Nutrition.* 32: 668-680.

29. Tiwana, U.S., and Thiwana, M.S. 1994. Effect of source, level and method of phosphorus application on the seed yield of clusterbean. *Indian J. Ecol.* 21:117-121.

30. Shivran, A.C., Kangarat, S.S., Shivran, P.L. and Gora, G.R. 1996. Response of clusterbean (*Cyamopsis tetragonoloba*) varieties to sulphur and phosphorus. *Indian J. Agron.* 41:340-342.

31. Bhati, D.S., Mathur, J.R. and Sharma, R.C. 1988. Response of moth bean to graded levels of N and P. *Ann. Arid Zone.* 27: 63-64.

32. Yadav, G.L. 1998. Response of moth bean varieties to row spacing and phosphorus levels. *Indian J. Pulse Res.* 11: 38-42.

33. Burman, U., Garg, B.K. and Kathju, S. 2007. Interactive effects of phosphorus, nitrogen and thiourea on clusterbean (*Cyamopsis tetragonoloba* L.) under rainfed conditions of the Indian arid zone. *J. Plant Nutr. Soil Sci.* 170:803-810.

34. Garg, B.K., Burman, U. and Kathju, S. 2007. Physiological basis of yield improvement in clusterbean in response to row spacings and phosphorus nutrition. *J. Plant Biol.* 34:1-8.

35. Singh, R.P. and Singh, K.C. 1980.Relative performance of rainfed moth bean varieties grown under different fertility levels. *Forage Res.* 6:161-164.

36. Chaudhary, M.I. and Fugita, K. 1998. Responses of nitrogen fixation and photosynthetic activity in moth bean (*Vigna aconitifolia*) cultivars under low P conditions. *Soil Sci. Plant Nutr.* 44:37-41.

37. Garg, B.K., Burman, U. and Kathju, S. 2004. The influence of phosphorus nutrition on the physiological response of moth bean genotypes to drought. *J. Plant Nutr. Soil Sci.* 167:503-508.

Effect of micronutrients on grain yield and quality of clusterbean

J.S. Yadav, B.D. Yadav*, U.N. Joshi* and S.S. Yadav

Regional Research Station, Bawal-123 501 (Haryana)

*Forage Section, CCS Haryana Agricultural University, Hisar-125 004 (Haryana)

ABSTRACT

A filed experiment was conducted for three consecutive years from *kharif* 2001 to 2003 at Regional Research Station, Bawal to know the response of zinc and iron. The soil of the experimental field was loamysand. The experiment consisted of nine treatments viz., control, 0.5% $FeSO_4$ and $ZnSO_4$ spray individually at 25, 45 and both at 25 and 45 DAS, respectively, 0.5% $FeSO_4$+$ZnSO_4$ spray at 25 DAS and soil applied 25 kg $ZnSO_4$/ha at sowing. Based on three years results, spray of 0.5% $FeSO_4$ alone either at 25 or 45 DAS and its two spray at 25 and 45 DAS did not cause any significant increase in grain yield, whereas 0.5% $ZnSO_4$ spray applied either the above stages or both the stages or in combination with 0.5% $FeSO_4$ at 25 DAS significantly increased the grain yield of clusterbean over control and spray of 0.5% $FeSO_4$ at any stage. There was no significant difference in grain yield when 0.5% $ZnSO_4$ spray was done either at 25 or 45 or both at 25 and 45 DAS. Application of 25 kg $ZnSO_4$/ha significantly increased the grain yield by 27.6 per cent over control. No significant differences in grain yield were observed between foliar and soil application of zinc sulphate. Spray of 0.5% $ZnSO_4$ either at 25 or 25+45 DAS exhibited significant increase in gum content over 45 DAS alone. All the treatments of zinc and iron spray recorded significantly higher gum and crude protein content over the control. Plant height remained at par under various treatments. Number of seed/pod, pods/plant and 1000-grain weight were significantly affected by 0.5% $ZnSO_4$ spray over control.

Key words: Clustern bean, micro-nutrients, quality, grain yield

INTRODUCTION

Clusterbean (*Cyamopsis tetragonoloba* (L.) Taub.) commonly known as guar, is a *kharif* arid legume grown for green fodder, vegetable, green manuring and seed purpose. It occupies an important place in the national economy especially among the dryland crops. The crop has a great role to play in nitrogen economy for the succeeding crop as it builds up soil fertility by fixing atmospheric nitrogen and by addition of organic matter. With the introduction of high yielding varieties and intensive cropping pattern the deficiency of micro-nutrients is increasing. Most of work on the nutrition of this crop has been related with major nutrients [1], whereas the significance of micro-nutrients is still ignored. Foliar spray of micro-nutrients depends upon the stage of plant growth. In general nutrition research, particularly on zinc and iron, has been confined to cereals and oilseeds and very little or no work has been reported on clusterbean. Of the several essential micro-nutrients zinc have been found to be showing positive response both as spray as well as soil application in clusterbean [2]. Hence, the above trial was conducted to study the response of zinc and iron on clusterbean.

MATERIALS AND METHODS

The field experiment was conducted for three consecutive years from *kharif* 2001 to 2003 at CCS Haryana Agricultural University's Hisar Regional Research Station, Bawal to study the effect of zinc and iron on growth, yield attributes, grain yield and its quality under rainfed conditions. The soil of the experimental site was loamysand, low in organic carbon (0.18%) and available phosphorus (8.34 kg/ha) with alkaline in reaction (pH 8.1); low in zinc (0.33 ppm DTPA) and iron (2.1 ppm DTPA). The experiment consisted of nine treatments viz., control, 0.5% $FeSO_4$ spray at 25, 45 and both at 25 and 45 DAS; 0.5% $ZnSO_4$ spray at 25, 45 and both at 25 and 45 DAS; 0.5% $FeSO_4$ + $ZnSO_4$ spray at 25 DAS; soil applied 25 kg $ZnSO_4$/ha at sowing. The trial was laid in randomized block design replicated thrice. In first year clusterbean variety HG 365 was sown in 30 cm row spacing using 15 kg seed/ha on June 23, 2001. In successive years the sowing was done on July 27, 2002 and July 9, 2003. The crop was raised under rainfed conditions during all the years except during 2002 where the crop was raised by applying pre-sowing irrigation because of failure of monsoon during the month of July. The crop received 357.0, 143.4 and 160.2 mm rainfall during the crop period of 2001, 2002 and 2003, respectively.

RESULTS AND DISCUSSION

The data presented in Table 1 revealed that plant height was not affected significantly by any of the treatments during all the years of studies. More plant height was obtained during 2001 because of high rainfall during the crop period and minimum height was achieved during 2002. Foliar spray of 0.5% $FeSO_4$ did not affect number of seeds/pod, pods/plant and 1000-grain weight but foliar spray of 0.5% $ZnSO_4$ gave better effect on different yield attributes under study. There was no significant differences in yield attributes among the different treatments of ferrous and zinc sulphate spray. With regards to all yield attributes soil application of 25 kg $ZnSO_4$/ha was found at par with foliar spray of 0.5% $ZnSO_4$ but showed significant differences over all treatments $FeSO_4$ and control.

Table 1. Effect of micronutrients on growth and yield attributes of clusterbean

Treatments	Plant height (cm)			Seeds/pod			Pods/plant			1000-seed wt. (g)		
	2001	2002	2003	2001	2002	2003	2001	2002	2003	2001	2002	2003
Control	131.0	92.2	108.5	8.20	7.27	7.95	46.77	36.33	54.36	27.04	26.60	28.12
0.5 % $FeSO_4$ spray 25 DAS	137.6	94.8	113.8	8.23	7.73	8.06	49.67	38.30	58.10	28.27	26.99	28.36
0.5 % $FeSO_4$ spray 45 DAS	129.5	94.3	115.7	8.00	7.67	8.02	44.23	38.23	56.30	28.30	26.98	28.65
0.5 % $FeSO_4$ spray 25 & 45 DAS	134.3	98.5	110.3	8.01	8.00	8.10	49.43	41.00	59.35	28.37	27.39	28.72
0.5 % $ZnSO_4$ spray 25 DAS	139.9	97.7	114.5	8.43	8.07	8.47	50.43	41.23	62.16	28.64	27.32	29.35
0.5 % $ZnSO_4$ spray 45 DAS	142.9	101.0	116.5	8.32	8.33	8.38	51.10	42.00	64.35	29.20	28.48	29.68
0.5 % $ZnSO_4$ spray 25 & 45DAS	141.2	98.8	112.6	8.43	8.37	8.65	50.43	41.87	68.70	28.26	28.56	29.82
0.5 % $FeSO_4$+$ZnSO_4$ spray 25 DAS	141.9	97.2	117.5	8.56	8.13	8.76	50.46	42.33	70.12	29.02	28.71	29.36
Soil applied 25 kg $ZnSO_4$ /ha	146.4	99.8	114.9	8.57	8.27	8.48	50.13	41.63	72.46	29.62	28.74	29.98
CD at 5 %	NS	NS	NS	0.45	0.40	0.35	NS	3.86	5.72	1.08	0.51	0.86

Results presented in Table 2 revealed that grain yield in general was higher during the year 2003 as compared to 2001, in spite of more rainfall received during 2001 (357.0 mm) when it was compared to year 2003 (160.2 mm). About 50% reduction in grain yield was observed during 2002 (142.0 mm) in comparison to 2001 and 2003. Low and erratic rainfall during 2002 may be the cause of decrease in grain yield of clusterbean. Based on mean of 3-years results it is revealed that 0.5% $FeSO_4$ spray at any of the stages could not increase grain yield significantly over control. No significant differences were observed when 0.5% $ZnSO_4$ spray was done either at 25 or 45 or two spray at 25 and 45 DAS. Two spray of 0.5% $ZnSO_4$ at 25 and 45 DAS and combination of 0.5% $FeSO_4$ and $ZnSO_4$ spray at 25 DAS produced significantly higher grain yield over $FeSO_4$ spray and control. Spray of $ZnSO_4$ applied either of the stages significantly increased grain yield over control. Similarly Gupta *et al.* [3] also got higher grain yield of clusterbean with foliar spray of $ZnSO_4$. Application of

25 kg $ZnSO_4$/ha significantly increased the grain yield by 27.6 per cent over control. There was no significant difference in grain yield between foliar and soil applied zinc sulphate. However, soil applied zinc sulphate produced higher grain yield over foliar application of $FeSO_4$. Among spray treatments $ZnSO_4$ spray proved to be more effective. These findings are in close agreement of the work of Nandwal *et al.* [4].

Table 2: Effect of micronutrients on grain yield and seed quality of clusterbean

Treatments	Grain yield (q/ha)				Gum (%)				Crude protein (%)			
	2001	2002	2003	Mean	2001	2002	2003	Mean	2001	2002	2003	Mean
Control	10.43	5.20	11.32	8.98	30.58	29.60	27.37	29.18	28.40	28.25	27.33	28.12
0.5 % $FeSO_4$ spray 25 DAS	10.86	5.47	12.08	9.47	30.95	30.24	27.61	29.60	29.49	28.44	28.44	28.79
0.5 % $FeSO_4$ spray 45 DAS	10.26	5.63	12.94	9.61	30.77	30.65	27.65	29.69	28.91	29.33	27.64	28.63
0.5 % $FeSO_4$ spray 25&45 DAS	10.52	5.50	12.96	9.66	31.66	30.67	28.25	30.19	28.88	30.44	28.79	29.37
0.5 % $ZnSO_4$ spray 25 DAS	11.69	6.77	13.27	10.57	32.70	31.43	28.60	30.91	29.68	28.38	28.73	28.93
0.5 % $ZnSO_4$ spray 45 DAS	11.78	6.60	13.92	10.76	31.45	30.58	28.42	30.15	29.79	29.56	28.73	29.36
0.5 % $ZnSO_4$ spray 25 & 45DAS	12.62	6.80	14.18	11.20	32.76	31.65	28.94	31.11	29.84	30.73	28.75	29.77
0.5 % $FeSO_4$+$ZnSO_4$ spray 25 DAS	12.05	7.10	14.82	11.32	33.57	31.65	28.39	31.20	29.81	30.93	28.71	29.81
Soil applied 25 kg $ZnSO_4$/ha	12.45	7.20	14.74	11.46	30.81	29.90	27.32	29.34	28.68	30.31	27.70	28.90
CD at 5 %	1.12	0.72	1.71	1.22	0.29	0.49	0.37	0.35	0.44	0.32	0.43	0.37

All the iron and zinc spray treatments recorded significantly higher gum and crude protein content over control. No significant differences were observed among different $FeSO_4$ spray treatments on grain quality. Spray of $ZnSO_4$ either at 25 or both at 25 and 45 DAS exhibited significant increase in gum content over spray of $ZnSO_4$ at 45 DAS. Soil applied $ZnSO_4$ could not increase gum content but increased crude protein value significantly over control. Two sprays of $FeSO_4$ at 25 and 45 DAS improved gum and crude protein content over $FeSO_4$ applied at 25 or 45 DAS alone. Spray of $FeSO_4$ in combination with $ZnSO_4$ showed significant improvement in gum content over all the treatments except $ZnSO_4$ spray at 25 + 45 DAS. Significant improvements in gum content by foliar spray of zinc sulphate in clusterbean have also been reported [5]. Thus it is conducted that zinc should be used @ 25 kg $ZnSO_4$/ha as basal dose in zinc deficient soils of Haryana. In case zinc is not use at the time of sowing, it may be applied by 0.5% spray of $ZnSO_4$ either at 25 or 25+45 DAS for obtaining higher grain yield and better grain quality of clusterbean.

REFERENCES

1. Dahiya, S.S., Khera, A.P., Tomer, N.K. and Panwar, B.S. 1986. Response of fertilization to guar under irrigation constraints. *Guar Res. Ann.* 4 : 18-20.
2. Gupta, V.K. and Potlia, B.S. 1988. Relative efficiency of zinc carriers on clusterbean and their residual effect on succeeding wheat crop. *Agri. Sci. Digest* 8(1) : 31-33.
3. Gupta, P.P., Yadav, B.D. and Joshi, U.N. 2007. Effect of micronutrients on major diseases of clusterbean (*Cyamopsis tetragonoloba* L. Taub.). *J. Arid Legumes,* 4 (1) : 1-5.
4. Nandwal, A.S., Dabas, S., Yadav, B.D. and Bharti, S. 1990. Effect of zinc on growth, yield, leghemoglobin and nitrogen of clusterbean nodule. *Guar Res. Ann.* 6 : 13-18.
5. Yadav, B.D., Joshi, U.N. and Joon, R.K. 2004. Effect of zinc and iron on growth, yield and quality of clusterbean. In : National Symposium on Arid Legumes for sustainable Agriculture and Trade from 5-7 November, 2004 at CAZRI, Jodhpur. Pp. 68.

Effect of host nutrition on mung bean yellow mosaic virus disease incidence in mothbean [*Vigna aconitifolia* (Jacq.) Marechal]

P.P. Gupta, Sunil Kumar, B.D. Yadav and Naresh Kumar*

Forage Section, Department of Plant Breeding,
CCS Haryana Agricultural University, Hisar-125 004
*Asstt. Plant Breeder, RRS Bawal (Rewari)-123 501 (Haryana)

ABSTRACT

An attempt was made in screen house during the years 2005 and 2006 to see MYMV disease incidence on susceptible cultivar (RMO 40) of moth bean using different (17+1 control) NPK combinations with four replications in nutrient deficient sandy soil. Six kg soil supplemented with different fertilizer combinations was taken in 12 inch earthen pots wrapped inside with polythene bags using CRD design. Twenty five seeds were sown per pot in first week of August. Irrigation was done with equal amount of water when needed. Three observations were recorded at 10 days interval starting from last week of August. The disease incidence and white fly population were recorded using standard method. On the basis of average of two years application of phosphorus and potassium (40 kgha^{-1} each) alone reduced the disease incidence from 49.4 % (in control) to 17.4% and 18.4% respectively, however, the application of 20 kg ha^{-1} nitrogen alone gave higher disease incidence (41.8%) in comparison to phosphorus and potassium application (40 kg ha^{-1} each alone). Highest reduction in MYMV disease incidence (71.05%) with respect to 20 kg ha^{-1} nitrogen application alone was recorded when $N_{20}P_{40}K_{40}$ kg ha^{-1} was used. Highest increase in MYMV disease incidence (80.80%) with respect to 10 kg ha^{-1} nitrogen and 40 kg ha^{-1} phosphorus application was recorded when nitrogen increased from 10-30 kg ha^{-1} with constant phosphorus dose. The number of white fly plant^{-1} on healthy and diseased plants differed significantly among the treatments. The grain yield plant^{-1} with respect to increase in disease incidence and white fly population decreased significantly. The disease incidence due to MYMV increased significantly with the increase in time interval. Hence

$N_{20}P_{40}K_{40}$ (kg ha^{-1}) is quite effective in reducing the MYMV disease incidence, white fly population and in increasing the grain yield.

Key words: *Vigna aconitifolia*, green gram, yellow mosaid virus, nutrition.

INTRODUCTION

Major pulse crops grown in the arid zone of Rajasthan and Haryana are moth bean [*Vigna aconitifolia* (Jacq.) Marechal], mung bean [*Vigna radiata* (L.) Witzec.] and cowpea [*Vigna unguiculata* (L.) Walp.]. Legumes are integral part of dry land farming system due to potent source of dietary protein. Besides, they are also used as fodder, feed, pasture and green manure purpose. They have the capabilty to fix atmospheric nitrogen. They are grown as a pure crop as well as in mixed/ inter crops. Moth bean is affected by several fungal, bacterial and viral diseases. It is a major commercial crop and mainly grown during *kharif* season where uneven conditions make it more vulnerable to the attack of diseases like mung bean yellow mosaic virus (MYMV), bacterial leaf spot etc. which cause heavy losses in yield. The precise estimation of the losses due to MYMV in moth bean has not been worked out. Maximum yield losses of 75.3 percent due to this disease were reported (1). The disease may appear at any stage of plant growth but if it appears at an initial stage of plant growth, then the plant may not flower and yield losses may be as high as 90 per cent (2). The virus is mainly transmitted through whitefly (*Bemisia tabaci*). It transmits the virus in non-persistent manner. The virus consisted of isometric particles (15-20 nm in diameter) forming loose aggregates in the nucleus of infected phloem cells. The nucleic acid identified was circular, single stranded DNA with a molecular weight of 8 x 10^5 Daltons. The thermal inactivation point of virus ranged from 40-50°C for 10 minutes. The dilution end point was 10^{-2} to 10^{-3} and the longevity *in vitro* 1-2 days at 20°C. Effect of fertilizer application on severity of MYMV disease on moth-bean has not been reported. However, effect of NPK treatments on severity of yellow mosaic virus in urdbean and mungbean was well studied. Sivaprakasam *et al.* (3) observed severe yellow mosaic virus disease in urdbean crop which received NPK treatments 50: 100:0 followed by 50: 100:50 kg/ha. Since, there are large variations in incidence of yellow mosaic virus disease due to application of different doses of NPK in other pulse crops. An attempt was therefore made in screen house condition during the years 2005 and 2006 to see MYMV disease incidence with variation of NPK application on susceptible cultivar of moth-bean in nutrient deficient sandy soil.

MATERIALS AND METHODS

The experiment was conducted during the *kharif* years 2005 and 2006 in 12 inch size earthen pots using CRD design. Polythene bags were wrapped inside the pots to check the leakage of water and nutrients from the pots. Pots were filled with nutrient deficient sandy soil collected from Gangwa Block. The soil was supplemented with different combinations of N, P and K. Twenty five seeds of susceptible cultivar (RMO 40) of moth were sown per pot with four replications for each treatment. Suitable control (no fertilizer) was also maintained. Equal amount of water was added in each pot when needed. Different N, P and K combinations (per hectare) used are as follows:

Combinations to note the effect of N : Combinations to note the effect of K:

(i) $N_0P_{40}K_0$ (iv) $N_{30}P_{40}K_0$ (i) $N_{20}P_{40}K_{10}$ (iv) $N_{20}P_{40}K_{40}$

(ii) $N_{10}P_{40}K_0$ (v) No application (ii) $N_{20}P_{40}K_{20}$ (v) $N_{20}P_{40}K_{50}$

(ii) $N_{20}P_{40}K_0$ (iii) $N_{20}P_{40}K_{30}$ (vi) N_0 P_0 K_{40}

Combinations to note the effect of P :

(i) $N_{20}P_0K_0$ (v) $N_{20}P_{40}K_0$

(ii) $N_{20}P_{10}K_0$ (vi) $N_{20}P_{50}K_0$

(iii) $N_{20}P_{20}K_0$ (vii) $N_{20}P_{60}K_0$

(iv) $N_{20}P_{30}K_0$

Observations were recorded at three successive intervals of 10 days each (29.08.05, 07.09.05and 17.08.05) after planting (04.08.05). Similar observations were recorded at intervals of 10 days on the same dates during 2006 after planting on 04.08.06. The incidence of mungbean yellow mosaic virus was recorded using modified 1-9 scale (4). To calculate the disease incidence, fifty leaves were selected from the top, middle and lower portion of the plants for finding the infected leaf area. The disease incidence due to MYMV was computed as mentioned:

$$\text{Disease incidence (\%)} = \frac{\text{Sum of all numerical ratings}}{\text{Highest Rating x No. of leaves observed}} \times 100$$

Whitefly population was recorded using the split cage method (5). Grain yield (g) per plant was also recorded.

RESULTS AND DISCUSSION

Phosphorus and potassium (40 kg/ha each) application alone reduced MYMV disease incidence from 49.4% (in control, no application of NPK) to 17.4% and 18.4% respectively (Table 1, Average of two years). The application of 20 kg/ha nitrogen alone gave higher disease incidence (41.8%) in comparison to phosphorus and potassium application (40 kg/ha each). When phosphorus dose (40 kg/ha) was kept constant and nitrogen dose varied from 0-30 kg/ha, the incidence of MYMV disease increased from 17.4% to 54.6%. When phosphorus application increased from 20 kg to 60 kg/ha keeping nitrogen application (20 kg/ha) as constant then MYMV disease incidence decreased significantly. On the other hand when potassium application increased from 20 kg to 50 kg/ha, keeping nitrogen application and phosphorus application constant i.e. 20 kg/ha and 40 kg/ha respectively, there was significant reduction in disease incidence and reduction in disease incidence was comparatively high with respect to increase in phosphorus application.

Table 1. Effect of host nutrition on MYMV disease progression, whitefly population and grain yield per plant in mothbean variety RMO-40 during the years 2005 and 2006 (Av. of two years)

Sr. No.	Treatments Fertilizer (kg/ha)			Disease score (1-9 scale)			DI (%) DAS			% Decrease in DI	WP			Y/P (in gm)
	N	P	K				25	35	45		25	35	45	
1	0	40	0	3	5	5	5.8 (13.86)	10.6 (19.12)	17.4(24.56)	67.78	1.3	1.6	4.6	5.2
2	10	40	0	5	5	7	10.3 (18.74)	18.3 (25.44)	30.2(33.24)	38.87	1.8	2.2	5.3	4.4
3	20	40	0	5	7	7	13.1 (21.45)	24.9 (29.86)	38.8(38.84)	21.46	2.1	2.6	6.6	3.4
4	30	40	0	5	7	9	18.8 (25.92)	35.6 (36.86)	54.6(47.74)	-10.53*	1.6	5.7	9.1	2.4
5	20	0	0	5	7	7	15.4 (23.24)	29.2 (32.68)	41.8(40.12)	15.38	1.0	3.8	7.6	3.6
6	20	10	0	5	7	7	16.4 (23.88)	30.8 (33.84)	46.2(42.76)	6.48	1.8	4.7	8.6	3.2
7	20	20	0	5	7	5	17.4 (24.82)	34.0 (35.84)	50.4(45.84)	-2.02*	1.9	4.1	7.4	2.9
8	20	30	0	5	7	7	15.6 (23.14)	31.0 (33.76)	47.1(43.36)	4.66	1.6	2.6	8.3	3.2
9	20	40	0	5	5	7	13.4 (21.64)	25.1 (30.34)	35.4(36.59)	28.34	1.8	3.7	6.6	3.8
10	20	50	0	5	7	7	14.5 (22.68)	25.7 (30.56)	36.4(37.12)	26.31	1.7	2.6	6.4	3.9
11	20	60	0	5	7	7	15.2 (22.92)	28.0 (31.82)	38.1(38.10)	22.87	2.1	2.7	6.8	3.7
12	20	40	10	5	7	7	14.3 (22.97)	26.5 (31.12)	36.6(37.15)	25.91	1.8	3.6	5.8	3.7
13	20	40	20	5	5	7	15.4 (23.24)	21.8 (27.36)	30.4(33.36)	38.46	1.3	5.7	5.4	4.2
14	20	40	30	3	5	5	7.9 (15.92)	15.4 (23.12)	22.6(28.15)	54.25	1.1	2.2	5.6	4.3
15	20	40	40	1	3	5	3.9 (11.82)	6.2 (14.42)	12.1(20.32)	75.51	1.0	1.6	3.2	5.8
16	20	40	50	1	3	5	4.2 (12.12)	6.8 (15.23)	13.4(21.46)	72.87	1.3	1.8	3.5	5.4
17	0	0	40	1	3	5	4.4 (12.68)	9.2 (17.72)	18.4(25.46)	62.75	1.4	2.7	5.1	4.7
18	0	0	0	5	5	7	15.7(23.82)	25.8 (31.42)	49.4(45.76)	-	2.2	3.2	8.6	3.1
SEm±							0.39	0.82	0.82		0.38	0.57	0.88	0.23
CD at 5%							1.16	1.42	2.34		NS	1.58	2.36	0.61

Figures in parenthesis are angular transformed values; DAS : Days after sowing; DI : Disease incidence; Y/P : Yield / Plant; * Per cent increase in disease incidence

Application of $N_{20}P_{40}K_{40}$ per hectare gave highest reduction (75.51%) in MYMV disease incidence over control and this reduction in disease incidence was highly significant. Increasing nitrogen dose from 0-30 kg/ha keeping phosphorus and potassium application (40 kg/ha each) constant, resulted in significant increase in MYMV disease incidence. On the other hand, keeping nitrogen application constant and increasing phosphorus application from 20 kg to 60 kg/ha and potassium application from 20 kg to 50 kg/ha there was significant decrease in disease incidence. The information on the effect of fertilizer application on the incidence of MYMV disease in moth bean is not available. However, the information of N P K treatments on severity of yellow mosaic virus in urdbean and mungbean is available. Our findings are in confirmation with the findings of Rajashekhar (6) who studied the effect of host nutrition on the incidence of MYMV disease on mungbean and found highest reduction in disease incidence (53.3%) with $N_{20}P_{40}K_{40}$. Application of 40 kg/ha each of phosphorus and potassium alone reduced incidence of MYMV disease significantly.

The number of white flies on healthy and diseased plants differed significantly among the treatments. White fly population increased with the increase in disease incidence. These findings are in confirmation with the findings of Murugeshan and Chelliah (7), Patel and Srivastava (8) and Nath and Saikia (9) who reported direct correlation between white fly population and mung bean yellow mosaic virus incidence in mungbean. Sahoo and Sahu (10) found similar effects in urdbean. The MYMV disease incidence and white fly population in mothbean increased significantly with the increase in time of observations. Such information appears to be reported for the first time in mothbean. As the data in support of above findings on mothbean with special references to fertilizer application have not been recorded from the available literature. The grain yield per plant with respect to increase in disease incidence and white fly population in mothbean decreased significantly. Effect of fertilizer application on severity of MYMV disease and increase in white fly population on the grain yield per plant in mothbean has not bean recorded from the available literature. Hence $N_{20}P_{40}K_{40}$ (kg/ha) may be used in reducing the MYMV disease incidence, white fly population and increasing the grain yield of mothbean.

REFERENCES

1. Lodha, S., Gupta, G.K. and Singh, S. 1986. Crop disease situation and some new records in Indian arid zone. *Ann. Arid Zone*. 25(4): 311-320.
2. Bhati, D.S., Mathur, J.R. and Sharma, R.C. 1987. "Jawala" a virus resistant mothbean. *Indian Farm.* 37(4): 7-14.
3. Sivaprakasam, K., Anabalagan, R. and Rajendran. T. 1983. Influence of NPK nutrition on incidence of yellow mosaic in urdbean. *Pulse Crop Newsletter*. 3: 55-56.

4 Singh, B.R., Singh, M., Yadav, M.D. and Dingar, S.M. 1982. Yield loss in mungbean due to yellow mosaic. *Sci. and Cult.* 48: 435-436.

5 Chhabra, K.S., Kooner, B.S. and Singh, G. 1979. Field resistance of certain cultivars of mung bean to whitefly (*B. tabaci*) and yellow mosaic virus. *J. Res. Punjab Agric Univ.* 16: 385-388.

6 Rajashekhar N. 2000. Studies on mungbean disease with special reference to mungbean yellow mosaic virus. M.Sc.Thesis. Hrayana Agric. Univ., Hisar. Pp. 1-49.

7 Murugeshan, S. and Chelliah, S. 1977. Influence of sowing time on the incidence of the vector *Bemisia tabaci* and yellow mosaic disease of green gram. *Madras Agric. J.* 64: 128-130.

8 Patel, M.B. and Srivastava, K.P. 1990. Field screening of some high yielding genotypes of mungbean, *vigna radiata* (Linnaeus). Wilczk of whitefly *Bemicia tabaci* (Gennadius) and yellow mosaic virus (YMV). *Indian J. Ent.* 52 (4): 547-551.

9 Nath, P.D. and Saikia, A.K. 1995. Effect of time of sowing on the incidence of mung bean yellow mosaic disease and whitefly (*Bemisia tabaci* Genn.) population in green gram. *Ann. Agric. Res.* 16 (4): 483-484.

10 Sahoo B.K. and Sahu P.N. 1991. Evaluation of promising blackgram varities against whitefly (*Bemisia tabaci* Genn.) and yellow mosaic. *Madras Agric. J.* 78 (1): 93-94.

Integrated nutrient management in clusterbean -Indian mustard cropping system in arid zone

Raj singh and Bhagwan singh

Central Arid Zone Research Institute, Jodhpur -342003

ABSTRACT

A field experiment was conducted during 2004-05 and 2005-06 at Satheen, Jodhpur, and Rajasthan, to find out the effect of nutrient management practices on the production potential of clusterbean-Indian mustard cropping system under arid zone condition. Significant increase in seed and straw yield of cluster bean was observed with the combined application of FYM @ 2.5 t ha^{-1}+50% RDF (10 kg N+20 kg P_2O_5 ha^{-1}) + bio fertilizer (*Rhizobium*+ PSB), followed by FYM @ 5 t ha^{-1}. The residual effect of nutrient management treatments significantly affected the growth, yield attributes and yield of succeeding Indian mustard. However, highest seed yield (2120 kg ha^{-1}) of Indian mustard was attained due to residual effect of FYM @ 5 t ha^{-1} applied in preceding cluster bean, which was 13.2 , 20.2 and 21.27% higher over the application of FYM @ 2.5 t ha^{-1}+50% RDF (10 kg N+20 kg P_2O_5 ha^{-1}) +biofertilizer (*Rhizobium*+ PSB), 100 % RDF (20 kg N+40 kg P_2O_5 ha^{-1}) and control, respectively. The highest mean mustard seed equivalent yield (2510 kg ha^{-1}) and net returns (Rs.22980 ha^{-1}) in cluster bean-mustard cropping systems were obtained, when cluster bean received FYM @ 5 t ha^{-1}, followed by FYM @ 2.5 t ha^{-1} +50% RDF (10 kg N+20 kg P_2O_5 ha^{-1})+bio fertilizer (*Rhizobium*+ PSB). Maximum improvement in physical and chemical properties of soil was also noticed with the application of FYM @ 5 t ha^{-1} and FYM @ 2.5 t ha^{-1}+50% RDF (10 kg N+20 kg P_2O_5 ha^{-1})+bio fertilizer (*Rhizobium*+ PSB).

Key words: Interacted nutritional management, clusterbean, indian mustard, cropping system, preceding crops.

INTRODUCTION

Cluster bean [*Cyamopsis tetragonoloba* (L.) Taub.] and Indian mustard [*Brassica juncea* (L.) Czern & Coss.] both are the important commercial crop of the arid zone of Rajasthan. The zone accounts about 78 and 67% to the total production and area of clusterbean and 30 and 37.15% to the area and production of mustard of the state, respectively. The introduction of short duration varieties of clusterbean has assured the feasibility of cluster bean-Indian mustard cropping system in arid zone of Rajasthan, where irrigation facilities are available. Of course raising the crops in sequences have become imperative to realize increased agricultural production per unit area and also meeting the ever-increasing demand for food, fiber and fuel. However, intensive cropping led to impoverishment of soil fertility and decline in crop productivity. Therefore to stabilize the productivity of the cropping system, judicious use of adequate nutrient management is of prime importance, as most of the soils of arid zone are deficient in organic matter, available N and phosphorus (1). Inclusion of legumes in the systems have many beneficial effects viz. increased nutrient availability, improvement in soil structure, reduced disease incidence and increased mycorrhizal colonization (2). Besides, a favourable effect of nutrient application on the growth and yield of clusterbean is well documented in arid zone condition (3, 4). Further, integrated nutrient management involving conjunctive use of organic, inorganic and bio fertilizer improved the total productivity and monetary benefit of the cropping systems due to direct and residual effect, as only part of its nutrient content is available to the first crop and succeeding crop benefits from the residual nutrients (5, 6). Keeping this in view, present investigation was therefore, undertaken to study the effect of nutrient management on the production potential of cluster bean-Indian mustard cropping system in arid zone of Rajasthan.

MATERIALS AND METHODS

A field experiment was conducted at Satheen, Jodhpur (Rajasthan) during *kharif* and *rabi* seasons of 2004-05 and 2005-06. The soil was sandy loam, with p^H 8.1, low in organic carbon (0.28%) and available N (148 kg ha^{-1}), medium in available P (13.6 kg ha^{-1}) and adequate in available K (268 kg ha^{-1}). Four nutrient management treatments [control, FYM @ 5 t ha^{-1}, 100% RDF (20 kg N+40 kg P_2O_5 ha^{-1}) and FYM @ 2.5 t ha^{-1}+ 50% RDF (10 kg N+ 20 kg P_2O_5 ha^{-1}) + bio fertilizer (*Rhizobium* + PSB)] were laid out in randomised block design with three replications. Cluster bean (RGC 936) was sown in rows 45 cm apart on 7 and 8 July during 2004 and 2005, respectively under rainfed condition and harvested at the physiological maturity of the crop. Well decomposed FYM containing 0.51%N, 0.26% P, and 0.60% K with 30:1 C: N ratio was applied 15 days before sowing, while full quantity of N and phosphorus was applied at the sowing time of cluster bean as per treatments. Seed treated with *Azotobacter* and PSB was sown in the respective plots.

In succeeding *rabi* season, the residual effect of treatments was studied in Indian mustard (Pusa Jai Kisan), which was sown in rows at 45 cm apart using 5 kg seed ha^{-1} on 28 and 21 October during respective seasons under irrigated condition. The recommended dose (60 kg N and 40 kg P_2O_5 ha^{-1}) was applied through urea and SSP, respectively. Half of N and full quantity of phosphorus were applied at the time of sowing as basal application, while remaining N was top dressed in 2 splits i.e. at 30 and 50 DAS. The crop was harvested on 27 and 25 February during 2005 and 2006, respectively. Mustard seed equivalent yield and economics were calculated considering the prevailing market prices of input and outputs. Soil was analyzed at the initial stage and after completion of the study to monitor the changes in nutrient status due to different treatments.

RESULTS AND DISCUSSION

Direct effect on cluster bean

Significant effect of different nutrient management treatments was noticed on growth, yield attributes and yield of cluster bean. (Table1). Combined application of FYM @ 2.5 t ha^{-1}+50% RDF +bio fertilizer significantly increased plant height and pods $plant^{-1}$ over rest of the treatments. Seeds pod^{-1} and 1000 seed weight were also recorded maximum with the combined application of FYM @ 2.5 t ha^{-1}+50 % RDF + biofertilizer and being statistically at par with 100% RDF and FYM @ 5 t ha^{-1} was significantly higher over control. On an average 22.51, 15.6 and 18.16% increase in plant height, 53.91, 38.8 and 9.94% increase in pods plant $^{-1}$; 22.95, 18.03 and 13.11% increase in seeds pod^{-1} and 6.52, 5.43 and 3.98% increase in test weight were registered with the application of FYM @ 2.5 t ha^{-1}+50 % RDF + bio fertilizer, 100% RDF and FYM @ 5 t ha^{-1}, respectively over control.

Table 1. Effects of nutrient management on the yield attributes and yield of cluster bean

Treatment	Plant height (cm)	Pods $plant^{-1}$	Seeds pod^{-1}	1000 seed weight (g)	Yield (kg ha^{-1})	
					Seed	Straw
Control	78.2	38.4	6.1	27.6	294	735
FYM 5 ton ha^{-1}	92.4	49.8	6.9	28.7	385	1078
100% RDF (20 kg N+40 kg P_2O_5 ha^{-1})	90.4	53.3	7.2	29.1	360	984
FYM @ 2. 5 ton +50% RDF (10 kg N+20 kg P_2O_5 ha^{-1}) +Biofertilizer (*Rhizobium* +PSB)	95.8	59.1	7.5	29.4	420	1142
C.D (P=0.05)	5.25	6.16	0.84	1.18	25.5	128

Significantly highest seed and Stover yield was recorded with the combined application of FYM @ 2.5 t ha^{-1}+50% RDF + bio fertilizer. The seed yield with this treatment was 14.3, 8.30 and 30.1% higher over 100% RDF, FYM @ 5 t ha^{-1} and control, respectively. Combined application of FYM @ 2.5 t ha^{-1}+50% RDF + bio fertilizer (*Rhizobium*+ PSB) also gave 13.83, 5.60 and 35.6% higher straw yield compared to 100% RDF, FYM @ 5 t ha^{-1}, and control, respectively. This might be attributed to the adequate supply of all required nutrients and better utilization by cluster bean crop due to application of FYM and inorganic nutrients (N and P). Further, seed treatment with bio fertilizers (*Rhizobium* and PSB) increased the supply of plant harmones by the microorganism. Thus it indicates that application of inorganic fertilizer in combination with organic manures is better than the single application. Gupta *et al* (7). also reported the beneficial effect of conjunctive use of organic and inorganic fertilizer than inorganic fertilizer alone in arid zone.

Residual effect on Indian mustard

Residual effect of nutrient management had significant influence on the growth, yield attributes and seed yield of succeeding Indian mustard except test weight (Table 2). FYM @ 5 t ha^{-1} applied in cluster bean registered maximum plant height, siliqua plant^{-1}, and seeds siliquae^{-1} of succeeding Indian mustard followed by combined application of FYM @, 5 t ha^{-1} +50 % RDF+ bio fertilizer and both these treatments found significantly superior over 100% RDF and control. However, no significant difference was observed on the test weight.

Table 2. Residual Effect of nutrient management treatments on the yield attributes and yield of Indian mustard

Treatment	Plant height (cm)	Siliquae plant^{-1}	Seeds Siliqua^{-1}	1000 seed weight(g)	Seed yield (kg ha^{-1})
Control	146	215	10.4	4.02	1669
FYM 5 ton ha-1	168	252	11.6	4.24	2120
100% RDF (20 kg N+40 kg P_2O_5 ha^{-1})	151	222	10.8	4.13	1692
FYM @ 2.5 ton +50% RDF (10 kg N+20 kg P_2O_5 ha^{-1}) +Biofertilizer (*Rhizobium* +PSB)	159	236	11.2	4.19	1840
C.D (P=0.05)	9.2	12.82	0.62	NS	129.16

Highest seed yield of Indian mustard (2120 kg ha^{-1}) was recorded with the application of FYM @ 5 t ha^{-1} in preceding cluster bean, which was significantly higher over rest of the treatments. This treatment gave 13.2, 20.2 and 21.27 % increase in the mean seed yield of Indian mustard over the application of FYM @ 2.5 t ha^{-1}+50% RDF +bio fertilizer, 100% RDF and control, respectively.

Combined application of FYM @ 2.5 t ha^{-1} +50% RDF + bio fertilizer being second best treatment showed 8.0 and 9.3% increase in seed yield over 100% RDF and control, respectively. This was owing to improvement in physical, chemical and biological properties of soil due to residual of FYM applied either alone or in combination during preceding season, which in turn led to increased the growth, yield attributes and yield of succeeding Indian mustard, indicating the residual effect of FYM either alone or in combination with fertilizer and bio fertilizer (8).

Changes in physio-chemical properties

The soil analysed after 2 years of field experimentation revealed, there was a gradual improvement in physical (reduction in soil pH and bulk density) and chemical properties (increase in the organic carbon, available N and P status) of the soil due to the application of nutritional treatments over initial value (Table 3). Application of FYM 5 t ha^{-1} exhibited maximum improvement in the physical properties of the soil by reducing soil pH and bulk density. This treatment also resulted in maximum improvement of organic carbon, available N and P status in the soil compared with other nutrient management practices. Further integration of FYM @ 2.5 t ha^{-1} with 50% RDF and bio fertilizer registered next best treatment for improving better physio-chemical properties of the soil compared with 100% RDF and control. The improvement in the physio-chemical properties of the soil could be attributed to the application of FYM.

Table 3. Effect of nutrient management on the Physio-chemical properties of the soil in cluster bean- Indian mustard cropping system

Treatment	Bulk density (g cm^{-3})	Soil pH	O.C (%)	Available N (kg ha^{-1})	Available P (kg ha^{-1})
Control	1.55	8.1	0.24	224	12.4
FYM 5 ton ha-1	1.52	7.9	0.29	265	14.2
100%RDF(20 kg N+40 kg P_2O_5 ha^{-1})	1.55	8.0	0.25	243	13.5
FYM @ 2.5 ton +50% RDF (10 kg N+20 kg P_2O_5 ha^{-1}) +Biofertilizer (*Rhizobium* +PSB)	1.54	7.9	0.27	252	13.8
Initial value	1.56	8.1	0.22	210	12.1

Economics

Economic analysis of the cropping system revealed that application of FYM @ 5 t ha^{-1} fetched the highest net return (Rs. 22980 ha^{-1}), followed by FYM @ 2.5 t ha^{-1} +50 % RDF+ bio-fertilizer (Table 4). The net profit was 18.29, 25.98 and 27.21% higher with the application of FYM @ 5 t ha^{-1} over combined application of FYM @ 2.5 t ha^{-1} + 50% RDF +bio fertilizer, 100% RDF and control, respectively. Integration of FYM @ 2.5 t ha^{-1} + 50% RDF +bio fertilizer also fetched 9.40 and 10.91% higher net return than 100% RDF and control, respectively. Application of 100% RDF was less remunerative compared to corresponding treatments. But benefit: cost ratio was maximum with the application of 100% RDF followed by FYM @ 2.5 t ha^{-1} + 50% RDF + bio fertilizer. However, application of FYM @ 5 t/ha resulted in lowest benefit: cost ratio. This could be attributed to the involvement of higher cost of the treatment, which caused low benefit: cost ratio.

Table 4. Effect of nutrient management on the mustard seed equivalent yield, monetary benefit and benefit: cost ratio in clusterbean - Indian mustard cropping system

Treatment	Mustard equivalent seed yield (kg ha^{-1})	Gross return (Rs ha^{-1})	Cost of cultivation (Rs.ha^{-1})	Net return (Rs.ha^{-1})	Benefit: cost ratio
Control	1996	35928	19200	16728	1.15
FYM 5 ton ha-1	2510	45180	22200	22980	0.97
100% RDF (20 kg N+40 kg P_2O_5 ha^{-1})	2120	38160	21150	17010	1.24
FYM @ 2.5 ton +50% RDF (10 kg N+ 20 kg P_2O_5 ha^{-1}) +Biofertilizer (*Rhizobium* +PSB)	2307	41526	22750	18776	1.21
C.D (P=0.05)	185	-	-	-	-

It was concluded that application of organic manure alone or in combination is beneficial to obtain more yield and monetary benefit by clusterbean-mustard cropping system in arid region of Rajasthan.

REFERENCES

2 Tripathi, K.P. and Praveen Kumar 2000. Phosphate adsorption in aridosols in relation to soil properties. *Annals of Arid Zone* 39:131-135

3 Georage, M. 1987. Studies on the effect of cereal and cereal-legume cropping systems on the productivity and fertility of soil. Ph.D. Thesis, Indian Agricultural Research Institute, New Delhi.

4 Singh Raj 2004. Response of cluster bean genotypes to nutrient management under arid condition of Rajasthan. *J. Arid Legumes* 1(1): 32-34.

5 Rathore, V.S., Singh, J.P., Soni, M.L., Yadava, N.D. and Beniwal, R.K. 2007. Productivity, quality and resource utilization by cluster bean (*Cyamopsis tetragonoloba*) as influenced by nutrient management. *Indian Journal of Agronomy* 52(3):243-246.

6 Singh R. 2002. Effect of integrated nutrient management in cluster bean (*Cyamopsis tetragonoloba* L. Taub.) – Wheat (*Triticum aestivum*) cropping system under western Rajasthan. *Indian Journal of Agronomy* 47(1):41-45.

7 Yadav, R.S., Yadav, P.C. and Dhama, A.K. 2003. Integrated nutrient management in wheat (*Triticum aestivum*) mungbean (*Phaseolus radiatus*) cropping sequence in arid region. *Indian Journal of Agronomy* 48(1):23-26.

8 Gupta, J.P., Joshi, D.C and Singh, G.B. 2000. Management of arid agro-ecosystem. In Natural Resource Management for Agricuture Production in India (Eds. J.S.P. Yadav and G.B. Singh) pp. 551-668. International Conference on Managing Natural Resource for Sustainable Agriculture Production in the 21ist Century, New Delhi.

9 Das Anup., Prasad, M. and Gautam, R.C. 2004. Residual effect of organic and inorganic sources of nitrogen applied to cotton (*Gossypium hirsutum*) on succeeding wheat (*Triticum aestivum*). *Indian Journal of Agronomy* 49(3):143-146.

Soil test based fertilizer recommendation under IPNS for clusterbean in *Torripsamments* of Western Rajasthan

I.J. Gulati and S.R. Yadav

STCR Project, Agricultural Research Station, Beechwal,
Rajasthan Agricultural University, Bikaner – 334006- India

ABSTRACT

Soil test crop response correlation studies were conducted with cluster bean (Var. RGC-986) under integrated plant nutrition system (STCR-IPNS) in *Torripsamments* of Rajasthan during *Kharif* 2006. The experiment was conducted in the field where three strips of fertility gradient were created during *Rabi* 2005-06. Three levels of each organic manure, nitrogen, phosphorus and potassium were taken in the augmented resolvable block design. Fertilizer adjustment equations under IPNS were formulated for cluster bean following Ramamoorthy's inductive-cum targeted yield approach. The nutrient requirement for producing one quintal of cluster bean was found to be 6.43, 2.15 and 3.98 kg of N, P_2O_5 and K_2O, respectively. The per cent nutrient utilization efficiencies from soil and fertilizer nutrients were found to be 35.83 and 95.99 for N, 46.23 and 21.71 for P_2O_5 and 13.56 and 58.72 for K_2O, respectively. Like wise the per cent organic nutrient contribution from farmyard manure (FYM) was 62.63 for N, 44.52 for P_2O_5 & 36.56 for K_2O, respectively. In STCR-IPNS technology, the fertilizer doses are tailored to the requirements of specific yield targets of cluster bean taking into account the contribution from soil, fertilizers and organics.

Key words: Clusterbean, soil test, fertilizer.

INTRODUCTION

The modern crop production technology emphasizes the need for integrated plant nutrient supply involving the judicious use of combination of organic and inorganic. The use of chemical fertilizers is a must for Indian agriculture but at least one third of the total nutrients must be in organic form for enhancing the efficiency of the inorganic component, cost effectiveness and reducing the burden of environmental damage (1). Fertilizers with instant ability to refurbish depleted nutrients in necessary quantities and forms have come to be recognized as a key component of soil fertility management. The fertilizer prices have gone up and hence their use in required amounts depends much upon the purchasing ability of the farmers. Accordingly, economic rationality dictates a more compressive approach for fertilizer use, incorporating soils tests, magnitude of crop response to applied nutrients and economic evaluation of the results (2). In these efforts, soil test based fertilizer recommendation plays a vital role in increasing the soil productivity. Research work done using soil test crop response approach in different parts of the country have shown that legume crops respond to the nutrients added through organics and chemical fertilizers. Therefore, in sandy soils of the poor fertility, it becomes pre-requisite to use organic manures along with fertilizers to achieve advancement in productivity of legume crops in an economically and ecologically sound manner.

Clusterbean (*Cyamopsis tetragonoluba L.Taub*) is an important arid legume grown in the arid and semi arid regions of India. Clusterbean is grown for seed, vegetable, fodder and green manure purposes. A main industrial important constituent of clusterbean seed is guar gum. So in order to step up the productivity of clusterbean in the state soil test based balanced fertilization is essential. Currently, a general recommendation of 20: 35: 0 kg of N, P_2O_5 and K_2O ha^{-1}, respectively, is being followed. Fertilization based on blanket recommendation results in either over use or under use of fertilizers, so balanced fertilization is a must for realizing higher efficiency and economy of fertilizer use (3). In fertilizing the crop existing soil fertility and crop requirement should be taken into account (4). This demands the maintenance of optimum balance between all essential nutrients, as per the crop requirement of the crop and their availability in the soil, keeping this in view, the present investigation was undertaken.

MATERIALS AND METHODS

A field experiment based on inductive methodology was conducted in *Torripsamments* of Bikaner during *Kharif* 2006 with clusterbean (Var. *RGC-986*). The soil of the experimental field is loamy sand in texture with pH 8.3 and non-saline (EC_2 0.3 dSm^{-1}). The initial $KMnO_4$-N, Olsen-P and NH_4OAc-K status were 104.00, 21.30 and 195.10 kg ha^{-1}, respectively. Following the inductive methodology of the Ramamoorthy *et al.* (4), three fertility gradients were created in the preceding season by dividing the experimental field into three equal strips. The first strip received no fertilizer, whereas second and third strips, one and two times the standardized dose of N, P and K, respectively. An exhaustive crop of barley (var. *RD 2508*) was grown prior to experimentation.

By growing the exhaustive crop, the operational range of soil fertility was created in the fertility strips, which was evaluated in terms of variations in yield, uptake and soil test values. After the harvest of the exhaustive crop, each strip of the fertility gradient was divided into three equal blocks across the strip for farmyard manure (FYM) levels. In all 18 fertilizer treatments were framed using three levels each of nitrogen, phosphorus and potassium as 20, 40 and 60 kg N/ha, 30, 60 and 90 kg P_2O_5/ha and 20, 40 and 60 kg K_2O/ha, respectively. All the 18 treatment combinations were divided into three groups, each group consisting of six treatment combination augmented with two absolute control treatments. Then each strip was divided into 24 plots and fertilizer treatments were distributed in such a manner that every treatment was accommodated in each gradient strip as well as in FYM block. The design may be viewed as an augmented resolvable block design with three replications.

Nitrogen was applied in the form of urea, phosphorus in the form of single super phosphate and potassium in the form of Muriate of potash. Clusterbean (RGC-986) was sown as test crop. The crop was harvested at maturity. The guar grain and stover yields were recorded plot wise. Composite grain and stover samples of each plot were collected for chemical analysis. Composite soil samples were also collected plot wise to analysis available N, P_2O_5 and K_2O.

Pre-sowing soil samples were collected from each plot before superimposition of the treatments and were analysed for alkaline $KMnO_4$-N (5), Olsen-P (6) and neutral N NH_4OAc-K (7). Clusterbean (Var. *RGC-986*) crop was grown with usual agro-nomic practices. After the harvest of crop, the grain and stover yields of cluster-bean were recorded plot wise. The plant samples from each plot were analysed for total N, P and K content (8) and total uptake was computed using clusterbean yield data.

Using the data of clusterbean yield, nutrient uptake, pre-sowing soil available nutrients and fertilizer doses applied, the basic parameters *viz.*, nutrient requirement (kg q^{-1}), contribution of nutrients from soil (Cs), fertilizer (Cf), and organics (Co) were derived and used for the formulation of fertilizer adjustment equations. Soil test based fertilizer recommendations were also prescribed in the form of a ready reckoner for desired yield target of clusterbean.

RESULTS AND DISCUSSION

Soil available nutriments and clusterbean yield

The range and mean values of clusterbean yield, uptake and available soil nutrients of treated and control plots are furnished in Table 1. The $KMnO_4$-N ranged from 89.43 to 161.22 kg ha^{-1} with a mean of 123.60 kg ha^{-1}, Olsen P ranged from 12.62 to 60.53 kg ha^{-1} with a mean of 31.70 kg ha^{-1} and NH_4OAc-K ranged from 165.43 to 248.27 kg ha^{-1} with a mean 203.90. The clusterbean yield in fertilizer treated plots ranged from 2.80 to 21.03 q ha^{-1} with a mean 11.81 q ha^{-1} and in control plots ranged from 3.21 to 9.23 q ha^{-1} with a mean 6.63 q ha^{-1}. The above data clearly indicate that a wide variability existed in the soil test values and clusterbean yield of treated and control plots, which is a pre-requisite for calculating the basic parameters and fertilizer adjustment equations for calibrating the fertilizer doses for specific yield targets Ramamoorthy *et al.* (4).

Table 1. Range and mean values of available nutrients in the pre-sowing surface soil and yield of clusterbean

Parameters	Range	Mean
	Soil test values	
$KMnO_4$-N	89.43-161.22	123.60
Olsen-P	12.62-60.53	31.70
NH_4OAc-K	165.43-248.27	203.90
	Clusterbean yield (q ha^{-1})	
Treated plots	2.80-21.03	11.81
Control plots	3.21-9.23	6.63
	Nutrient uptake (kg ha^{-1})	
	Treated plots	
N uptake	17.07-144.16	76.04
P uptake	5.26-47.45	25.43
K uptake	10.50-90.64	47.00
	Control plots	
N uptake	16.58-67.79	40.58
P uptake	6.35-18.86	13.38
K uptake	12.13-34.98	26.15

Basic parameters

The basic data, *viz.,* nutrient requirement for producing one quintal of clusterbean yield, the per cent nutrient utilization efficiency for soil and fertilizer have been calculated and presented in Table 2. The nutrient requirements of N, P_2O_5 and K_2O obtained were 6.43, 2.15 and 3.98 kg q^{-1} of clusterbean, respectively. The per cent nutrient utilization efficiencies from soil (Cs) and fertilizer nutrients (Cf) were found to be 35.83 and 95.99 for nitrogen, 46.23 and 21.71 for phosphorus (P_2O_5) and 13.56 and 58.72 for potassium (K_2O).

Similarly the per cent nutrient utilization efficiencies for N, P_2O_5 and K_2O obtained from organics were 62.63, 44.52 and 36.56, respectively. The data on C_s and C_f indicated that the per cent utilization efficiencies from fertilizer source were greater than that from soil in case of N and K_2O.

Table 2. Nutrient requirement, per cent contribution from soil, fertilizer and organics for clusterbean

Parameters	N	P_2O_5	K_2O
Nutrient requirement (kg q^{-1})	6.43	2.15	3.98
Soil nutrient utilization efficiency (%)	35.83	46.23	13.56
Fertilizer nutrient utilization efficiency (%)	95.99	21.71	58.72
Organic nutrient contribution (%)	62.63	44.52	36.56

Fertilizer adjustment equations for desired yield targets of clusterbean

Soil test based fertilizer models equations for targeted yield of clusterbean were formulated using initial soil test values and the data on NR, Cs ,Cf and Co are furnished in Table 3.

Table 3. Soil test based fertilizer prescription equations for targeted yield of clusterbean

$$FN = 6.70\,T - 0.37\,SN - 0.65\,ON$$

$$FP = 9.90\,T - 2.13\,SP - 2.05\,OP$$

$$FK = 6.78\,T - 0.23\,SK - 0.62\,OK$$

Note: FN, FP and FK – Fertilizer N, P_2O_5 and K_2O in kg ha^{-1}, respectively; T – Yield target in q ha^{-1}; SN, SP and SK are $KMnO_4$-N, Olsen- P (P_2O_5) and NH_4OAc-K (K_2O) in kg ha^{-1}, respectively. ON, OP and OK denote percent N, P_2O_5 and K_2O content, respectively in organic manure.

The yield targeting equations are valid under the following situations

1. These should be used for similar soils occurring in a particular –Agro climatic zone.
2. Targets chosen should not be unduly high or low and should be within the range of experimental yields obtained.
3. Adjustment equations must be used within the experimental range of soil test values and can not be extrapolate.
4. Good and recommended agronomic practices need to be followed while raising crops.
5. Other micro and secondary nutrients should not be yield limiting.

On the basis of these equations, a ready reckoner was prepared for range of soil test values and for a yield target of 12.00 q ha^{-1} with and without FYM/compost (Table 4) under different soil test values. For producing 12.00 q ha^{-1} of clusterbean grain yield in *Torripsamments*, the fertilizer doses required without FYM for the average soil test values of 110, 40 and 220 kg ha^{-1} N, P_2O_5 and K_2O, respectively would be 39.7, 33.6 and 30.8 kg N, P_2O_5 and K_2O respectively. But with with 5 t ha^{-1} FYM these doses would be 17.6, 8.6 and 15.9 (maintenance dose) kg ha^{-1} N, P_2O_5 and K_2O, respectively. This shows that application of FYM also contributed for particular nutrients i.e nitrogen, phosphorus and potassium. The findings are in line with those reported by Reddy *et al.* (9, 10) and Santhi *et al.* (11).

Table 4. Estimates of soil test based fertilizer recommendations for 12 q ha^{-1} grain yield of clusterbean (kg ha^{-1}).

Soil test values (kg ha^{-1})			Fertilizer for 12 q ha^{-1} without FYM			Fertilizer for 12 q ha^{-1} with 5 t ha^{-1} FYM		
N	P_2O_5	K_2O	N	P_2O_5	K_2O	N	P_2O_5	K_2O
90	10	160	47.1	97.5	44.6	25.0	54.4	23.5
100	20	180	43.4	76.2	40.0	21.3	33.1	18.9
110	30	200	39.7	54.9	35.4	17.6	11.8	14.3 (15.9)
120	40	220	36.0	33.6	30.8	13.9 (15.0)	- (8.6)	9.9 (15.9)
130	50	240	32.3	12.3	26.2	10.2 (15.0)	- (8.6)	5.1 (15.9)
140	60	260	28.6	0 (8.6)	21.6	6.5 (15.0)	- (8.6)	0.5 (15.9)
150	70	280	24.9	0 (8.6)	17.0	2.8 (15.0)	- (8.6)	- (15.9)
160	80	300	21.2	0 (8.6)	12.4 (15.9)	- (15.0)	- (8.6)	- (15.9)

Average N, P_2O_5 and K_2O content in FYM taken as 0.68, 0.42 and 0.68 percent, respectively. Figures in parentheses indicate maintenance doses of fertilizer N, P_2O_5 and K_2O

On the basis of a large amount of data obtained from different soil agro-climatic regions of the country under the All India Coordinated Research Project on Soil Test Crop Response Correlation, Subba Rao and Srivastava (12) reported similar results. The present study amply demonstrates that the STCR based recommendations lead to balanced supply of nutrients avoiding either under or over uses of fertilizer inputs, as in the technology fertilizer doses are tailored to the requirements of specific yield targets of crop taking into account the contribution from soil and fertilizers (13). Hence, there will be a balanced supply of nutrients avoiding either under or over uses.

REFERENCES

1. Sankaram, A. 1997. The Hindu Survey of Indian Agriculture. P. 169.
2. Katyal, J.C. 2001. Fertilizer Use Situation in India. *J.Indian Soc. Soil Sci.* 49, 570-592.
3. Velayutham, M and Reddy, K.C.K. 1990. *Soil Fertility and Fertilizer Use. Vol. IV, IFFCO*. New Delhi.
4. Ramamoorthy, B., Narasimham, R.L. and Dinesh, R.S. 1967. Fertilizer application for specific yield target of *Sonora-64* wheat. *Indian Fmg.*, 17,43.
5. Subhiah, B.V. and Asija, G.L. 1956. Rapid procedure for estimation of available nitrogen in soils. *Curr. Sci.* 25, 259.
6. Olsen, S.R., Cole, C.V., Watanabe, F.S. and Dean, L. 1954. Estimation of available phosphorus in soil by extraction with sodium biocarbonate. *U.S.D.A. Circ. 939, U.S. Govt. Printing Office*, Washington D.C.
7. Hanway, J.J and Heidal, H. 1952. Soil Analysis Methods used in Iowa State College Soil Testing Laboratory, *Iowa Agronomy*, 57,1.
8. Piper, C.S. 1950. *Soil and plant analysis*. Hans Publications, Bombay.
9. Reddy, K.C.K., Velayutham, M. and Maruthi Sankar, G.R. 1994. Soil Test Based Fertilizer Prescription for yield targets and crops-Extension Bulletin of All India Coordinated Soil Test Crop Response Correlation Project, CRIDA, Hyderabad.
10. Naidu, L.G.K, Hunsigi, G and Mohan, N.G.R. 1999. Soil test based fertilizer recommendation to sugarcane- A rational Approach. *Fert. News* 44 (8): 29-36.
11. Rao, V.N., Rego, J.J. and Meyers, R.J.K. 1997. Balanced fertilizer use in black soils. *Fert. News,* 42 : 35-45
12. Subba Rao, A. and Srivastava, S. 1999. Experiences on current status of crop responses to fertilizers in different agro climatic zones as learnt from AICRP on Soil Test Crop Response Correlation. *Fert. News* 44 (4): 83-95.
13. Santhi, R., Natesan, R. and Selva Kumari, G. 2002. Soil test crop response correlation studies under irrigated plant nutrient ion system for onion (*Allium cepa* L. var. *aggregatum*) in Inceptisols of Tamil Nadu. *J. Indian Soc. Soil Sci.* 50 : 489-492.

Role of arid legumes in mobilization of resin phosphorus

Indira Rathore, Poonam Joshi and J. C. Tarafdar

Central Arid Zone Research Institute, Jodhpur – 342 003, Rajasthan

ABSTRACT

A field experiment was conducted in Central Research Farm, Central Arid Zone Research Institute, to investigate the changes in resin-P status in the rhizosphere of three important arid legumes (clusterbean, moth bean, mung bean) due to residue incorporation, tillage operation and inoculation with phosphatase/ phytase releasing/producing bacteria *Bacillus coagulans*. The results were also compared with the most important cereal (pearl millet) under arid environment. Both basic anion exchange and chelating resin was used below 15-18 cm in the rhizosphere during the critical period of growth of different crops. Resin P (the P available by the plant during the growing period) was, in general, 26% higher in untilled plots than tilled ones, which was more under chelating resins after six year of cropping. Application of residue @ 4t ha^{-1} increased the resin-P status by 12 to 54% than corresponding control. Chelating resins showed higher P trapping than basic anion exchange resin especially under untilled condition. Inoculation of legumes with P-mobilizing bacteria *Bacillus coagulans* showed 84% higher resin-P mobilization in the rhizosphere than uninoculated control.

Key words: Anion-exchange resin, arid crops, *Bacillus coagulans*, chelating resin, crop residue, resin-P, tillage

INTRODUCTION

Ion exchange resins are highly ionic, covalently cross-linked, insoluble poly-electrolytes supplied as beads. The use of resins for measuring soil nutrient availability in plots of limited size is especially attractive because of their nondestructive nature [1]. Several studies have been attempted to measure nutrient dynamics under field conditions with resin filled nylon bags or capsules. An ion exchange membranes (AEMS) have been used successfully to measure inorganic soil P fluxes under field conditions [2, 3] and employment of ion-exchange resins to quantify labile soil P constituted a major methodological breakthrough. The resin method overcomes the disadvantage of chemical extraction that are inherently static and do not account for the kinetics of nutrient release and transport [4] and may mobilize nutrient forms other than those that are truly plant available [5].

P in soil has a varying labiality, i.e. availability for potential mineralization by microorganisms or phosphohydrolyzes [6]. The labiality may vary according to the nature of the compound, its association and other soil constituents, and its physical position inside or outside the soil micro-aggregates. In general, old, highly weathered soils are thought to contain little P in the most labile forms, whereas in partially weathered soils and soils high in organic matter, the labile fractions are present in higher concentrations, and represent a greater percentage of the total P [7].

This paper reports the results of field experiments, which investigated the changes of resin P status due to crop rotation, residue application, tillage and inoculation with P mobilizing bacteria *Bacillus coagulans* under different arid crops.

MATERIALS AND METHODS

***Field experiment*:** The experiments were conducted at Central Research Farm of Central Arid Zone Research Institute, Jodhpur, Rajasthan, India, located at 26°18' N latitude and 73°01' longitudes. The climate of the region is arid and a mean annual precipitation of 360 mm. Rainfall occurs mostly between July and September. The annual mean temperature is 26.8°C. The soil is loamy sand. The arid crops grown were three legumes such as clusterbean, moth bean, mung bean and most important cereal pearl millet in *kharif* season (July – September) in triplicate split plot design with an equal plot size of 20 m^2 (4 m × 5 m) each.

Construction of resin bags: Rectangular resin bags (3.5 × 4.0 cm) were constructed from terylene-net in a similar manner to that of Sibbesen [8]. The terylene had a mesh size (0.3 mm) sufficient to retain the resin without loss and allow through-movement of moisture. The bags were sewan with non-rotting polyester thread and heat sealed. Before sealed the bags, 0.4 g of strongly basic anion exchange resin (INDION® GS 400) in Cl^- form or chelating resin were put into each bag.

Regeneration of the resin bags: To regenerate the resin bags prior to use the bags were initially washed with distilled water. Each bag was placed in a 50 mL beaker containing 20 mL of 0.5M $NaHCO_3$ and shaken for 60 minutes at 150 rpm. This was repeated for three times. After the final wash and further rinse with distilled water the bags were ready for use.

Placement of resin bags in the rhizosphere: Three resin bags of each replicate in all genotypes were placed 15 to 18 cm below the soil surface into the root zone at 4 weeks of plant age. The bags were removed after 3 weeks of placement (i.e. at 7 weeks of crop age).

Elusion of resin bag: Each resin bag was initially rinsed in distilled water to clean off attached soil and root material. Subsequently each resin bag was shaken for 60 minutes at 150 rpm with 20 mL 5 %(v/v) HCl and the aliquot was collected for estimation of P by ammonium molybdenum blue colour method [9].

Inoculation with Bacillus coagulans: All the seeds were sterilized (in case of inoculated treatment) with acidified 0.05% $HgCl_2$ for 2 minutes followed by surface sterilization with alcohol for 5 minutes. Thereafter, washed with sterilized deionized water for 5 to 6 times. The inoculation with *Bacillus coagulans* was carried out in the slurry of carrier based culture in sterilized (20%) jaggery (gur) solution and seed was treated with *Bacillus coagulans* (10 mL 100 g^{-1} seed) having 2×10^8 mL^{-1} population. The spore coated seeds were air dried before sowing.

Statistical analysis: Standard errors of means were calculated and analysis of variance and least significance difference (LSD) were carried out by means of standard procedure [10].

RESULTS AND DISCUSSION

The effect of tillage on labile P status under both anion exchange and chelating resins were presented (Table 1). The accumulation of P under legumes was 3.5 to 5.1% higher than cereals. Resin-P was found more in chelating resin than anion-exchange resin used for study. In general, 26% higher resin-P was observed under untilled plots than tilled ones.

Table 1. Accumulation of resin P (mg kg^{-1}) under tilled and untilled condition

	Tilled	Untilled
(A) Anion exchange resin		
Cereals	6.49	6.47
Legumes	6.73	6.71
% improvement in legumes	3.5	3.7
(B) Chelating resin		
Cereals	6.73	8.35
Legumes	7.02	9.20
% improvement in legumes	4.3	5.1

The application of residues showed positive influence on resin-P accumulation both under tilled and untilled condition (Table 2). The effect of residues was more under untilled (52.3 to 54.2%) than tilled condition (12.4 to 12.6%). The effect under legumes was slightly more than cereals.

Table 2. Effect of residues on accumulation of P (mg kg-1) by ion exchange resin*

	Tilled		Untilled	
	Control	Residue	Control	Residue
Cereals	6.49	7.41 (12.4)**	6.47	13.57 (52.3)
Legumes	6.73	7.70 (12.6)	6.71	14.64 (54.2)
LSD (p = 0.05)	0.29	0.50	0.63	1.93

* Average of both anion exchange and chelating resins
** Figures in parenthesis indicates the % improvement due to residue application

Resin-P (labile P) is thought to available to microbial and plant communities for a short period, because it is rapidly disappeared from the surface of soil particles. Non-labile P fractions are thought to be tightly bound to soil particles and unavailable to plants. The high levels of resin-P polls due to cultivation of legumes and residue application indicate an ample P supply for crop growth. Our result suggests the rhizosphere influence the resin-P status.

Accumulation of resin P under different crops (Fig. 1) clearly indicated significant effect of inoculation under different crops. Resin P accumulation was up to 133% higher due to inoculation than control treatment. Resin-P was estimated more under inoculated condition followed by residue application than control soil. In general, inoculation resulted 84% higher resin-P accumulation than in control soil.

Many mechanisms have been proposed to account for the increased mobilization of soil phosphorus in plant rhizosphere. These include the exudation of reducing substances, the production of mucilage and sloughed off root cells, which provide the energy substrate for microbial activity in the rhizosphere, which subsequently influences the hydrolysis of minerals and organic P [11]. Our results suggest that the regulation of resin P pools is neither closely related to vegetation demand, nor it is strongly correlated by plant uptake. Extent of actual availability of each P fraction to plants is somewhat uncertain but it appears that the pools of resin P thought to be available to plants are considerably larger than the annual demand for P. Tissen *et al.* [12] used multiple regression and path analysis to explore the relationships they postulated in their conceptual model of P flows. They found that bicarbonate-Po was most closely correlated with resin P, and suggested that the mineralization of Po fractions was supporting the resin-P pools. Our study demonstrated that legume rotation, residue incorporation and inoculation with P mobilizing bacteria many increase the resin-P (labile P) status in the soil.

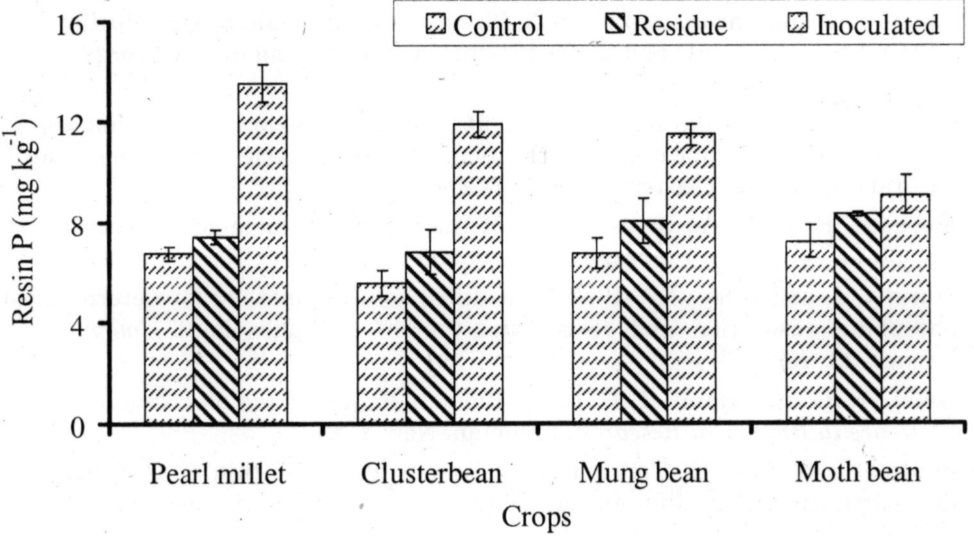

Fig. 1. A comparison of P accumulation by ion exchange resins under different arid crops (average of two different resins). I indicate the standard errors of mean.

ACKNOWLEDGEMENT

The authors are grateful to Dr. K. P. R. Vittal, Director, CAZRI, for his keen interest of the work and ICAR for funding the National Fellow Scheme under which the present work has been carried out.

REFERENCES

1. Johnson, D.W., Hungate, B.A., Dijkstra, P., Hymus, G. and Drake 2001. Effects of elevated CO_2 on soil in a Florida oak ecosystem. *Journal of Environmental Quality*, 30: 501-507.

2. Abrams, M.M. and Jarrel, W.M. 1992. Bioavailability index for phosphorus using ion exchange resin impregnated membranes. *Soil Science Society of American Journal*, 56: 1532-1537.

3. Cooparband, L.R. and Logan, T.J. 1924. Measuring *in-situ* changes in labile soil phosphorus with anion exchange membranes. *Soil Science Society of American Journal,* 58: 105-114.

4. Curtin, D., Syars, J.K. and Smillie, G.W. 1987. The importance of exchangeable cations and resin-sink characteristics in the release of soil phosphorus. *Journal of Soil Science*, 38: 711-716.
5. Menon, R.G., Hammond, L.L. and Sissing, H. 1988. Determination of plant-available phosphorus by iron-impregnated filter paper (Pi) soil test. *Soil Science Society of American Journal*, 53: 110-115.
6. Yadav, R.S. and Tarafdar, J.C. 2003. Phytase and phosphatases producing fungi in arid and semi-arid soils and their efficiency in hydrolyzing different organic P. *Soil Biology and Biochemistry*, 35: 745-751.
7. Cross, A. F. and Schlesinger, W.H. 1995. A literate review and evaluation of the Hedley fraction: applications to the biogeochemical cycle of soil phosphorus in natural ecosystems. *Geoderma,* 64: 197-214.
8. Sibbesen, E. 1978. An investigation of the anion-exchange resin method for soil phosphate extraction. *Plant and Soil,* 50: 305-321.
9. Murphy, J. and Riley, J.P. 1962. A modified solution method for determination of phosphate adsorption and resin extraction of soil phosphate. *Journal of Soil Science*, 33: 709-717.
10. Sokal, R.R. and Rohlf, F.J. 1981. Biometry- *The Principles and Practices of Statistics in Biological Research*, Freeman, New York.
11. Panwar, J., Saini, V.K., Tarafdar, J.C., Kumar, P. and Kathju, S. 2005. Changes in labile P status under different cropping systems in an arid environment. *Journal of Arid Environments,* 61: 137-145.
12. Tiessen, H., Steward, J.W.B. and Cole, C.V. 1984. Pathways of phosphorus transformations in soils of differing pedogensis. *Soil Science Society of America Journal*, 48: 853-858.

Alleviation of water stress effects in mungbean by phosphorus application

B.K. Garg, Uday Burman and S. Kathju
Division of Soil-Water-Plant Relationships
Central Arid Zone Research Institute, Jodhpur 342 003

ABSTRACT

Mungbean [*Vigna radiata* (L.) Wilczek] plants grown under two levels of phosphorus (0 and 40 kg ha^{-1}) were subjected to different intensities of water stress at pre-flowering stage by withholding of irrigation for 0, 3, 6 and 9 days in a pot culture trial. Increasing water stress progressively and significantly decreased plant water potential, relative water content, leaf area, net photosynthetic rate, levels of total chlorophyll, starch and soluble protein as well as nitrate reductase activity under both the P levels. However, mung bean plants grown with P application displayed better photosynthetic efficiency, higher levels of different leaf metabolites and a higher NR activity as compared to no application of phosphorus under all the intensities of water stress, particularly upto 6 days of water deprivation. Furthermore, recovery in all the metabolic parameters was comparatively higher in P-fertilized than unfertilized plants. Therefore, P application enhanced both dry matter and seed yield at all stress levels. On an average P-fertilized plants had 11.2% higher dry matter, 11.7% higher pod weight and 13.4% higher seed yield over unfertilized plants. The results indicate the possibility of water stress alleviation by P application, at least upto moderate levels of water stress in mungbean. The beneficial effects of P application on different physiological processes and plant performance under water limited conditions have been discussed in the light of the published reports.

Keywords: Mungbean, water stress, phosphorus application, photosynthesis, nitrogen metabolism, seed yield.

INTRODUCTION

Mungbean [*Vigna radiata* (L.) Wilczek] commonly known as 'green gram' is a short season summer and *kharif* pulse legume cultivated predominantly under dry land conditions throughout the tropics and subtropics. India is the largest producer of mungbean in the world and accounts for 65% of mungbean area and 54% of world's production. In Rajasthan, mungbean occupy 5.15 lac ha area with the productivity of 163 kg ha^{-1} only, primarily due to abiotic and biotic stresses (1, 2). The crop is exposed to varying timing and severity of water deficits due to erratic nature of monsoon rains and variation in stored soil water at the time of sowing, resulting in low and unstable yields in arid regions (3). Furthermore, yields are also low due to poor soil fertility of arid soils. Therefore, addressing the prevailing fertility constraints can reportedly alleviate adverse effects of water stress on crop productivity under arid environments (4, 5). The supply of phosphorus in particularly critical as declining water availability causes reduced P availability and absorption by mungbean under such conditions (6, 7).

Individual effects of P and drought on plant water relations, leaf metabolism and yield have been reported in several crops but less work has been carried out on their possible interactions. At higher P supply, drought stressed white clover plants maintained a higher leaf water potential without showing moisture stress symptoms and extracted more soil water than P-deficient plants that wilted rapidly and failed to recover on rewatering (8). Furthermore, P deficiency induced decline in leaf growth and photosynthetic rate in spinach leaves has been ascribed to a reduction in stomatal conductance and RUBP carboxylase regeneration capacity (9). Phosphorus deficiency is also known to reduce the uptake of nitrate and its assimilation by nitrate reductase (10). Several field studies have also indicated positive interactions between P and water availability on yield of crops in arid and semi-arid regions (4, 7, 11). However, information is lacking on the responses of mungbean to P nutrition under different intensities of drought. Therefore, interactive effects of P nutrition and drought intensities were studied at the critical pre-flowering stage of mungbean to determine the level of water deficit which could be alleviated by P application.

MATERIALS AND METHODS

This study was carried out in the pot culture during 2004 under net house conditions. Mungbean [*Vigna radiata* (L.) Wilczek cv.K-851] plants were grown in earthen pots (2 plants per pot) containing 10 kg sandy loam soil, composed of 7.1% clay, 5.6 silt, 63.1% fine sand and 24.1% coarse sand. The soil contained 0.28% organic carbon, 0.023% total nitrogen, 80 kg ha^{-1} available N, 12 kg ha^{-1} available P and 120 kg ha^{-1} available K. Plants were supplied with 0 or 40 kg ha^{-1} P (designated as P_0 and P_{40}) applied as single super phosphate at sowing. Soil moisture in the pots was maintained close to field capacity (10%) by regular watering until 40 days after sowing. Thereafter, watering was withheld in different sets of pots (both for P_0 and P_{40} treatments) for 3, 6 and 9 days in a staggered fashion so that drought could be simultaneously terminated on the 49th dry of sowing. One set of undroughted plants, both under P_0 and P_{40} levels, served as controls. Just before rewatering on the 49th day after sowing and again 2 days after rewatering the following observations were recorded in quadruplicate, in all the eight treatments. The plant water potential was determined using a pressure chamber (PMS instrument company, USA) and the relative water content (RWC) was determined according to the standard procedure (12).

The rate of net photosynthesis was measured in two uppermost fully expanded leaves of intact plants using a LICOR-6200 portable photosynthetic system. The two uppermost fully expanded leaves of eight plants from 4 pots in each treatment were analyzed for the estimation of total chlorophyll and the alcohol extract (80%) was used for the estimation of reducing sugars while starch was released by dissolving the pellets in 26% perchloric acid and then estimated by anthrone method. Soluble protein was determined following the method of Lowry et al. (13) while nitrate reductase activity was estimated *in vivo* from leaf discs using KNO_3 as substrate. All these estimations were made following the procedures described earlier (5). Data on seed yield and dry matter of above ground biomass were recorded at harvest from 10 pots in each treatment. Data were subjected to analysis of variance and mean separation was done by LSD at 5% probability.

RESULTS AND DISCUSSION

Plant water status

Increasing water stress progressively and significantly decreased plant water potential (ψ_{plant}) and relative water content (RWC) under both the P levels (Table 1). However, drought induced decrease in ψ_{plant} and RWC was more in P-fertilized plants as compared to unfertilized plants at all stress intensities, probably due to higher leaf area and more plant growth in the P-fertilized plants. More reduction in plant water status in fertilized than unfertilized plants under water stress conditions has also been reported in other crops like cotton (14), pearl millet (15) and clusterbean (16). Rewatering of stressed plants led to fast recovery within 48 hours. However, extent of recovery was dependent upon stress experienced and plants stressed for 9 days were not able to recover fully, whereas recovery was almost complete upto 6 days of water stress (Table 1). Thus severity of stress experienced by the plants has an important role for normalization of leaf turgor on irrigation and water availability due to rains (3).

Leaf area and net photosynthesis

Drought induced decrease in ψ_{plant} and RWC resulted in significant decrease in leaf area and net photosynthetic rate (P_n) under both the P levels

(Fig. 1). Leaf area decreased from 10.7 to 34.1% in P_0 and from 4.5 to 33.7% in P_{40} plants under increased water stress intensity. Likewise P_n decreased from 20.1 to 54.5% in P_0 and from 19.6% to 52.1% in P_{40} plants. Furthermore, both leaf area and P_n were significantly higher in P-fertilized than unfertilized plants under all the stress intensities. P application, on an average increased leaf area by 8.6% and P_n by 16.6%. Furthermore, the recovery in P_n and leaf area upon rewatering was higher in P-fertilized plants, particularly upto 6 days of water stress (Fig. 1). A number of reports indicate higher photosynthesis at high than low levels of P in water stressed plants (16, 17). Similarly the role of P application in enhancement of leaf area is well documented (11, 16). Garg *et al.* (5) reported that under water stress condition, moth bean genotypes grown with adequate P had higher net photosynthesis than those grown without P application. In sugarbeet, P induced enhancement in photosynthesis was attributed to high RuBP carboxylase activity and CO_2 fixation (18).

Table 1. Effect of different intensities of water stress on plant water potential & relative water content of mungbean grown at two levels of phosphorus.

Days without water	ψ plant (- MPa)						RWC (%)					
	Stressed			Rewatered			Stressed			Rewatered		
	P_0		P_{40}	P_0		P_{40}	P_0		P_{40}	P_0		P_{40}
0	1.13		1.23	1.15		1.20	88.0		85.3	84.6		83.4
3	1.70		1.73	1.25		1.30	78.7		74.9	83.0		83.1
6	1.97		2.07	1.35		1.35	69.1		67.2	82.6		81.0
9	2.07		2.47	1.40		1.60	62.2		57.7	81.9		76.7
Mean	1.72		1.88	1.29		1.36	74.8		71.1	83.0		81.1
LSD	D	P	D x P	D	P	D x P	D	P	D x P	D	P	D x P
(p=0.05)	0.08	0.06	0.11	0.10	NS	0.15	2.5	1.0	3.5	1.1	0.8	1.6

D- Drought; P- Phosphorus

Chlorophyll and starch

Increasing intensity of water stress progressively and significantly decreased total chlorophyll and starch contents under both the P levels (Table 2). The reduction in chlorophyll and starch levels was slightly less in P-fertilized than unfertilized plants under all intensities of water stress. Besides, their concentrations as such were significantly higher in P_{40} than P_0 plants and on an average P application enhanced chlorophyll content by 4.5% and starch content by 7.9%. Furthermore, P supplied plants also showed higher recovery on rewatering in both chlorophyll and starch contents at various stress intensities.

However, the recovery was not complete in plants stressed for 6 and 9 days irrespective of P levels. Favourable influence of P nutrition on chlorophyll and starch contents has earlier been reported in mulberry plants (17) and in moth bean genotypes (5) under water stress condition. Thus results indicate that P nutrition helps in maintenance of higher chlorophyll and starch contents even under water stress condition.

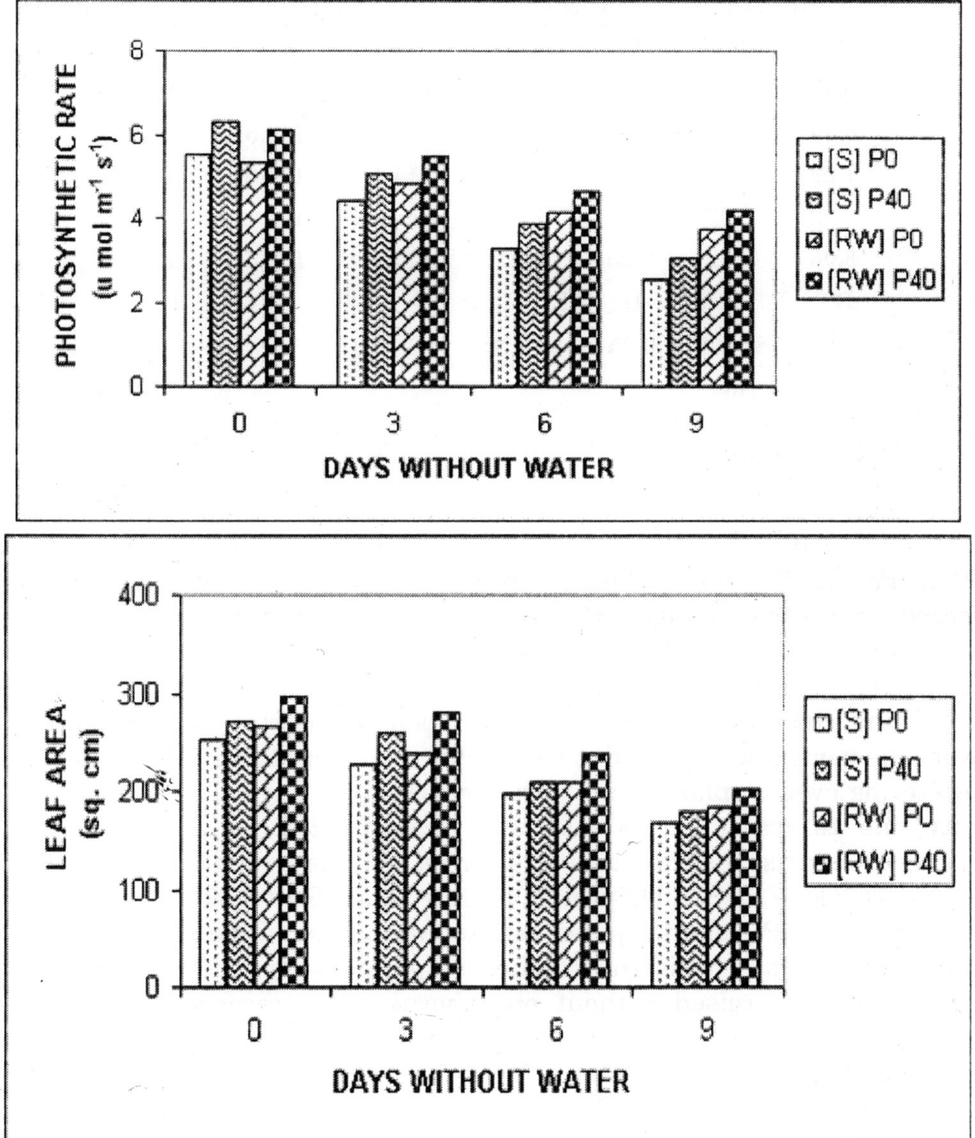

Fig. 1. Effect of phosphorus application on net photosynthetic rate and leaf area of mungbean under different intensities of water stress and 2 days after rewatering

Table 2. Effect of P application on total chlorophyll and starch contents of mung-bean under different intensities of water stress and 2 days after rewatering

Days without water	Total chlorophyll ($mg\ g^{-1}$ dw)						Starch ($mg\ g^{-1}$ dw)					
	Stressed			Rewatered			Stressed			Rewatered		
	P_0		P_{40}	P_0		P_{40}	P_0		P_{40}	P_0		P_{40}
0	9.52		9.74	9.41		9.72	158.2		165.6	162.6		178.8
3	8.39		8.87	8.67		8.90	131.0		143.6	156.1		170.1
6	7.89		8.25	8.07		8.63	114.1		120.9	149.2		164.6
9	6.42		6.83	6.72		7.13	95.3		100.1	133.5		148.5
Mean	8.06		8.42	8.25		8.60	122.7		132.5	150.3		165.5
LSD	D	P	D x P	D	P	D x P	D	P	D x P	D	P	D x P
(p=0.05)	0.26	0.18	0.37	0.25	0.18	NS	5.0	3.5	NS	5.0	3.6	7.1

D- Drought; P- Phosphorus

Nitrogen metabolism

Increasing water stress drastically reduced the nitrate reductase (NR) activity under both the P levels (Table 3). The reduction in NR activity was 40.6% in P_0 and 37.8% in P_{40} plants even after 3 days of water deprivation as compared to control plants. NR activity was reduced by more than 70%, irrespective of P levels after 9 days of water stress. This indicated very high sensitivity of nitrate reductase to soil moisture deficits as also reported in moth bean (5) and clusterbean (16). However, upon rewatering there was a rapid recovery in enzyme activity and it was restored to more than 75% within 2 days of rewatering even in plants that were stressed for 9 days. The decrease in NR activity due to water stress was also accompanied by a significant and progressive decrease in the concentration of soluble protein under increased water stress (Table 3). Soluble protein declined from 6.1 to 16.6% in P_0 and from 3.8% to 17.5% in P_{40} plants under various intensities of water stress. Soluble protein content was consistency higher in plants supplied with phosphorus, compared to those raised without phosphorus, after recovery from stress of varying intensity. Thus plants supplied with P maintained higher NR activity and soluble protein content than unfertilized plants under all intensities of water stress as well as upon rewatering (Table 3). On an average P application enhanced NR activity by 17.9% and soluble protein by 6.5% in water stressed plants. Similar trend was observed in rewatered plants with the P application. The maintenance of higher NR activity and soluble protein content by P application indicates efficient N metabolism in P-treated plants as plant's N balance, growth and NR activity are closely related (19). Positive relationship between P level, plant growth and N metabolism in sorghum and bean plants (20) further strengthens the view that in legumes role of P nutrition is critical. Application of P is reported to enhance NR activity and mobilize reduced N from leaves in chickpea also (21).

Table 3. Effect of P application on nitrate reductase activity and soluble protein content of mungbean under different intensities of water stress and 2 days after rewatering

Days without water	Nitrate reductase ($\mu g\ NO_2\ g^{-1}\ dwh^{-1}$)						Soluble protein ($mg\ g^{-1}\ dw$)					
	Stressed			Rewatered			Stressed			Rewatered		
	P_0		P_{40}	P_0		P_{40}	P_0		P_{40}	P_0		P_{40}
0	210.9		234.8	205.1		221.9	74.10		78.82	70.30		78.10
3	125.3		146.0	190.7		208.7	69.62		75.80	68.70		79.00
6	102.4		118.2	180.4		192.7	66.92		70.50	67.70		73.50
9	44.4		67.2	152.9		169.9	61.82		65.02	64.60		69.12
Mean	120.0		141.5	182.3		198.3	68.11		72.54	67.72		74.93
LSD	D	P	D x P	D	P	D x P	D	P	D x P	D	P	D x P
(p=0.05)	11.1	7.9	NS	11.4	8.1	16.1	1.79	1.27	NS	1.46	1.03	2.06

D- Drought; P- Phosphorus

Biomass and seed yield

Increasing intensities of water stress caused a progressive decline in growth and seed yield (Table 4). Total plant biomass as well pod and seed yields were significantly reduced even after 3 days of water stress at both the P levels. At the highest drought intensity, the reduction in seed yield was 38.9% while shoot dry matter decreased by 26.6% as compared to control plants. On an average P fertilized plants had 11.2% higher DMP, 11.7% higher pod weight and 13.4% higher seed yield over unfertilized plants. This indicates that application of P can help in obtaining sustainable yields of mungbean even under water limited conditions. There are number of reports on the adverse effects of water stress on mungbean growth and seed yield (2, 3, 22, 23, 24). However, few studies have been undertaken on the application of P for yield enhancement of mungbean, particularly under water limited conditions (6,7, 25, 26). In this regard, several studies indicate that P nutrition under water deficits increased drought resistance and yield of plants (11, 27). Garg et al. (5) reported similar positive effects of P application in moth bean genotypes under water stress condition. The present study further suggests that beneficial effects of P application may be achieved in mungbean upto moderate levels of water stress.

Table 4. Effect of phosphorus application on seed yield, pod weight and shoot dry mater of mungbean under different intensities of water stress.

Days without water	Seed yield (g plant^{-1})			Pod weight (g plant^{-1})			Shoot dry matter (g plant^{-1})		
	P_0		P_{40}	P_0		P_{40}	P_0		P_{40}
0	10.33		11.63	16.31		18.21	33.13		36.70
3	9.24		10.37	13.85		15.59	29.21		32.33
6	7.58		8.79	12.39		13.88	26.71		29.98
9	6.29		7.13	10.26		11.36	24.24		27.00
Mean	8.36		9.48	13.23		14.76	28.32		31.50
LSD (p=0.05)	D	P	D x P	D	P	D x P	D	P	D x P
	1.21	0.85	NS	1.62	1.15	NS	2.78	1.96	NS

D- Drought; P- Phosphorus

REFERENCES

1. Singh, D.P. 2000. Advances in Management of Biotic and Abiotic Stress in Pulse Crops. Proceeding of National Symposium on, "Management of Biotic and Abiotic Stresses in Pulse Crops", held at Kanpur, June 26-28, 1998. pp.93-98.
2. Poehlman, J.M. 1991. The Mungbean. Oxford and IBH Publishing Co. Pvt. Ltd., New Delhi. 375 p.
3. Thomas, M.J., Robertson, S.F. and Peoples, M.B. 2004. The effect of timing and severity of water deficit on growth, development, yield accumulation and nitrogen fixation of mungbean. Field Crops Research. 86:67-80.
4. Payne, W.A., Drew, M.C., Hosner, L.R., Lascano, R.J., Onken, A.B. and Wendt, C.W. 1992. Soil moisture availability and pearl millet water use efficiency. Crop Science. 32:1010-1015.
5. Garg, B.K., Burman, U. and Kathju, S. 2004. The influence of phosphorus nutrition on the physiological response of moth bean genotypes to drought. Journal of Plant Nutrition and Soil Science. 167:503-508.
6. Sharma, B.M., Yadav, J.S.P. and Rajput, R.K. 1984. Effect of irrigation and phosphorus application on available phosphorus and grain yield of moong. Indian Journal of Agronomy. 29:613-620.
7. Kathju, S., Aggarwal, R.K. and Lahiri, A.N. 1987. Evaluation of diverse effects of phosphate application on legumes of arid areas. Tropical Agriculture. 64:91-96.

8. Singh, D.K., Sale, P.W.G. and Mckenzie, B.M. 1997. Water relations of white clover (*Trifolium repens* L.) in a drying soil as a function of phosphorus supply and defoliation frequency. Australian Journal of Agriculture Research. 48:675-681.
9. Brooks, A. 1986. Effects of phosphorus nutrition on ribulose – 1, 5-biphosphate carboxylase activation, photosynthetic quantum yield and amount of some calvin cycle metabolites in spinach leaves. Australian Journal of Plant Physiology. 13:221-237.
10. Pilbeam, D.J., Cakmak, I., Marschner, H. and Kirkby, E.A. 1993. Effect of withdrawl of phosphorus on nitrate assimilation and PEP carboxylase activity in tomato. Plant and Soil. 154:111-117.
11. Gutierrez-Boem, F. and Thomas, G.W. 1999. Phosphorus nutrition and water deficits in field grown soyabeans. Plant and Soil. 207:87-96.
12. Barrs, H.D. 1968. Determination of water deficit in plant tissues. In: T.T. Korlowski (ed.) Water Deficits and Plant Growth. Vol.I.pp.236-368. Academic Press.
13. Lowry, O.H., Ropsenbrough, N.J., Farr, A.Z. and Randall, R.J. 1951. Protein measurement with Folin-phenol reagent. Journal of Biological Chemistry. 193:245-275.
14. Radin, J.W. and Parker, L.L. 1979. Water relations of cotton plants under nitrogen deficiency II Environmental interaction on stomata. Plant Physiology. 64:499-501.
15. Kathju, S., Burman, U. and Garg. 2001. Influence of nitrogen fertilization on water relations, photosynthesis, carbohydrate and nitrogen metabolism of diverse pearl millet genotypes under arid conditions. Journal of Agricultural Sciences (Cambridge) 137:307-318.
16. Burman, U., Garg, B.K. and Kathju, S. 2004. Interactive effects of thiourea and phosphorus on clusterbean under water stress. Biologia Plantarum. 48:61-65.
17. Sharma, P.N. 1995. Water relations and photosynthesis in phosphorus deficient mulberry plants. Indian Journal of Plant Physiology. 38:298-300.
18. Rao, I.M. and Terry, N. 1995. Leaf phosphate status, photosynthesis and carbon portioning in sugarbeet IV Changes with time following increased supply of phosphate to low-phosphate plants. Plant Physiology. 107:1313-1321.
19. El-Komy, H.M., Hamida, M.A. and Abdel-Baki, G.K. 2003. Nitrate reductase in wheat plants grown under water stress and inoculated with *Azospirillum* spp. Biologia Plantarum. 46-281-287.
20. Al-Karaki, G.N., Clark, R. B. and Sullivan, C.Y. 1996. Phosphorus nutrition and water stress effects on proline accumulation in sorghum and bean. Journal of Plant Physiology. 148:745-751.
21. Thapar, S., Sekhon, B.S., Atwal, A. and Singh, R. 1991. Phosphorus nutrition and nitrogen assimilation in chickpea (*Cicer arietinum*). Journal of Plant Physiology. 1378:111-114.
22. Kubar, R.A., Kumbhar, A.M., Jamro, G.H. and Abro, B.A. 2004. Effect of water stress on growth and yield of mung bean. Indus Journal of Biological Sciences. 1:200-205.

23. Pandey, R.K., Herrera, W.A.T. and Pendleton, J.W. 1984. Drought response of grain legumes under irrigation gradient. I. Yield and yield components. Agronomy Journal. 76:549-553.

24. Garg, B.K., Burman, U. and Kathju, S. 2005. Comparative water relations, photosynthesis and nitrogen metabolism of arid legumes under water stress. Journal of Plant Biology. 32:83-93.

25. Agarwal, S.K., Behl, N.K. and Moolani, M.K. 1976. Response of summer mung to levels of phosphorus and irrigation under different dates of planting. Indian Journal of Agronomy. 21:290-291.

26. Prasad, R., Bhendia, M.L. and Bains, S.S. 1968. Response of grain legumes to levels and sources of phosphorus on different soils. Indian Journal of Agronomy. 13:305-309.

27. Singh, D.K. and Sale, P.W.G. 1998. Phosphorus supply and growth of frequently defoliated white clover (*Trifolium repens* L.) in dry soil. Plant and Soil. 205:155-162.

Assessing soil productivity through potassium permanganate oxidizable carbon – a new perspective

Praveen-Kumar and Uday Burman

Central Arid Zone Research Institute, Jodhpur 342 003

ABSTRACT

Indian agriculture is currently in pressure from two sources viz. (i) ever increasing demand of food and (ii) declining availability and increasing cost of fertilizers. These challenges can be overcome, only with site-specific assessment of soil quality and application of fertilizer. Soil organic carbon is an important indicator of quality and productivity. But its quantity is very low in arid soils and display high level of heterogeneity. As a result, large numbers of samples need to be collected and analysed for reliable estimations. Further, conventional methods of SOC estimation are based on use of highly corrosive reagents, expensive equipments and need high level of technical skill and are thus unsuitable for a field kit approach. A new method is proposed for estimation of SOC and for soil productivity. The method is based on reactivity of C compounds of SOC with potassium permanganate. Estimation of permanganate oxidizable C (POC) through the proposed method truly reflects the changes in soil productivity observed after treatments of tillage, application of manures and cultivation of legumes. Changes SOC by the soil management intervention like tillage, manure application and crop rotation reflected in SOC were also reflected in POC in similar way. Thus the new method is suitable both for the assessment of soil organic carbon and also for assessment of soil productivity.

Key words: Crop rotation, soil organic carbon, soil management, tillage, yield.

INTRODUCTION

More than ninety-nine percent of the food requirement in our country is met through terrestrial resources. Out of total 329 m ha of available land in India, around 139 m ha was cultivated which produced about 212 m tons food grain in 2007-08. But even this production level fell short of target envisaged in Ninth Five Year Plan document (1997-2002). Low fertility status of most Indian soils is major bottleneck for increasing productivity. The organic carbon (C) content is low and thus status of available N range from medium to low in 89% of the area. Similarly 80% area for phosphorous, 50% for potassium, 40% for sulphur, 48% for zinc, 33 % for boron, 12% for iron and 5% for manganese is categorized as low to medium [1, 2]. Lal [3] estimated that 29.4 M ha of our soils are experiencing decline in fertility and net negative balance of 8-10 m tons of nutrients per annum has become a constant feature from the onset of green revolution. Tandon [4] reported that during 1998-99 16.6 m ton of fertilizer nutrients were added against the removal of 25 m tons plant nutrients. The current estimated average depletion per ha is about 16 kg N, 11 kg P_2O_5 and 42 kg K_2O. With negative nutrient balance deficiencies may become more widespread, extent of deficiency may become more acute and adversely affect fertilizer use efficiency.

Over past fifty years our country has witness unprecedented increase in population. Population increase is also leading to decline in size and increase in number and fragmentation of farm holdings. The size of farm holdings in India fell from 2.3 ha in 1970/71 to 1.47 ha in 1995/96 (Table 1).

Table 1. Number, area covered and average size of landholdings

Category of holding	Number (million)		Area (million ha)		Average size (ha)	
	1970/71	1995/96	1970/71	1995/96	1970/71	1995/96
Marginal	35.7	71.2	14.5	28.1	0.41	0.40
(< 1 ha)	(50.6)	(61.6)	(9.0)	(17.2)		
Small	13.4	21.6	19.3	30.7	1.44	1.42
(1 – 2 ha)	(19.1)	(18.7)	(11.9)	(18.8)		
Semi-medium	10.7	14.3	30.0	39.0	2.81	2.73
(2 – 4 ha)	(15.2)	(12.3)	(18.5)	(23.8)		
Medium	7.9	7.1	48.2	41.4	6.08	5.84
(4 – 10 ha)	(11.3)	(6.1)	(29.8)	(25.3)		
Large	2.8	1.4	50.1	24.2	18.10	17.21
(> 10 ha)	(3.9)	(1.2)	(30.9)	(14.8)		
Total	70.5	115.6	162.1	163.4	2.30	1.41

( ) = Percentage share of various categories to total number and area.

About 62 percent of the farm holdings are of <1 ha and large holdings >10 ha constitute only 1.2 percent of the total number. Farmers with smallholdings lack resources to maintain soil fertility and as a result soils get degraded, their crop yields decline and resources are further depleted. Thus, farmers are caught in the vicious cycle of poverty (Figure 1).

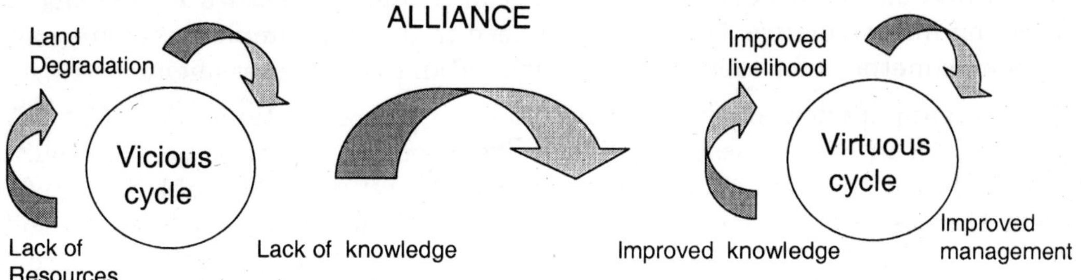

Figure 1. Vicious cycle of poverty and way to get out of it

Apart from negative nutrient balance these developments are further threatening our food security and food shortage is becoming a first major crisis of Indian agriculture ever since Green Revolution. The only way to get out of this crisis is to treat each field as a unit for soil fertility management. This calls for massive investments in improving knowledge level of farmers and also in their resource inventory to take them out of this crisis situation.

Thus, food security of our nation depends on the maintenance and improvement of soil quality by small landholders through field specific soil quality assessment and management. Frequent assessment of soil quality, which is necessary for field specific soil fertility management, cannot be taken care of by the existing soil testing infrastructure. A field kit type of diagnostic approach can offer the viable option.

In this regard, soil organic carbon (SOC) is most important indicator of soil quality and productivity as it effects several critical functions including aggregation, infiltration, compaction, moisture holding capacity, resistance to erosion, bulk density and nutrient retention. Its estimation is pivotal in assessing soil quality and health. Soil organic matter is composed of a variety of fractions like raw plant residues, micro-organisms and humus. Soil organic matter consists a large pool of recalcitrant fractions and a small active pool. The active fraction of soil organic matter is more directly related to important biological processes in soil and affects the soil productivity. Measurement of total soil organic matter though provides important information, but high portion of the recalcitrant carbon mask slight changes in active pool due to alterations in soil management. Changes in this component may provide a better indication of management impact of soil quality. Further, estimation of

SOC by conventional Walkley & Black's [5] method is time consuming, expensive, involves corrosive reagents and thus requires technical skills. Modern C-N analysers too are very expensive and can only be operated by trained technicians. In arid region, SOC displays high level of heterogeneity. Thus for estimation of soil quality through SOC, large number of samples must be collected and analysed periodically which can potentially saturate the existing soil testing infrastructure Thus, there is need to develop simple, less expensive and reliable methods for labile C estimation for an on-site assessment.

Soil organic carbon oxidized with a range of concentrations 33 to 333 mM has been used in past to assess influence management practices on soil quality [6,7,8,9,10]. But the amount of C extracted from soil with 333 mM $KMnO_4$ far exceeds the theoretical amount in the soil [11] or the CO_2 evolution from soil. Stine and Weil [12] & Weil *et al.* [13] hypothesized that labile SOC fractions associated with soil quality might better be represented by more dilute concentrations of $KMnO_4$. The method was further refined by Gruver [14] to detect effect of soil management practices like cropping rotation. However, total C in soils used by these workers ranged from 2 to 6%. In arid soils where SOC level is very low (0.1 to 0.2%) these methods are not adequate and thus authors felt the need to modify the method to make it suitable for assessing soil quality having very low organic carbon soils. The method is extremely simple, non-corrosive, less expensive and allows processing of two to three times more samples as compared to conventional SOC estimation within a given time.

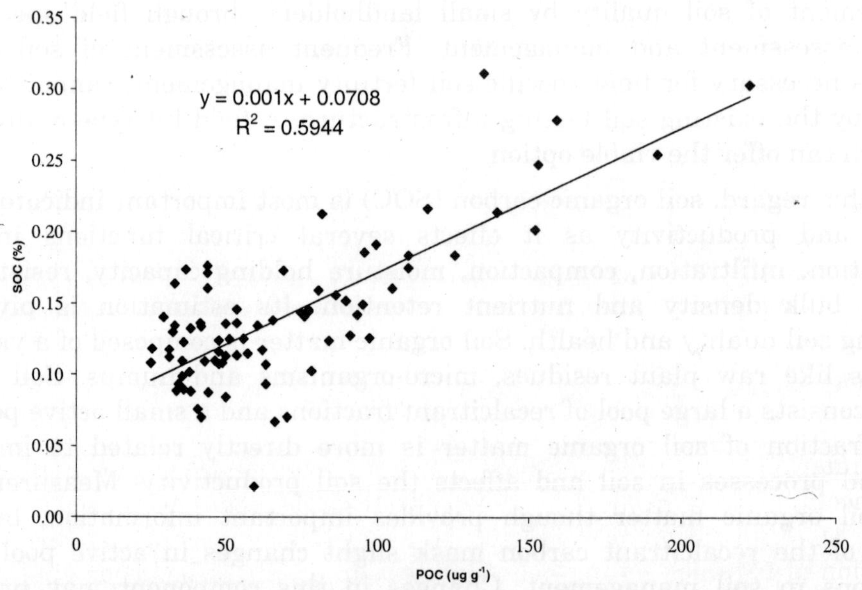

Figure. 2. Correlation between labile-C and soil organic carbon (SOC)

Five g soil is weighed in 150 ml conical flask and 50 mL aqueous solution of 5mM $KMnO_4$ is added to it. Contents are shaken for an hour on a platform shaker and allowed to settle for an hour. Contents are then filtered through Whatman 42 filter paper. 2mL of filtrate is the pipette out in 100mL volumetric flask. The absorbance is read on 560 nm. Decline in absorbance in comparison to control is used to calculate POC.

Potassium permanganate oxidizable carbon (POC) estimated by the new modified method showed significant correlation with total SOC (n=76; Figure 2).

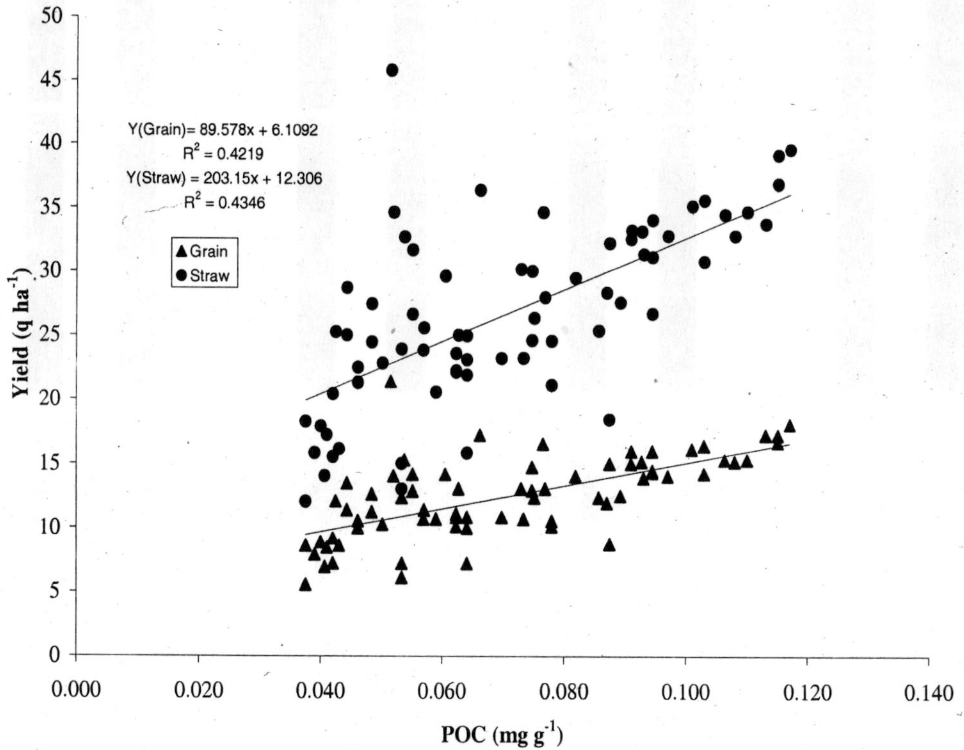

Figure 3. Correlation between pearl millet yield and status of POC in soil.

As the correlation is high it is possible to use labile-C for assessment of SOC. POC was also significantly correlated with pearl millet yield (n=65; Figure 3).

Estimation of POC in samples of a long-term tillage experiment revealed that concentration of POC pool (as percent of total C pool) decreased with tillage (Figure 4). Application of fertilizer N also slightly decreased this pool, but cultivation of legumes (mung bean, moth bean or clusterbean) or application of manure increased the labile-C.

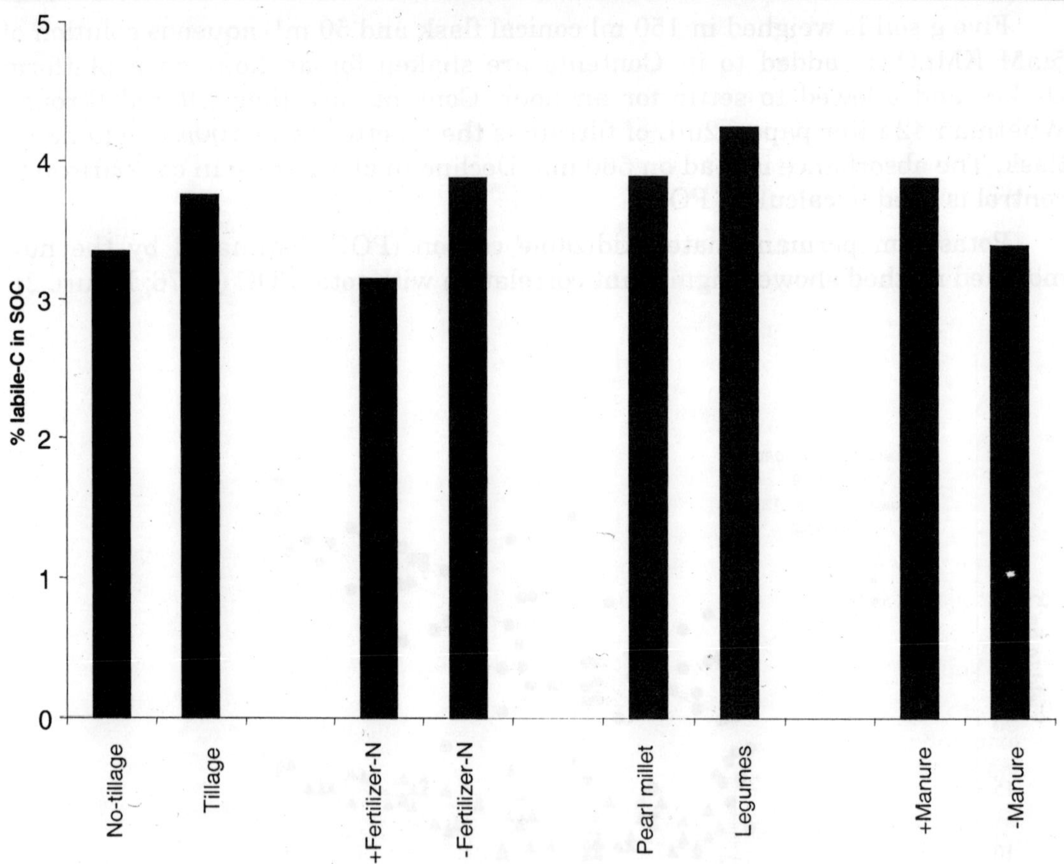

Figure 4. Effect of tillage, cultivation of legumes, application of fertilizer-N and incorporation of manure on content of labile-C in SOC.

Our results with POC clearly showed it to be highly correlated with SOC. Also the effect of soil management practices like tillage, legume based crop rotations and manure applications on soil productivity were reflected in a similar way both in SOC and POC. Thus, POC is suitable both for the estimation of SOC and also for assessing productivity of soil. Further, simple procedure, being less expensive and rapid make POC estimation highly suitable for field kit especially in arid soils where extent of heterogeneity is high.

REFERENCES

1. Motsara, M.R. 2002. Available nitrogen, phosphorus and potassium status of Indian soils as depicted by soil fertiliser maps. *Fertilizer News* 47(8): 15–21.
2. Singh, M.V. 2001. Evaluation of current micronutrient stocks in different agro-ecological zones of India for sustainable crops production. *Fertilizer News* 42(2): 25–42.

3. Lal, R. 2004. Soil carbon sequestration impacts on global climate change and food security. *Science* 304:1623-1627.

4. Tandon, H.L.S. 2004. Fertilizers in Indian Agriculture- From 20th to 21st century. Fertilizer Development and Consultation Organisation, New Delhi

5. Blacks C.A. 1965. Soil Chemical Analysis Pt II. American Society of Soil Science, Wisconsin, USA

6. Loginow W., Wisniewski W., Gonet S.S and Ciescinska B. 1987. Fractionation of organic C based on susceptibility to oxidation. *Polish Journal of Soil Science.* 20:47-52.

7. Blair, G.J., Lefroy, R.D.B. and Lisle, L. 1995. Soil carbon fractions based on their degree of oxidation and the development of a carbon management index for agricultural systems. *Aust. J. Agric. Res.* 46: 1459-1466.

8. Blair, G.J.,. Lefroy, R.D.B., Singh B.P and. Till, A.R. 1997. Development and use of a carbon management index to monitor changes in soil C pool size and turnover rate. p. 273-281. *In* G. Cadisch and K.E. Giller (ed.) Driven by nature: Plant litter quality and decomposition. CAB International, Wallingford, UK.

9. Blair, G.J., Lefroy, R.D.B. Whitbread, A. Blair, N. and Conteh A. 2001. The development of the $KMnO_4$ oxidation technique to determine labile carbon in soil and its use in a carbon management index. p. 323-337. *In* R. Lal *et al.* (ed.) Assessment methods for soil carbon. Lewis Publishers, CRC Press, Boca Raton, FL.

10. Tirole-Padre, A. and Ladha, J.K. 2004. Assessing the reliability of permanganate oxidizable carbon as an index of soil labile carbon. *Soil Science Society of Anmerica Journal* 68: 969-978.

11. Parton, W. J., and Rasmussen, P. E. 1994. Long-term effects of crop management in wheat/fallow: II. Century model simulations. *Soil Science Society of America Journal* 58 :530-536

12. Stine M.A. and Weil, R.R. 2002. The relationship between soil quality and crop productivity across three tillage systems in south central Honduras. *American Journal of Alternative Agriculture* 17: 2-8.

13. Weil R.W., Islam K.R., Stine, M., Gruver, J.B. and Samson-Liebig S.E. 2003. Estimating active carbon for soil quality assessment: a simplified method for laboratory and field use. *American J. Alt. Ag.* 18 (1): 3-17.

14. Gruver, J. 2004. quality Control for a new permanganate oxidizable C method. 26th Southern Conservation Tillage Conference, Raleigh, North Carolina, June 8-9, 2004.

Leguminous plants of ethnomedicinal value in the Indian arid zone

Suresh Kumar, Farzana Parveen, Aruna Chauhan & Sangeeta Goyal

Central Arid Zone Research Institute, Jodhpur

ABSTRACT

Floristic elements in the desert districts viz. Barmer, Bikaner, Jaiselmer and Jodhpur were surveyed to know their local uses in healthcare and other economic activities. In each district, 6-8 villages in each tehsil (block) were selected of which 2-3 villages were of small, medium and big size each in terms of population and households. Thus a total of 128 villages were sampled. In each village, local vaidyas, elders, herbal healers and 'Dais' were interviewed to know medicinal uses of these plants, and the way of its usage. From this wealth of data collected, we present here the ethnomedicinal uses of Leguminous plants of all three families i.e., Fabaceae, Mimosaceae and Caesalpiniaceae.

A total of 22 species have been found to have medicinal value. Of these, 13 belong to family Fabaceae, seven Mimosaceae and two Caesalpiniaceae. These include nine trees, five shrubs and eight herbs. Plant parts reportedly useful included bark, flowers, fruits, pods, galls, gum, leaves, seeds, stem, thorns and in some cases whole plant. People use these plants to cure as many as 32 diseases, using them in a large variety of ways. Some important ailments include arth-ritis, constipation, debility, dysentery, flatulence, fractured bone, skin burn and urinary problems. Since *Acacia nilotica, A. senegal, Butea monosperma* (all for gum), and *Glycyrrhiza glabra* (Mulethi) are imported, these could be taken up for large scale plantation while *Cassia angustifolia* is an export crop. In view of their economic, ecological and social value, many of these can be inculcated in the farming system for its diversification.

Key words: Ethnomeidicnal, legume, arid zone, diversification.

Introduction

Fabaceae, Caesalpiniaceae and Mimosaceae together have 28 genera and 64 species, 5 genera and 10 species and 7 genera and 14 species, respectively in the

Indian arid zone (1). These leguminous species are known to be traditional source of food and fodder. Some attempts have been made in the past to document their other uses (2,3,4,5,6,7). However, a detailed field based appraisal of ethno medicinal value of legumes in arid zone of India was lacking and hence this study was undertaken and results are described in this paper.

Material and Methods

The study covered the districts of Jodhpur, Jaisalmer, Bikaner and Barmer in Rajasthan state lying between 24°40'-29°3'N and 69°7'-74°7'E, approximately 100 m above the sea level. The area receives very low, erratic annual rainfall of 150-300 mm, mainly during the months of July to September. It faces extremes of diurnal and annual mean temperature, ranging between 24°C and 2.2°C during January in winter and between 26.7°C and 48°C in the months of May-June in summer. Vegetation is sparse, stunted, with low canopy cover, comprising mainly small trees and shrubs, many of which are spinous. Due to over-exploitation and misuse, the vegetation is extremely degraded, with best expressions only at places where it is protected for religious reasons.

A total of 128 selected villages in four districts were visited and a wide range of people were interviewed first in groups and then individually. Local herbal healers and livestock herders (Raikas) were also contacted. Voucher specimens were identified with the help of respondents and their botanical identities were confirmed through "Flora of the Indian Desert" by Bhandari (1990). Voucher specimens have been deposited in the herbarium of Central Arid Zone Research Institute, Jodhpur.

The plants are arranged alphabetically according to their botanical name followed by respective family in bracket. It is followed by local name, part used, ailments and mode of administration which are summarized in table 1.

Result and Discussion

A perusal of listed uses of different leguminous plants (Table 1) revealed that a total of 22 species are being used for treating 30 different ailments. Their habit wise classification (Table 2) revealed that in Fabaceace, herbaceous species are dominantly medicinal while in Mimosaceae and Caesalpiniaecae, ligneous species are medicinal. These plants are used both externally as well as internally. External applications include use of its paste or boiling it in water and using this water for bathing. Internal consumption has been by way of using its juice, decoction, filtrate, dry powder, cakes and boiled, cooked and eaten raw. Leaves of maximum species (31.1%) are used as medicine (Fig. 1) followed by seeds (19.0%), roots (16%), flowers (9%), bark (9%), fruits (8%), gum (5%), and remaining (pods, stem, whole plant, gall) as 1%.

Table 1. Ethno Leguminous-Medicinal Plants in Indian arid zone.

Plant name & Family	Local name	Ailments	Part used	Mode of administration
Acacia nilotica (L.) Del. *Mimosaceae*	Desi Babool	Asthma	Flower	Fresh flowers are dried in shade and ground to form powder. One teaspoon powder mixed with honey is given orally to the patient during asthmatic attack.
			Gum	Gum is boiled in water and is allowed to cool and taken orally to get relief in asthma.
		Blisters	Leaves	Aqueous paste of leaves is applied on blisters.
		Boils	Leaves	Aqueous paste of fresh leaves is applied on boils to hasten suppuration.
		Cough	Fruit	Dried powdered fruit mixed with sugar and one tea spoon mixture is taken orally with water twice a day to get relief in cough.
		Debility	Gum	Sweet cakes are prepared by mixing its gum with sugar, butter oil, 'wheat' flour (*Triticum aestivum*) and dry fruits. It is given to mother after delivery as a tonic.
			Leaves	One teaspoon powdered leaves is taken orally with goat's fresh milk twice a day as a tonic.
		Eye pain	Leaves	Few drops of extract of leaves are put into eyes to remove the redness and to get relief in pain.
		Eye irritation	Leaves	Fresh leaves are kept on earthen pot for the whole night. Next morning these leaves are placed on eyelids to get relief from irritation.
		Eye problem	Flower	Aqueous paste of fresh flowers is applied over eyes as a coolant.
		Fever	Leaves	Aqueous paste of leaves is taken orally with water.
		Flatulence	Thorns	Thorns are boiled in water and filtrate is given orally to relieve muscular pain in the body.
		Fracture	Bark	Crushed bark is boiled in water. A cotton cloth soaked in this filtrates tied on fractured bone for early healing and pain relief. It is repeated many times a day for early cure.
		Gynecological disorder	Flower	Dried flowers are crushed into powder and mixed with sugar candy and some water. It is taken orally four times a day. It gives relief in pain during menstruation.
		Jaundice	Flower	Fresh flowers are boiled in water and its filtrate is given orally twice a day till the patient gets cured.
			Fruit	Dried powdered fruits mixed with fresh goat's milk and sweetened with sugar are given orally.
		Kidney problem	Flower	Fresh flowers are soaked in water for an hour and ground to form paste. It is mixed with fresh goat's milk in equal quantity and taken orally empty stomach in the morning.
		Skin irritation	Bark	Aqueous paste of bark applied on affected part, gives cooling effect.
		Stomach ache	Fruit	Fresh fruits are fried in butter and consumed as such to reduce stomach pain.
		Throm-	Bark	Fresh bark is soaked in water for 15 days and mashed. Its extract

		bosis		is given orally to prevent blood clotting in the body..
		Typhoid	Thorns	Thorns are boiled in water and its filtrate is used for bathing.
			Bark	Crushed bark is boiled in water and its filtrate is used for bathing.
			Flower	Fresh flowers are boiled in water and its filtrate is taken orally.
			Gum	Gum is boiled in water and is allowed to cool and given orally to the patient.
		Wound	Bark	Aqueous paste of bark is applied on wounds for early healing.
Acacia senegal (L.) Willd. *Mimosaceae*	Kumat	Debility	Gum	Cakes are prepared by mixing its gum with sugar, butter oil, 'wheat' flour (*Triticum aestivum*) and dry fruits. It is given to mother after delivery as a tonic
			Leaves	Fresh leaves are boiled in water and its filtrate is taken orally to regain body strength
Acacia tortilis (Forsk.) Hayne *Mimosaceae*	Kikar	Fever	Thorns	20-25 thorns are boiled in water and its filtrate is taken orally to reduce body temperature
Albizia lebbeck (L.) Willd. *Mimosaceae*	Sares	Injury	Bark	Aqueous paste of bark is applied on affected portion for early healing
Butea monosperma (Lamk.) Taub. *Fabaceae*	Palash	Debility	Flower	Sweet cakes are prepared by mixing its flowers, butter oil, sugar and gum of 'kumat' (*Acacia senegal*). It is consumed as such by mother after delivery
		Fever	Flower	Aqueous paste of flowers mixed in bentonite paste is applied on whole body to cure fever caused due to heat stroke
		Stomach ache	Flower	Fresh flowers are tied on stomach with the help of a cloth to get relief in stomachache
Cassia angustifolia Vahl. *Caesalpiniaceae*	Sona-mikhi	Asthma	Leaves	Powdered leaves mixed with sugar taken orally with water once a day before going to bed for one month
		Constipa tion	Leaves	One teaspoon powdered leaves is taken orally with water before going to bed to cure acute problem of constipation
		Flatule-nce	Leaves	One teaspoon powdered leaves is taken orally with water before going to bed
		Piles	Leaves	Extract of fresh leaves is taken orally with water
		Stomach ache	Leaves	A mixture is prepared from powdered leaves and seed of 'ajwain' (*Trachyspermum ammi*) and taken orally with water
Cicer arietinum L. *Fabaceae*	Chanas	Jaundice	Seed	Seeds are slightly warmed on a pan and consumed as such with sour curd till jaundice is cured
		Skin burns	Seed	Powdered seed mixed with cow's butter milk and fine sand and applied on burnt skin for early healing
Crotalaria burhia Buch-Ham *Fabaceae*	Sinia	Boils	Root	Aqueous paste of roots is applied on boils to hasten suppuration
		Cold	Root	Fresh crushed roots are boiled in water and its filtrate is taken orally
		Constipa tion	Root	Crushed roots are boiled in water and its filtrate is taken orally
		Cough	Root	Roots boiled in water and its filtrate is taken orally
		Diuretic	Root	Decoction is prepared by adding its crushed roots and sugar candy.

				It is filtered and given orally
		Ear pain	Root	Few drops of extract of roots is put into ears to get relief in ear pain
		Fever	Root	Roots are boiled in water and its filtrate is taken orally to cure fever
		Flatulence	Root	Decoction is prepared by adding crushed roots and sugar candy. It is filtered and then the filtrate is taken orally
		Nasal bleeding	Root	Crushed roots are boiled in water and its filtrate is taken orally to stop nasal bleeding
		Stomach ache	Root	Crushed roots are boiled in water and its filtrate is taken orally in stomachache
		Stone problem	Root	One teaspoon powdered roots is taken orally with water to cure stone problem in urinary bladder
		Urinary infection	Root	Decoction is prepared by adding its crushed roots and sugar candy. Its filtrate is taken orally
Cyamopsis tetragonoloba (L.) Taub. *Fabaceae*	Gwar-phali	Back-ache	Pods	Fresh pods are soaked in water for 2-3 hours and then ground to paste. This paste is used in preparation of sweet dish with other ingredients like butter oil, and sugar. Patient having backache is advised to consume it twice a day to get relief in pain
		Cuts	Leaves	Aqueous paste of leaves is applied on cuts to stop bleeding and for early healing
		Dysentery	Seed	Powdered seed are mixed with curd and taken orally in dysentery
Dalbergia sisso Roxb.Roxb. *Fabaceae*	Shisham	Cough	Leaves	Ten fresh leaves are chewed daily in early morning to cure cough
Glycyrrhiza glabra L. *Fabaceae*	Mulethi	Constipation	Leaves	Powdered leaves is taken orally with water
		Typhoid	Whole plant	Whole plant is boiled in water and its filtrate is taken orally to cure typhoid
Indigofera cordifolia Heyne.ex Roth *Fabaceae*	Bek-ariya	Cold	Seed	Decoction is prepared by adding its powdered seeds with common salt and sugar candy. Its filtrate is taken orally
		Eye sight	Seed	Sweet cakes are prepared by adding its powdered roasted seeds, 'singhara' flour (*Trapa natans)* sugar and butter oil. It is consumed daily in morning to improve eye sight
		Fever	Seed	A dish is prepared by adding its powdered seeds with 'chawal' (*Oryza sativa)* and 'moong' seeds (*Vigna radiata*) and is boiled in water. It is consumed as such with butter oil for four days
		Insanity	Seed	Equal amounts of its powdered seeds and 'dhamasa' seed (*Fagonia indica*) are boiled in water and its filtrate is taken orally for four days
		Malaria	Seed	A sweet dish is prepared by adding its 100 g powdered seeds with 1/2 kg gum of 'kumat' (*Acacia senegal*), 1/2 kg 'singhara flour (*Trapa natans*), 100 g powdered roots of 'satawari' (*Asparagus racemosus*), butter oil and sugar. It is consumed for a few days

		Typhoid	Seed	Seed are boiled in water. Sugar is added in its filtrate and taken orally for three days
Medicago sativa L. Fabaceae	Rizka	Injury	Leaves	Aqueous paste of leaves is applied on injured portion for early healing
Mimosa hamata Willd. Mimosaceae	Jinjani	Asthma	Stem	Crushed stem is boiled in water and filtrate is given orally to patient for immediate relief
Pisum sativum L. Fabaceae	Matar	Jaundice	Seed	Few seeds are consumed as such while reciting some mantras. It is said that repeating this for five days will cure jaundice
Prosopis cineraria (L.) Druce Mimosaceae	Khejri	Antidote	Bark	In snake bite extract of inner bark is taken orally so that the poison comes out through vomiting
		Blisters	Bark	Aqueous paste of bark is applied on blisters for early healing
		Boils	Bark	Heat the bark in fire for one or two minutes and put it immediately into water. Its aqueous paste is applied on boils to hasten suppuration
		Boils	Leaves	Aqueous paste of leaves is applied on boils to hasten suppuration
		Boils	Stem	Ash of stem is applied on boils to hasten suppuration
		Cataract	Leaves	Juice of leaves added with sugar candy is taken orally to cure cataract in the eyes
		Debility	Gum	Sweet cakes are prepared by adding its gum with sugar, butter oil and 'wheat' flour (*Triticum aestivum*). It is given orally to mother after delivery
		Dysentery	Leaves	2 g of its fresh leaves and sugar candy is boiled in water and its filtrate is taken orally after two hours of interval for one day. It stops bleeding during dysentery
		Malaria	Fruit	A mixture is prepared by mixing its powdered Pod with flowers of 'keir' (*Capparis decidua*), 'neem' (*Azadirachta indica*) and 'safed aak' (*Calotropis gigantea*). This mixture is boiled in water till the whole water gets evaporated. The residue is then ground with cane sugar and consumed orally
		Pneumonia	Fruit	Equal amount of its fruit and seeds of 'moong' (*Vigna radiata*) were boiled in water and its filtrate is taken orally twice a day for three days
		Skin burns	Galls	Aqueous paste of galls is applied on burnt skin for early healing
		Thrombosis	Leaves	Fresh green leaves are warmed on a pan and then tied as a bandage with the help of a cloth on affected portion to prevent blood clotting
		Tumor	Fruit	Aqueous paste of fresh pods is applied on tumors for early healing
		Typhoid	Fruit	Old fruits are cooked as vegetable and consumed as such
Prosopis juliflora DC. (Swartz) Mimosaceae	Jangli kikar	Asthma	Leaves	Powdered leaves are filled in traditional cigars locally called as 'chilams'. The smoke is inhaled to get relief in asthma
		Boils	Leaves	Aqueous paste of leaves is applied on boils to hasten suppuration

		Cataract	Leaves	Juice of leaves is mixed with sugar candy and is taken orally to cure cataract problem
		Eye irritation	Leaves	Aqueous paste of leaves is applied over eyelids to get relief in eye irritation
		Stomach ache	Leaves	Aqueous paste of leaves is applied on abdomen region to cure stomachache
Tamarindus indica L. Caesalpiniaceae	Imli	Eye sight	Leaves	Aqueous paste of leaves is applied over eyelids with the help of a cloth to improve eyesight
		Fever	Fruit	Fruits boiled in water and its filtrate is taken orally
		Injury	Leaves	Aqueous paste of leaves is applied on injured portion for early healing
Tephrosia purpurea (L.) Pers. Fabaceae	Biyani	Antidote	Root	In snake bite the extract of fresh roots is applied on bit portion to neutralize the toxic effect of poison
		Blisters	Leaves	Aqueous paste of leaves is applied on blisters for early healing
		Boils	Leaves	Aqueous paste of leaves is applied on boils to hasten suppuration
		Boils	Seed	Aqueous paste of Seed is applied on boils to hasten suppuration
		Dental problem	Root	Fresh roots are used as tooth brush having antibacterial property
		Ear pain	Leaves	Few drops of leaf extract put in to ears to get relief in pain
		Fever	Leaves	Leaves are boiled in water and its filtrate is taken orally
		Flatulence	Root	Aqueous paste of roots is applied on affected portion
		Flatulence	Seed	Powder of 2-4 seeds is mixed with butter and consumed orally
		Ringworm	Leaves	Aqueous paste of leaves applied on affected portion
		Stomach ache	Root	Roots are boiled in water and its filtrate is taken orally
Trigonella foenum-graecum L. Fabaceae	Methi	Arthritis	Seed	Seed of 3-4 years old plant are sauté in butter oil and then it is mixed with sugar and consumed as such for a few days
		Cough	Seed	Decoction is prepared by adding its seeds, 'dhania' seed (*Coriandrum sativum*) and sugar. Its filtrate is taken orally
		Eye problem	Seed	Seeds are boiled in water and its filtrate is applied on eyelids with the help of a cloth to cure the redness of eyes
Vigna aconitifolia (Jacq.) Marechal Fabaceae	Moth	Cancer	Seed	One g ash of its seed is taken orally with water twice a day for one month to cure throat cancer
		Constipation	Seed	3-4 years old seeds are sauted in butter oil. Sweet cakes prepared from these seeds, sugar and 'wheat' flour (*Triticum aestivum*) are consumed for a few days.
		Debility	Seed	Sweet cakes prepared from its seed flour mixed with butter oil and 'wheat' flour (*Triticum aestivum*) act as a tonic.
		Typhoid	Root	Crushed roots are boiled in water and its filtrate is taken orally.
			Seed	Decoction is prepared by adding its 25 g seeds, 'dhania' seed (*Coriandrum sativum*) and 'methi' Seed (*Trigonella foenum-graecum*). Its filtrate is taken orally twice a day for a few days.

		Flatu-lence	Seed	Chapatti's are prepared from seed flour and tied on affected portion.
Vigna radiata (L.) Wilczek Fabaceae	Mung	Typhoid	Seed	25 gm seeds boiled in 2.5 liter of water till it gets reduced to 250 ml. Its filtrate is taken orally twice a day for seven days

Table 2. Habit wise classification

Fabaceae	
Trees	*Butea monosperma*
	Dalbergia sisso
Shrubs	*Crotalaria burhia*
Herbs	*Cicer arietinum*
	Cyamopsis tetragonoloba
	Glycyrrhiza glabra
	Indigofera cordifolia
	Medicago sativa
	Pisum sativum
	Tephrosia purpurea
	Trigonella foenum-graecum
	Vigna aconitifolia
	Vigna radiata
Ceasalpiniaceae	
Trees	*Tamarindus indica*
Shrubs	*Cassia angustifolia*
Mimosaceae	
Trees	*Acacia nilotica*
	Acacia senegal
	Acacia tortilis
	Albizia lebbeck
	Prosopis cineraria
	Prosopis juliflora
Shrubs	*Mimosa hamata*

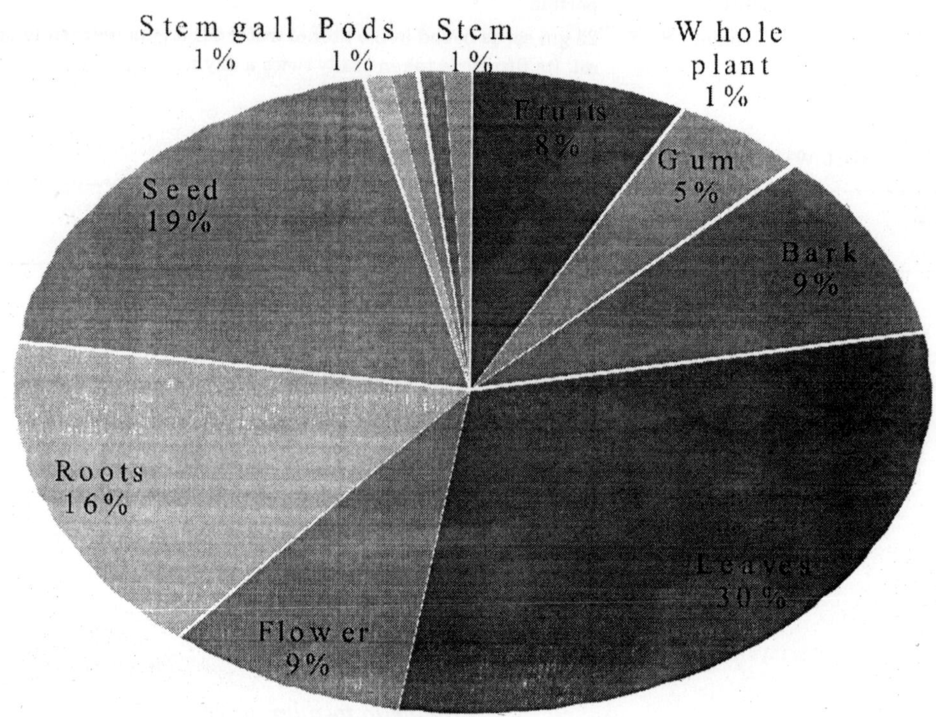

Fig. 1.

It is evident from above that many plants are used for curing one ailment throughout the desert, Like wise, a single plant can be used for curing many ailments, for example: *Acacia nilotica* being useful in 18 diseases. Almost all the plants listed occur commonly in grazingland, wastelands, croplands, water bodies and around habitations. Besides cultivated plants, some of these species are also weeds e.g. *Indigoflora cordifolia, Crotalaria burhia, Tephrosia purpurea P. juliflora* and *Mimosa hamata* all having immense medicinal value. Analysis of the ways of using herbs revealed that most of these are being used as filtrates taken empty stomach in the morning. The trading is unorganized and need institutional support. In view of their immense value, *Acacia nilotica, Butea monosperma, Glycyrhiza glabra* and *Cassia angustifolia*, having established and validated medicinal uses, could be considered for crop diversification in the Indian arid zone.

REFERENCES

1. Bhandari, M. M. 1990. *Flora of Indian Desert* (Revised Ed.) MPS Repros, Jodhpur 435p.

2. Anonymous, 1972. *The wealth of India, Raw Materials*. Vol. 2. Publication & Information Directorate. CSIR New Delhi
3. Jain, S.K. 1997. *Contribution to Indian Ethnobotany*. Scientific Publishers. Jodhpur
4. Lancelot D' Cruz, 2007. Ethnobotanical studies on Leguminous plants of Dedipada forest. *Ethnobotany* 19, 73-77.
5. Pande, P.C. and Pande, H.C. 2004. Traditional usage of 'Urd' (*Vigna mungo*) in Kumaon Himalaya. *Ethnobotany* 16: 96-98.
6. Pandey, R.K. and Shrivastava, J.L. 1991. Wild medicinal plants of *Fabaceae* in Madhya Pradesh : An Ethno botanical approach. *J. of Tropical Forestry* 7(3) : 220-.227
7. Singh, V. and Pandey, R.P. 1998. *Ethnobotany of Rajasthan, India*. Scientific Publishers, Jodhpur 367p.

Occurrence of triacontanol - a plant growth regulator in Senna (*Cassia angustifolia*)

Hamid A. Khan, B.K. Garg and Z.D. Kavia
Central Arid Zone Research Institute, Jodhpur - 342003

ABSTRACT

Cassia senna Linn. (*leguminosae*) var. senna (*Cassia angustifolia*) is an important commercial medicinal crop. Initially confining to Tamil Nadu, its cultivation spread to Maharashtra, Andhra Pradesh, Karnataka, Gujarat and Rajasthan. The plant is highly drought hardy and is cultivated on marginal lands. It is particularly suitable for the desert and its cultivation is spreading in the western Rajasthan. Senna is valued for its cathartic properties and is specially useful for habitual constipation. Both leaves and pods of the plant contain sennosides and constitute the drug. India is the major supplier of the drug to the world market. Occurrence of triacontanol in senna stems - a by-product in senna processing industry, has already been reported. Triacontanol is an effective commercial plant growth regulator which increases growth and water uptake of rice seedlings and also the growth of corn and barley. Triacontanol was also found effective in promoting germination and stubble sprouting. Occurrence of triacontanol in the stems of senna is partly responsible for the better yield of Bajra and wheat observed when senna stems are ploughed back into the soil. Senna dust, another by-product available in senna processing industry has also been found to contain triacontanol as high as 0.0439% which is almost four times to the presence of triacontanol (0.0103%) in *senna* stems. Occurrence of triacontanol in *senna* stems (reported earlier) and in senna dust enhances over-all importance of *senna* as a medicinal crop as the left-over stems and senna dust in senna processing industry, after taking leaves and pods for medicinal purposes, may be utilized for increasing the productivity of the field.

Key words: *Cassia angustifolia*, senna, triacontanol, plant growth regulator.

INTRODUCTION

Cassia senna linn. (*Leguminosae*) Var. senna (*Cassia angustifolia*) is an important commercial medicinal crop and finds wide use in Ayurveda, Unani, Allopathy and other traditional systems of medicine mainly because of the laxative property of its leaves and pods which contain sennosides A and B and form the drug. Senna is specially useful for habitual constipation [1]. Besides Tamil Nadu, its cultivation spread to Maharashtra, Andhra Pradesh, Karnataka, Gujarat and Rajasthan. The plant is highly drought hardy and its cultivation is particularly suited to western Rajasthan. India is the major supplier of the drug to the world market [2] and large quantities of dried senna leaves and pods are exported to many European countries, USA, Japan and Australia. We undertook the analysis of different parts of *Cassia angustifolia* when it came to our notice that ploughing of stems of *Cassia angustifolia* back into the soil increases the growth and yield of Bajra and wheat. However, senna stems and senna dust considered waste in senna processing industry have been found to contain triacontanol - a plant growth regulator. Triacontanol increases the growth and water uptake of rice seedlings and also the growth of corn and barley [3]. Triacontanol was also found effective in promoting germination and stubble sprouting [4].

MATERIALS AND METHODS

Senna dust available as by-product in senna processing industry was procured from a farmer in Jodhpur District. Senna dust was extracted in a soxhlet apparatus with petroleum-ether (60-80°). When the extraction was complete, the extract was concentrated and chromatographed over a column of silica gel. The eluant with 2% ethyl acetate-petroleum ether (60-80°) mixture afforded a mixture of compounds. Fractional crystallization of the mixture with chloroform afforded crystals of triacontanol. The identity of triacontanol was confirmed by its m.p. m.m.p. and co-TLC with an authentic sample.

RESULTS AND DISCUSSION

We earlier undertook the analysis of the stems of *Cassia angustifolia* when it came to our notice that ploughing of its stems back into the soil increases the growth and yield of Bajra and wheat. The stems of authentic senna were collected by us from the field of a farmer dealing in senna trade in Jodhpur district in Rajasthan. The stems were chemically analysed and it was found that the stems contained a plant growth regulator, namely, triacontanol in 0.0103% yield [5]. Triacontanol is an effective and commercial plant growth regulator which increases growth and water uptake of rice seedlings and also the growth of corn and barley. Triacontanol was also found effective in promoting germination and stubble sprouting. Occurrence of triacontanol in the stems of senna may be partly responsible for the reported better yield of Bajra and wheat crops when senna stems are ploughed back into the soil. The effect of the presence of triacontanol as plant growth regulator in senna stems is clearly manifested as the application of senna stems significantly increased the leaf area and fresh and dry weight of pearl-millet seedlings within twenty days of growth which cannot be ascribed to crop residue effect only [6]. Further, soil application of senna stems to wheat crop before sowing either alone or in combination with nitrogen could significantly improve plant growth and grain yield [7].

Senna (*Cassia angustifolia*) dust available as another by-product in senna processing industry has also been chemically analyzed now. Senna dust (551g) was extracted in a soxhlet apparatus with petroleum-ether (60-80^0). When the extraction was complete, the extract was concentrated and chromatographed over a column of silica gel. The eluant with 2% ethyl acetate- petroleum ether (60-80°) mixture afforded a mixture of compounds. Fractional crystallization of the mixture with chloroform afforded crystals of triacontanol, m.p. 85° in 0.0439% yield. The identity of triacontanol was confirmed by its m.p., m.m.p. and co-TLC with and authentic sample. Triacontanol is known to be an effective and commercial plant growth regulator. Chemical analysis of senna dust has thus revealed that it contains triacontanol (0.0439%) which is almost four times to the presence of triacontanol (0.0103%) in senna stems. Occurrence of triacontanol in senna stems (reported earlier) and in senna dust enhances over-all importance of senna as a medicinal crop as the left-over stems and senna dust, after taking leaves and pods for medicinal purposes, may be utilized for increasing the productivity of the field.

REFERENCES

1. Chopra, I.C., Abrol, B.K. and Handa, K.L. 1960. Medicinal plants of the arid zone, UNESCO, pp. 27.
2. Anonymous 1992. Wealth of India, CSIR, New Delhi, 3: 362.
3. Reis, S.K., Sweeley, C.C. and Leavit, R.A. 1977. Triacontanol: A New Naturally Qccurring Plant Growth Regulator. Science, 95: 1339.
4. Anonymous 1987. Annual Progress Report (1986-87), All India Coordinated Research Project on under utilized and under- exploited plants, National Bureau of Plant Genetic Resources, New Delhi, pp. 51.
5. Khan, Hamid, A., Kavia, Z.D. and Garg, B.K. 2004. Occurrence of triacontanol - a plant growth regulator in senna stems. Paper presented at National Seminar on Recent Advances in Analytical Chemistry (RAAC 2004), JNV University Jodhpur, pp. 37.
6. Garg, B.K., Khan, Hamid, A. and Kavia, Z.D. 2005. Influence of senna stems on growth, chlorophyll and nitratereductase activityof pearl millet. Annals of Arid Zone, **44(1)**: 79 - 82.
7. Garg, B.K., Khan, Hamid, A. and Kavia, Z.D. 2007. Interactive effects of senna stems and nitrogen fertilization on net photosynthesis, carbohydrate and nitrogen metabolism and yield of wheat. J. Plant Biol., **34(1)**: 5 -12.

Seed yield of different accessions and selection of best plants from exotic collections of *Acacia senegal*

Amit Tak, S.K. Jindal and Anjly Pancholy

Central Arid Zone Research Institute, Jodhpur-342 003

ABSTRACT

Acacia senegal (L.) Willd. most drought- hardy plant is well known for its multi purpose uses in Indian hot desert. Its seeds rich in protein are used by the local people in the form of vegetable known as "Panchkuta". To identify high seed yielding accessions and genotypes, plantation of six exotic and one indigenous collections planted in 1988 were evaluated for seed yield in 2003 and 2007. Highest number of plants yielding seeds was from EC87/7493, which was found to have maximum survival after 19 years of establishment. Maximum average seed yield of seed yielding plants was 1.17 kg for Sudan collection in 2003, and 0.92 kg for EC87/7493 in 2007. Individual trees of Sudan material yielded the maximum seed in 2003 and in 2007. There was not much change in the average seed yield of seed yielding trees of 2003 and 2007. Positive and significant relationship of seed yield of 2003 and 2007 showed that high yielding genotypes could be selected at early stage. Six plants were selected on the basis of seed yield out of which three were from Sudan material and others were from EC87/7490 (Niger), EC87/7493 (Mali) and EC87/7497(Mali) and seed of these genotypes can be used for large-scale plantations.

Key words: *Acacia senegal,* accessoris, exotic, selection.

INTRODUCTION

Acacia senegal (L.) Willd. most drought- hardy plant is well known for its multipurpose uses in Indian hot desert. Locally it is known as kumat /kher and is a major source of gum arabic. Its seeds rich in protein are boiled, dried and used as one of the ingredients of desert delicacy by local people in the form of vegetable known as "Panchkuta". Selection of right seed source of species for a given region is necessary for maximum productivity [1]. Exploration of naturally occurring variability in seeds and seed yield tells the effect of environment and insect pests on pod and seed yield. For any improvement programme assessment of degree and pattern of variation, and collection and conservation of this variation is utmost necessary. In crop plants the importance of broad genetic base is well known and conservation of germplasm is given a high priority. Relationship of geographical diversity with genetic diversity is well studied and documented in annual crop plants. With this approach, genetic material from the areas where *A. senegal* is naturally grown was procured, planted in 1988 and evaluated for seed yield in 2003 and 2007 to identify high seed yielding genotypes, the results of which are presented here.

MATERIALS AND METHODS

Seeds of five accessions from Niger (EC 87/7490), Mali (EC 87/7493, EC 87/7497 and EC 87/7499) and Senegal (EC 87/7500), were procured from CIFT Nogent Sur Marne (Table 1). Seedlings of these accessions along with of Sudan and Rajasthan were transplanted in 1988 at Central Research Farm of Central Arid Zone Research Institute, Jodhpur. Pods were collected from each surviving tree during December of 2003 and 2007 and were thrashed tree wise to record seed yield of individual trees. Mean, SD values for each accession were worked out as per standard statistical procedures.

Table 1. Source of seed of *Acacia senegal* planted at CAZRI in 1988

S.No	Accession	Origin	Longitude	Latitude	Altitude (m)	Precipitation (mm)
1	EC87/7490	Niger	-	-	-	-
2	EC87/7493	Mali	11°35'W	14°57'N	100	560
3	EC87/7497	Mali	10°22'W	1°437'N	150	650
4	EC87/7499	Mali	7°10'W	15°08'N	270	450
5	EC87/7500	Senegal	14°49'W	15°40'N	60	400
6	Sudan	Sudan	-	-	-	-
7	Rajasthan	Rajasthan	73°08'W	26°18'N	241	317

RESULT AND DISCUSSION

Rainfall was 419.7 mm during 2003 and 323.7 mm during 2007 at Jodhpur and most of it was during July in 2003 and well distributed over 4 months June-September in 2007 (Table 2). The average annual rainfall of this area was 367.9 ± 39.4 mm during 1988-2007 ranging between 50.6 mm (2002) to 844 mm (1990) with 20.8 ± 1.06 number of rainy days.

Table 2. Rainfall during 2003 and 2007 at Jodhpur

Month	Rainfall (mm)	
	2003	2007
Jan	0.7	0
Feb	23.1	22
Mar	2.7	29.1
April	0	14.8
May	0.4	1
June	65.6	22.3
July	260.3	73.4
August	65.3	77.2
Sep	1.6	83.7
Oct	0	0
Nov	0	0
Dec	0	0.2
Total	419.7	323.7

The number of trees bearing seed, mean seed yield per tree, minimum and maximum seed yield of individual trees during 2003 and 2007 are given in table 3. Highest number of plant yielding seed was from accessions EC87/7493 in 2003 and from Sudan in 2007. Accession EC87/7493 was also found to have maximum number of surviving trees after 19 years of establishment. Maximum average seed yield of seed yielding plants was 1167.8 g for Sudan collection in 2003 and 918.1 g for EC87/7493 in 2007. Among the individual trees Sudan material yielded the maximum seeds (2576 g) in 2003 and 2510 g in 2007. There was not much difference in average seed yield of 2003 (645 g) and 2007 (669.7 g). But accessions EC87/7490, EC87/7493, EC87/7499 and EC87/7500 yielded more in 2007 than in 2003 and the only significant change from year to year was in case of only EC87/7493 where its seed yield was more than double in 2007 (446.9 g in 2003, 918.1 g in 2007) (Chart 1). In 2003, only 2 accessions EC87/7497 (t value =2.572* at 14 df) and Sudan (t value =3.021** at 25 df) had significantly higher seed yield than accession Rajasthan. In 2007 four accessions namely EC87/7490 (t value =2.549* at 23 df), EC87/7493(t value =4.039** at 32 df), EC87/7499(t value =3.856** at 24 df) and Sudan(t value =2.823** at 34 df) gave significantly higher seed yield than the accession Rajasthan. Only accession Sudan performed significantly better than accession Rajasthan during both the years. The coefficient of variation was more than 50% for all the accessions during both the years and it was very high i.e., more than 100.0% in case of EC87/7500 showing that there is significant variation within the accessions also. This high variation shows that selecting the individual trees for high seed yield from and within the accessions will be most effective.

Table 3. Seed yield per plant (g) of seed bearing plants of different accessions of *A. senegal* during 2003 & 2007

Accessions	n	Seed yield	Min.	Max.	SD	CV%
2003						
EC87/7490	14	580.1	57.7	2129.2	508.9	87.7
EC87/7493	19	446.9	83.0	984.7	246.0	55.0
EC87/7497	3	1105.5	450.9	1722.6	636.7	57.6
EC87/7499	13	672.7	122.8	2120.6	570.4	84.8
EC87/7500	6	323.4	84.6	970.3	331.0	102.4
Sudan	14	1167.8	153.2	2576.0	784.0	67.1
Rajasthan	13	455.9	106.3	1341.5	337.5	74.0
Population	82	645.0	57.7	2576.0	560.4	86.9
2007						
EC87/7490	9	723.3	30.0	1995.0	676.1	93.5
EC87/7493	18	918.1	20.0	1980.0	618.6	67.4
EC87/7497	5	532.0	20.0	1285.0	628.8	118.2
EC87/7499	10	778.5	130.0	1850.0	454.7	58.4
EC87/7500	12	562.5	15.0	1850.0	596.2	106.0
Sudan	20	802.3	50.0	2510.0	749.1	93.4
Rajasthan	16	249.7	20.0	1070.0	247.2	99.0
Population	90	669.7	15.0	2510.0	615.8	92.0

A. senegal is highly cross pollinated due to its self- incompatibility operating inside the embryo sac. Under natural conditions fruit set is very low and 0.36% flowers set into fruits [2]. Insufficient pollination is the main cause of low fruit set. There are various factors to affect the seed yield of a tree. These factors are genetically and environmentally regulated. Genetically means that all accessions have different genetic structure so these are different in pod and seed yield as this tree is basically cross-pollinated. Within the same accession also, there is a huge difference in seed yield. This can be due to environmental factors. Rainfall and temperature are the most important factors for flower setting. Both factors also affect the insect breeding. Other than the environmental factors there are genetic differences, which is confirmed by the positive and significant relationship of seed yield of individual trees in 2003 with of 2007 (correlation coefficient = 0.470 at 95 degree of freedom P<0.01). Seed yield is a complex character and it is a function of many parameters like seed size, number of seeds per pods etc. Though this variation was not recorded.

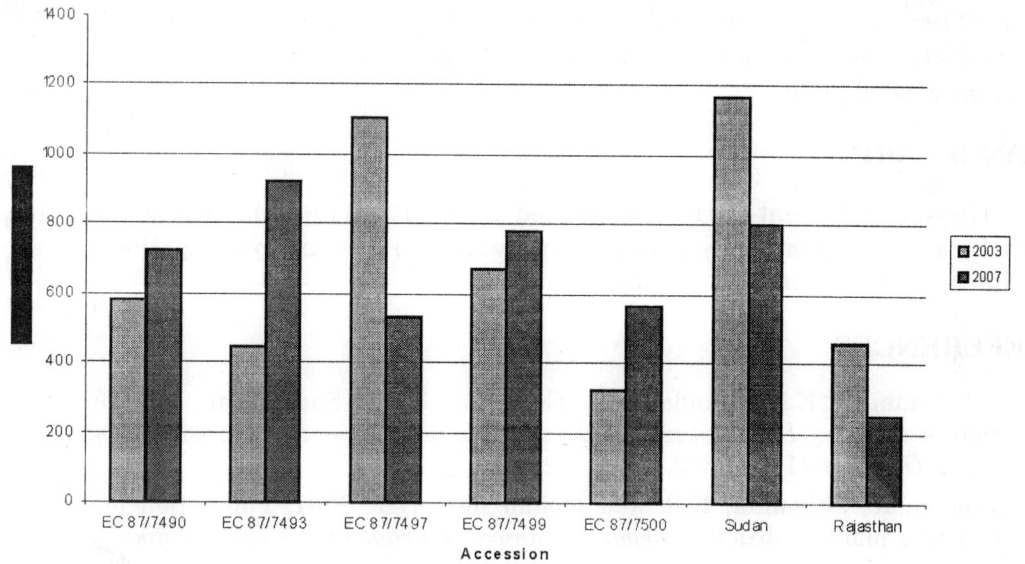

Chart 1. Seed yield/tree (g) in different accessions in 2003 and 2007.

Table 4. Six best seed yielding plants of *A. senegal*

Source/Acc. No.	Tree No.	Seed yield (g)	
		2003	**2007**
Sudan	31	1854	1415
Sudan	37	1936	2220
Sudan	28	2069	1490
EC 87/7490	67	918	1020
EC 87/7493	223	985	1320
EC 87/7497	166	1723	1150

The positive relationship of rainfall of the sites (from where the accessions were collected) with seed yield of different accessions in 2003 (r=0.722 at 3 df) and seed yield in 2007 (r=0.512 at 3 df) showed that seed collected from better rainfall area gives better seed yield when evaluated under Indian desert conditions.

The significance of source variation in seed and seedling traits has been reported in number of tree species, viz, *A. mangium* [3; 4; 5], *Dalbergia sissoo*

[6], *Eucalyptus tereticornis* [7] and *Prosopis cineraria* [8] which is confirmed here in *A. senegal.*

On the basis of best seed yield, six plants were selected (Table 4). These were tree numbers 28, 31 and 37 from Sudan material; tree number 67 from EC87/7490; tree number 223 from EC87/7493 and tree number 166 from EC87/7497. Seeds of these genotypes can be used for large-scale plantations for maximum seed yield.

CONCLUSION

There is sufficient variation for seed yield within and among the accessions, which can be exploited by selecting individual trees and putting them in seed orchard for future use.

REFERENCES

1. Subramanium, K.N., Mandal, A.K., Govindraj, P. and Sasidharan, K.R. 1992. Provenance trials in *Eucalyptus grandis* and its implications to forestry programmes. *Silvae Genetica* 41: 239-242.
2. Tandon, R., Shivanna, K.R. and Mohan Ram H.Y. 2001. Pollination biology and breeding plant of *Acacia senegal*. *Botanical Journal of the Linean Society* 135 (3): 251-262.
3. Salazar, R. 1989. Genetic variation of 16 provenances of *Acacia mangium* at nursery level, Turrialba. Costa Rica. *Commonwealth Forestry* 68: 263-272.
4. Bagchi, S.K. and Dobriyal, N.D. 1990. Provenance variation in seed parameters of *Acacia nilotica*. *Indian Forester* 116: 958-961.
5. Bagchi, S.K., Joshi, D.N. and Rawat, D.S. 1990. Variation in seed size of *Acacia* spp. Silvae Genetica, 39: 107-110.
6. Vakshasya, R.K. Rajora, O.P. and Rawat, M.S. 1992. Seed and seedling traits of *Dalbargia sissoo* Roxb., seed sources variation studies among ten sources of India. *Forest Ecology and Management* 48: 271-275.
7. Ginwal, H.S., Pradip Kumar, Sharma, V.K., Mandal, A.K. and Harwood, C.E. 2004. Genetic variability and early growth performance of *Eucalyptus tereticornis* Sm. in provenance cum progeny trials in India. *Silvae Genetica* 53: 148-153.
8. Kackar, N.L., Solanki, K.R. and Jindal, S.K. 1986. Variation in fruit and seed characters of *Prosopis cineraria* Mac Bride, in Thar desert. *Indian Journal of Forestry* 9: 113-115.

Germination among different accessions of *Prosopis cineraria* – an important leguminous tree of the Indian arid zone

V.K. Manga[1] and D.N. Sen

Department of Botany, Jai Narain Vyas University, Jodhpur 342 001

1. Principal Scientist (Plant breeding) Central Arid Zone Research Institute, Jodhpur

ABSTRACT

Prosopis cineraria is widely distributed in the arid and semi-arid tracts spread over north and north-west India. It is a multipurpose tree and is a source of fuel, timber and vegetable to human beings and green fodder to animals. Through generations of experience, farmers have understood the usefulness of this tree species, consequently it is encouraged in the cultivated fields and village grazing lands. Seeds of *P.cineraria* have mechanical dormancy due to impervious and hard seed coat, consequently need some pre-sowing treatment to enhance germination quality of seedlings. In the present experiment germination of 51 accessions of *Prosopis cineraria* from Rajasthan, Gujarat and Haryana, was studied in laboratory at 27 ± 2°C under 12 hour light and 12 hour darkness, first without any seed treatment (control) and then following scarification with concentrated sulphuric acid for 10, 15, 20 and 25 minutes. Germination among accessions ranged from 0 to 63% under control (no scarification). Twenty five accessions did not germinate. Thus level of dormancy varied among accessions. Following scarification with concentrated sulphuric acid, increase in germination percentage was observed. Maximum germination percentage was observed following 15 minutes seed treatment with concentrated sulphuric acid. Accessions NIC-1848, NIC-1937, NIC-1936, IC-104194, NIC-1961, NIC-1908, NIC-1920, NIC-1966, NIC-13549, NIC-13580, and NIC-13555 showed 100 per cent germination. For remaining accessions, germination percentage ranged from 30 to 96.67%. Seeds of all accessions were also sown in field in polythene bags, following 15 minute scarification. Germination in field ranged from 6.67 to 90%. Visnagar showed the maximum (90%) germination, followed by Hansot (82.22%), NIC-1848 (80%), NIC-1966 (80%), and NIC-13580 (80%). The study revealed that germination among different accessions of *Prosopis cineraria* can be significantly enhanced by seed scarification with conc. Sulphuric acid for 15 minutes. Further information obtained in lab on germination percentage can be used to predict germination under field conditions.

Key words: *Prosopis cineraria,* seed germination, germplasm.

INTRODUCTION

Seed may appear simple externally but has a complex ecophysiology for further resumption of growth, primarily its germination (17). A number of environmental factors together with the genetic makeup of the seed affect the germination phenomenon.

Seed dormancy among plants growing in arid zone is of great advantage if germination and growth can be managed during the relative brief period of rainfall in these areas. Thus the seeds that remain viable but dormant until sufficient water is available have good chance for survival. Seed dormancy therefore is an ecological adaptation by which a species is able to delay its germination in an ecosystem till the favourable environmental conditions are resumed. The pattern of seed germination in a species has been found to vary according the seed source (1), family (2,3), parental nutrition (4), seed maturity (5), environmental pre-conditioning during seed development (6,7), and seed size (8,9). Germination is further influenced by maternal effects in most plants, since a major portion of the seed components (seed coat (2n), endoperm (1n), and half of the embryo (1n)) are maternally contributed (10,11). Seeds of *P. cineraria* have poor germination mainly because of mechanical dormancy due to impervious and hard seed coat. According to Nambier (12), Martin and Alexander (13), Chatterji and Mohnot (14), impervious teguments are the major cause behind seed dormancy in the genus *Prosopis*. Several methods like filing, heat treatment, hot water treatment, concentrated sulphuric acid treatment etc. have been tried to break dormancy in other species. Sulphuric acid has however been found to be the most suitable one. No such information is available in case of *P. cineraria*. In the present study germination was studied in different accessions of *P. cineraria* to find out: (i) differences in germination percentage among different accessions of this species and (ii) optimum duration of seed scarification with concentrated sulphuric acid for obtaining maximum germination. Improving the seed lot for higher germination would require lower seed rate and lesser seedbed area reducing seed sowing costs. Finally uniform germination with better seedling vigour will provide a uniform and better plantable stock with less management cost and time for large scale plantation.

MATERIAL AND METHODS

Materials for the present study consisted of seeds of fifty one accession of *Prosopis cineraria* (L.) MacBride, forty seven of which were from different districts of Rajasthan, four from Haryana and four from Gujarat. Five treatments were given to the seeds before placing in petri plates for germination study in lab (a) Control (no scarification), (b) Seeds treated with concentrated (95%,GR) sulphuric acid for ten minutes, (c) fifteen minutes treatment with concentrated sulphuric acid, (d) Twenty minutes treatment with concentrated sulphuric acid, (e) twenty five minutes treatment with concentrated sulphuric acid. After scarification with acid, seeds were washed thoroughly in running tap water for thirty minutes and finally with distilled water. Ten seeds of each accession for each treatment were placed in triplicate in 10 cm diameter petri plates lined with ordinary filter paper, and kept regularly moist with distilled water. These petri plates were placed in seed germinator at 27±2°C under 12 hour light and 12 hour darkness. Recording of germination data was initiated 24 hours after placing petri plates in the germinator. Number of seeds that germinated everyday were recorded for each accession up to ten days. Germination percentage was calculated as percentage of seed kept for germination. Seeds of each accession were also planted in polythene bags in nursery after treatment with concentrated sulphuric acid for 15 minutes. Data obtained on germination percentage in different accession were subjected to analysis of variance.

RESULTS AND DISCUSSION

Data obtained on germination percentage of different accessions following different durations of acid scarification were subjected to analysis of variance. Analysis of variance (Table 1) revealed highly significant differences among treatments for germination percentage, indicating the effectiveness of seed scarification with concentrated sulphuric acid and opportunity of selecting the treatment giving highest seed germination. Similarly, significant differences were found to exist among accessions. This showed that accessions of *Prosopis cineraria* differed in germination percentage, as a result it should be possible to choose accessions showing higher seed germination. The treatment x accessions interaction was also significant. This implied that different accessions showed different expression for germination percentage following different treatments. Consequently accessions ranked differently for germination percentage under different treatments. Perusal of table 1 revealed that treatments accounted for the largest portion of variation (95%), while variation among accessions accounted for a small portion of total variation (3.30%). The treatments x accessions interaction contributed 0.05% to the total variation. To understand the effect of accessions data obtained on germination in Lab under 15 minute treatment were further subjected to analysis of variance. A study of table 2, revealed highly significant differences among accessions for germination percentage. As is evident accessions accounted for 85 percent of the total variation.

Table 1. Analysis of variance for germination percentage (lab) in *Prosopis cineraria* following different duration of scarification with concentrated H_2SO_4

Sources of variation	df	Mean squares
Replication	2	667.00**
Treatment (T)	4	74394.40**
Accession (A)	50	2583.10**
T x A	200	419.51**
Error	508	64.48
Variance components (%)		
Replication		0.85
Treatment (T)		95.22
Accession (A)		3.30
T x A		0.05
Error		0.58

** Significant at 1% level of probability

Mean values of germination percentage obtained under different treatments are presented in Table 3. Germination ranged from 0.00 to 63.33% in control (no seed treatment). It was found that out of 51 accessions, twenty five did not show germination under control. This showed that level of dormancy varied from accession to accession. Similar observations have been made in Sitka spruce by Chaisurisri *et al.* (15), in Douglas fir by Allen (1961) and El-Kassaby *et al.* (1992) and Virginia pine (1, 3). Following scarification with concentrated sulphuric acid, significant increase in germination was observed (plate 1). The increase in germination percentage following 10 minutes scarification was sharp (table 3), it shot up from mean germination of 9.8% (control) to 65.3%. Further increase in germination following longer durations of scarification was however not as sharp as the previous one, but the mean germination percentage increased from 65.3% (10 minute scarification) to 89.67% (15 minute scarification). A further increase in duration of scarification led to a decline in germination. The downtrend which started following 20 minutes scarification continued for 25 minutes treatment also. This reduction in germination percentage could probably be due to the damaging effect of sulphuric acid on the embryo, which might have entered seed following seed coat softening. Hermosilla and Rosales (16), however did not report any harmful effect of different durations (0 to 24 minutes) of concentrated sulphuric acid treatment on four species of *Prosopis*. In the present study most of the accessions showed maximum germination percentage following 15 minutes scarification. Accessions NIC-1848, NIC-1937,

NIC-1936, IC-104194, NIC-1961, NIC-1908, NIC-1920, NIC-1966, NIC-13549, NIC-13580, and NIC-13555 showed 100 per cent germination after 15 minutes scarification. For remaining accessions, germination percentage ranged from 30 to 96.67%. Only 17.64% of the accessions showed less then 60% germination.

Table 2. Analysis of variance for germination percentage in *Prosopis cineraria* (Lab) following 15 minutes treatment with concentrated H_2SO_4

Sources of variation	df	Mean squares
Replication	2	8.62
Accession (A)	50	747.67**
Error	100	117.45
Variance components (%)		
Replication		0.98
Accession (A)		85.57
Error		13.45

** Significant at 1% level of probability

Table 3. Analysis of variance for germination percentage (field) in *Prosopis cineraria*

Sources of variation	df	Mean squares
Replication	2	870.75**
Accession (A)	50	546.33**
Error	100	49.61
Variance components (%)		
Replication		59.36
Accession (A)		37.25
Error		3.38

** Significant at 1% level of probability

After it was established in Lab study that seed scarification with concentrated sulphuric acid for 15 minutes gave the highest germination, seeds of all accessions were sown in soil filled polyethylene bags following 15 minutes of acid scarification. Data obtained on germination percentage were subjected to analysis of variance (Table 3). Significant differences were observed among accessions for germination percentage. The germination percentage in field ranged from 6.67 to 90%. Visnagar showed the maximum (90%) germination, followed by Hansot (82.22%), NIC-1848 (80%), NIC-1966 (80%), and NIC-13580 (80%).

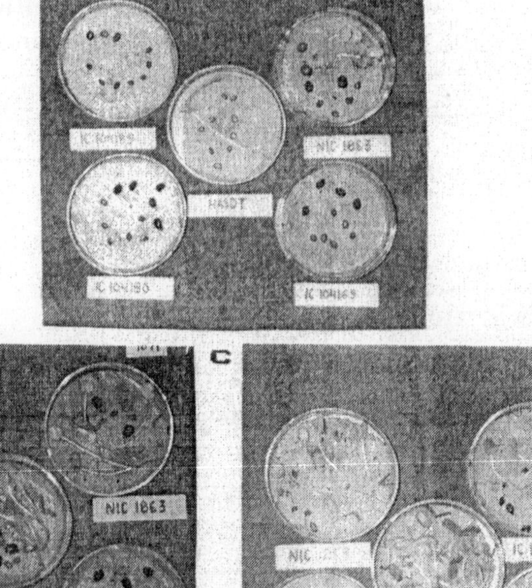

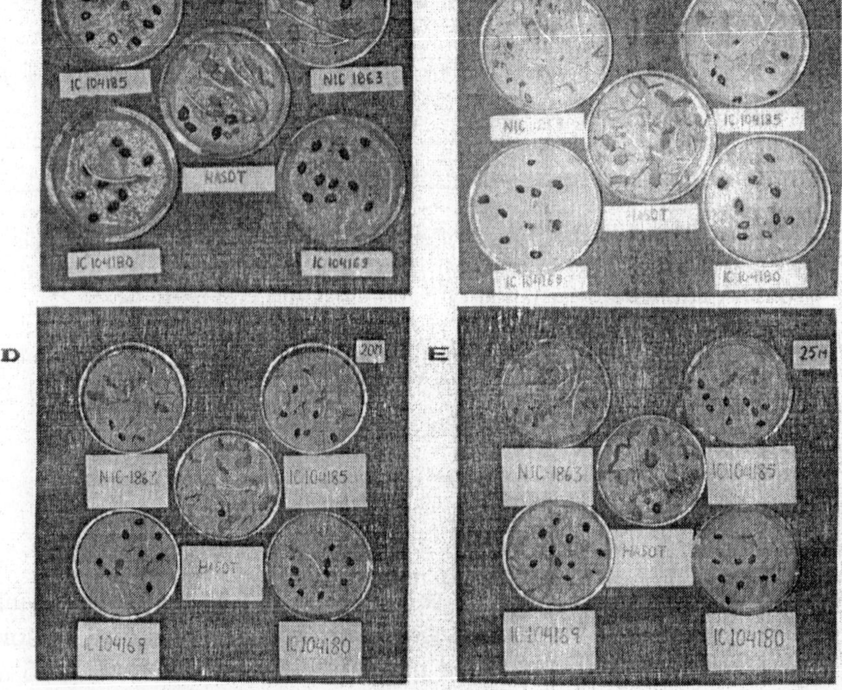

Seed germination in control (A), after 10 (B), 15 (C), 20 (D) and 25 (E) minutes with H_2SO_4 scarification, in some genotypes of *P. cineraria*.

Plate 1.

Table 3 Mean and range of germination percentage (lab) following different duration of acid scarification in lab, and comparison of lab and filed germination following 15 minute of acid scarification.

Treatment (scarificatin with conc. H_2SO_4)	Mean germination (%)		Range (%)	
	Lab	Field	Lab	Field
Control	9.80	-	0.00 - 63.33	-
10 minute	65.29	-	6.67 - 100.00	-
15 minute	82.16	55.42	30.00 - 100.00	6.67 – 90.00
20 minute	72.09	-	13.33-100.00	-
25 minute	64.12	-	6.67 - 96.67	-

Results obtained following Lab and field study are presented in Table 4. Range for percent germination was higher in field in comparison to that obtained in Lab. Mean germination percentage, was however, higher in Lab. This could be because of better germination conditions in Lab as compared to field. To find out whether Lab study can be utilized for predicting germination in field or not, simple correlation was worked out between germination percentage obtained in Lab following 15 minute acid treatment, with those obtained in field. A high value of positive correlation (0.79) between these two revealed that information obtained in Lab on germination percentage could be used confidently to predict germination behaviour of different *Prosopis* accessions in field. As such it can be concluded from the study that a). differences for percent germination existed among different accessions of *Prosopis cineraria*, b). germination percentage can be improved by scarifying seeds of *Prosopis cineraria* with concentrated sulphuric acid, and the optimum duration is 15 minutes for getting maximum germination and c). the information obtained from lab study on germination percentage can be used to predict germination behaviour in field.

REFERENCES

1. Allen, G.S. 1961. Testing Douglas-fir seed for provenance. *Proc. Int. Seed Test. Assoc.* 26: 388-403.
2. Bramlett, D.L., Dell, T.R. and Pepper, W.D. 1983. Genetic and maternal influences on Virginia pine seed germination. *Silvae Genet.* 32: 1-4.
3. El-Kassaby, Y.A., Edwards, D.G.W. and Taylor, D.W. 1992. Genetic control of germination parameters in Douglas-fir and its importance for domestication. *Silvae Genet.* 41: 48-54.

4. Allen, G.S. 1960. Factors affecting the viability and germination behaviour of coniferous seed. IV. Stratification period and incubation temperature, *Pseudotsuga menziesii* (Mirb.) Franco. *For. Chron.* 36: 18-29.

5. Edwards, D.G.W. and El-Kassaby, Y.A. 1988. Effect of flowering phenology, date of cone collection, cone-storage treatment and seed pretreatment on yield and germination of seeds from a Douglas-fir seed orchard. *For. Ecol. Manage.* 25: 17-29.

6. Sawhney, R. and Noylor, J.M. 1979. Dormancy studies in seed of *Avena fatua* 9. Demonstration of genetic variability affecting the response to temperature during seed development. *Can. J. Bot.* 57: 59-63.

7. Sawhney, R. and Naylor, J.M. 1980. Dormancy studies in seed of *Avena fatua*.12. Influence of temperature on germination behaviour of nondormant families. *Can. J. Bot.* 58: 578-581.

8. Naylor, J. M. 1983. Studies on the genetic control of some physiological processes in seeds. *Can. J. Bot.* 61: 3561- 3567.

9. Wulff, R. 1972. Intrapopulational variation in the seeds in *Hyptis suavelens. Ecology* 54: 646-649.

10. Hellum, A.K. 1990. Seed ecology in a population of *Acacia holoserica. Can. J. For. Res.* 20: 927-934.

11. Ellner, S. 1986. Germination dimorphism and parent-offspring conflict in seed germination. *J. Theor. Biol.* 123: 173-185.

12. Nambier, K. 1946. A novel method of improving the germination of *Prosopis juliflora* seeds. *Indian Forester* 72: 193-195.

13. Martin, C. and Alexander, R. 1974. Agricutural Handbook. 450, US Forest Service, Washington. D.C. pp. 656-657.

14. Chaisurisri, K., Edwards, D.G.W. and El-Kassaby, Y.A. 1992. Genetic control of seed size and germination in Sitka Spruce. *Silvae Genet.* 41: 348-355.

15. Chatterji, U.N. and Mohnot, K. 1969. Eco-physiological investigations on the imbibition and germination of seeds of *Prosopis juliflora* Linn. *Recent Advances in Tropical Ecology, Symposium*, pp. 261-268.

16. Hermosilla, J.L. and Rosales, B.A. 1988. Germination improvement treatments for seeds of four Chilean *Prosopis* species. In: *The Current State of Knowledge on Prosopis juliflora. II Intl. Conference on Prosopis.* Recife, Brazil, 25-29 Aug. 1986. Mario A. Habit (ed.). pp. 495-501.

17. Sen, D. N. 1977. *Environment and Seed Germination of Indian Plants*. The Chronica Botanica Co., New Delhi.

Biotecnological interventions in pulses: future prospects

N.P. Singh and Indu Singh Yadav

Indian Institute of Pulses Research,Kanpur-208004

Pulses playa significant role in world agriculture because they provide a protein rich diet to poor and marginal sections of population especially in developing nations. The world population is predicted to reach 8 billion by the year 2010. To feed 3 billion additional people in the next 20 year will require dramatic increase in crop productivity a formidable task by any standard. Despite a number of positive interventions such as high yielding varieties, extension of irrigated areas and use of agrochemical, pulse productivity in India and elsewhere remains low and stagnant for the past several years. Conventional plant breeders and other agricultural scientists have worked diligently and skillfully to upgrade quality and raise the yields by employing various crop improvement techniques. Not withstanding the impressive gains in productivity so far, there are limitations to conventional plant breeding technology, either due to the limited gene pool or to the restricted range of organisms between which genes can be transferred due to species barriers. Although these methods have proved to be useful, they depend ultimately upon reliable testing and proper selection and do not involve the identification and manipulation of genetic targets as defined in molecular terms. Successful augmentation of traditional approaches to plant breeding based on advances in applied biotechnology has assumed utmost significance and considered crucial for future realization of yield potential and value addition. In recent days, areas of molecular research: transgenics, marker aided selection and genomics significantly contribute towards pulse improvement.

With the advent of recombinant DNA methods and genetic transformation procedures, it is possible to transfer genes into crop plants from unrelated plants, microbes and animals. Most of the modifications being carried out or envisaged are for disease pest or herbicide resistance. Plants with new traits, such as resistance to herbicides, insect pests and viruses have been genetically engineered using genes from unrelated organisms. Introducing new genes or manipulating endogenous gene expression via genetic transformation generates new phenotypic variations useful for crop improvement. In general new phenotypic variations are useful for crop improvement. In general legume species are much more difficult to transform than others chiefly due to the poor regenerability and lack of availability of compatible gene delivery methods. Thus developing protocols to enhance regeneration still remains an important goal in grains legumes besides the objective of producing single copy transgenes expressing the transgene stably and capable of transmitting it to the progeny. Transformation of plants involves the stable introduction of DNA sequences usually into the nuclear genome of cells capable to giving rise to a whole transformed plant. Transformation without regeneration and regeneration without transformation are of limited value. In recent years many of the shoot regeneration protocols have been developed specifically for the purpose of genetic manipulations. Several techniques for DNA delivery into plant cells are available, ranging from the Agrobacterium -mediated gene transfer, direct gene transfer through electroporation into protoplasts/ intact tissues either by PEG method or by uptake of DNA into naked protoplasts by injection and the use of microprojectile bombardment mediated transformation to introduce DNA into intact tissues. The most prevalent mode of regeneration in grain legumes is via direct organogenesis from cotyledonary node explants. In general, the callus mediated regeneration systems are amenable to Agrobacterium mediated transformation. Agrobacterium-mediated transformation which has been successful in other crops, has registered only limited success with respect to grain legumes because all the four modes of regeneration are not generally available. In pigeonpea and chickpea, preliminary reports of transformation of cotyledonary nodes and shoot apices are available but confirmed transgenics are yet to be realized. Enhanced transformation frequency has been reported in pea using a new binary vector: pGreen. Particle bombardment has been used for genetic transformation of many

grains legumes such as soybean, and Vigna radiata. Although, at present all the transgenic varieties of soybean under commercial cultivation have been generated via particle bombardment, but no reports of successful biolistic transformation of pigeonpea and chickpea are available so far. The major problem in the development of transformation system is providing induced Agrobacterium with access to cells capable of dedifferentiation followed by regeneration. In species such as tobacco and brassica napus, this combination can be achieved with relative ease, however it is often difficult to combine transformation competence with totipotency. Tissue culture independent transformation systems have also been demonstrated in various crops such as soybean, sunflower, safflower and peanut. A number of positive reports describing legume transformation systems that require reduced or no tissue culture such as electroporation of nodal axillary buds and seedlings and flower filtration using A. tumefaciens for medicago truncatula have been observed. Alternatively legume specific non-tissue culture systems could be a significant boon to legume transformation. In pulses, there are a dozen laboratories in the country who have capabilities to transfonn important pulse crops like chickpea, pigeonpea, mungbean and fieldpea. Institutions like ICRISAT, AAU, Jorhat, IARI, New

Delhi, IIPR, Kanpur, NBRI, Lucknow, Bose Institute, Kolkata are at different stage of development of transgenic varieties of chickpea and pigeonpea using CrylAc, CrylAb,CrylAa gene constructs. However, there is need to generate large number of transfonned plants with different events and screen and identify an event which gives optimum expression level to provide satisfactory control to Helicoverpa. Besides, pod borer resistance in chickpea and pigeonpea, there are also efforts to develop transgenics resistance to fungal pathogens involving PGIP gene in chickpea, resistance to viral diseases resistance using coat protein and replicase genes in blackgram and herbicide tolerance using dianthin and barnase gene in blackgram. In future, the deficiency in essential amino acids and nutritional quality as a whole and tolerance to drought and salinity in particular need biotechnological interventions.

During the next decade traditional approaches will be greatly aided by newer molecular techniques such as DNA markers and genomics, enabling rapid generation of cultivars. Marker assisted selection or marker aided selection is a boon whereby a marker (morphological, biochemical or one based on DNAIRNA variation is used for indirect selection of a genetic determinant or determinant of a trait of interest (ie., productivity, disease resistance, abiotic stress tolerance and or quality). The primary resource of a plant breeding programme is the genetic variability available within germplasm. In general marker assisted breeding strategies are pursued for three basic purposes (i) Identification of suitable parental lines for improvement with respect to a specific target trait. (ii) Tracing favourable alleles (dominant or recessive) across generations. (iii) Identifying the most suitable individuals among the segregating progeny, based on allelic composition across a part or the entire genome.

Genomics is likely to bring in profound gains in pulse production and nutritional enhancement. Genetic modification can contribute significantly in enhancing the nutritional value of grain legumes. Attempts to alter storage protein composition, carbohydrate and mineral content and elimination! reduction of anti-nutritional factors in legume seeds have particularly gained attention in recent days. Genome wide analysis of gene function and gene expression are beginning to yield valuable information in many areas of biological research and these genomic tools are now being applied in grain legumes. DNA sequencing of cDNA libraries to generate sets of expressed sequence tags (ESTs) are allowing gene compendiums for crop diseases to be compiled. Annotation of such data collection is also providing a wealth of functional genomics era will be to employ large scale techniques to knock out or silence genes in order to synthesize gene specific mutants for phenotypic analysis and to use micro-array methodology to analyze global gene expression protein turnover processing. Application of these technologies promises to accelerate the pace that biological infonnation relevant to crop protection accures. There has been substantial progress made in legume genomics over the past decade. Marker techniques were adopted or newly designed for the major legumes. The diverse molecular markers (spanning from dominant RAPD, DAF, and microsatellite fingerprint markers to co-dominant RFLP or simple sequence repeat markers) were used to detect genetic diversity in lines and populations, and to establish genetic maps for nearly all the more agronomically important legumes. It comes as no surprise that soybean genomics is presently most advanced, given the number of laboratories working on this crop and the support being granted for soybean research. In spite of all these advances, legumes as a whole are still under-researched, the research generally underfinanced and the genomics of these important crops necessarily underdeveloped, especially if compared with other crops such as rice, maize or potato, or even the so-called model plant Arabidopsis thaliana. However, it is expected that legumes research will intensify in near future and probably catch up with research on other crop plants.

Isolation and insecticidal activity of *Bacillus thuringiensis* from arid soil against *Corcyra cepholonica*

Anjly Pancholy, Nisha Patel, S.K. Jindal and Manjit Singh

Central Arid Zone Research Institute, Jodhpur-342003 (Rajasthan)

ABSTRACT

Bacillus thuringiensis Berliner (Bt), a crystalliferous, spore-forming bacterium, is active against a wide variety of insect pests. These pests as defoliators and pod borers cause great loss to the arid legumes. To isolate and identify Bt, 5 g samples were collected in September, 2006 from 5-10 cm deep soils from the fields of *Vigna unguiculata* (L.) Walp and *Vigna radiata (L.) Wilczek* (where these crops are being grown for the last ten years) and stored in the refrigerator until further use. *Bacillus* cultures were isolated from 500 mg of soil samples using Acetate Selection Method with a slight modification. Cultures from each soil samples were tested against larvae of *Corcyra cepholonica*, a stored grain pest. Four cultures from soils of *V. unguiculata* and two from *V. radiata* fields showed promising larvicidal effects. These cultures were also analysed by Polymerase Chain Reaction (PCR) method using universal primers: Un1, Un2, Un3, Un4, Un7, 8 for the presence of crystal protein genes (*cry* genes). Five *Bacillus* cultures gave PCR products representing *cry*1 (~ 270 bp) with the universal primer Un1 and one culture gave PCR product (~700 bp) representing *cry*2 with primer Un2, respectively. These *Bacillus* strains can be explored for the management of lepidopteron insects of the legume crops of arid zone.

Key words: *Bacillus thuringiensis*, isolation, arid zone soil

INTRODUCTION

Bacillus thuringiensis, commonly found in different niches such as soil, dead larvae, grains, rice dust and snow (1), has been valued for its specificity for target insects, its low development cost, and its environmental compatibility. *B. thuringiensis* also constitutes a gene pool for engineering insecticidal activity into other organisms, such as crop plants. Though, it is one of the most widely used bio-pesticide in the world (2), in India it is still restricted to experimental field trials with few crops only. Commercial formulations of Bt are available, which are not effective against insect larvae at high temperature during summer season in arid region. Native strains already acclimatized to the environmental conditions of the desert can be more effective bio-control agents against various insect pests of this area. Crystal proteins produced by Bt as the products of *cry* genes and active ingredients against target insects, have a highly specific insecticidal effect. Therefore, searching for newer, native strains of Bt with broad-spectrum activity is imperative for cost effective biological control. Although, the occurrence of Bt is random but soil has been reported to be the major source for the isolation of Bt (3, 4, 5). Therefore, an attempt has been made to isolate and characterize native strains of Bt from local arid region soil by using a selective medium followed by PCR screening and insect bioassay.

MATERIAL & METHODS

Collection of soil samples: Soil samples (5 g) were collected in September, 2006 at the depth of 5-10 cm from fields of mung bean (*Vigna radiata*) and cowpea (*V. unguiculata*) located in Central Research Farm, CAZRI, Jodhpur, where the soil temperature goes up to 55°C. The two pulse crops are being grown for the past 10 yrs in the said fields. The soil of this area is sandy loam (camborthid) with a pH 8.1, has low nutrient level, with 0.23% organic carbon, 0.03% nitrogen and 0.02% phosphorous (6). All samples were stored in refrigerator in sterile polythene bags (10 X 5 cm) until further use.

Isolation of Bt: This was carried out by acetate selection protocol (7). 500 mg of soil was mixed with 10 ml of Luria broth (contains per liter - Casein enzymic hydrolysate, 10 g; yeast extract, 5 g; sodium chloride, 5 g and final pH 7.0) buffered with 0.25 M sodium acetate, rotated at 250 rpm for 4h at 30°C and heat-shocked at 80°C for 10 minutes. Serial dilutions (10^{-1} to 10^{-4}) were made in sterile, distilled water and 100μl of each dilution was spread on Luria agar plates followed by incubation at 30°C for 48h. There were three replications for each dilution. All the colonies were further purified by single colony isolation and maintained on Luria agar at 4°C until further use. For sporulation and protein crystal formation, isolated colonies were picked up and streaked on T3 medium (contains per liter – Tryptone, 3 g; tryptose, 2 g; yeast extract, 1.5 g; sodium phosphate, 0.05M, pH 6.8; $MnCl_2$, 0.005 g) plates and incubated at 30°C for 72h.

Anti-biotic resistance of the isolated cultures: All the bacillus cultures were screened for their antibiotic resistance using antibiotic susceptibility discs of Hi Media. For this, each culture was spread on Luria agar plate to form a lawn of bacteria and the discs of different antibiotics such as ampicillin (25μg), trimethprim (25μg), streptomycin (25μg), erythromycin (15μg), kanamycin (5μg), tetracycline (5μg) and novobiocin (30μg), were placed on the bacterial lawn at a distance of approximately 3-4 cm. Five such discs were used per plate and observations on the formation of zones of inhibition were made after 24h.

Molecular characterization of the isolated cultures: This was done by PCR amplification of the crystal protein genes or the *cry* genes. For this, the isolates were grown on Luria agar medium for 24h. A loopful of cells was transferred to an eppendorf tube containing 100μl of sterile, distilled water and kept in boiling water bath for 10 min for disruption of cells. The disrupted cell suspension was then centrifuged and supernatant was used in PCR assay. Five pairs of universal primers (8) were used having sequences as follows: (i) *cry*1 - Un1 (d), 5' CAT GAT TCA TGC GGC AGA TAA AC 3' & Un1 (r), 5' TTG TGA CAC TTC TGC TTC CCA TT 3'; (ii) *cry*2 - Un2 (d), 5' GTT ATT CTT AAT GCA GAT GAA TGG G 3' & Un2 (r), 5' CGG ATA AAA TAA TCT GGG AAA TAG T 3'; (iii) *cry*3 - Un3 (d), 5' CGT TAT CGC AGA GAG ATG ACA TTA AC 3' & Un3 (r), 5' CAT CTG TTG TTT CTG GAG GCA AT 3'; (iv) *cry*4 - Un4 (d), 5' GCA TAT GAT GTA GCG AAA CAA GCC 3' & Un4 (r), 5' GCG TGA CAT ACC CAT TTC CAG GTC C 3'; and (v) *cry*7 & cry8 - Un7,8 (d), 5'-AAG CAG TGA ATG CCT TGT TTA C-3' & Un7,8 (r), 5' CTT CTA AAC CTT GAC TAC TT 3'. The PCR reaction mixture (50 μl) contained 5μl cell supernatant, 0.5 μl each of cry gene primers (100 pmole / μl stock), 2 μl dNTPs mix (10mM stock), 5 μl 10X Taq buffer (without $MgCl_2$), 3 μl $MgCl_2$ (25mM stock), 0.5 μl Taq DNA polymerase and 33.5 μl sterile, distilled water. Amplification was done in a DNA thermocycler by using the program: initial denaturation at 94°C for 5min., 30 cycles of denaturation at 94°C (30 sec.), annealing at 50°C (30 sec.), extension at 72°C (30 sec.) and single cycle of final extension at 72°C for 7 min. Following amplification, the PCR reaction products were analyzed by agarose gel electrophoresis on 1% gel in 1 x TBE (Tris-borate-EDTA) buffer, stained with ethidium bromide and viewed and photographed under UV illumination using Gene Genius bio-imaging system.

Insect Bioassay: Insecticidal activity of the isolates was determined against second instar larvae of rice meal moth, *Corcyra cepholonica*, of the order lepidoptera. Cultures of Bt were grown for 48h on Luria agar plates and cell suspension (~10^7cells/ml) prepared in sterile, distilled water. 0.5, 1.0 and 1.5 ml each of the above Bt suspensions were added to 50g of jowar grains and mixed thoroughly in 100ml glass beakers. Ten larvae each of *C. cepholonica* were put in each of these beakers. Three replications were made for each concentration. Observations were recorded on inactivity / mortality of larvae after 24, 48 and 96 h. Sterile- distilled water was used as control.

RESULTS AND DISCUSSION

Isolation of Bt from soil: Isolated 44 bacillus cultures (18 cultures from *V. unguiculata* and 26 from *V. radiata* fields) using Acetate selection protocol, a method selective for the isolation of *B. thuringiensis*. Plate1 shows a Bt culture growing on Luria agar medium isolated in this study.

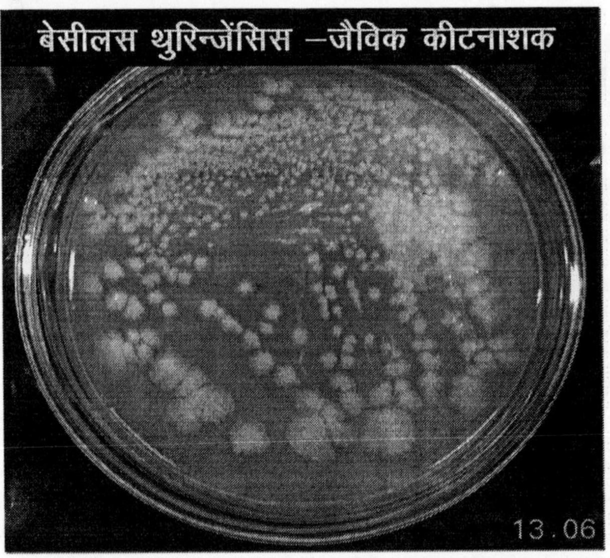

Fig.1. Locally isolated *Bacillus thuringiensis* culture

Antibiotic resistance: Antibiotic resistance of the isolated cultures was studied to identify easily selectable markers. The results presented in Table1 showed that 29.5% of the isolated cultures were resistant and 70.5% were susceptible to ampicillin (25μg) and erythromycin (15μg). 31.8% of the cultures were resistant to Streptomycin (25μg) and 27.2% were resistant to tetracycline (5μg). None of the cultures were resistant to trimethprim (25μg), Kanamycin (5μg) and novobiocin (30μg).

Table 1. Response of isolated bacillus cultures to different antibiotics

S.No.	Antibiotics	Resistant (%)	Susceptible (%)
1	Ampicillin (25μg)	29.5	70.5
2.	Trimethoprim (25μg)	Nil	100
3.	Streptomycin (25μg)	31.8	68.2
4.	Erythromycin (15μg)	29.5	70.5
5.	Kanamycin (5μg)	Nil	100

6.	Tetracycline (5μg)	27.2	72.8
7.	Novobiocin (30μg)	Nil	100

Molecular characterization of the isolated cultures: The polymerase chain reaction (PCR) method, a powerful tool for identification of the specific insecticidal genes carried by Bt was used to characterize newly isolated, native *B. thuringiensis* isolates. The universal oligo-nucleotide primers used in this study were from a region that is highly conserved in 20 *cry1*, 3 *cry2*, 4 *cry3*, 2 *cry4*, 2 *cry7*, and 3 *cry8* genes (8) and identification was done on the basis of the size of the PCR products obtained. In the present study, out of 44 bacillus cultures isolated, five (two from *V radiata* field and three from *V. unguiculata* field) gave PCR products (~270 bp) with *cry*1 universal primers and one (from *V. unguiculata* field) gave PCR product (size ~ 700 bp) with *cry*2 universal primers (Table 2) (Fig. 2). Native strains of *B. thuringiensis* isolated from different ecosystems of Columbia giving amplification mostly with *cry*1 general primers were also isolated by Ben-Dov *et al.* (8). Rest 38 bacillus cultures did not respond to the universal primers used in the present investigation. The absence of PCR products in these *Bacillus* isolates does not necessarily imply that these strains were devoid of genes coding for insecticidal polypeptides. Either these new isolates might be containing genes from known strains which have not been sequenced yet as discussed by (9) or these could be having novel cry genes, which may not be detected by using the *cry* specific primers used in the present study.

Table 2. PCR product profile of Bacillus isolates

Universal primers used*	Cry group genes from which derived	PCR product obtained	Approximate PCR product size (bp)	No. of Isolates giving PCR product	
				From *V. radiata* field	From *V. unguiculata* field
Un1	*cry*1	Yes	270	2	3
Un2	*Cry*2	Yes	700	-	1
Un3	*Cry*3	No	-	-	-
Un4	*Cry*4	No	-	-	-
Un7,8	*Cry*7,8	No	-	-	-

*Ben-Dov et al. (1997)

Insect Bioassay with the isolated cultures: *Bacillus* isolates were tested for their activities against insects of the order Lepidoptera by using the stored grain pest, *Corcyra cepholonica* (Table 3). Larvae of this insect were fed on jowar grains treated with different doses of the isolated Bt cultures. As is clear from the Table 3 that 2% and 3% of the Bt doses were effective in causing death of *C. cepholonica* larvae. But in either case the mortality was not more than 20-30%. One reason for low mortality of insect larvae could be that the treated larvae were a mix of different in star stages. The younger larvae were affected by the Bt treatment whereas the older ones that were about to enter into pupation were not affected by the above treatment. Also, the activity of delta endotoxin is generally confined to the digestive tract of any insect. Susceptible insects have high midgut pH that is required for crystal dissolution and subsequent action of proteolytic enzymes to produce endotoxins. Reducing substances also play a role in crystal dissolution (10). In some insects such as *Popillia japonica* ingested food regulated pH of the digestic tracts (11). In the present study also, low mortality of the treated larvae could be due to unfavorable midgut pH. This needs to be experimentally proved by further research.

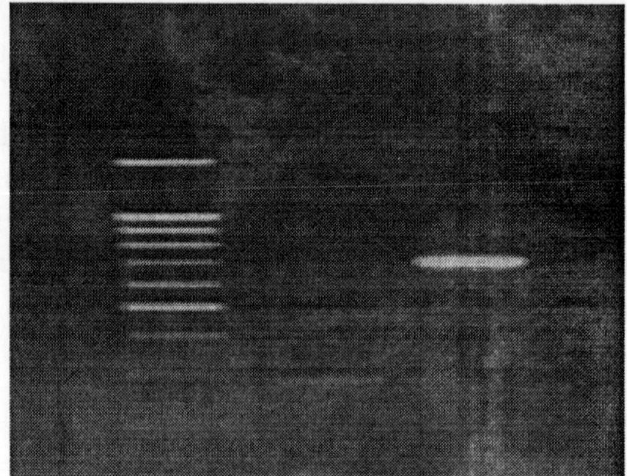

Fig. 2. Molecular Characterization of Bt cultures

Table 3. Insect (*Corcyra cepholonica)* bioassay with the isolated cultures

Treatment	Inactivity / mortality (%) after		
(10^7cells/ml)	24h	48h	96h
Control	nil	nil	nil
1%	nil	nil	nil
2%	nil	nil	20
3%	nil	30	30

CONCLUSION

The soil of this area seems to be richer in Bt bacteria carrying *cry*1 gene followed by *cry*2 gene, although the possibility of presence of other types of cry genes in the soil under study can not be ruled out. *The B. thuringiensis* cultures isolated in the present study will not only enrich the existing insecticidal strains, but can also be multiplied as Bt formulations against some of the lepidopteron insects of the arid zone crops after extensive testing in the field.

REFERENCES

1. Martin PAW and Travers R.S. 1989. Worldwide abundance and distribution of *Bacillus thuringiensis* isolates. Appl. Environ. Microbiol., 55:2437-2442.
2. Van Frankenhuyzen, K. 1993. The challenge of *Bacillus thuringiensis*, p, 1-35. *In* P. F. Entwistle, J.S. Cory, M.J. Bailey and S.R. Higgs (ed.), *Bacillus thuringiensis*, an environmental biopesticide: Theory and practice. John Wiley & Sons, Ltd., Chichester, United Kingdom.
3. Bernhard, K., Meadows, M., Butt, J., Ellis, D.J., Roberts, G.M., Pauli, S., Rodgers, P. and Berges, H.D. 1997. Natural isolates of *Bacillus thuringiensis*: worldwide distribution, characterization and activity against insect pests. J. Invert. Pathol., 70: 59-68.
4. Hongyu, Z., Yu, Z. and Deng, W. 2000. Isolation, distribution and toxicity of *Bacillus thuringiensis* from Warehouses in China. Journal of Crop Protection. 19: 449-454.
5. Asokan, R. and Puttaswamy, 2007. Isolation and characterization of *Bacillus thuringiensis* Berliner from soil, leaf, seed dust and insect cadaver. J. Biol. Control. 21(1): 83-90.
6. Dhir R.P. 1984. Soils of arid regions, their characteristics and properties. In: *Agroforestry for Arid and Semi-Arid Zones* (ed.) K.A. Shankarnarayan, CAZRI, Jodhpur, pp. 20-29.
7. Travers, R.S., Martin, P.A.W. and Reichelderfer, C.F. 1987. Selective process for efficient isolation of soil *Bacillus* species. Appl. Environ. Microbiol., 53: 1263-1266.
8. Ben-Dov, E, Zaritsky, A., Dahan, E., Barak, Z., Sinai, R., Manasherob, R., Khamraev, A., Troitskaya, E., Dubitsky, A., Berezina, N. and Margalith, Y. 1997. Extended screening by PCR for seven *cry*-group genes from field-collected strains of *Bacillus thuringiensis*. Appl. Environ Microbiol., 63(12): 4883-4890.
9. Carozzi, N.B., Kramer, V.C., Warren, G.W., Evola, S. and Koziel, M.G. 1991. Prediction of insecticidal activity of *Bacillus thuringiensis* strains by polymerase chain reaction product profiles. Appl. Environ Microbiol., 57:3057-3061.
10. Tanada, Y. and Kaya, H.K. 1993. Insect Pathlogy, Academic Press, Inc, California. Pp. 104-105.
11. Sharpe, E.S. and Detroy, R.W. 1979. Susceptibility of Japanese beetle larvae to *Bacillus thuringiensis*: Associated effects of diapause, midgut pH, and milky disease. J. Invert. Pathol., 34: 92-94.

Association among traits at different growth stages in *Prosopis cineraria* (L.) McBride

V.K. Manga*[1] *and David N. Sen

Botany Department, J.N.V. University, Jodhpur, 342 001
1. Principal Scientist (Plant Breeding),
Central Arid Zone Research Institute, Jodhpur, Rajasthan 342 003

ABSTRACT

Prosopis cineraria is widely distributed in the arid and semi-arid traits spread over north and north west India. It forms prominent constituent of the Desert Thorn Forest with very open crop of scattered trees. This species holds an important place in the rural economy in the north west arid region of Indian Sub-continent. Variability and interrelationships among various traits were studied in forty accessions of *Prosopis cineraria* (L.) MacBride, from different districts of Rajasthan, Haryana and Gujarat. The experiment was conducted in a randomized complete block design with three replications. Ten seedlings of each accession in each of the three replications were randomly tagged and data on various morphological traits were recorded in two months, four months, six months, nine months and one year old seedlings. Seedling height, internode length pinna length, pinna width, number of pinnules/pinna were recorded at all growth stages, while data on number of branches and collar diameter were recorded at four month and later growth stages. Dry shoot weight was recorded in four month old and in one year old seedlings. Data so obtained were subjected to analysis of variance and correlation coefficients were worked out among different traits at different growth stages. Highly significant differences were observed among different genotypes, for all traits studied at all the growth stages. Seedling height, internode length, pinna length and number of pinnules showed moderately high heritability at various growth stages. Seedling height associated positively and significantly with internode length, number of pinnules/pinna, number of branches, collar diameter and dry shoot weight at all growth stages. Similarly internode length correlated positively with pinna length at all growth stages, and with number of pinnules in four out of five stages. Pinna length also associated positively with pinna width as well as number of pinnules, while pinna width and number of number of pinnules did not show consistent association with other traits. Looking to these associations and heritability of traits, seedling height was the most appropriate trait, selection for which can bring about improvement in most of the traits including dry shoot weight.

Key words: *Prosopis cineraria*, variability, heritability, association, correlation, inter-relationships

INTRODUCTION

Prosopis cineraria is widely distributed in the arid and semi-arid traits spread over north and north west India. It forms prominent constituent of the Desert Thorn Forest with very open crop of scattered trees. This species holds an important place in the rural economy in the north west arid region of Indian Sub-continent. It is the only leguminous tree that grows well against all odds of climatic conditions. It is a multipurpose tree and is a source of fuel, timber and vegetable to human beings and green fodder to animals. Through generations of experience, farmers have realized the usefulness of this tree species, consequently it is encouraged in the cultivated fields and village grazing lands. Despite being an important multipurpose tree of the arid zone, very little efforts have been made to improve this species using genetic principles and breeding methods. *Prosopis cineraria* is a multipurpose tree species highly suitable for agroforestry and afforestation of desert wastelands. Studies conducted (1,2,3) revealed that large amount of genetic variability existed for various morphological traits. This offers scope for selection of desirable types. However in a tree improvement programme, knowledge of inter-relationship among different traits is essential for identifying an easily manipulative trait that is associated with a complex trait of interest. Such an association if remain strong enough even with increase in age of seedling could be used for making selection at an early age with confidence. In the present study attempt has been made to study inter-relationship among traits in seedlings of *Prosopis cineraria* at different ages.

MATERIAL AND METHODS

Materials for the present study consisted of seeds of forty accession of *Prosopis cineraria* (L.) MacBride, from different districts of Rajasthan, Haryana and Gujarat. The experiment was conducted in March 1993 in randomized complete block design with three replications. Thirty bags of each accession were seeded in each replication for getting sufficient number of seedlings for recording data. The bags were irrigated daily in summer, and as and when required in rainy and winter seasons. During the duration of the experiment, hoeing and weeding was done regularly and insecticide sprayed to keep seedlings healthy. Ten seedlings of each accession in each of the three replications were randomly tagged and data on various morphological traits were recorded in two months, four months, six months, nine months and one year old seedlings. Seedling height, internode length pinna length, pinna width, number of pinnules/pinna were recorded at all growth stages, while data on number of branches and collar diameter were recorded at four month and later growth stages. Dry shoot weight was recorded in four month old and in one year old seedlings. Data so obtained were subjected to analysis of variance and correlation coefficients were worked out among different traits at different growth stages.

RESULTS AND DISCUSSION

Highly significant differences were observed among genotypes, for all traits studied at all the growth stages, revealing that sufficient variability was available in the material, which meets the pre-requisites of a breeding programme. Range, mean and heritability of various traits at different growth stages is given in table 1. For the purpose of selection, seedling stage that showed high range and high heritability, would be more appropriate. A study of the table revealed that for seedling height, four months old seedling stage showed the maximum range as well heritability. For internode length, nine months seedling stage showed the maximum heritability, though range for this trait was next only to four month old seedlings. Similar was the case for pinna length and pinna width, for which nine month seedling stage formed the most appropriate stage for selection of material. For number of pinnules/pinna, six months seedlings stage was suitable as seedlings at this stage showed high heritability. For number of branches, twelve months seedlings showed maximum range, while heritability for this trait was moderate at four months stage, hence direct selection for this trait may be useful at four months stage. Range for collar diameter was almost same at different seedling stages except for two months old seedlings. Heritability for this trait was however higher in four months old seedlings, hence selection for collar diameter would be effective at this stage. For dry shoot weight, though four months stage showed high range in twelve months old seedlings, heritability was high at four month stage.

Table 1. Mean, range and heritability in broad sense of various morphological traits at different growth stages in *Prosopis cineraria*

Growth stage	Parameter	Seedling height (cm)	Inter-node length (cm)	Pinna length (mm)	Pinna width (mm)	No. of pinnules	No.of branches	Collar diameter (mm)	Dry shoot weight (g)
Two	Mean	17.72	8.83	18.86	9.01	12.70	0.51	1.83	Not

month	Range	12.05-23.30	6.63-11.40	15.83-21.97	7.87-10.50	10.87-16.0	0.00-2.13	1.44-2.14	recorded
	Heritability	0.51	0.47	0.42	0.30	0.76	0.25	0.47	
Four	Mean	21.61	11.61	19.63	10.11	12.34	2.46	2.47	1.16
month	Range	13.63-34.20	9.53-14.67	16.20- 23.7	9.00-11.73	9.60- 15.60	0.87- 5.00	2.13-3.12	0.60-2.33
	Heritability	0.66	0.24	0.49	0.28	0.58	0.47	0.43	0.41
Six	Mean	32.54	11.42	20.60	9.75	13.08	8.00	3.47	Not
month	Range	27.02-39.90	9.57-13.60	17.60-23.80	8.97-10.83	11.00-15.6	5.47-15.2	2.99-4.01	recorded
	Heritability	0.56	0.32	0.50	0.48	0.72	0.40	0.37	
Nine	Mean	38.05	11.31	20.18	9.39	12.95	13.25	3.88	Not
month	Range	29.50-47.87	9.27-13.93	15.87-23.63	8.13-10.53	10.93-15.3	8.57-19.6	3.24-4.44	recorded
	Heritability	0.48	0.55	0.67	0.55	0.67	0.40	0.35	
Twelve	Mean	39.90	11.38	13.36	6.81	11.23	18.82	4.04	3.77
month	Range	32.07-48.72	9.57-13.33	11.20-15.97	5.83-8.3	9.87-13.2	12.03-29.0	3.52-4.64	2.59-6.31
	Heritability	0.59	0.54	0.23	0.24	0.36	0.32	0.30	0.23

Phenotypic and genotypic correlation coefficients obtained between different combination of traits at 2 month growth stage (Table 2) revealed that seedling height associated significantly and positively with internode length, pinna length, number of pinnules/pinna and collar diameter. Internode length associated positively with pinna length and number of pinnules/pinna, while pinna length with pinna width and number of pinnules/pinna. No association of number of branches with other traits was observed at this growth stage. Genotypic correlation coefficients of plant height with internode length, number of pinnules/pinna and collar diameter were higher than the phenotypic values. Genotypic association of internode length with pinna width was slightly lower than phenotypic association. Pinna width that had negative and non-significant association with number of branches and number of pinnules/pinna at phenotypic level showed high negative association with number of branches, and moderate positive correlation with number of pinnules/pinna.

Association among various traits at four month growth stage are presented in Table 3. Seedling height showed high and positive association at phenotypic as well as genotypic level with internode length, pinna length, number of pinnules/pinna, number of branches and collar diameter. Pinna width did not show any association with seedling height. Internode length had significant and positive association with pinna length and pinna width, while no association was observed with number of pinnules/pinna, number of branches and collar diameter. Pinna length showed significant and positive association with pinna width, number of pinnules/pinna and collar diameter. High association of these traits with internode length was also observed at genotypic level. Number of branches which had no association with pinna length at phenotypic level showed high value of genotypic correlation coefficient, revealing the influence of environment. Pinna width did not show association with number of pinnules, number of branches and collar diameter. Number of branches showed high positive correlation with collar diameter both at phenotypic and genotypic level. Dry shoot weight showed highly significant and positive correlation coefficient both at phenotypic and genotypic level with seedling height, pinna length, number of pinnules, number of branches and collar diameter. In *Pinus roxburghii,* Khosla *et al.* (4) reported positive phenotypic and genotypic association of shoot dry weight with shoot length and collar diameter. Similarly, Jindal *et al.* (5), observed significant association of dry shoot weight with its components in 3 month old seedlings of *Acacia senegal.* Internode length which had non-significant association with dry shoot weight at phenotypic level, showed high value of correlation coefficient at genotypic level indicating high environmental influence on this trait. Pinna width which had non-significant association with dry shoot weight also had low values of correlation coefficient at genotypic level.

Table 2. Phenotypic (upper diagonal) and genotypic (lower diagonal) correlation coefficients among morphological traits in two month old seedlings of *Prosopis cineraria*

Character	Seedling height	Internode length (cm)	Pinna length(mm)	Pinna width (mm)	No. of pinnules	No.of branches	Collar diameter (mm)
Seedling height		0.642**	0.415**	0.077	0.464**	0.247	0.526**
Internode length (cm)	0.702		0.578**	0.173	0.414**	0.039	0.216
Pinna length (mm)	0.243	0.451		0.497**	0.426**	-0.240	0.266
Pinna width (mm)	-0.238	-0.175	0.331		-0.113	-0.146	0.215
No. of pinnules	0.487	0.447	0.488	-0.469		-0.104	0.169
No.of branches	0.203	-0.189	-0.898	-0.513	-0.351		0.188
Collar diameter (mm)	0.593	0.077	0.182	0.258	0.096	-0.124	

*, ** significant at five and one percent level respectively

Table 3. Phenotypic (upper diagonal) and genotypic (lower diagonal) correlation coefficients among morphological traits in four month old seedlings of *Prosopis cineraria*

Character	Dry shoot weight	Seedling height	Internode length (cm)	Pinna length (mm)	Pinna width (mm)	No. of pinnules	No.of branches	Collar diameter (mm)
Dry shoot weight		0.728**	0.281	0.453**	0.164	0.438**	0.636**	0.648**
Seedling height	1.003		0.476**	0.578**	0.176	0.510**	0.505**	0.601**
Internode length (cm)	0.847	0.657		0.550**	0.434**	0.264	0.069	0.271
Pinna length (mm)	1.020	0.765	0.557		0.442**	0.611**	0.241	0.553**
Pinna width (mm)	0.452	-0.007	0.446	0.458		0.079	-0.042	0.182
No. of pinnules	0.628	0.545	0.384	0.765	0.121		0.227	0.482**
No.of branches	0.748	0.735	0.232	0.700	0.013	0.492		0.552**
Collar diameter (mm)	1.004	0.737	0.371	0.839	0.431	0.599	0.638	

*, ** significant at five and one percent level respectively

In six months old seedlings, seedling height associated significantly and positively with internode length, number of pinnules/pinna, number of branches and collar diameter at phenotypic level (Table 4). Values of genotypic correlation coefficients of these traits with seedling height were higher than the phenotypic values. Internode length associated significantly and positively with pinna length, pinna width and number of pinnules, while number of branches and collar diameter had no association with internode length at phenotypic level. However at genotypic level these traits had positive and high values. Interestingly pinna width had positive significant association with internode length at phenotypic level but showed no association at genotypic level. Pinna length had high and positive association with pinna width and number of pinnules, both at phenotypic and genotypic level. Number of pinnules/pinna did not show any association with number of branches and collar diameter. Association of number of branches with collar diameter was significant only at phenotypic level.

Phenotypic and genotypic correlation coefficients between various traits at nine month seedling stage are presented in Table 5. Seedling height was positively associated both at phenotypic and genotypic level with internode length, pinna length, pinna width, number of pinnules, and number of branches and collar diameter. Highest association was observed with internode length, followed by number of branches and collar diameter. Internode length also had positive association with pinna length, pinna width and number of pinnules both at phenotypic and genotypic level. Pinna length was highly associated with pinna width and number of pinnules/pinna, while no association of this trait was observed with number of branches and collar diameter. Pinna width did not show any association with number of pinnules/pinna, number of branches and collar diameter, both at phenotypic and genotypic level. Phenotypic and genotypic correlation coefficients of number of pinnules/pinna with number of number of branches were very low and non-significant.

Table 4. Phenotypic (upper diagonal) and genotypic (lower diagonal) correlation coefficients among morphological traits in six month oldseedlings of *Prosopis cineraria*

Character	Seedling height	Internode length (cm)	Pinna length (mm)	Pinna width (mm)	No. of pinnules	No.of branches	Collar diameter (mm)
Seedling height		0.352*	0.284	-0.136	0.398*	0.411*	0.514**
Internode length (cm)	0.781		0.581**	0.307*	0.345*	-0.074	0.084
Pinna length (mm)	0.365	0.582		0.632**	0.455**	-0.208	0.190
Pinna width (mm)	-0.207	-0.007	0.625		0.042	-0.273	0.042
No. of pinnules	0.457	0.478	0.499	-0.102		-0.109	0.200
No.of branches	0.480	0.325	-0.330	-0.439	-0.149		0.497**
Collar diameter (mm)	0.712	0.732	0.419	0.130	0.429	0.186	

*, ** significant at five and one percent level respectively

Table 5. Phenotypic (upper diagonal) and genotypic (lower diagonal) correlation coefficients among morphological traits in nine month old seedlings of *Prosopis cineraria*

Character	Seedling height	Internode length (cm)	Pinna length (mm)	Pinna width (mm)	No. of pinnules	No.of branches	Collar diameter (mm)
Seedling height		0.517**	0.424**	0.345*	0.355*	0.512**	0.505**
Internode length (cm)	0.663		0.368*	0.349*	0.332*	0.129	0.069

Pinna length (mm)	0.480	0.539		0.763**	0.505**	-0.060	0.178
Pinna width (mm)	0.424	0.483	0.817		0.237	-0.180	0.121
No. of pinnules	0.354	0.521	0.598	0.232		-0.045	0.166
No.of branches	0.491	0.355	-0.136	-0.176	-0.207		0.590**
Collar diameter (mm)	0.740	0.440	0.214	0.281	0.096	0.353	

*, ** significant at five and one percent level respectively

In twelve month old seedlings, like other growth stages genotypic correlation coefficient in general were higher than phenotypic correlation coefficients (table 6). Phenotypic as well as genotypic association of seedling height with internode length, number of pinnules/pinna, number of branches and collar diameter was high and positive. Pinna length that showed low association at phenotypic level exhibited high value of genotypic correlation coefficient with seedling height. Internode length associated significantly and positively with pinna length, and number of pinnules/pinna at phenotypic as well as genotypic level. Pinna width, number of pinnules/pinna, number of branches and collar diameter did not show any association with internode length. Pinna length showed significant positive association with pinna width and number of pinnules/pinna. Value of genotypic correlation coefficient of number of pinnules with pinna length was slightly less than the phenotypic value. Pinna length that did not have any association with collar diameter showed high association with pinna length at genotypic level. At phenotypic level, pinna width showed positive association with number of pinnules/pinna, while this association at genotypic level was very low and negative. Conversely pinna width did not associate with collar diameter at phenotypic level, while this association was moderate at genotypic level. No association of pinna width both at phenotypic and genotypic level was observed with number of branches. Phenotypic correlation coefficient of pinnules/pinna number of branches and collar diameter were low and non-significant. Number of branches had high association with collar diameter. This showed good inherent relationship among traits. Shoot dry weight possessed high and positive values of both phenotypic and genotypic correlation coefficients with seedling height, number of pinnules/pinna, number branches and collar diameter. Internode length, pinna length and pinna width which had low and non-significant phenotypic correlation with shoot dry weight showed high association at genotypic level, indicating the influence of environment on these associations.

Table 6. Phenotypic (upper diagonal) and genotypic (lower diagonal) correlation coefficients among morphological traits in twelve month old seedlings of *Prosopis cineraria*

Character	Dry shoot weight	Seedling height	Internode length (cm)	Pinna length (mm)	Pinna width (mm)	No. of pinnules	No.of branches	Collar diameter (mm)
Dry shoot weight		0.786**	0.242	0.159	0.080	0.383*	0.707**	0.786**
Seedling height	1.034		0.469**	0.289	0.137	0.377*	0.567**	0.711**
Internode length (cm)	0.822	0.668		0.377*	0.276	0.307*	0.140	0.250
Pinna length (mm)	0.975	0.615	0.625		0.349*	0.480**	0.037	0.088
Pinna width (mm)	0.521	0.225	0.382	0.895		0.351*	-0.120	0.067
No. of pinnules	0.534	0.474	0.540	0.333	-0.045		0.139	0.299
No.of branches	0.749	0.701	0.353	0.311	0.001	0.262		0.551**
Collar diameter (mm)	0.899	0.848	0.461	0.747	0.472	0.314	0.675	

*, ** significant at five and one percent level respectively

A comparison of association among traits at various growth stages revealed that seedling height associated positively and significantly with internode length, number of pinnules/pinna, number of branches and collar diameter at all growth stages. Seedling height also showed highly significant and positive association with dry shoot weight at different growth stages. This showed the stability of these associations, which could be due to tight linkage between these traits. Internode length on the other hand correlated positively with pinna length at all growth stages, and with number of pinnules in four out of five stages. Pinna length associated positively in all stages with pinna width as well as number of pinnules. Pinna width and number of number of pinnules did not show consistent association with other traits. Number of branches associated positively with collar diameter at all stages. Dry shoot weight showed positive association with seedling height, number of pinnules, number of branches and collar diameter, both in four months and twelve months old seedlings. Looking to these associations and heritability of traits (table 1), seedling height was the most appropriate trait, selection for which can bring about improvement in most of the traits including dry shoot weight.

REFERENCES

1. Jatasara, D.S. 1982. Germplasm collection of *Prosopis cineraria* (L.) McBride from Indian Desert. *Forage Research* 26: 82-87.
2. Jindal, S.K., Kackar, N.L. and Solanki, K.R. 1987. Interrelationship between shoot dry weight and components in Acacia Senegal (L.) Willd. Progeny. *NFTRRS*: 16-17.
3. Khosla, P.K., Sehgal, R.N., Dhall, S.P. and Chauhan, K. 1985. Path coefficient analysis of correlation between dry shoot weight and other characters in *Pinus roxburghii* progeny. *Journal of Tree Science* 4: 15-19.
4. Kackar, N.L., Solanki, K.R. and Jindal, S.K. 1986. Variation for morphological characters in Prosopis cineraria in natural stands. *My forest* 22: 129-134.
5. Singh, Manjit, Kackar, N.L., Jindal, S.K. and Solanki, K.R. 1993. Variation in morphological and seed related traits of Prosopis cineraia in natural stands. *Annals of Arid Zone* 32: 142.

Growth and yield of senna in association with neem

S.K. Jindal, Anjly Pancholy, Nisha Patel and Manjit Singh
Central Arid Zone Research Institute, Jodhpur 342 003

ABSTRACT

Senna (*Cassia angustifolia* Linn.), an important medicinal plant with laxative properties, is becoming a popular crop in large areas in Rajasthan. Its ability to grow in wastelands and non-palatability of leaves has contributed towards its success. It is grown as a single crop as well as in association with trees. To study the effect of distance of neem trees on senna crop, eighteen accessions were sown between distances of 9 to 23 m from neem plantation (more than 20 years) in second week of August 2006 as three strips of equal width of 4m each (with a gap of 1m between the strips) parallel to the trees. Data were recorded for plant height, number of primary branches, bunches/plant, pods/plant; fresh and dry weight of leaf, stem and pods; pod length and seeds per pod in the first week of November, 2006 after 88 days of seed sowing. Similar observations were recorded in second week of October 2007. In 2006, there were significant differences among the three treatments (strips), for all the characters except dry weight of pods, pod length and seeds/pod and values for such traits were almost half in strip-I (strip nearest to neem trees) as compared to strip III. In the next year (2007), even though the pattern of change in values with the distance from neem was same but the differences in the values of individual traits got diminished and were statistically non-significant. It appears that there is competition of the plants with the trees in the first year when the plants have to establish and it diminishes with the age of the crop.

Key words: Senna, growth, neem.

INTRODUCTION

Senna (*Cassia angustifolia* Linn.), an important medicinal plant is cultivated for its leaves and immature pods which contain dianthone glycoside A and B possessing strong laxative properties. It has been in cultivation in Tamil nadu since late eighteenth century (1). Now it is becoming a popular crop in large areas in Rajasthan. Its ability to grow in wastelands and non-palatability of leaves has contributed towards its success. It is grown as a single crop as well as in association with trees. Growing of crops with woody species provide physical and ecological benefits of a multi-species and multiple land – use production systems capable of providing positive returns in uncertainties. The overall returns in such systems are better than the pure crops under rainfed conditions (2, 3 and 4). Objective of this experiment was to study the effect of distance of neem (*Azadirachta indica* A. Juss) trees on senna crop.

MATERIAL AND METHODS

Eighteen accessions were sown between distances of 9 m to 23 m from neem plantation of more than 20 years, in second week of August 2006 as three strips of equal width of 4 m each (with a gap of 1m between the strips) parallel to the trees. The three strips were actually the blocks and the blocking was done in such a way to minimize the residual variance. Row to row distance was 50 cm. Data were recorded on randomly selected five plants each from individual 18 plots in each strip for plant height, number of primary branches, bunches/plant, pods/plant; fresh and dry weight of leaf, stem and pods; pod length and seeds per pod in the first week of November 2006 after 88 days of seed sowing. Similar observations were recorded in second week of October 2007. The climate of this area is typically arid characterized by dry summer, exceptionally hot, sub humid monsoon and cold dry winters. The soil is loamy sand with pH 8.1 and low nutrient levels 0.02% phosphorus, 0.03% nitrogen and 0.23% organic carbon. The soil depth of the experimental site is 2.25 m. Data was analyzed as per standard statistical procedures for analysis of variance and Student's t-distribution (5).

RESULTS AND DISCUSSION

The analysis of variance for various traits is senna were recorded in both the years (Table 1). There were highly significant differences among distances of senna from neem, both during the year of establishment (2006) in subsequent year (2007) for all the morphological traits like plant height and primary branches. Among the pod characteristics, bunches per plant and pods per plant showed significant differences with distances during both the years. During the establishment year, the differences for pod length and seeds per pod were non-significant. However, differences in the second year started expressing and were found to be significant. The differences among the distances for plant morphology and pod characteristics were reflected in the fresh and dry weight of leaves, stem and pods as indicated by significant to highly significant differences in weight of leaves, stem and pods during both the years except dry weight of pods per plant in the year of establishment.

Table 1. Analysis of variance (mean square) for various traits in senna

Source	D.F.	Mean square					
		Plant height	Primary branches/ plant	Bunches/ plant	Pods/ plant	Pod length	Seeds/ pod
				2006			
Distance	2	996.4**	12.0**	22.2**	561.0**	0.030	0.057
Accessions	17	29.6	1.4	3.1	89.2	0.098	0.441
Error	34	43.0	1.6	2.9	89.4	0.094	0.601
				2007			
Distance	2	99.0*	24.3**	46.0**	471.7**	0.367*	5.164*
Accessions	17	20.6	2.9	2.7	37.5	0.263	1.483
Error	34	18.8	4.0	4.7	41.7	0.089	1.453
		Fresh weight/plant			Dry weight /plant		
		Leaves	Stems	Pods	Leaves	Stems	Pods
				2006			
Distance	2	57.8**	55.6**	63.4*	5.0**	7.0**	1.73
Accessions	17	5.5	5.3	15.9	0.4	0.7	0.95
Error	34	7.7	6.0	13.9	0.7	0.7	0.75
				2007			
Distance	2	27.0**	64.8**	48.3**	3.2**	8.1*	2.77**
Accessions	17	3.8	9.6	3.3	0.5	1.5	0.33
Error	34	4.9	11.0	6.8	0.6	1.7	0.47

* $p<0.05$, **$p<0.01$

It is clearly evident that neem affects the growth and development of senna in its vicinity and the negative effect declines as the distance increases from the trees. In the first year even though plants got affected but pod characters were not affected as indicated by non significant effect on pod length and seeds per pod. Even during subsequent years, the effect is only significant as compared to highly significant effect on some of the other traits like primary branches, bunches and pods per plant. It appears that overall pod characteristics showed relatively less change under changed environments.

Number of pods per plant is expected to be related to plant height and primary branches per plant. Significant differences in number of pods per plant as compared to non significant differences in dry weight of pods per plant during the year of establishment suggests that partitioning of biomass towards pod weight is more in plants that are more stressed. This higher weight, however, does not mean more length of pods or more number of seed per pod. During the second year the differences in pods per plant over three distances started getting reflected in pod length and seeds per pod as indicated by significant difference in these two traits.

Mean height of the senna crop changed from 15.5 cm to 25.0 cm to 30.1 cm in 2006 and it was 24.5 cm, 27.1 cm, 29.2 cm in case of 11m (strip I), 16m (strip II) and 21 m (strip III) in 2007 (Table 2). This was the situation for most of the traits except pod length and seeds per pod during both the years (Table 2 & 3).

Table 2. Effect of different distances of plantation on plant height, primary branches, bunches, pods per plant, pod length and seeds per pod in senna

Distance from neem plantation		Plant height	Primary branches/ plant	Bunches/ plant	Pods/ plant	Pod length	Seeds/ pod
m		cm	No.	No.	No.	cm	No.
2006							
11	Mean	15.5	1.18	2.07	8.9	3.88	6.54
	SD	4.86	0.95	1.16	6.26	0.21	0.44
16	Mean	25.0	2.38	3.70	17.7	3.80	6.66
	SD	4.85	1.24	1.75	10.79	0.25	0.64
21	Mean	30.1	2.73	4.19	19.3	3.85	6.58
	SD	8.27	1.44	2.12	10.60	0.42	1.02
Mean		23.5	2.10	3.32	15.3	3.84	6.59
SE mean		1.55	0.29	0.40	2.23	0.07	0.18
CD 1%		5.63	1.07	1.46	8.12	0.26	0.67
CD 5%		4.28	0.81	1.11	6.18	0.20	0.51
2007							
11	Mean	24.5	4.98	5.29	16.8	3.55	5.11
	SD	3.70	1.34	1.98	5.95	0.26	1.11
16	Mean	27.1	6.09	7.98	25.9	3.79	5.19
	SD	4.63	1.12	2.21	6.92	0.40	1.19
21	Mean	29.2	7.30	5.13	17.2	3.81	6.07
	SD	4.80	2.79	1.80	6.13	0.46	1.32
Mean		26.9	6.12	6.13	20.0	3.72	5.46
SE mean		1.02	0.47	0.51	1.52	0.07	0.28

CD 1%	3.72	1.71	1.85	5.54	0.26	1.03
CD 5%	2.83	1.30	1.41	4.22	0.19	0.79

In the first year, the differences between II and III strips for primary branches per plant, bunches per plant and pods per plant were non significant and the values for these traits were significantly low in strip I as compared to strip II and III. Same was the trend for fresh and dry weight of leaves, stem and pods. But in the second year, the values for bunches per plant, pods per plant, fresh and dry weights of leaves, stems and pods were more in strip II.

Table 3. Effect of different distances of plantation of neem on fresh and dry weight of leaves, stem and pods per plant in senna

Distance from neem plantation (m)		Fresh weight/plant (g)			Dry weight /plant (g)		
		Leaves	Stems	Pods	Leaves	Stems	Pods
			2006				
11	Mean	2.94	2.23	3.66	0.96	0.80	1.06
	SD	1.70	1.30	2.59	0.55	0.48	0.73
16	Mean	5.58	4.98	6.84	1.77	1.77	1.64
	SD	2.02	2.45	4.38	0.66	0.79	1.09
21	Mean	6.37	5.51	6.97	1.95	1.97	1.54
	SD	3.72	3.10	4.22	1.02	1.11	0.86
Mean		4.96	4.24	5.82	1.56	1.51	1.41
SE mean		0.65	0.58	0.88	0.20	0.20	0.20
CD 1%		2.38	2.11	3.20	0.71	0.71	-
CD 5%		1.81	1.60	2.43	0.54	0.54	-
			2007				
11	Mean	6.44	8.33	5.16	2.37	3.53	1.56
	SD	2.44	3.13	2.12	0.87	1.25	0.63
16	Mean	6.89	11.87	7.91	2.43	4.68	2.24
	SD	2.46	2.89	2.57	0.80	1.10	0.62
21	Mean	4.58	8.91	5.00	1.67	3.51	1.55
	SD	1.28	3.66	2.42	0.56	1.43	0.71
Mean		5.97	9.70	6.02	2.16	3.91	1.78
SE mean		0.52	0.78	0.61	0.18	0.31	0.16
CD 1%		1.90	2.85	2.24	0.66	1.11	0.59
CD 5%		1.45	2.16	1.70	0.50	0.85	0.45

(-) as differences were non-significant

Leaves and pods are the major economic traits as these have higher amounts of sennosides than the stem. In the first year, dry biomass per plant was 2.82 g in case of plants in strip I and there was more than 80% increase in the other two strips (Table 4, Fig 1). In second year, dry biomass per plant was 74.8% more than in the first year which might be due to the benefit of established crop. The significant differences for dry biomass/ plant between the years in case of 11m ($t=6.65$, $df=34$, $p<0.01$), 16 m ($t=5.49$, $df=34$, $p<0.01$) and grand mean ($t=6.85$, $df=106$, $p<0.01$) showed the neem plants affected more in the first year i.e. the year of establishment. Non-significant differences between years for 21 m distance ($t=1.49$, $df=34$, NS) revealed that the effect of neem plantation on the crop diminishes from this distance. Same were the trend for significant differences when dry leaves and pods per plant was considered together at 11 m ($t=4.74$, $df=34$, $p<0.01$), 16 m ($t=2.63$, $df=34$, $p<0.01$) and 21 m ($t=0.59$, $df=106$, NS) distances. Ratio of economic yield with total dry biomass in first year was more (66.3%) than in the second year (50.2%) showing that it is more woody in the second year (Fig. 2).

Table 4. Effect of distance of neem plantation on dry biomass, and leaves & pods per plant in senna

Distance from neem plantation (m)		Dry biomass / plant (g)	Dry leaves and pods/plant (g)
		2006	
11	Mean	2.82	2.02
	SD	1.53	1.09
16	Mean	5.18	3.42
	SD	2.43	1.66
21	Mean	5.46	3.49
	SD	2.76	1.67
Grand mean		4.49	2.98
SD		2.55	1.62
		2007	
11	Mean	7.47	3.94
	SD	2.54	1.33
16	Mean	9.35	4.67
	SD	2.12	1.15
21	Mean	6.73	3.22
	SD	2.35	1.02
Grand mean		7.85	3.94
SD		2.55	1.30

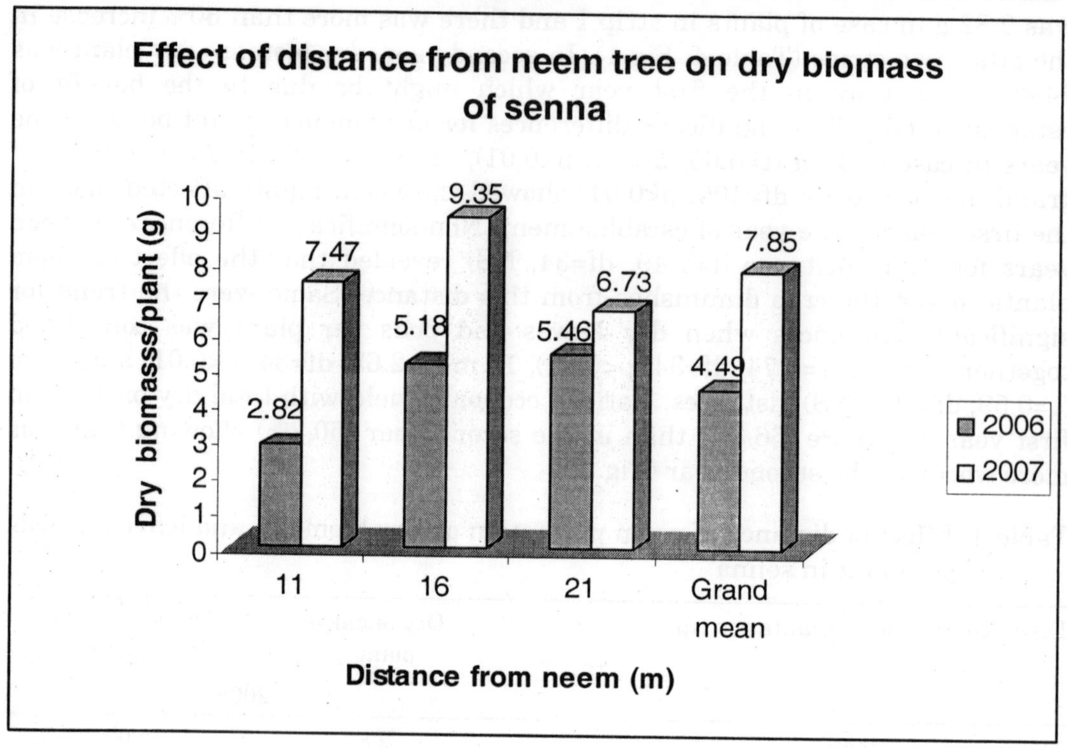

Fig 1

The increase in yield of associated crops with increasing distance from trees have also been reported in maize and finger millet (6), wheat (7) and in rainfed crops (8), but higher grain yield of wheat near tree line under irrigated conditions has been reported (9). Arid and semi arid lands have special characteristics of location and accessibility, climate and soils, human social and economic structures, and products and services required (10). To meet these, crops are grown in association with shrubs and trees. No doubt, trees /shrubs affect the associated crops especially when there is a competition for moisture where the rains are erratic. Though yields of crops are less but overall returns from such systems are more in comparison to growing only crops. Even leaf extracts of neem have been reported to have deleterious effect on seed germination and seedling growth in arid zone crops (11). *C. angustifolia* as inter-crop provides support to the farming system by way of conferring stability and generating assured income (12). Different varieties and crops respond differently with different tree species (13). Care must be taken by laying out field experiments for maximizing the block effect by keeping them parallel to the plantation .Such data where differences due to accessions were non significant some information still could be worked out from such trials.

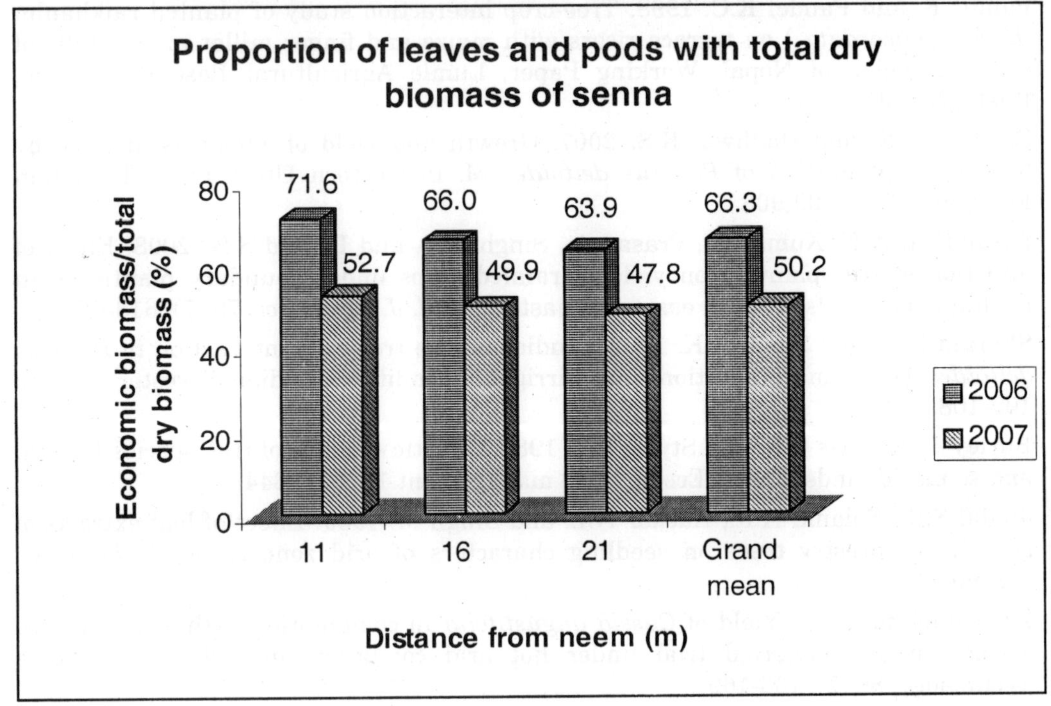

Fig. 2

Plantation of neem reduces the biological yield of senna upto 16 m distance and the yield increases with increase in the distance. Competition of the crop plants with trees decreases with the age of the senna crop.

References

1. Sastry K.P., Mehta V.K., Radhakrishnan K., Saleem S.M. and Sushil Kumar 2000-01. Cultivation of Senna (*Casssia angustifolia)* in the southern districts of the Indian state of Tamilnadu for export purposes. Journal of Medicinal and Aromatic Plant Sciences 22-23 : 674- 678.
2. Prasad A. and Prasad S.N. 1997. Effect of white papinoc (*Leucaena leucocephala*) on field crops and nutrient addition in soil under agro forestry system. Ind. J. Agric. Sci. 67(11): 523-528.
3. Narain P., Singh R.K., Sindhwal N.S. and Joshie P. 1998. Agroforestry for soil and water conservation in the Western Himalayan Valley region of India. Agroforestry Systems 39: 191-203.
4. Vyas S.P. 2003. Effect of *Prosopis cineraria* (L.) Macbride on growth and productivity of Senna. Range Management and Agro forestry. 24 (2): 159-160.
5. Panse V.G. and Sukhatme P.V. 1978. Statistical Methods for Agricultural Workers. ICAR New Delhi pp 347.

6. Pandit R. and Paudel K.C. 1998. Tree-crop interaction study of planted raikhanim (*Ficus semicordata*) on terrace risers with maize and finger millet in mid hills of western region of Nepal. Working Paper, Lumle Agricultural Research Centre, Kaski, Nepal.
7. Sharma N.K. and Dadhwal K.S. 2007. Growth and yield of wheat as affected by boundary plantation of *Populus deltoides* M. in western Uttar Pradesh. Indian Forester 133(7): 899-908.
8. Parandiyal A K, Kumar A., Prasad A., Singh K.D. and Prasad S.N. 2008. Effect of direction of tree planting on yield of rainfed crops under boundary plantation in medium black soils of south eastern Rajasthan. *Ind. J. Agric. Sci.* 78 (7): 618-621.
9. Sharma K.K. and Singh R.K. 1992. Studies on the tree-crop interaction in *Populus deltoides* 'G-3' bund plantation under irrigated conditions. Indian Forester 118 (2): 102-108.
10. Burley J , Hughes C.E. and Styles B.T. 1986. Genetic systems of tree species for arid and semiarid lands. Forest Ecology and management.16 :317 -344.
11. Jindal S.K., Solanki K.R., Kackar N.L. and Singh M. 1989. Effect of leaf extracts of some agro forestry trees on seedling characters of arid zone crops. J. Tree Sci. 8(2):90-91.
12. Arya Ranjana 2003. Yield of *Cassia angustifolia* in combination with different tree species in a silvi-herbal trial under hot arid conditions in India. Bioresource Technology, 86 (2): 165-169.
13. Jindal S.K., Solanki K.R., Kackar N.L. and Singh M. 1990. Seed yield of clusterbean, cowpea and mothbean varieties in association with three arid zone tree species. Ann. Arid Zone. 29(2): 145-146.

Genetic variability, correlation and path analysis in moth bean [*Vigna aconitifolia* (Jacq.) Marechal]

H.R. Mahla*[1] *and D. Kumar

Central Arid Zone Research Institute, Jodhpur – 342 003
[1]Present address: CAZRI, Regional Research Station, Post Box # 63, Jaisalmer (Rajasthan) – 345 001

ABSTRACT

Evaluation of two hundred germplasm accessions of moth bean revealed wide range of variability for yield and yield attributing traits. Extreme values of GCV (41.9) and PCV (132.0) were observed for yield per plant while heritability and genetic advance were low. Branches per plant, clusters per plant and total pods per plant had close association between the values of GCV and PCV accompanied with higher heritability and genetic advance as per cent of mean. All the component traits except test weight showed significant positive correlation with yield per plant. The maximum values were from total pods per plant (0.755) followed by number of clusters per plant (0.598), plant height (0.424) and branches per plant (0.372). The direct and indirect effect of nine quantitative characters indicated that the total pods per plant had maximum direct positive effect on seed yield per plant (1.059) while rest of the traits did not contribute directly much.

Key words: Heritability, genetic advance, correlation, variability, path analysis, moth bean.

INTRODUCTION

Moth bean [*Vigna aconitifolia* (Jacq.) Marechal] is an important pulse crop of arid region having diversified uses. It grows well in the area receiving less than 300 mm of rainfall on the sandy soils of Indian Thar desert. The world famous Bikaneri *Bhujia* and *Papad* are the products of moth bean. The efficiency of any crop improvement programme mainly depends on the extent of genetic variability present in the germplasm and heritability of the traits under selection. Yield is regarded as a complex trait, which is influenced by many component traits both in positive and negative directions. Generally, direct selection for yield is not sufficiently effective due to low heritability. Therefore, it is desirable to know the mutual relationship between various component traits on which selection can be made for improvement in yield. If the correlation between yield and character is due to the direct effect of the character, it reflects a true relationship between them and selection can be practiced for such traits in order to improve yield. The path analysis unravels whether the association of these traits with yield is due to their direct effect on yield or it is consequence of their indirect effect via some other trait (s).

MATERIALS AND METHODS

The experimental materials comprised of 200 germplasm accessions received from NBPGR Regional Research Station, CAZRI, Jodhpur. All these accessions were sown in Augumented Block Design having five blocks in single row plots of 5 m spaced 30 cm apart. In each block 3 check varieties (CZM – 1, RMO – 225 and RMO – 257) were also sown in order to avoid block effect. The data were recorded on 5 randomly selected competitive plants for all the traits (listed in the tables). The analysis of variance was done following the method suggested by Federer (1) and procedure outlined by Sharma (2). Genotypic and phenotypic coefficients of variation (GCV and PCV) were calculated as per procedure given by Burton (3) while heritability (bs), expected genetic gain and correlations were worked out using the formulae of Johnson et. al. (4). The path analysis was performed as per Dewey and Lu (5).

RESULTS AND DISCUSSIONS

Evaluation of 200 germplasm accessions showed wide range of variability for yield and yield attributing traits. The extreme values of GCV (41.9) and PCV (132.0) for yield per plant indicate that environment played a significant role for this complex trait. The low heritability (10.1) and low genetic gain (27.5) does not permit direct selection for yield *per se*. Further, the higher coefficient of variability (35.30) questions the consistency of yield. But branches per plant, cluster per plant and total pods per plant had close association between the values of GCV and PCV accompanied with higher heritability and genetic advance as per cent of mean (Table 1). This clearly indicates that these traits were not much influenced by the environment and selection may be rewarding for improvement of yield in moth bean for these traits.

The degree of association between different characters was estimated. All the component traits except test weight showed significant positive correlation with yield per plant (Table 2). The maximum values were from total pods per plant (0.755), followed by number of clusters per plant (0.598), plant height (0.424), branches per plant (0.372) and low values were observed for pod length, days to flowering and days to maturity. Days to flowering and maturity had positive correlations (0.508) with each other. Both branches per plant and number of clusters per plant showed significant positive correlations with each other as well as other traits except test weight. Plant height, seeds per pod and pod length had negative associations with days to flowering, days to maturity and test weight. Sihag *et al.* (6) observed that grain yield was positively and significantly associated with number of clusters per plant, number of pods per plant and 100 grain weight but negatively associated with plant height and days to flowering. Positive association between yield and its attributes have also been reported by Tikka *et al.* (7, 8).

Table 1. Genetic parameters in moth bean for different quantitative traits.

Characters	GCV (%)	PCV (%)	h^2 (bs) (%)	GA as % of mean	Mean ± SEm	CV (%)
Plant height (cm)	17.83	78.47	5.00	8.06	16.66 ± 2.64	14.46
Branches per plant	25.82	29.88	74.70	45.97	5.37 ± 0.71	12.07
Clusters per plant	37.20	38.49	93.40	74.40	11.05 ± 1.61	13.34
Pod length (cm)	12.20	56.94	4.60	5.37	3.58 ± 0.82	20.87
Seeds per pod	10.46	11.63	80.90	19.38	6.00 ± 0.70	10.75
Pods per plant	37.61	37.80	99.00	77.10	19.94 ± 5.37	24.58
Yield per plant (g)	41.93	132.00	10.10	27.40	2.18 ± 0.84	35.27
Test weight (g)	10.80	10.98	96.90	21.91	27.85 ± 2.81	9.21
Days to flowering	5.79	7.31	62.70	9.44	38.51 ± 1.20	2.84
Days to maturity	8.34	8.56	95.00	16.74	69.31 ± 2.80	3.69

Table 2. Correlation coefficients of different quantitative characters.

Characters	1	2	3	4	5	6	7	8	9	10
1. Yield per plant (g)	1.000	0.199**	0.179*	0.424**	0.372**	0.598**	0.276**	0.294**	0.755**	-0.016
2. Days to flowering		1.000	0.508**	-0.012	0.305**	0.168*	0.058	-0.036	0.138	-0.215**
3. Days to maturity			1.000	0.167	0.273**	0.306**	-0.058	-0.286**	0.220**	-0.227**
4. Plant height (cm)				1.000	0.316**	0.573**	0.270**	0.232**	0.589**	-0.043

5. Branches per plant	1.000	0.571**	0.243**	0.270**	0.445**	0.103
6. Clusters per plant		1.000	0.291**	0.286**	0.877**	-0.603**
7. Pod length (cm)			1.000	0.274**	0.327**	0.085
8. Seeds per pod				1.000	0.295**	0.108
9. Pods per plant					1.000	-0.136
10. Test weight (g)						1.000

Table 3. Direct (Bold figures) and indirect effects of various traits on yield per plant in moth bean.

Characters	1	2	3	4	5	6	7	8	9	Correlation
1. Days to flowering	0.106	0.054	-0.001	0.032	0.018	0.006	-0.004	0.015	-0.023	0.199**
2. Days to maturity	0.031	0.060	0.010	0.016	0.018	-0.003	-0.017	0.013	-0.014	0.179**
3. Plant height (cm)	0.000	0.000	-0005	-0.002	-0.003	-0.001	-0.001	-0.003	0.000	0.424**
4. Branches per plant	0.016	0.014	0.016	0.051	0.029	0.012	0.014	0.023	0.005	0.372**
5. Clusters per plant	-0.069	-0.126	-0.237	-0.236	-0.414	-0.120	-0.118	-0.363	0.025	0.598**
6. Pod length (cm)	0.000	0.000	0.000	0.000	0.000	-0.001	0.000	0.000	0.000	0.276**
7. Seeds per pod	-0.003	-0.027	0.022	0.026	0.027	0.026	0.095	0.028	0.010	0.294**
8. Pods per plant	0.146	0.233	0.624	0.471	0.930	0.347	0.312	1.059	-0.144	0.755**
9. Test weight (g)	-0.027	-0.028	-0.053	0.128	-0.007	0.106	0.013	-0.017	0.124	-0.016

The direct and indirect effect of nine quantitative characters indicated that the total number of pods per plant had the maximum direct positive effect on seed yield per plant (1.059), while rest of the traits did not contribute much directly (Table 3). The negative direct effect (-0.414) of clusters per plant was compensated through total pods indirectly. Further the total number of pods has maximum positive effect for all the traits except test weight. Sharma and Mathur (9) reported that correlations between seed yield and pods per plant, branches, peduncle length, plant height and seeds per pod were significant except for test weight. Tikka *et al.* (10) reported that pod number, test weight and seeds per pod were of vital importance for maximization of seed yield. Misra *et al.* (11) observed high direct effects of days to maturity and branches per plant for seed yield.

In the present study keeping GCV, PCV, heritability, genetic gain as per cent of mean, direct and indirect effects and correlations, branches per plant, clusters per plant and total number of pods may be considered as selection criteria for improvement of seed yield in moth bean.

REFERENCES

1. Federer, W.T. 1956. Augmented Designs. Hawain Planters Record. **55** : 191-207.
2. Sharma, J.R. 1999. Statistical and Biometrical techniques in Plant Breeding. Tata McGraw-Hill Publishing Company, New Delhi. Pp. 19 -29.
3. Burton, G.W. 1952. Quantitative inheritance in grasses. Proceedings of 6th International Grassland Congress, **1**: 277-283.
4. Johnson, H.W., Robinson, H.F. and Comstock, R.E. 1955. Estimates of genetic and environmental variability in soybeans. *Agronomy Journal*. **47** : 314-318.
5. Dewey, D.R. and Lu, K.H. 1959. A correlation and path coefficient analysis of components of crested wheat grass and seed production. *Agronomy Journal,* **51**: 515-58.
6. Sihag, S.K., Khatri, R.S. and Singh, J.V. 2005. Correlation and path analysis for grain yield in moth bean [*Vigna aconitifolia* (Jacq.)]. *Journal of Arid Legumes*, **2**: 306 – 308.
7. Tikka, S.B.S., Asawa, B.M. and Kumar, S. 1977. Correlated response to selection in moth bean. *Gujarat Agricultural University Research Journal* **3**: 1-4.
8. Tikka, S.B.S., Asawa, B.M., Sharma, R.K. and Kumar, S. 1980. Selection indices in moth bean. [*Vigna aconitifolia* (Jacq.)]. *Tropical Grain Legumes Bulletin*, **17-18**:46-48.
9. Sharma, R.C. and Mathur, J.R. 1991. Correlation and path analysis studies in moth bean. *Rajasthan Agricultural Research Journal*, **5** : 103 -106.
10. Tikka, S.B.S., Yadevendra, J.P., Bordia, C.P. and Kumar, S. 1973. Variation in moth bean. *Rajasthan Agricultural Research Journal* **4**: 50-55.
11. Misra, S.P., Mehta, T.R. and Rao, K.V.S. 1981. Studies on path coefficient analysis in moth bean. *Agriculture Science Digest*. **1** : 43-45.

Response of plant extracts against major root pathogens in clusterbean [*Cyamopsis tetragonoloba* (L) Taub.]

P.P. Gupta, Shekhar Kumar and Sher Singh

Forage Section, Department of Plant Breeding, CCS Haryana Agricultural University, Hisar-125 004

ABSTRACT

With the increase in awareness towards toxic hazards of fungicides different plant extracts viz; *neem* (*Azadirachta indica*) leaves, *eucalyptus* (*Eucalyptus globulens*) leaves, *datura* (*Datura alba*) leaves, rhizomes of *ginger* (*Zingiber officinale*), cloves of *garlic* (*Allium sativum*) and bulbs of *onion* (*Allium cepa*) were prepared in 1:1 ratio(w/v) after filteration through muslin cloth. These solutions were sterilized by passing through millipore filter (0.40 μm) and considered as standard (100%). Different concentrations (5, 10 and 20%, respectively) of these extracts were tested *in vitro* and *in vivo* during the years 2004 and 2005 against *Rhizoctonia bataticola*, *Fusarium solani* and *Sclerotium rolfsii* to manage root rot diseases in clusterbean. In *in vitro* studies desired conc. (5, 10 and 20% respectively) were tested using poisoned food technique. Discs measuring 5mm diameter were cut from the margin of 7 days old culture of test organism (fungi) with the help of cork borer and placed in the centre of solidified PDA and incubated in BOD incubator at 28 ± 2°C for one week and observations on radial growth of the pathogens were recorded at 24 h interval. It has been observed that all the tested plant extracts were inhibitory to all the three fungi and 20% garlic extract exhibited maximum viz; 84.50, 87.50 and 90.25% inhibition of *Rhizoctonia bataticola*, *Fusarium solani* and *Sclerotium rolfsii,* respectively. In *in vivo* studies seeds of clusterbean cv. FS-277, PNB and HG-75 were soaked in these extracts for 30 min., shade dried and sown in earthen pots along with untreated check. Observations on seedling mortality were recorded daily up to one month. At 20% conc. garlic extract exhibited minimum seedling mortality of clusterbean cv FS-277 having *Rhizoctonia bataticola* (13.5%), *Fusarium solani* (12.0%), *Sclerotium rolfsii* (8.5%) and their combination (15.75%). Similar results were obtained in cv. PNB and HG-75. Hence garlic extract (20%) was most effective in inhibition of growth of root rot fungi and in reducing seedling mortality of clusterbean caused by these fungi.

Key words: clusterbean, plant extracts, root pathogens.

Introduction

Clusterbean [*Cyamopsis tetragonoloba* (L) Taub.] is an important arid legume crop grown for feed, green fodder, vegetable, green manuring and gum purpose. The gum present in its seed endosperm has diversified uses in many industries like cosmetics, gramophone, pharmaceuticals, mining and textile industries, etc. Numerous soil borne fungal pathogens infect clustebean roots in Northwestern India including *Rhizoctonia bataticola* [*Macrophomina phaseolina* (Tassi) Goid], *Fusarium solani, Sclerotium rolfsii, Rhizoctonia solani, Fusarium caeruleum and, Neocosmospora vasinfecta* (1). The fungi belonging to three genera of *Rhizoctonia, Fusarium* and *Sclerotium* has been found responsible for root rot complex of clusterbean in Haryana (2). The yield losses range from 21.6-90.0 per cent (2,3). In recent years, constraints in disease control especially those involving fungicides, due to their acquired resistance to plant pathogens, ecological hazards and their high cost have led to the necessity of finding alternate source. Plant derivatives are potential sources of fungicides against many plant pathogens causing serious diseases (4). With the increase in awareness towards toxic hazards of fungicides to crops, consumers and environment due to their phytotoxic, residual and pollution effects, the importance of plant products/plant extracts in plant disease control has been emphasized by many workers. Keeping in view the importance of disease and available information, the present investigation was undertaken with the objective of testing the efficacy of botanicals/plant extracts against three major root rot pathogens on clusterbean in Haryana.

MATERIALS AND METHODS

Screening of plant extracts against root rot causing fungi *in vitro*

For evaluating the antifungal activity of different plant extracts, the fresh leaves of *neem* (*Azadirachta indica*), *eucalyptus* (*Eucalyptus globulens*), *datura* (*Datura alba*), rhizomes of *zinger* (*Zingibir officinale*), cloves of *garlic* (*Allium sativum*) and bulbs of onion (*Allium cepa*) were collected during the years 2004 and 2005 and washed 2-3 times with sterilized distilled water. The parts of each plant species were crushed in distilled water in 1:1 ratio (w/v) and filtered through muslin cloth. This formed the standard plant extract solutions (100%). The solutions thus obtained were passed through a Millipore filter (0.40μm) using filter adapter attached to a glass syringe (20 ml) to get sterilized extracts.

These extracts were immediately used for experiment. The desired concentrations 5, 10 and 20 percent were made by adding appropriate amount of standard solution of plant extracts to 15 ml sterilized Potato Dextrose Agar (PDA) medium, which was poured into sterilized Petri dishes. The inhibition of radial growth of three fungi,viz *Rhizoctonia bataticola, Fusarium solani* and *Sclerotium rolfsii*, on PDA was studied using poisoned food technique. Each treatment was replicated four times. The Petri dishes without plant extract but with test fungi served as control. All plates were incubated at 28 ± 2°C in BOD incubator for one week and observations on radial growth of the pathogens were recorded at 24 hr interval.

Evaluation of plant extracts against root rot pathogens in screen house

Efficacy of plant extracts namely, *neem, eucalyptus, datura, zinger, garlic* and *onion*, were tested as seed treatment @ 5, 10 and 20 per cent concentration in 1:1 ratio (w/v) against root rot pathogens under screen house conditions during the years 2004 and 2005. The inoculums of root rot fungi were mixed in the soil and allowed to stabilize for two days in pots before sowing. The plant extracts of *neem* (*Azadirachta indica*), *eucalyptus* (*Eucalyptus globulens*), *datura* (*Datura alba*), rhizomes of *zinger* (*Zingibir officinale*), cloves of *garlic* (*Allium sativum*) and bulbs of *onion* (*Allium cepa*) were prepared in cold sterilized distilled water as described earlier. The plant extracts were passed through a Millipore filter (0.40μm) using filter adapter attached to a glass syringe (20 ml) to get sterilized extracts. Hundred seeds of clusterbean varieties FS-277, Pusha Nav Bahar (PNB) and HG-75 were soaked in these extracts separately for half an hour and air dried under shade and sown in pots in four replications and untreated check was also maintained. Pots were irrigated regularly. Observations on seedling mortality were recorded daily up to one month.

RESULTS AND DISCUSSION

It is clear from table 1 (Av. of two years) that out of six plant extracts, *eucalyptus* leaves showed minimum area of inhibition of *Rhizoctonia bataticola* viz 13.25, 29.75 and 67.75 per cent at 5, 10 and 20 percent concentration, respectively while at same concentration extract of *garlic* cloves resulted in maximum area of inhibition of 29.75, 46.00 and 84.50 per cent of *Rhizoctonia bataticola*. The extracts from rhizomes of *zinger*, bulbs of *onion* and leaves of *neem* were also found equally effective. The extract from *datura* leaves was found to be less effective. Similar type of inhibition of area of *Fusarium solani* and *Sclerotium rolfsii* was recorded by these plant extracts. The data from Table 1 also reveals that with the increase in concentration of plant extracts, there was decrease in radial growth of all the three fungi. However a marked inhibition in all the three fungi was observed by the extract of garlic cloves at all the concentration level which was followed by *zinger, neem, onion* and *datura. Eucalyptus* leaves extract was found to be least effective in the inhibition all the three fungi at all the three concentrations.

Table 1. Effect of plant extracts on the inhibition of radial growth of fungi in vitro during the years 2004 and 2005 (Av. of two years)

Plant Extracts	Inhibition (%)											
	Rhizoctonia bataticola				*Fusarium solani*				*Sclerotium rolfsii*			
	5%	**10%**	**20%**	**Mean**	**5%**	**10%**	**20%**	**Mean**	**5%**	**10%**	**20%**	**Mean**
Neem	26.50*	43.00	82.25	50.58	29.50	47.50	85.25	54.08	37.25	51.50	87.50	58.75
Eucalyptus	13.25	29.75	67.75	36.91	16.25	34.75	70.25	40.41	24.75	39.75	75.50	46.66
Datura	18.50	34.75	71.75	41.66	21.50	39.50	74.75	45.25	29.25	43.75	78.75	50.58
Zinger	28.25	43.50	83.50	51.75	31.50	48.50	86.75	55.58	38.00	53.75	89.25	60.33
Garlic	29.75	46.00	84.50	53.41	32.50	51.00	87.50	57.00	40.25	55.00	90.25	61.83
Onion	25.25	40.26	81.00	48.83	27.25	45.25	84.00	52.16	35.25	49.50	86.25	57.00
Mean	23.58	39.54	78.45		26.41	44.41	81.41		34.12	48.87	84.58	
C.D. at 5%	1.403	1.137	1.412		1.542	1.304	1.434		1.332	2.139	1.429	

*Average of four replications

It is clear from table 2 that clusterbean seed (cv. FS 277) soaking in *garlic* cloves extract (20%) for half an hour induced minimum seedling mortality by *Rhizoctonia bataticola*, i.e., 13.50 per cent while seed soaking in *eucalyptus* leaves extract induced maximum seedling mortality (23.50%) over control (72.75%). Seed soaking with *neem, datura, zinger* and *onion* extracts resulted in seedling mortality of 16.25, 22.00, 15.75 and 19.00 per cent, respectively. The trend for seedling mortality in varieties PNB (Table 3) and HG-75 (Table 4) were similar to that of FS-277. However, the average seedling mortality was maximum (26.11%) in FS-277 and minimum (10.68%) in HG-75. The average seedling mortality was found to be intermediate (24.03%) in PNB. Similar trend of seedling mortality by *Fusarium solani, Sclerotium rolfsii* and combination of all the three fungi (Table 2, 3 and 4 Av. of two years) was recorded by these plant extracts when the seeds of PNB (Table 3, Av. of two years) and HG-75 (Table 4) were soaked in them. The data from Table 2, 3 and 4 also reveals that garlic extract was most effective followed by *zinger, neem* and *onion* extracts where as *eucalyptus* and *datura* leaves were least effective. However, the maximum seedling mortality in all the three fungi individually and in combinations was observed with *eucalyptus* extract. Maximum seedling mortality was found on variety FS 277 followed by PNB and HG 75. Seedling mortality was highest at 5 per cent concentration when compared with 10 per cent and 20 per cent concentration but trend for all plant extracts in three clusterbean varieties were similar for all the three fungi.

Table 2: Effect of seed treatment with plant extracts on the incidence of root-rot in clusterbean cv. FS 277 under screen house condition during 2004 and 2005 (Av. of two years)

Plant Extracts	Inhibition (%)											
	Rhizoctonia bataticola			*Fusarium solani*			*Sclerotium rolfsii*			All three fungi		
	5%	10%	20%	5%	10%	20%	5%	10%	20%	5%	10%	20%
Neem	50.00*	33.25	16.25	43.50	29.25	14.25	28.25	19.75	11.50	55.25	39.50	19.25
	(44.98)**	(35.19)	(23.74)	(41.25)	(32.72)	(22.14)	(32.23)	(26.36)	(19.79)	(47.99)	(38.91)	(25.99)
Eucalyptus	58.00	46.00	23.50	50.75	40.25	20.50	34.00	27.50	15.50	64.25	49.25	28.25
	(49.59)	(42.69)	(28.94)	(45.41)	(39.36)	(26.85)	(35.64)	(31.61)	(23.16)	(53.26)	(44.55)	(32.06)
Datura	55.25	40.00	22.00	48.50	36.25	18.25	32.50	24.50	13.75	61.50	46.25	25.75
	(47.99)	(39.21)	(27.92)	(44.12)	(37.00)	(25.26)	(34.73)	(29.64)	(21.73)	(51.63)	(42.83)	(30.48)
Zinger	46.25	31.75	15.75	43.00	27.00	13.75	27.50	18.50	10.50	51.50	35.50	19.00
	(42.83)	(34.26)	(23.36)	(40.94)	(31.27)	(21.73)	(31.61)	(25.44)	(18.87)	(45.84)	(36.55)	(25.81)
Garlic	41.50	27.25	13.50	36.00	22.00	12.00	23.75	15.50	8.50	46.50	29.25	15.75
	(40.08)	(31.43)	(21.51)	(36.84)	(27.95)	(20.24)	(29.14)	(23.15)	(16.91)	(42.97)	(32.71)	(23.35)
Onion	51.25	34.75	19.00	45.25	30.50	16.50	31.75	21.00	12.75	58.25	40.25	22.50
	(45.69)	(36.10)	(25.82)	(42.25)	(33.51)	(23.91)	(34.27)	(27.23)	(20.88)	(49.73)	(39.36)	(28.27)
Control	75.25	71.50	72.75	67.50	64.50	66.00	47.25	45.75	45.25	81.25	78.50	79.25
	(60.19)	(57.71)	(58.54)	(55.24)	(53.41)	(54.32)	(43.50)	(42.54)	(42.26)	(64.33)	(62.37)	(62.89)
Mean	53.93	40.64	26.11	47.78	35.68	23.03	32.18	24.64	16.82	59.78	45.50	29.96
	(46.62)	(39.64)	(29.98)	(43.72)	(36.46)	(27.78)	(34.45)	(29.42)	(23.37)	(50.82)	(42.47)	(32.69)
C.D. at 5%	(2.618)	(1.732)	(2.455)	(2.746)	(1.468)	(2.427)	(2.045)	(2.019)	(1.817)	(1.959)	(2.635)	(2.346)

* Average of four replications and two years; ** Figures in parenthesis are angular transformed values

Table 3: Effect of seed treatment with plant extracts on the incidence of root-rot in clusterbean cv. Pusa Nav Bahar under screen house condition during 2004 and 2005 (Av. of two years)

Plant Extracts	Inhibition (%)											
	Rhizoctonia bataticola			*Fusarium solani*			*Sclerotium rolfsii*			All three fungi		
	5%	10%	20%	5%	10%	20%	5%	10%	20%	5%	10%	20%
Neem	46.25*	28.05	15.25	38.25	24.00	13.50	22.50	16.75	11.00	50.00	32.50	18.25
	(42.83)**	(32.24)	(22.97)	(38.17)	(29.31)	(21.51)	(28.29)	(24.14)	(19.35)	(44.98)	(34.73)	(25.24)

Eucalyptus	53.75	41.00	22.00	44.50	35.00	18.50	28.25	23.75	15.00	58.25	45.75	23.75
	(47.13)	(39.79)	(27.95)	(41.82)	(36.25)	(25.42)	(32.09)	(29.14)	(22.74)	(49.73)	(42.54)	(29.14)
Datura	51.25	36.25	19.75	42.50	30.50	17.50	26.50	22.75	15.25	56.00	40.25	21.50
	(45.69)	(37.00)	(26.36)	(40.67)	(33.49)	(24.71)	(30.96)	(28.46)	(22.91)	(48.43)	(39.36)	(27.56)
Zinger	42.75	26.00	12.75	35.25	21.00	11.75	21.25	15.75	10.75	46.25	29.75	14.50
	(40.81)	(30.64)	(20.90)	(36.39)	(27.23)	(20.03)	(27.42)	(23.35)	(19.04)	(42.83)	(33.03)	(22.36)
Garlic	37.75	20.75	11.25	31.25	17.75	10.25	18.50	12.50	9.00	42.00	23.00	13.00
	(37.89)	(27.06)	(19.56)	(33.95)	(22.89)	(18.66)	(25.42)	(20.65)	(17.41)	(40.38)	(28.62)	(21.08)
Onion	47.75	30.50	16.50	39.50	25.75	15.25	26.00	18.75	12.75	52.25	34.00	18.75
	(43.69)	(33.50)	(23.94)	(38.91)	(30.47)	(22.94)	(30.64)	(25.59)	(20.84)	(46.27)	(35.65)	(25.62)
Control	75.50	71.50	70.75	61.00	60.00	59.25	37.25	36.25	42.50	75.75	74.75	73.50
	(58.35)	(57.71)	(57.25)	(51.34)	(50.75)	(50.32)	(37.59)	(37.00)	(40.67)	(60.48)	(59.82)	(59.00)
Mean	50.28	37.79	24.03	41.75	30.57	20.86	25.75	20.93	16.61	54.36	40.00	26.18
	(45.69)	(36.85)	(28.42)	(40.18)	(33.19)	(26.23)	(30.34)	(26.90)	(23.28)	(47.59)	(39.11)	(30.00)
C.D. at 5%	(1.742)	(1.802)	(1.717)	(2.492)	(1.959)	(2.079)	(1.871)	(2.081)	(2.583)	(1.359)	(2.032)	(2.433)

*Average of four replications and two years; ** Figures in parenthesis are angular transformed values

Table 4: Effect of seed treatment with plant extracts on the incidence of root-rot in clusterbean cv. HG 75 under screen house condition during 2004 and 2005 (Av. of two years)

Plant Extracts	Inhibition (%)											
	Rhizoctonia bataticola			*Fusarium solani*			*Sclerotium rolfsii*			All three fungi		
	5%	10%	20%	5%	10%	20%	5%	10%	20%	5%	10%	20%
Neem	11.50*	10.75	9.50	10.50	9.25	8.00	9.75	8.50	7.00	11.75	10.75	9.75
	(19.81)**	(19.12)	(17.94)	(18.87)	(17.68)	(16.41)	(18.17)	(16.91)	(15.32)	(20.03)	(19.12)	(18.14)
Eucalyptus	15.75	15.75	15.50	14.50	14.50	13.75	13.25	14.00	12.75	16.75	16.25	16.50
	(23.35)	(23.37)	(23.17)	(22.37)	(22.36)	(21.74)	(21.32)	(21.96)	(20.89)	(24.14)	(23.76)	(23.93)
Datura	14.25	13.25	12.00	13.50	12.25	11.50	12.75	11.75	10.50	14.75	13.50	12.25
	(22.16)	(21.33)	(20.25)	(21.14)	(20.47)	(19.81)	(20.90)	(20.03)	(18.87)	(22.57)	(21.54)	(20.47)
Zinger	9.75	9.50	7.25	8.75	8.75	7.00	8.25	8.00	5.75	10.25	10.25	7.50
	(18.17)	(16.91)	(15.61)	(17.19)	(17.14)	(15.27)	(16.66)	(16.38)	(13.86)	(18.66)	(17.18)	(15.85)
Garlic	7.00	8.25	3.50	6.25	7.25	3.25	5.75	6.75	3.25	7.75	9.50	4.75
	(15.32)	(14.71)	(10.75)	(14.46)	(15.59)	(10.29)	(13.83)	(15.03)	(10.17)	(16.15)	(14.75)	(12.46)
Onion	11.75	11.00	9.75	10.75	10.25	9.00	10.00	9.50	8.25	12.25	11.75	10.25
	(20.03)	(19.35)	(18.17)	(19.12)	(18.65)	(17.44)	(18.42)	(17.92)	(16.66)	(20.48)	(20.03)	(18.65)
Control	18.25	17.25	17.25	16.75	17.25	16.50	15.00	14.75	14.00	19.25	18.25	18.00

	(25.28)	(24.90)	(24.52)	(24.14)	(24.53)	(23.94)	(22.75)	(22.57)	(21.94`)	(26.01)	(25.27)	(25.08)
Mean	12.61	12.32	10.68	11.57	11.36	9.86	10.68	8.80	8.78	13.25	12.89	11.82
	(20.59)	(19.96)	(18.63)	(19.61)	(19.49)	(17.84)	(18.86)	(18.69)	(16.82)	(21.15)	(20.24)	(19.23)
C.D. at 5%	(1.204)	(1.439)	(1.043)	(1.086)	(1.551)	(1.682)	(1.547)	(1.563)	(1.977)	(0.833)	(1.134)	(2.070)

*Average of four replications and two years ** Figures in parenthesis are angular transformed values

Several higher plants and their constituents have shown success in controlling plant diseases and proved to be harmless and non-phyto toxic (5). All plant extracts under study inhibited the growth/spread of *Rhizoctonia bataticola* but garlic extract was found to be most effective on all the three varieties both in vitro and screen house conditions. These findings are supported by the findings of Dubey and Dwivedi (5) and Sindhan *et al.* (6). They reported that the plant extracts were inhibitory to *Rhizoctonia bataticola.* However, fungus was more sensitive to extracts of *onion, zinger, neem* and *garlic* followed by *mint, eucalyptus*, but it was most effective at higher concentration. It has been found that garlic extract was most effective and other plant extracts were inhibitory to *Fusarium solani,* These findings are supported by the findings of Philip and Sharma (7) and Bianchi *et al.* (8) who investigated the effect of *neem* and *garlic* extracts on the mycelial growth and root rot development by *Fusarium solani* and found that mycelial growth was strongly inhibited at the maximum concentration of the aqueous extract tested @ 100 ml per litre. They also found inhibitory effect on root rot development in different crops.

In the present findings, all the five plant extracts were found to be effective against *Sclerotium rolfsii* as well as against all the three fungi under study. Ram *et al.* (9) reported that soil application of neem leaf extract reduced both pre and post emergence collar rot in chickpea caused by *Sclerotium rolfsii.* Similarly Yossry *et al.* (10) studied the fungitoxic properties of plant extracts especially datura and river red gum (*Eucalyptus camaldulensis*) and found that these extracts were having inhibitory effect against the growth of soil borne fungi. They also found that increased concentration of plant extracts reduced growth of *Rhizoctonia bataticola (Macrophomina phaseolina), Fusarium oxysporum, Sclerotium rolfsii* and *Rhizoctonia solani.* Hence to avoid toxic hazards of synthetic chemicals to clusterbean crop, on human beings and to reduce the spread of root rot complex clusterbean seed must be soaked in garlic cloves extract (20%, 1:1 w/v stock solution) for half an hour before sowing.

REFERENCES

1 Gupta P.P., Kumar, S. and Yadav B.D. 2007. Effect of soil temperature and moisture levels on root rot of clusterbean. *J. Arid Legumes* 4(2): 95-99.

2 Gupta P.P., Kumar, S. and Singh, S. 2007. Influence of inoculum age, density and soil pH on seedling mortality of clusterbean caused by root rot pathogens. Lead

paper presented in national symposium organized by IPS (Central zone) & RAU, Bikaner on "Innovative Approaches for Disease Management in Arid Crops for Enhancing Export Potentials. Held at ARS, Mandor, Nov. 2-3, 2007 Abstract S-9 pp 45.

3 Gupta P.P. and Rohilla H.R. 2008. Plant Protection in Arid Legumes. In : "Souvenir" Published by D. Kumar in 3rd National Symposium on "Enhancing Productivity, Nutritional Security and Export Potential through Arid Legumes" held at CAZRI, Jodhpur June 28-30, 2008, pp 127-137.

4 Kumar, S. 2005. Studies on Epidemiolgy and Management of Root rot of Clusterbean. M.Sc. Thesis. Haryana Agric. Univ., Hisar pp 1-126.

5 Dubey, R.C. and Dwivedi, R.S. 1991. Fungi toxic properties of some plant extracts against vegetative growth and sclerotial viability of *Macrophomina phaseolina. Indian Phytopath*. 44: 411-413.

6 Sindhan, G.S., Hodda, Indra and Parashar, R.D. 1999. Effect of some plant extracts on the vegetative growth of root rot causing fungi. *Indian J. Mycol. Pl. Pathol*. 29 : 110-111.

7 Philip, T. and Sharma, D.D. 1997. *In vitro* evaluation of leaf and oil cake extracts of *Azadirachta indica* and *Pongamia glabra* on mulberry root rot pathogens. *Indian J. Mycol. Pl. Pathol*. 36: 150-152.

8 Bianchi, A., Zambonelli, A., D`- Aulero, A.Z. and Bellesia, F. 1997. Ultrastructural studies on the effects of *Allium sativum* on phytopathogenic fungi *in vitro*. *Plant Dis*. 81: 1241-1246.

9 Ram, D., Tewari, V.P. and Ram, D. 1994. Control of chickpea collar rot (*Sclerotium rolfsii)* with soil application of plant products. *J. Appl. Biol*. 4: 38-40

10 Yossry, A.A., El, A.S. and El, I.S. 1998. Fungitoxic properties of some plant extract against growth of soil borne disease fungi. *Proc. Seventh Agrl. Development Res. Conference, Cairo, Egypt Annal. Agrl. Sci. Cairo*.3: 891-909.

Diseases and their management in arid legumes: Present status

Satish Lodha, Rajendra Kumar and Ritu Mawar
Central Arid Zone Research Institute, Jodhpur – 342 003

Arid legumes, due to their inherent drought hardy characteristics, are mostly pushed to late monsoon conditions or in a piece of land not suitable for cultivation of cereals. This trend has changed to a little extent in these years when growers have realized the economic returns from cultivation of guar, moth bean and cowpea compared to pearl millet. Addition to soil fertility is yet another advantage that has gained attention of growers in adopting cereal-legume cultivation in Indian arid region. Once the crops get due attention, growers also become concern about securing maximum profitability from their lands. Occurrence of disease on these legumes is a major constraint in realizing maximum output. In the past two decades, research organizations have also paid serious attention in developing strategies against important diseases. In all India coordinated research project, several multilocational experiments have been conducted in order to develop management strategies suitable for all the arid legumes growing regions or for a specific zone. Diseases of arid legumes, their epidemiology and management aspect have been recently reviewed by Pande (2008). An attempt has been made in this article to discuss the recent advances made in understanding various aspects of important diseases of arid legumes and future strategies.

I. Blights

Blights caused by different pathovars of *Xanthomonas axonopodis* have continued to remain most serious problem in all the arid legume growing regions. No doubt, due attention has been given in these years to identify resistant sources and to breed disease resistant variety; still large area remains under cultivation of local strains. These strains are highly susceptible to blight, often causing 35-40% yield losses. The extent of losses depends on stage of the crop, cultivar and prevailing environmental conditions. In arid districts of Rajasthan, bacterial blight on guar occurs mostly after cluster initiation, thus number of pods are severely affected with increasing disease intensity (Lodha and Gupta, 1981), while in Haryana, it occurs before cluster initiation (Gandhi and Chand, 1985), therefore prophylactic measures needs to be applied at vegetable stage of the crop i.e. 25-30 days after sowing. Recently this disease has become a serious problem in Karnataka also (Chakarvarty *et al.*, 2005).

Bacterial blight is favoured by spattering rains, high humidity (80-90%) and moderate temperature, which are usually available during August-September in arid regions (Singh and Swarup, 1987).

Efforts made at various research organizations have resulted in developing resistant genotypes. GAUG 63, an unbranched type was the best entry of guar showing moderate resistance to bacterial blight and high resistance against powdery mildew (Pandy *et al.*, 1993). Since bacterium is seed borne, hot water treatment at 56°C for 10 min or seed treatment with streptocycline (0.025%) was found highly effective (Singh and Swarup, 1986; Lodha and Ananthram, 1993). Secondary spread of the disease can be checked by two sprays of streptocycline (0.01%) at 35 and 49 days after sowing.

These studies have shown that physical, chemical and biological methods of control are now available for managing bacterial blight. However, there exists ample scope to develop resistant varieties so as to minimize dependence on chemicals. Guar gum is used in various food industries by developed countries. Their requirement now a days is to import guar seed produced without the use of chemicals. In view of increasing export demand, there is a need to develop eco-friendly measures to control bacterial blight. Use of effective botanicals, bio-control agents, proper sanitation, crop rotation and other agronomic practices clubbed with fairly resistant variety can reduce yield losses.

II. Dry root rot

Low and erratic rainfall, short and long durations of soil moisture stress, low microbial population of sandy soils alongwith concurrent heat stress are few of the factors, which favors occurrence of dry root rot caused by *Macrophomina phaseolina* in mild to severe farm on all the arid legumes. Other than legumes, the pathogen has a wide host range in many other cultivating plants. Therefore, it may cause infection in succeeding crops from one field to another. Considerable variation has been observed in these legumes for susceptibility to *Macrophomina* under similar soil moisture stress. Cowpea is more prone to dry root rot development compared to guar, while moth bean is relatively resistant.

This variation explains the legume growing conditions where growers are cultivating moth bean in extremely low (> 250 mm), guar in medium (< 350 mm) and cowpea in better rainfall regions (< 450 mm). Thus, often in spite of availability of sufficient inoculum load of *M. phaseolina*, development of dry root-rot is mainly dependent on critical stress condition of a crop or its variety. In arid regions, 15 days old seedlings were more susceptible to *Macrophomina* than 30 days old plants (Singh and Lodha, 1986; Lodha, 1998).

Considerable work has been done on various management approaches at the Central Arid Zone Research Institute for the past 2-3 decades. Rotation with less susceptible crops like moth bean and Pearl millet reduced the population of *M. phaseolina* (Lodha *et al.*, 1990). Certain genotypes like RGC 471 and Kutch 8 were found to be moderately resistant. Mulching of the soil with a layer of pearl millet Stover (3.5 ton ha^{-1}), low plant population (1.6 lakh ha^{-1}) and farmyard manure (10 ton ha^{-1}), singly or in combination effectively conserved the soil moisture and decreased the sclerotial counts and dry root rot incidence (Lodha, 1996). Amendment of soil with composts reduced *M. phaseolina* population and increased the antagonistic actinomycetes (Lodha *et al.*, 2002). Soil solarization is also a very highly effective technique but appeared too expensive for arid legumes. However, use of biological control agents like *Trichoderma* and *Bacillus firmus* holds promise in augmenting disease control if applied with other management strategies. However, efforts are needed to develop a sound schedule of steps encompassing different management strategies to minimize losses due to this most important disease. Crop rotation, selection of healthy seeds, soil amendment with Biocontrol agents, composts, effective weed residues or farm yard manure, adoption of moisture conservation practices, etc. are few of the proven management approaches, which can find place in rainfed agriculture.

Apart from these management approaches, there is a need to select relatively tolerant genotypes of arid legumes. In this research efforts are needed to select short duration genotypes or those lines, which have more tolerance to moisture stress at seedling stage. In regions, where irrigation is available for growing winter season crops, use of *Brassica* residues with summer irrigation is one of the practical approaches for managing *Macrophomina* and *Fusarium* induced diseases. It is now established that VAM may improve plant growth through increased uptake of phosphorus especially in soils of low fertility (Gupta *et al.*, 1999). In nutrient deficient soil, *G. fasciculatum* has been found effective for growth of cowpea, indicating the need for inoculation of VAM for desiring maximum benefit for plant growth.

III. Leaf spots

Arid legumes also suffer from attack of several types of leaf spots. The intensity and losses may vary from crop to crop, variety to variety and season to season. In guar, Alternaria leaf spot caused by *Alternaria cyamopsidis* is seed borne in nature and survive as mycelium in debris and crop stubbles. Under conducive environment, pathogen infects susceptible guar crop. Continuous rainfall, 80% RH, 25-31°C favors the development of disease. The disease is more prevalent in Punjab, Haryana and north Rajasthan. Many resistant varieties have been identified. Efforts should be made to keep the field clean from infected plant debris of last year or crop rotation should be practiced. Seed borne infection can be minimized by use of healthy seeds or treating seeds with fungicides. Secondary spread of disease can be checked by spray of mancozeb or other contact fungicides. In the regions where both the diseases are important a combination of contact fungicide with streptocycline is recommended for control of bacterial blight also.

In view of great demand of organically produced guar, use of effective botanicals is also an area, which requires attention of research workers. At present neem or compost extract are effective in checking *Alternaria* species.

IV. Powdery mildew

This disease is also of considerable importance in certain areas and is favored by warm temperature (33°C and above), low humidity and bright sunshine. The disease is caused by pathogens like *Leveillula taurica*, *Erysiphe polygonii* and *Sphaerotheca fuliginea*. High water content of the fungal conidia enables then to withstand harsh conditions of Indian arid zone. In certain arid districts of Rajasthan like Jodhpur, Nagour, Jaisalmer, Barmer and Pali, legumes mature by end of September or in early October, thus escape the attack of powdery mildew. However, in north Gujarat and north Rajasthan, where rainy season crops mature by end of October, occurrence of powdery mildew in mild to sever form can be invariably observed on guar and other legumes.

Resistant cultivars have been identified in case of all the legumes particularly from Gujarat. Control of powdery mildew by sulphur dusting is also a common practice adopted by the farmer, as spray of sulphur fungicides is not possible due to scarcity of water. Early sown crop has been observed to escape the disease. A Biocontrol agent *Cicinnobolus cesatii* has been reported to be a natural fungal hyper parasite of powdery mildew (Venkatarayan, 1946).

In addition to Alternaria leaf spot, several other leaf spots also occur on arid legume in mild to severe form. The pathogen includes *Cercospora*, *Myrothecium*, *Curvularia*, Anthracnose, etc. Among these, Cercospora leaf spot caused by two species viz., *C.cancescens* and *C. cruenta* are more important. In India, *C. cruenta* is more serious, which usually occurs during humid weather. The pathogen can survive in plant debris. Use of health seeds, seed treatment with fungicides and destruction of infected crop debris can minimize disease incidence to considerable extent.

IV. Viral diseases

Among arid legumes, viral diseases are major constraint in cowpea and moth bean. Seed borne viruses have epidemiological, ecological and quarantine implications. Some of the viral diseases of economic importance in cowpea includes cowpea mosaic, cowpea yellow mosaic, cowpea mild mottle, cowpea cucumo viruses, southern bean mosaic, cowpea tobamo viruses, etc. This group of diseases occurs frequently at different growth stages of the crop. Large number of genotypes have been found to possess resistance against one or another viral disease. In addition to main legumes, certain weeds also act as alternative host. Therefore, weed free cultivation is recommended. Spraying the crop with endosulphan (0.2%) may check the insect vector.

In moth bean, occurrence of yellow mosaic virus (YMV) in severe form is a regular feature in the entire moth been growing regions, lack of resistance in available germ plasm is the major constrain in this crop. The disease is transmitted by a white fly (*Bemisia tabaci*). Often due to late onset of rains, farmers grow moth bean in their fields in place of cereals. On late plantation, the vegetative stage of the crop coincides with multiplication of vector, which spread YMV on susceptible cultivars. These cultivars are planted in large areas as single crop favoring rapid spread of vector transmitted virus particularly at vulnerable stage. Intercropping or mixed cropping is one practical way to minimize losses. Moderately resistant cultivars should be cultivated.

Table 1. Major diseases of arid legumes

Diseases	Pathogen	Crop
Bacterial blight	Xanthomonas axonopodis	Guar
	pv. *cyamopsidis*	Cowpea
	X. a. pv. *vignicola*	Moth bean
Dry root rot	Macrophomina phaseolina	Guar
		Cowpea
		Moth bean and horse gram
Powdery mildew	*Leveilula taurica,*	Guar, Horse gram, Moth bean
	Erysiphe polygoni,	Guar, cowpea
	Sphaerotheca fuliginea	Guar

Leaf spots	*Alternaria cyamopsidis*	Guar
	Myrothecium roridum	Guar
	Curvularia lunata	Guar
	Colletotrichum capsici f. *cyamopsicola*	Guar
	Cercospora canescens	Cowpea
	C. cruenta	Horse gram
	C. dolichi	Cowpea
	Septoria vignicola	Horse gram
		Cowpea
Anthracnose	*Colletotrichum capsici f. cyamopsicola*	Guar
	C. lindemuthanium	Cowpea, horsegram
Viral	Mosaic, yellow mosaic, mild mottle , cucumo viruses, southern bean mosaic,	Cowpea
	Tobamo viruses, yellow mosaic	Moth bean, horse gram

CONCLUSION

The foregoing reveals that arid legume cultivation is confronted with losses caused by fungal, bacterial and viral diseases. In water deficient regions, limited choice of cultivation of drought hardy, unavailability of water for foliar sprays, increasing inoculum of soil borne plant pathogens, inherent nature of sandy soils are the major reasons for occurrence of these diseases in mild to severe form. Research investigations carried at various research organizations have led to development of host of ecofriendly management strategies against important diseases. However, proper dissemination of these strategies to the level of farmers has not occurred so far. This is an area which requires almost attention of researchers, policy planners and extension personnels, simultaneously, there is a need to develop low input sustainable technologies, which can be adopted by small and marginal farmers of arid region. Forewarning of weather dependent diseases like blights, powdery mildew, etc is also needed to minimize losses if farmers adopt prophylactic measures against weather dependent diseases.

REFERENCES

1. Chakravartry C N, Krishnappa M and Thippeswamy B (2005). Investigation on bacterial blight (*Xanthomonas axanopodis* pv. *cyamopsidis)* of clusterbean {*Cyamopsis tetragonoloba* (L.) Taub.} and *in vitro* control. *Journal of Mycology and Plant Pathology* Symposium Abstract pp. 119.
2. Gandhi S K and Chand J N (1985). Yield losses in guar due to bacterial blight caused by *Xanthomonas campestris* pv. *cyamopsidis*. *Indian Phytopathology* 38 516-519.

3. Gupta P P, Kumar R and Jalali B L (1999). Response of different vesicular arbuscular mycorrhizal fungi on cowpea. *Plant Disease Reporter* 14: 25-30
4. Lodha S and Gupta G K (1981). The relationship between bacterial blight and loss in yield components of guar (Abstr.) In proceeding *Third International Symposium on Plant Pathology* held at New Delhi. Dec. 13-17, 94-95.
5. Lodha, S. (1984). Varietal resistance and evaluation of seed dressers against bacterial blight of guar (*Cyamopsis tetragonoloba* L. Taub.). *Indian Phytopath-ology* 37: 438-440.
6. Lodha, S. (1998). Effect of sources of inoculum on population dynamics of *Macrophomina phaseolina* and disease intensity in clusterbean. *Indian Phytopathology* 51: 175-179.
7. Lodha, S. (1996). Influence of moisture conservation techniques on *Macrophomina phaseolina* population, dry root rot and yield of clusterbean. *Indian Phytopat-hology* 49: 342-349.
8. Lodha, S. and Anantha Ram, K. (1993). Effectiveness of streptocycline spray schedules on the bacterial blight intensity and yield of clusterbean. *Annals of Arid Zone* 32: 237-240.
9. Lodha, S., Sharma, S. K. and Aggarwal, R. K. (2002). Inactivation of *Macrophomina phaseolina* propagules during composting and efficacy of composts on dry root rot and seed yield of clusterbean. *European Journal of Plant Pathology* 108(3): 253-261.
10. Pande R N (2008). Legumes pp 133-196 In *Disease Management in Arid Land Crops* (Eds. Lodha *et al.*) Scientific Publishers (India), Jodhpur 430 pp
11. Pandey R N, Pathak A R, Kher H R and Bharodia P S (1993). Evaluation of clusterbean (*Cyamopsis tetragonoloba*) germplasm for resistance to bacterial blight (*Xanthomonas campestris* pv. *cyamopsidis*) and powdery mildew (*Leveillula taurica*). *Indian Journal of agricultural Sciences* 63 : 68-70.
12. Singh J and Swarup J (1986). Control of bacterial blight of clusterbean by seed treatment and foliar spray. *Pesticides* 20: 23-24.
13. Singh J and Swarup J (1987). Epidemiology of bacterial blight of clusterbean caused by *Xanthomonas campestris* pv. *Cyamopsidis*. *Farm Science Journal* 2: 91-92
14. Singh, S. and Lodha, S. (1986). Varietal resistance of cowpea to *Macrophomina phaseolina* (Tassi.) Goid. causing dry root rot and its control. *Indian Journal of agricultural Sciences.* 56: 552-555.
15. Venkatarayan S B (1946). A natural fungus parasite of powdery mildew on C. psoraloides DC. *Current Science* 15: 319.

Management of insect pests and diseases of pulses with special reference to arid legumes

B.S. Chhillar

Director of Research,CCSHAU, Hissar

Arid legumes are grown in about 50 lakh ha area and producing approximately 21 lakh tones of grain annually in India. The major arid legumes are, clusterbean, cowpea, mothbean and horse gram which are predominantly grown in arid and semi-arid tracts of Indian sub-continent. Since, these are generally grown in arid areas where drought is a rule rather exception, that's why the average productivity is poor. Therefore, it is imperative to search tolerance potential towards major biotic and abiotic stresses besides developing strategies for stability of production and productivity through improved management practices. The clusterbean and mothbean commonly grown in kharif season in the North-Western dry zone of India, whereas cowpea and horse gram are commonly grown in small quantity in arid ecosystem of various parts of the country.

Despite the use of quality seed of high yielding varieties, fertilizers and possible irrigation, the farmers are unable to harvest good crop. Insect pests and diseases are the major constraints in increasing the productivity of these crops. Out of a large numbers of insect pests and diseases attacking these crops in India a few are destructive to these crops. The extent of damage caused depends upon the residual pest population and source of primary disease inoculum, environmental conditions and control measures adopted. Integrated pest and disease management is a combined approach to reduce the crop losses. The management of major insect pests and diseases of economic importance in various part of the country on clusterbean, cowpea and mothbean have been discussed.

Among more than 3 dozen of insect pests attacking arid legumes, the leafhopper, Empoasca kerri Pruthi and black aphid, Aphis craccivora Koch and among diseases bacterial leaf blight, alternaria leaf spot, root rot complex, wilt, powdery mildew in clusterbean; cowpea yellow mosaic, root rot/charco I rot/ seedling blight, anthracnose, cercospora leaf spot, powdery mildew, fusarium wilt, bacterial blight, cowpea mosaic in cowpea and yellow mosaic virus (YMV), bacterial leaf spot (BLS), seed rot, root rot and seedling blight, brown web blight and colletotrichum leaf spot in mothbean are important.

Insect pest management

The insect pest attacking arid legumes can be effectively control by the use of various safer insecticides / biopesticides in addition to the use of some cultural practices like early sowing, intercropping with non-legume crops removal of infested plants, early maturing and short duration cultivars and balanced use of fertilizer are the ecofriendly approaches. The pest tolerant varieties such as GC 8927 and CS 88 of cowpea; HG 75 and HGS 563 of clusterbean are fairly tolerant to the incidence of leafhopper and pod borer. On the basis of information available various practices have been suggested to evolve strategy to develop IPM approaches against major pests. Pest wise information about their biology, bionomics, nature of damage and management approaches have been discussed in detail.

Disease management

Lines with moderate degree of resistance in clusterbean identified from genetic stocks include HG 75, HG 258, RGC 990, HGS 365, HGS 119, HG 182 and HFG 504. Infection rate has been reported to be much lower in Cvs. HG 75, HFG 119, HG 182 and HFG 340. Seed treatment with 1000 ppm streptomycin/ streptocycline and thiram (2000 ppm) is most effective to check pre- emergence mortality, whereas Bavistin (0.2%) is best to check post-emergence seedling mortality. Spray, streptocycline @ 150 ppm thrice at an interval of 10 days starting from onset of BLB. Spray Dithane M-45 @ 0.2% thrice at an interval of 10 days starting from onset of disease. In cowpea grow CS 88 (against CYMV and leafhopper), V 130, V 240 and GC 3 are the resistant varieties. Use CS 105, CS 106, CS 107, CS 108, BL 2 and UPC 5286 as resistant sources to yellow mosaic in breeding programmes. Control insect vectors by spraying endosulfan (0.07%) in grain crop. Use Captan (3 g/kg seed) or Carbendazim (2 g/kg seed) alongwith chlorophyriphos @ 4 ml/kg seed to check the presence of insect pests and fungi. In mothbean grow CAZRI Moth-I, Jawala and early maturing ones, like RMO 40, RMO 257, RMO 225 resistant against MYMV. Soil solarisation in the month of May is most effective to check soil borne pathogens. Spray the crop with wettable sulphur (0.3%)/wettable Ceresan and Cupravid @ 1 kg/ha at an interval of 15 days starting from onset of disease to control powdery mildew. Treat the seed with Benlate and Brassicol @ 2 g/kg seed to eliminate seed born infection of brown web blight. Use bio-control agent like *Trichoderma harzianum* to check incidence of *Macrophomina phaseolina* in mothbean.

Bruchids infestation in arid legumes, assessment of losses and their management

Satya Vir

Central Arid Zone Research Institute, Jodhpur-342003
Email: svir@cazri.res.in

Abstract

Bruchids responsible for most infestation in arid legumes (mothbean, mungbean and cowpea etc.) are *Callosobruchus maculatus* and *C. chinensis*. The pest attack nearly mature and dried pods with even an one or two pair of insects could result in major damage (100%) within 4-6 months. Mungbean is the most preferred host followed by mothbean, arhar, blackgram, cowpea and dolichos. Oviposition is preferred by color, texture and volume of seed with bright colored smooth surfaces with greater seed volume most preferred but the development and emergence of adults is not affected on grains preferred for oviposition. Use of comparative least susceptible lines is recommended for storage at the moisture contents below 9% level. The use of neem and karanj oil is the best for protection of leguminous seeds for period of six months. The grain should be dried sufficiently to reduce the moisture contents and dried periodically at 2-3 months interval in gunny bags and again inspection after every 3 months. Use of neem oil 5ml/kg seed can protect all leguminous seeds for a year. An economical and safer method for the protection of arid legumes is also suggested.

Key words : Bruchids, arid legumes, multiplication in stores, losses, resistant/ susceptible cultivars, control

Introduction

Bruchids are a major and growing problem in stored arid legumes (mungbean, mothbean, cowpea etc.). Infestation of bruchids starts in the field at the time of maturity that is too low to be detected at harvest. This infestations are most often not detected until seed has been stored for a reasonable period (e.g. for longer than three months). Bruchids multiply fast in storage and by the time they are detected, the infested grain is usually unmarketable. The bruchid responsible for most infestations in arid legumes are *Callosobruchus maculatus* (Fab.) and *Callosobruchus chinensis* (Lin). The adults of these insects are small (3 mm long) brown beetles with a tear-shaped body (i.e. taper towards the head). Both the bruchids are typically orange-brown with dark markings, but colour can vary and they are easily confused with each other. Adults are strong fliers. The eggs are small and white, despite being only 0.5-0.7 mm long, are readily visible. The larvae look like short, cream-coloured maggots, and are rarely seen as they feed within grains. Eggs are laid on maturing pods in the field, but in storage are laid directly on grains and if the number is more the eggs are also laid on the walls that do not mature. Hatching larvae burrow into the seed and complete their develop-ment (including pupation) within a seed before emerging as adults. Both the species breed rapidly in stored seed and can complete a lifecycle in as little as 24 days (at 30°C). Adults live only 10-15 days and females lay about 60-100 eggs.

Symptoms and losses: The pest attack nearly mature and dried pods. The round exist hole and the white eggs on the pod walls are conspicuous. Infested stored seeds can be recognised by the eggs on the seed surface and the round exhibit holes with the 'flap' of seed coat. Actual figures on losses to the growing crop are very spare. In 1961, Prevett recorded that pest complex affected 3.1 to 11% of pods collected as they are dried. In 1958, Phelps and Oosthuzens recorded only 1.9% damage of freshly picked cowpea pods *by C. chinensis.* However, such levels of pod infestation by potential storage pests represented a serious threat.

Even an extremely one or two pair of insects could result in major damage (100% of grains damaged) within 4-6 months. Because infestations are often difficult to detect at harvest, carefully check for live insects and the characteristic grain damage at regular intervals during storage. Regular checking of bulk and bagged pulses stored is a critical bruchid management strategy. Both the species are tropical breeding ceases below 20 °C. All storage structructures in tropical region are a good insulator and seeds remain at the temperature they were at harvest. Consequently growth and development of bruchids continues unabated but unseen during the winter.

Oviposition response and development of *C. maculatus* on different host species

Oviposition response and development of *C. maculatus* was studied on eight seed materials viz. moongbean, mothbean, arhar, cowpea, dolichos, blackgram and guar (Satya Vir and Jindal 1981). The study revealed that arhar (80.66 eggs) was most preferred for oviposition followed by dolichos (54.66), cowpea

(50), blackgram (46.0), mothbean (33.33), moongbean (22.33), ricebean (22.0) and guar (16.66). The grubs failed to penetrate the seed coat of guar and rice bean and there was no development on these seed materials. The development period was minimum on moongbean and mothbean and maximum in case of dolichos. Emergence of beetles was maximum on cowpea followed by moongbean, mothbean, blackgram, arhar and dolichos. The study has concluded that moong is the most preferred host followed by mothbean, arhar, blackgram, cowpea and dolichos.

Our study also revealed that beetle was found to be guided in its oviposition on preference by colour, texture and volume of seed grain (Satya Vir 1980). Bright coloured smooth surfaces with greater seed volume were much preferred for oviposition. Egg laying was minimum on wrinkled surfaces. Female tended to avoid adding second egg as long as there were unused seeds. The development and percentage emergence of adults was not affected on the grain preferred for oviposition. The study on the comparison of data on nutritional value of food and seed consumed revealed that the amount of food consumed per grub increase with the increase in protein value of seed. The protein value of different varieties of cowpea varied from 21 to 31.5 per cent. The comparative resistant varieties contained only 21 to 24.5 per cent protein as compare to most susceptible varieties containing 28 to 31.5 per cent protein which concluded that it is not the morphological characters of the seeds but the nutritional value of seeds that governs the mechanism of resistance in cowpea to the attack of *C. maculatus*. The data obtained on sex ratio were non-conclusive.

Multiplication of bruchids and loss in weight of mothbean and cowpea seeds

Multiplication of *C. maculatus* on mothbean and cowpea seeds, percentage infestation of seed and loss in seed weight was studied at four levels of infestation for a period of six months (Satya Vir, 1989). Population on both the seeds increased sharply after one months of storage (Fig1.). Increase in population on mothbean seed was 510, 768, 996, and 1258 with the infestation of 21, 40, 68, and 81 per cent seeds and a loss of 24, 39, 49, and 58 per cent seed weight while in cowpea it was 417, 805, 912 and 1027 insects with the infestation of 19, 21, 30, 38 per cent seed weight after six months of storage initially infested with one, two, three and five pair of insects, respectively (Fig 2). Caswell (1961, 1970) showed that unthreshed cowpea put into 'rumbus' immediately the crop becomes ripe sustained 24% damage after six months, if left in the field and harvested at a later date. The losses after six months in storage in 'rumbus' was 33%. The damage to threshold cowpea was 87% after nine months of storage. The total loss in storage of cowpea material per annum was calculated to be approximately 7000 tonnes which represents 5% of the production of 8,20,000 tonnes.

Fig.1

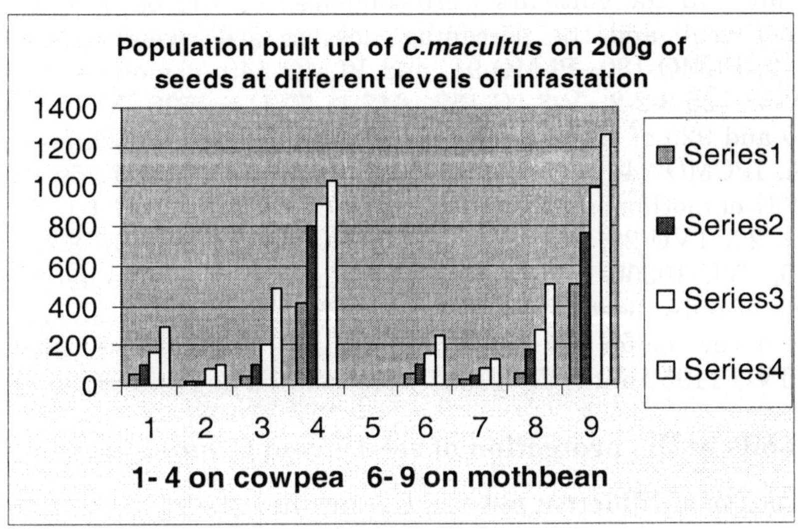

Series 1,2,3 4 Initial infestation with 1,23 and 5 pair of adults

Fig 2.

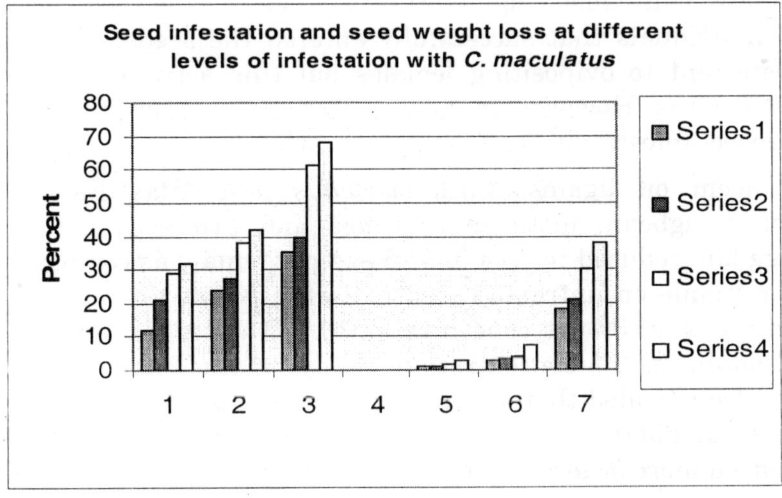

Screening of cultivars to pulse beetle attack:

Nearly 1000 lines of mothbean and cowpea cowpea were screened for their relative susceptibility to the attack of *C. maculates* (Satya Vir, 1981, 1982a, 1982b, 1982c, 1983, 1984). The results revealed that there was no significant difference among the cultivars in relation to the average number of eggs laid and the percentage infestation of seeds. Cultivars reaction in regard to food consumed per grub, percentage loss in seed weight and the emergence of beetles was significant. All the cultivars were compared on the basis of average food consumed per grub and the percentage loss in seed weight as a combined criterion. T-3, PLMO-130, PLMO-91, and PLMO-149 of mothbean and J.C. 5, Sel-1476, V-38, 25-8-2-2, V-3, C-152, 647-1, TVU-6, 526, TVU-373-1-1, EC-107134, 820 and 920 of cowpea were found least susceptible while IPCMO-926, PLMO-84-A, IPCMO-344, G-1, PLMO-169, Jhalawar-1, JMM-259, IPCMO-844 and ipcmo-937 of mothbean and AV-7, Copusa-2, Culture-2, 26-4-1, PTB-1, 5269, C.G. 5, H.G. 22, TVU-22, 901, 665, CG68, Pb-208, 3183-1, EC-8203, 304, 534, 628, EC-455, TVU-10350-1, and 16954 of cowpea proved to be most susceptible to the attack of pulse beetle. Singh *et al.* 1985 screened more than 8000 germ-plasms lines of cowpea for resistance to stored seed damage by *C. maculates* and TVU 2027, TVU 11952 and TVU 11953 were isolated as resistant lines.

Efficacy of oils in the protection of seeds from *C. maculatus*:

Application of six different oils viz. Karanz, neem seed, coconut, groundnut, sawflower, mustard, neem leaves powder and malathion dust was compared in protecting the cowpea seeds from the attack of *C. maculatus* Fab. Effectiveness was compared on the basis of the rate of oviposition, population growth and extent of damage. Oils causes high mortality of eggs on the seed surface but had no effect on individuals that successfully entered the seeds. Oil treated seeds were also deterrent to ovipositing females but this activity was limited to a period of 2-4 months. However, malathion was found most effective and could prevent the insect infestation for 8 months period.

Use of neem oil against *Callosobruchus maculatus* Fab. studied for protection of mungbean, mothbean, cowpea and urd seed material reports number of egg laid reduced to 25.6 to 38.8 egg per female as compared to 81.0 to 90.6 eggs per female on untreated seed material(Satya Vir and Yadav, 2004). Adult toxicity was 93.3% at concentration 0.75% after 72 hrs and complete mortality of adults at 2% concentration. Study on the comparative efficacy of different oils also revealed that karanj and neem oil had better efficacy as for as the discriminating quotient is concerned while coconut, mustard, safflower and mustard behaved more or less at par in reducing the oviposition (Table 1). Neem leaves powder did not show any deterrence efficacy while the use of malathion resulted quick mortality of adults thereby resulting laying a few non viable eggs on the seeds. The chi-square test between the egg laying in treated and control was highly significant on treatment of malathion, karanj and neem oils ($p > 0.001$) while the use of coconut, mustard, safflower and mustard oil were statistically significant and at par ($p > 0.01$). The use of karanj and neem oil gave equal protection to seeds even after six months of storage as has been found in malathion treatment.

Table 1. Comparative efficacy of oils in protecting cowpea seeds against *C. maculatus*

Treatment	Control/ Treated	Av. Eggs laid	Adults emergence after 3months	Adults emergence after 6months
Karang oil	C	530	65.0	135.0
	T	240	0.00	9.5
Neem oil	C	570	68.0	195.0
	T	152	0.00	0.00
Coconut oil	C	520	62.0	214.0
	T	330	2.0	27.3
Mustard oil	C	545	59.0	159.0
	T	320	3.0	22.8
Saflower oil	C	597	62.0	164.0
	T	370	22.0	43.2
Groundnut oil	C	588	68.0	168.0
	T	340	29.0	35.8
Neem leaves powder	C	576	68.0	170.0
	T	505	38.6	77.5
Malathion	C	585	60.0	162.0
	T	10	0.00	0.00

Comparative bio-efficacy of different neem formulation against pulse beetle *Callosobruchus maculatus* Fabricius.

The bio-activity of different neem formulations with varied azadirachtin concentrations were evaluated for their comparative performance for protection of mungbean seeds against Pulse beetle *C. maculatus.* This performance was evaluated on the basis of average number of eggs laid per female and the percentage emergence of adults from seeds (Fig 3 & 4). The average eggs laid was reduced to 2.33 eggs per female when treated with 2% aza of 10,000ppm as compared to 12-18 eggs per female on treating with same concentration of 1500ppm and neem oil. But there was no emergence of adults at this concentration irrespective of 1500 or 10,000 ppm azadiractin or neem oil. The doses of 0.1 and 0.50% were found to be ineffective irrespective of the azazdirachtin concentration present in the formulation.

Fig. 3.

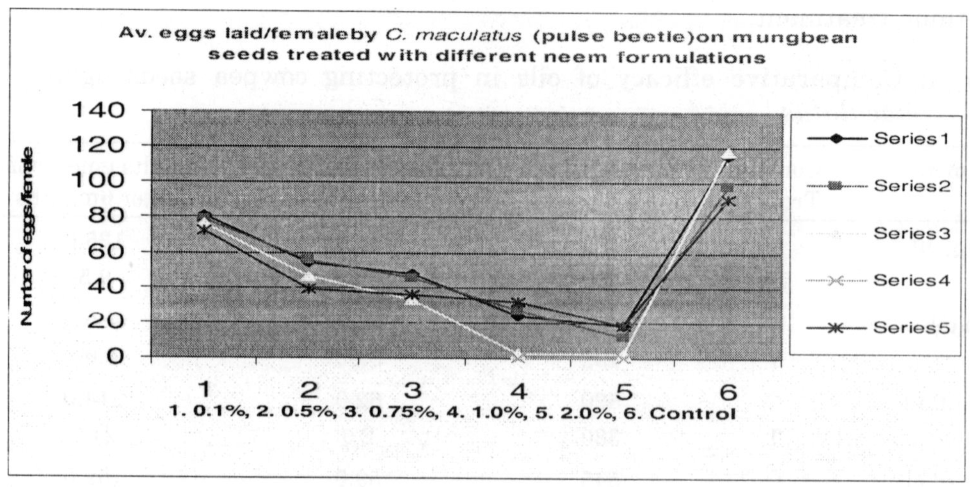

Treatments: Series 1. Neem EC 1500 ppm, Series 2. Achook 1500ppm, Series 3. Neemazal 1500 ppm, Series 4. Econeem 1500 ppm, Series 5. Neem oil

Fig. 4.

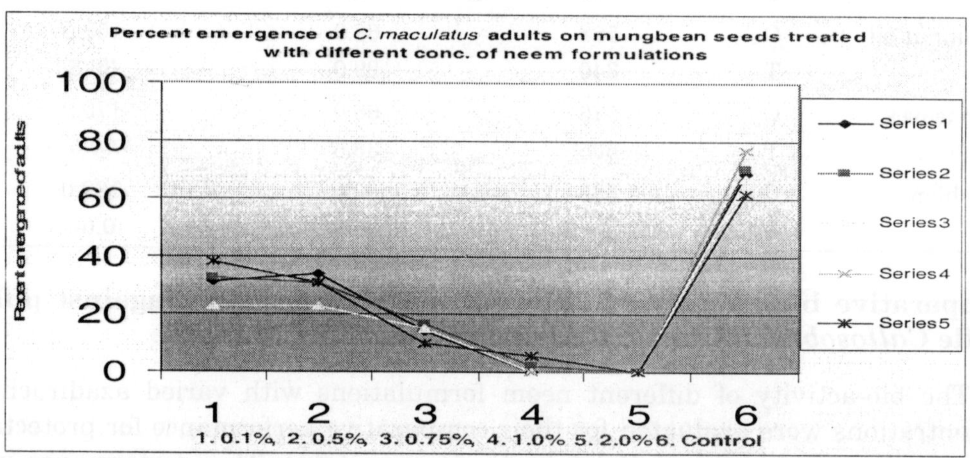

Treatments: Series 1. Neem EC 1500ppm, Series 2. Achook 1500ppm, Series 3. Neemazal 500 ppm, Series 4. Econeem 1500 ppm, Series 5. Neem oil

Control

The grain should be dried sufficiently to reduce the moisture content as far as possible before storage. The grain may be dried periodically at 2 months interval. Grain should be inspected prior to storing and also at an interval of 2 months after storage. It should be stored preferably in earthen containers with sealed earthen lid or a top cover of sand or ash (7.5 cm thick) or in gunny bags with polythene linning and closed tightly. If there is occurence of beetle or eggs on seeds are observed in any of the sampling, control measures should be adopted immediately fumigated in an air tight container is recommended. Treatment with neem oil or linseed oil or karanj oil 5 ml/kg, followed by storage in sealed polythene bags or in jute bags provided the bag surface is sprayed with decis 2.5 ml/m just after storage and again at an interval of 6 months. None of the treatments affected the germination of cowpea (Rahman, 1991). Pulse grain stored in earthen containers sealed with cowdung, coaltar or gunny bags with polythene lining and mud remained free from external infestation. A cover or a layer of sand over the stored grains in earthen containers also prevented external infestation of pulse beetle (BARI 1984). Beans should be inspected for insects once a fortnight in summer and once a month in winter. Controlling bruchids in the field is not an effective option and would not guarantee immunity from storage damage. Anything less than 100% in-field control of bruchids could still result in significant bruchid damage in storage.

At farm level storage and warehouses, the application of Neem derivatives to bags and stored grains has provided protection against insect pests. Seeds of different legumes viz. green gram, chick pea, cowpea and pea when mixed with powdered Neem kernel at 1 or 2 parts to 100 parts can protect them from damage caused by pulse beetle, *Callosobruchus* spp. for 8 to 11 months (Satya Vir and Yadav, 2004). Neem kernels not only protected different legumes against *C. maculates* and *C. chinensis,* but there was also no progeny development even 12 months after insects were released on treated lentil seeds. Similarly, chick pea and pigeon pea seeds remained undamaged up to 12 months after treating with 2 g Neem kernel powder per 100g seeds. The best way to store arid legumes in big boxes has been evolved by putting 2kg of malathion powder in the muslin cloth at the bottom of box that can store 1 to 2 quintals of seed and putting another muslin cloth bag with 1 kg of malathion powder without mixing on the top of the seeds. But in such treatments the seeds should be properly dried inspected and than stored.

Reference

1. BARI (Bangladesh Agricultural Research Institute). 1984. Pulse beetle incidence in stored pulses and control measures. Pages 6-13 in Legumes Post harvest technology Project. Annual Report 1983-84. Joydebpur, Gazipur, Bangladesh.
2. Caswell. G.H. 1961. The infestation of cowpea in the western region of Nigeria. Trop. Sci. 3:154-158.
3. Caswell. G.H. 1970. The storage of cowpea in the Northern States of Nigeria. Trop. Sci. 3:154-158. Samaru Res. Bull.120. Institute for Agric. Research Samaru, Ahmadu Bello Univ., Zaria, Nigeria.

4. Rahman, M.M. 1991. Control measures for important insect pests of major pulses. Advances in Pulse Research in Bangladesh. Proc. of IInd National Workshop on Pulses, 6-8 June, 1989. (Pub. BARI & ICRISAT) Jaydebpur, Bangladesh.

5. Satya Vir. 1980. Oviposition response and development of *Callosobruchus maculatus* Fab. On different varieties of cowpea. *Bull. Grain Tech.* 18 (3): 200-203.

6. Satya Vir. 1981. Note on the relative susceptibility of different varieties of cowpea to *Callosobruchus maculatus* Fab. *Indian J. Agric. Sci.* 51 (11): 813-815.

7. Satya Vir. 1982a Varietal preference in moth (*Vigna aconitifolia* Jacq.) for the pulse beetle *Callosobruchus maculatus* Fab. *Bull. Grain Tech.* 20 (1): 3-7.

8. Satya Vir. 1982b. Varietal preference in moth and cowpea by the pulse beetle, *Callosobruchus maculatus* Fab. *Sci. & Cult.* 48 (10): 346-347.

9. Satya Vir. 1982c. Relative susceptibility of cowpea varieties to the pulse beetle *Callosobruchus maculatus*. *Pestology*. 6 (7): 9-11

10. Satya Vir. 1983. Varietal resistance and susceptibility of cowpea to *Callosobruchus maculatus* Fab. *Indian J. Ent.* 45 (3): 213-217.

11. Satya Vir. 1984. Screening of mothbean *Vigna aconitifolia* Jacq. varieties against the pulse beetle *Callosobruchus ma culatus* Linn. *Proc. Indian Acad. Sci. (Animal Sci.)* 93 (1):25-28.

12. Satya Vir. 1984. Screening of some cowpea varieties against the pulse beetle *Callosobruchus chinensiuss* Linn. Seeds and Farms 9 (6): 29-31

13. Satya Vir. 1989. Multiplication of *Callosobruchus maculatus* Fab. Infestation and loss in weight of mothbean and cowpea seeds at different level of infestation. *Bull. Grain Tech*. 27 (1): 21-25

14. Satya Vir and Jindal, S.K. 1981. The oviposition and development *of Callosobruchus maculatus* Fabricius (Coleoptera: Bruchidae) on different host species. *Bull. Grain Tech*. 19 (3): 180-189.

15. Satya Vir and Jindal, S.K. 1982. Assessment of yield loss due to insect pest of pulses in cowpea, *Vigna unguiculata* (L) Walp. *Seeds and Farms*. 8 (8 & 7): 59-60.

16. Satya Vir and Yadav, S. 2004. Bio-efficacy of neem oil as seed protectant of mungbean, mothbean, cowpea and urd against *Callosobruchus maculatus* Fabricius *J. Arid Legumes*. 1(2): 125-128.

17. Saxena, H.P. 1978. Pest of grain legumes and their control in India. In pest of grain Legumes: Ecology and Control, Eds: Singh, Emden and Taylor. Academic Press pp: 15-24

18. Singh, B.B., Singh S.R. and Adjadi, O. 1985. Bruchid resistance in cowpea. *Crop Science 25* (5): 736-739.

Effect of some plant extracts on the ovipositional behaviour of *Callosobruchus maculatus* infesting stored moth bean (*Vigna aconitifolia*)

Lekhu Gehlot, P.M. Singhvi and Dushyent Gehlot

Department of Zoology, J.N.Vyas University, Jodhpur - 342005, Rajasthan

ABSTRACT

Insect attacks all the parts of the plants, however, the important stored grains insect pest *Callosobruchus maculatus* is also found to have infestation in stored moth bean seeds. After attacking the seeds, this insect put eggs on the healthy seeds for the multiplication of progenies, so it is necessary to control the ovipositional behaviour of *C.maculates*. For this purpose, experiments were conducted at Entomology lab, Department of Zoology, J.N.Vyas University Jodhpur to evaluated some plant extracts viz., *Azadirachta indica* (Neem), *Barleria acanthoides* (Bajradanti) *Ocimum sanctum* (Tulsi), *Balanites aegyptiaca* (Hingota), *Eucalyptus globulus* (Safeda), *Salvadora Persica* (Khara Jaal/ Pilu), *Maytenus emarginata* (Kankera), *Commiphora wightii* (Guggal), *Fagonia cretica* (Dhamasa), and *Prosopis cineraria* (Khejari), for their biopesticidal activity against the ovipositional behaviour of *C.maculatus* to protect the stored moth bean seeds of Rajasthan. The plant extracts were taken @ 1.0ml., 2.5ml. and 5.0 ml. extract/100gm seeds. All the treatments caused significant reduction in oviposition by test insect in comparision to control. The minimum number of egg were laid on seeds treated with 5.0 ml. extract/100 gm seeds in case of all the plant extracts. *Azadirachta indica* was found the most effective at all the doses. Hingota, Guggal Safeda, Dhamasa Bajradanti and Tulsi also showed effective result. Maximum oviposition observed in case of seeds treated with extract of *Prosopis cineraria* @ 1ml./100gm, seeds. Analysis of variance due to different plant species and concentration were also highly significant (at 99, probabilitie level).

Key words: Plant extract, callosobruchus, biopesticides.

Introduction

Insects are the most important and cause tremendous qualitative and quantitative losses to pulses. Amongst insect pests of stored pulses, the members of the family Bruchidae have long been reported to destroy the seeds of the leguminous plants. The pulse beetle *Callosobruchus maculatus* severely damage the moth bean during storage conditions. Moth bean is one of the important pulse crop of the arid zone of Rajasthan. It is a common food grains of villagers as well. Due to unfavorable environmental conditions in the arid zone, the yield of moth bean is limited. Hence, users of it are storing the moth bean for long duration, which leads to infestation of several microorganisms and stored grain insect pests. To minimize losses during storage, in modern agriculture the insect pests are controlled by the use of chemicals. These chemicals are often associated with problems such as toxic residues, workers' safety, development of resistant strains and ever increasing cost of purchase. In view of hazardous effect of chemical insecticides in present study, an attempt was made to test some common desert plants of Rajasthan for their biopesticidal properties.

MATERIALS AND METHODS

(A) Rearing of test culture

The laboratory culture of *Callosobruchus maculatus* maintained at the Department of Zoology, J.N.V. University, Jodhpur. The culture obtained from naturally infested moth bean seeds from local market. The test insect reared on clean uninfested moth bean seeds in bacalite jars, covered with muslin cloth (to avoid escape of adults) at 28% ± 2°C and 70% ± 5% RH. New generations of the bruchids were obtained from this stock culture by continuously replacing the devoured and infested seeds with fresh un-infested seeds and introduction of male and female adults of *Callosobruchus maculatus* into the jar.

(B) Method of Ovipositional behaviour

To study the ovipositional behaviour of *Callosobruchus maculatus* on moth bean seeds, appropriate amount of extract was pipetted from the stock solution into a jar containing moth bean seeds. The extracts were applied at doses of 1.0, 2.5 and 5ml. per hundred grams of moth bean seeds. The seeds and the extracts were thoroughly mixed with the aid of a glass rod to ensure an even coating of extracts on seeds. Treated seeds were allowed to dry for about one hour at room temperature. Twenty extract treated and twenty control (untreated) seeds for each concentration were placed into a 'petridish'. Five pairs of newly emerged *C. maculatus* were introduced into each petridish and covered with mulsin cloth. Each experiment was performed in triplicate. The experiment was terminated after seven days by removing the adult insects. The number of eggs laid was recorded on both the extracts treated and untreated moth bean seeds.

Results and discssion

All the treatments caused significant reduction of oviposition by *C. maculates* in comparision to the control Minimum number of eggs were laid on the seeds treated with *Azadirachta indica*, where the oviposition rate was (8.49%), followed by *Commiphora wightii* (20.58%), *Balanites aegyptiaca* (20.58%), *Fagonia cretica* (21.56%), *Eucalyptus globulus* (26.14%), *Barleria acanthoides* (26.14%), *Ocimum sanctum* (33.66%), *Maytenus emarginata* (36.92%), *Salvadora persica* (37.90%), *Prosopis cineraria* (44.44%). @ 2.5ml extract/100g seeds, it was observed that minimum eggs were laid on the seeds treated with *Azadirachta indica* (1.96%) and other plant extracts range in betw-een 10.78% to 32.67%. @ 5ml extract/100g seeds no egg laying was observed in case of *Azadirachta indica* and other plant extracts range in between 3.26% to 23.85%.

Analysis of variance ascertained that the variance due to plant species are highly significant (99% probability). The variance due to concentration of the plant extracts are also highly significant, therefore, the interaction effect of species and concentration will also be highly significant. The contribution to variance by replicate revealed no significance indicating the lesser experimental error (Table 1).

Table 1. Effect of Plant Extracts on the Ovipositional Behaviour of *C. maculatus*

Plant Species	Doses (ml ext./100g seeds)	Total No. of eggs on 20 seeds (Replicates)			Mean	%
		R1	R2	R3		
Barleria acanthoides	1.0	8	8	8	8.00	26.14
	2.5	5	6	5	5.30	17.32
	5.0	2	3	3	2.60	8.49
Ocimum sanctum	1.0	10	11	10	10.30	33.66
	2.5	8	7	8	7.60	24.83
	5.0	5	5	5	5.00	16.30
Balanites aegyptiaca	1.0	7	6	6	6.30	20.58
	2.5	4	3	3	3.30	10.78
	5.0	1	1	1	1.00	3.26
Eucalyptus globulus	1.0	7	8	9	8.00	26.14
	2.5	5	5	5	5.00	16.33
	5.0	2	1	2	1.60	5.22
Salvadora persica	1.0	11	12	12	11.60	37.90
	2.5	9	9	9	9.00	29.41
	5.0	6	7	6	6.30	20.50

Maytenus emarginata	1.0	12	11	11	11.30	36.92
	2.5	10	9	9	9.30	30.39
	5.0	6	6	5	5.60	18.30
Azadirachta indica	1.0	3	2	3	2.60	8.49
	2.5	1	0	1	0.60	1.96
	5.0	0	0	0	0.00	0.00
Commiphora weightii	1.0	6	7	6	6.30	20.58
	2.5	3	4	3	3.30	10.89
	5.0	1	1	1	1.00	3.26
Fagonia cretica	1.0	7	6	7	6.60	21.56
	2.5	5	4	5	4.60	15.00
	5.0	2	2	2	2.00	6.53
Prosopis cineraria	1.0	14	13	14	13.60	44.44
	2.5	11	10	9	10.00	32.67
	5.0	8	7	7	7.30	23.85
Control		30	32	30	30.60	

Das (1) reported that total number of eggs laid by *Callosobruchus chine-nsis* on seeds treated with neem oil @ 6,8 and 10 ml/kg seeds was significantly lower than control. Kumar and Singhvi (2) stated that the oviposition of *Tribolium castaneum* on bajra seeds treated with neem cake powder in various ratios is significantly lower than any of the other cake powder mixed with grains. Petroleum ether extracts of 4 plant species of Egypt were found to have ovipositional deterrency against *Callosobruchus chinensis* Ghar *et al.* (3). Laboratory experiments were conducted by Parsai *et al.* (4) to investigate the efficacy of mustard, sesame and soybean oil and it was reported that these oil reduce the oviposition of *Callosobruchus chinensis*. Ogunwolu and Odunlanu (5) stated that Zanthoxylum root bark powder was an effective oviposition suppressant of *Callosobruchus maculatus* and more effective than neem seeds powder. Malatu and Gebremedhin (6) studied the oviposition deterrent effect of various botanicals on the azuki bean beetle, *Callosobruchus chinensis*. Singh (7) studied the effect of edible and non edible oils against *C. chinensis* at 8 ml/kg seeds. All the oils proved highly effective preventing egg laying and population of the beetle.

REFERENCES

1. Das, G.P. (1986). Effect of different concentrations of neem oil on the adult mortality and oviposition of *Callosobruchus chinensis*, *Ind. J. Agric. Sci.*, 56 (10) : 743-744.
2. Kumar, S and Singhvi, P.M. (1993). Evaluation of certain cake powders as protectant against Tribolium castaneum, *Asian J. Appli. Sci*, 7:49-52.

3. El-Ghar, GESA; El-Sheikh, A.E.; Ghar, GESA-El and Sheikh, AE-El (1987). Effectiveness of some plant extracts as surface protectants of cowpea seeds against the pulse beetle, *Callosobruchus chinensis*, *Phytoparasitica,* 15(2): 109:113.

4. Parsai, S.K.; Verma, R.S.; Shaw, S.S. and Badaya, A.K. (1994) Efficacy of edible oils on the incidence of *Callosobruchus chinensis* Linn. and their effect on cooking quality of pigeonpea, *Ind. J. Pulses Res.*, 7(2):148-152.

5. Ogunwolu, E.O. and Odunlanu, A.T. (1996). Supression of seeds ruchid (*Callosobruchus maculatus*) development and damage on cowpea with Zanthoxylum zanthoxyloides root bark powder when compared to neem seed powder and pirimiphosmethyl, *Crop Prot.,* 15:603-607.

6. Malatu, B. and Gebremedhin, T. (2000). Oviposition detrrent and toxic effects of various botanicals on the azukibean beetle, *Callosobruchus chinensis*(L), Insect Sci. Appl., 20:33-38.

7. Singh, P.K. (2003), Effect of some oils against pulse beetle, *Callosobruchus chinensis* in infesting pigeon pea. *India J. Ent.*, 65(1):55-58.

Effect of dates of sowing and varieties on development of bacterial blight on clusterbean at Bikaner

N.K. Khatri
Agricultural Research Station, Rajasthan Agricultural University, Bikaner-334006

ABSTRACT

The studies on dates of sowing compressing three clusterbean varieites were carried out to see the development if bacterial blight dsiease in Bikaner regions of arid western Rajasthan. The results revealed that the clusterbean sown in second dates of serving (12th July) gave the minimum incidence of disease and resulted in better yield levels.

Key words: Clusterbean, sowing dates, bacterial blight.

INTRODUCTION

Clusterbean [*Cyamopsis tetragonaloba* (L.) Taub.] popularly known by its vermacular name "guar" is an important leguminous crop of India. It is cultivated mainly under arid and semi regions of the country. Guar is mostly grown in Rajasthan, Gujarat, Hartana, Punjab and Uttar Pradesh. Rajasthan ranks first in respect of area and production. It is attacked by a number of bacterial, fungal and viral pathogens. Among these, bacterial blight caused by *Xanthomonas axonopodis* pv. *cyamosidis* (1,2) is one of the economically important disease. Yadav *et al.* (3) reported that the highest disease incidence (58.3%) was observed when the crop was sown in the last week of July. Gandhi and Chand (4) concluded that bacterial blight of clusterbean caused significant decreases in the numbers of clusters per plant, pods per plant, 1000-grain weight and also grain yield per plant. Despite of its importance, very little attention has been paid to study the intensity of bacterial blight and losses in yield of clusterbean as affected by dates of sowing and varieties in arid region of Rajasthan. Therefore, a field experiment was conducted under natural field condition to determine the effect of dates of sowing on incidence and development of bacterial blight on clusterbean and level of resistance of three popular varieties of clusterbean.

MATERIAL AND METHODS

A field experiment was conducted during *kharif*-2007 at Agricultural Research Station, Bikaner to study the performance of three popular varieties of clusterban *viz.,* M-83, RGC-986 and RGC-1017 under three dates of sowing at an interval of 15 days i.e. 26 June, 2007, 12 July, 2007 and 27 July, 2007. The experiment was laid out in Randomized Block Design with three replication. The net plot size was 4.0 x 3.0 m. All the agronomical recommended practices were followed as per package of practices. Sowing was done keeping spacing 30 x 10 cm. The disease intensity was recorded using 1-9 scale (1=0 %, 3=0.1 – 10.0%, 5= 10.1-25.0%, 7=25.1-50.0%, 9= >50% disease intensity).

RESULTS AND DISCUSSION

The results present in Table 1 revealed that maximum 40.00 per cent infection of bacterial blight was observed in 1st date of sowing i.e. 26.06.2008 and yield was 5.05 q ha^{-1} which was minimum among three dates of sowing. The second date of sowing i.e. 12.07.2007 was the best among three dates of sowing, where disease intensity was 21.11 per cent and yield was 6.13 q ha^{-1}. The higher grain yield in second date of sowing may be due to higher yield attributing characters *viz.,* plant height, primary branches, number of pods per plant.

If we compare the performance of the varieties, although variety RGC-1017 gave maximum higher yield (5.81 q ha^{-1}), it may be due to its more number of pods per plant and 100-seed weight, followed by M-83 (5.65 q ha^{-1}) and RGC-986 (5.59 q ha^{-1}) but these three varieties did not differ significantly. The interaction between dates of sowing and varieties had no significant difference. Butter *et al.* (5) found significant result on development of bacterial blight on different varieties sown under various dates of sowing in Punjab during 2001-03.

Hence IInd dates of sowing (i.e. 12.07.07) was most suitable for clusterbean in Bikaner region.

Table 1: Effect of dates of sowing and varieties on development of bacterial blight on clusterbean

S.No.	Treatment	Disease intensity (%)	Yield (q ha^{-1})
1.	Date of sowing		
	(i) 26.06.2007	40.00	5.05
	(ii) 12.07.2007	21.11	6.13

	(iii) 27.07.2007	26.11	5.87
	S.Em±	2.73	0.14
	C.D. (5%)	8.19	0.42
2.	Varieties		
	(i) M-83	37.78	5.65
	(ii) RGC-986	22.22	5.59
	(iii) RGC-1017	27.22	5.81
	S.Em±	2.73	0.14
	C.D. (5%)	8.19	NS

REFERENCES

1. Patel, M.K., Dhande, G.W. and Kulkarni, Y.S. 1953. Bacterial leaf spot of clusterbean. *Curr. Sci.,* 22 : 183.
2. Vauterin, L., Haste, B., Kersters, S.K. and Swings, J. 1995. Reclassificatin of *Xanthomonas. Int. J. Syst. Bacteriol.* 45 : 472-489.
3. Yadav, S.C., Nath, R. and Yadav, R.K. 2005. Occurance of bacterial blight (*Xanthomonas axonopodis* pv. *Cyamosidis*) on clusterbean. *J. Mycol. Pl. Pathol.,* 35(3) : 496.
4. Gandhi, S.K. and Chand, J.N. 1985. Yield losses in guar due to bacterial blight caused by *Xanthomonas axonopodis* pv. *Cyamosidis. Indian Phytopath.* 38(3) : 516-519.
5. Buttar, G.S., Singh, S. and Kaur, A. 2008. Performance of different varieties of guar sown under various dates of sowing in Punjab Condition. Abstracts of 3rd Nation Symposium of Arid Legumes, 2008. pp. 64.

Estimation of diversity among various isolates of *Xanthomonas axonopodis* pv. *cyamopsidis* using molecular markers

***G.L. Kakraliya, N.K. Khatri*[1] *& R.A. Sharma*[2]**

College of Agriculture, [1]Agriculture Research Station,
[2]Plant Biotechnology Centre, Rajasthan Agricultural University,
Bikaner - 334006 (Rajasthan)

ABSTRACT

In the present study four primers used, had generated a significant information. The number and size of amplicons depended on primers and strains used. Though, most amplicons were smaller than 3kb in size but one of the amplification product was larger than 5kb. Our study showed that the regional *Xanthomonas axonopodis* pv. *cyamopsidis* population had a high degree of genetic diversity. In this study, strain Nokha was not clustered with other strains either in the phytogenetic cluster or the RAPD analysis. Nokha and Bikaner isolates belonging to equally distant places were most diverse one. In conclusion it can be said that the *Xanthomonas axonopodis* pv. *cyamopsidis* depicts a very high diversity in harsher climate of Rajasthan probably for its survival.

Key words: Clusterbean, bacterial blight strains, diversity, molecular market.

INTRODUCTION

Clusterbean [*Cyamopsis tetragonoloba* (Linn.) Taub.] locally known as 'Guar' is a legume of Indian origin is attacked by more than 30 fungal, bacterial and viral pathogens. Among them bacterial blight caused by *Xanthomonas axonopodis* pv. *cyamopsidis* (1), is the most important and wide spread disease. It appears all most every year and may cause considerable damage depending upon the variety and environment conditions. In case of heavy infection, it may cause 58-68 per cent losses in grain yield (2,3). Thus there is an urgent need to search for economic and eco-friendly alternative strategies for the management of this disease. Recently greater emphasis on stable source of resistance and other useful control measures *viz.,* biological, cultural and botanicals have been laid. However their efficient employment would require information on genetic diversity present in the pathogen.

The traditionally applied techniques for diversity studies based on morphological and biochemical traits need tedious culture and analytical procedures. More recently following

DNA fingerprinting methods are considered as ideal methods and have been widely used for specific identification of many organisms, including bacteria (4,5). The PCR based random amplified polymorphic DNA (RAPD) method, one of the most frequently used methods of DNA fingerprinting, can be used to generate distinctively amplified patterns for different tested species.

RAPDs have been successfully used to study bacterial blight communities (6). Looking to the possibility that RAPD markers can be successfully used for characterization of bacterial blight communities, present investigation was taken up to find out effect of annual plant species like cluster bean on bacterial diversity in arid regions of Rajasthan. An attempt has been made in the present study to determine the genetic diversity present in the *Xanthomonas axonopodis* pv. *cyamopsidis* in a vide cluster bean grown region.

MATERIALS AND METHODS

For the isolation of causal bacterium, young diseased portions/lesions were cut into small bits and rinsed with spirit and immediately transferred into mercuric chloride solution (0.01%) for 15 seconds and were finally passed through 3-changes of sterile distilled water for surface sterilization. The bits were then transferred individually on flame sterilized, coded glass slide containing drop of sterilized water under aseptic conditions. The lesions were cut into 2 halves with heat sterilized razor blade and the slide was left for few minutes (2-5 min) to permit the diffusion of bacteria from tissue into the water drops. A loopful of suspension was streaked - quadrantly on previously sterilized petriplates containing nutrient agar (NA) and potato sucrose peptone agar (PSPA) media with the help of sterilized inoculation needle. The inoculated plates were incubated at 27± 2°C for over night. The bacterial colonies appeared on the surface of the medium like straw yellow droplets. The single fresh and pure colony was picked up under a stereoscopic microscope, with sterilized inoculation needle and transferred on other petriplates having NA and PSPA medium. The process was repeated until all the colonies developed on the medium had similar cultural appearance. Such culturally identical single colony was then transferred on yeast glucose chalk agar storage medium slants, and maintained by preserving in a refrigerator for all future experiments, and making periodic transfers after every fortnight. The stock culture was kept in refrigerator after covering culture slants with sterilized mineral oil.

The experimental material for the present investigation consisted of six strains of isolates of *Xanthomonas axonopodis* pv. *cyamopsidis* which were collected from different locations e.g. Bikaner, Hisar, Durgapura (Jaipur), Udaipur, Jobner and Nokha.

RAPD ANALYSIS : RAPD analysis was conducted using purified DNA from each accessions, in order to determine genetic diversity among genotypes taken for study.

DNA isolation: Bacterial colonies were Stretched from peteriplates and dissolved in 1 ml $T_{10}E_1$ in appendrof. Centrifuged at 5000 rpm for 3 minutes, at 40 C. Added 800 -1000 μl (60°C) DNA isolation buffer [2x CTAB DNA etraction buffer-100 mM Tris 20 mM EDTA, 1.4 M NaCl, 2% CTAB and 2 μl/ml ß-mercaptoethanol]. To the isolation buffer 100 μl of 20% SDS was also added. Incubated for 1 hr at 60°C with occasional mixing by gentle swirling in water bath. After removing from water bath one volume of chloroform : Isoamyl alcohol (24:1) was added and mixed by inversion for 15 minutes to ensure emulsification of the phase. It was then centrifuged at 15000 rpm for 10 minutes to recover aqueous phase in another tube. Ice cold 0.6 vol. of isopropanol was added to a aqueous phase to precipitate DNA. DNA-CTAB complex was precipitated as a fibrous network. It was lifted by Pasteur pipette and was transferred to washing solution. 20 ml of 70% alcohol was added to pellet and was kept for 20 minutes with gentle agitation. The pellet was collected by centrifugation at 10,000 rpm for 5 minutes at 20°C. The tubes were inverted and drained on a paper towel. The pellet was dried over-night after covering with parafilm with tiny holes. The pellet was re-dissolved in 100 µl of $T_{10}E_1$ buffer by keeping over night at room temperature.

Purification of DNA : RNA was removed from DAN by treating the sample with DNase free RNase, protein including RNase was removed by chloroform: Isoamyl alcohol (24 :1) treatment. Presence of single compact band at the corresponding band of λ phage DNA indicates high molecular weight of isolated DNA on 0.8% agarose gel.

The quantification of DNA was done by observing it at 260 nm wave lengths by using a spectrophotometer (UV visible from UNICAM). The quantified DNA was diluted to final concentration of 25 ng/µl in $T_{10}E_1$ buffer (10 mM Tris Cl, 1 mM EDTA, pH 8.0).

RAPD analysis: Random amplification of polymorphic DNA was done by using 20 primers of OPG series obtained from "OPERON TECHNOLOGIES" (Inc. Alameda, Calif). PCR reaction were performed in final volume of 25 µl containing 10 X Assay Buffer (Bangalore Genei) 0.5 units of Taq. DNA polymerase (Bangalore Genei), 200 μM each of dNTPs (Bangalore Genei), 10 pmols/ reaction of random primer's (OPERON TECHNOLOGIES) and 50 ng of tempelate DNA. The PCR was performed in 'Biometra Thermocycler' using the following cycling parameters.

Cycle 1: Denaturation (94°C) 5 minutes. Primer annealing (37°C) 1 minute. Primer Extension (72°C) 2 minutes, Cycle 2-43 : Denaturation (94°C) 1 minute. Primer annealing (37°C) 1 minute. Primer Extension (72°C) 2 minutes, Cycle 44: Denaturation (94°C) 1 minute. Primer annealing (37°C) 1 minute. Primer Extension (72°C) 7 minutes

Following the amplification, the PCR products were loaded on 1.2% Agarose gel (Merk) which was prepared in 1 x TBE buffer containing 0.5 μg/ml of the ethidium bromide. The Amplified products were electrophoresed for 4-5 hrs at 50 V with cooling. After separation the gel was viewed under UV transillum-inator and photographed by digital camera.

RESULTS AND DISCUSSION

Total genomic DNA was isolated with the CTAB method described by Ausubel *et al.* (7) with some modification and treated with RNase to elimi-nate RNA. DNA concentration was measured by UV absorbance method and the quantity of DNA ranged from 32 μg/ml to 108 μg/ml of culture (Table 4.5). The integrity of isolated DNA was verified by visualization of DNA on agarose gel (0.8%) with DNA standard uncut lambda DNA. The appearance of single band corresponding to the size of lambda uncut confirms high molecular size of DNA isolated (Plate 4). The quality of DNA was further determined for the presence of contaminants as the ratio of A_{260}/A_{280}, ranged from 1.73 to 2.00 which is indicative of good quality bacterial DNA and the ratio was almost consistent irrespective of the *Xanthomonas axonopodis* pv. *cyamopsidis* strains.

The total genomic DNA isolated using SDS supplemented (TAB method yielded satisfactory quantity, 32 μg/ml to 108 μg/ml from one petri plate). The DNA was of high molecular weight mostly visible in single band corresponding to λ DNA and A_{260}/A_{280} ratio was also between 1.73 to 1.90 indicating contamination free preparation.

All the six strains of *X. axonopodis* pv. *cyamopsidis* were examined for random amplified polymorphic DNA (RAPD) genetic markers with 20 decamer primers (Operon Technology, USA) (Table 4.6). Primers of OPG series were screened taking two strains, only four primers giving consistent results were used for final amplification (OPG-3, OPG-4, OPG-5, OPG-19). All the four primers were repeated at least twice to confirm the polymorphism. This is because the stoichastic nature of primer annealing has been known to change the band and banding pattern of DNA amplification with RAPD. Reproducibility of the banding pattern has been found to change.

Finally, only those bands were considered as polymorphic, which did not amplify in at least one strain on repetition. The banding pattern generated by each primer was primer and strain dependent and varied from 7-13 at 37C annealing temperature. A total of 38 amplicons were obtained with 4 primers with an average of 9.5 bands per primer. Highest number of bands (13) were generated by primer OPG-5, while OPG-4 produced least (7) bands. Most of the amplicons were below 3 kb size abut as large as 5148 kb was also observed in OPG-3. All the 38 bands were found to be polymorphic and the level of polymorphism was 100 per cent (Table 1, Plate 1).

Table 1 : List of primers and polymorphic amplicons generated

Primers	Sequence (5'-3')	Total No. of bands (a)	No. of polymorphic bands (b)	Polymorphism b/a x 100 (%)
OPG-3	GAGCCCTCCA	8	8	100
OPG-4	AGCGTGTCTG	7	7	100
OPG-5	CTGAGACGGA	13	13	100
OPG-19	GTCAGGGCAA	10	10	100
	Total	38	38	100

Identification markers were generated for all of the isolates by various primers but OPG-4 and OPG-5 generated maximum isolate specific amplicons (Table 3).

Table 2 : Primer that gave strain specific marker

Primer	Strains	Marker size (bp)	Present (+) or absent (-)
OPG-3	Job	5148	+
	Nok	564	+
OPG-4	Udr	3514	+
	Dur	2351	+
	His	820	+
	Job	564	+
OPG-5	Dur	1620	+
	Nok	1400, 230, 110	+
	Bik	640, 400, 150	+
	His	310	+

OPG-19	Bik	1904	-
	Udr	1010	+

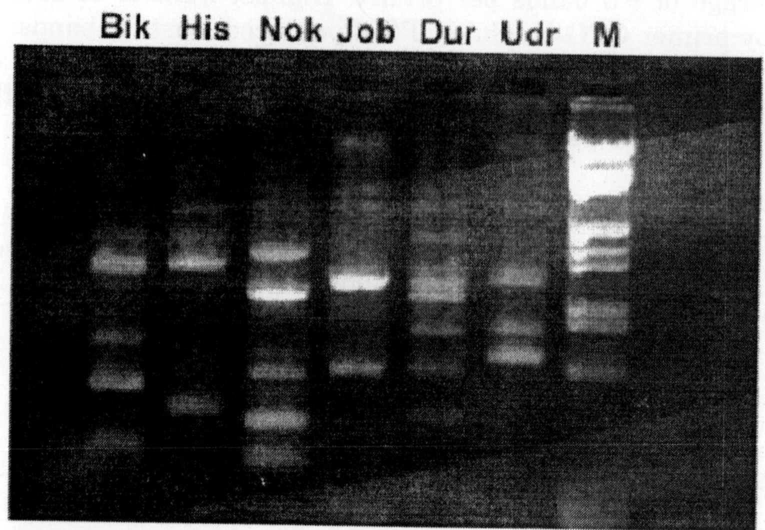

Plate-0 RAPD profile generated by primer OPG-5 in six *Xanthomonas axonopodis p.v. cyamopsidis* strains.

The Jaccard's pair wise similarity coefficient values on the basis of RAPD patterns for six *Xanthomonas axonopodis* pv. *cyamopsidis* isolates revealed a high degree of diversity among the isolates (Table 4). The range of diversity was observed to be 57% between Job and Dur) to 92% among Bik and Dur isolates while average being 79%. The highest similarity coefficient 0.43 between Job and Dur was followed by 0.40 among His and Udr isolates. However, least similar isolates were Bik and Dur (0.08) followed by Nok-His and Nok-Udr (0.09).

Table 3: Jaccard's similarity coefficient among six isolates of *Xanthomonas axonopodis* pv. *cyamopsidis* based on RAPD

Isolates	Bik	His	Nok	Job	Dur	UDR
Bik	1.00					
His	0.19	1.00				
Nok	0.11	0.09	1.00			
Job	0.17	0.28	0.17	1.00		
Dur	0.08	0.29	0.13	0.43	1.00	

Udr	0.21	0.40	0.09	0.25	0.26	1.00

Genetic similarity estimates (Jaccard's coefficient) were further used to group the isolates using UPGMA (Un-weighted Pair Group Method for Arithmetic Average) algorithm. The genetic relation ships (clusters) so obtained have been presented in the form of dendrogram (Fig. 3). The isolates could not form obvious groupings. Four of the isolates (His, Udr, Dur and Job) obtained from relatively higher rainfall regions were comparatively more related than the two isolates Bik and Nok, belonging to dryer regions. Nevertheless, similarity within group of four isolates was much less (~28%). The isolates from Nokha and Bikaner were most diverse.

The number and size of amplicons depended on primers and strains used. Though, most amplicons were smaller than 3kb in size one of the amplification product was larger than 5kb. RAPD marker system has been indicated to amplify fragments smaller than 3 kb by various authors (Kaur *et al.,* 2005) and are known to depend on primer and species used (Johnson *et al.,* 1994).

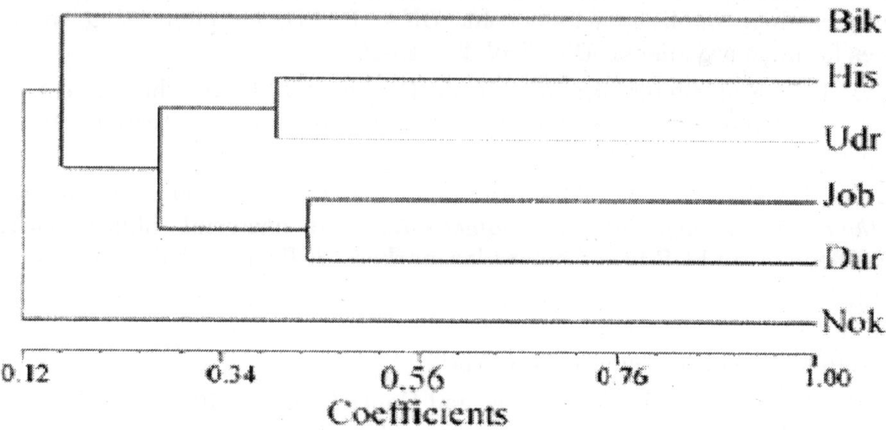

fig. 3 Dendogram generated on six Xanthomonas axonopodis pv. cyamopsidis strain using UPGMA cluster analysis based on Jaccard's similarity coefficients for RAPD data

The diverse origin of isolates from various places was reflected in very high diversity (average being 79%) obtained in the form of Jaccard's similarity coefficients. Though a high level of diversity among isolates belonging to different regions have been reported such a high diversity level has not been reported in *Xanthomonas* spp (8). Job and Dur, the isolates collected from near by locations i.e. Jobner and Durgapura respectively though showed more closeness but it was not the general trend followed in the present study. Nok and Bik isolates belonging to equally distant places were most diverse one. Overall, clustering was not conspicuous as within group diversity was more than between group diversity as evident by dendrogram. However, little similarity between isolates belonging to places having better rainfall (His, Udr) was evident. Nevertheless as stated earlier Nok and Bik isolates belonging to drier regions of Rajasthan (Nokha and Bikaner respectively) were most diverse ones. In conclusion it can be said that the *Xanthomonas axonopodis* pv. *cyamopsidis* depicts a very high diversity in harsher climate of Rajasthan probably for its survival.

REFERENCES

1. Patel, M.K., Dhande, G.W. and Kulkarni, Y.S. 1953. Bacterial leaf spot of clusterbean. *Curr. Sci.* **22** : 183
2. Gupta, D.K. 1977. Evaluation of antibiotics and fungicides against bacterial blight of guar *(Cyamopsis tetragonoloba* (L) Taub*) Veg. Sci.* **4** : 25-27.
3. Gandhi, S.K. and Chand, J.N. 1985. Yield losses in guar due to bacterial blight caused by *X. campestris* pv. *cyamopsidis*. *Indian Phytopath.* **38** (3) : 516-519.
4. Fani, R., Damiani, G., Diserio, C., Gaillari, E., Friioni, A. and Bazzicalupo, M 1993. Use of random amplified polymorphic DNA (RAPD) for generating specific DNA probes for microorganisms. *Mol. Ecol.* **2** : 243-250.
5. Louws, F.J., Rademaker, J.L.W. and de Bruijn, F.J. 1999. The three DS of PCR based genomonic analysis of phytobacteria, diversity detection and disease diagnosis. *Annu. Rev. Phytopatho.* **37** : 81-125.
6. Kaur, B., Purkayastha, S., Dilbaghi, N. and Chaudhary, A. 2005. Characterization of *Xanthomonas axonopodis* pv. *cyamopsidis*, the bacterial blight pathogen of clusterbean using PCR-based molecular markers. *J. Phytopathol.* **153** : 470-478.
7. Ausubel, F.M., Brent, R., Kigdom, R.E., Moore, D.D., Smith, J.A., Sidemon, S.G. and Strutid, K. 1992. Current protocols in molecular biology. Greene Publishing Associates, Wiley Interscience, New York.
8. Manulis, S., Valinsky, L., Lichter, A. and Gabriel, D.W. 1994. Sensitive and specific detection of *Xanthomonas campestris* pv. *pelargonii* with DNA primers and probes DNA analysis. *Appl. Environ. Microbiol.* **60** : 4094-4099.

To identify resistant source against yellow mosaic virus in mungbean

O.P. Khedar, R.V. Singh, Kuldeep Sharma and Mahesh Shrimali

Rajasthan Agriculture University, Agricultural Research Station, Durgapura, Jaipur-302018

ABSTRACT

A study was conducted at hot spot of YMV, ARSS Hanumangarh, Rajasthan to screen most effective crosses for transferring diseases resistance (YMV) in mungbean. Keeping this view in mind a field experiment was conducted to screen 270 F_3 lines of 44 crosses under the natural field condition during *kharif* 2006. The results revealed that none of the progeny showed resistance against YMV, whereas, 66.67% progenies of Pusa vishal x SML 589, 50% of MUM 2 x Pusa vishal, 28.57% of PDM 11 x SML 589, 25% of RMG 344 x PDM 54 and SML 589 x RMG 344, 16.67% of RMG 62 x KM 2170, 13.33% of RMG 344 x TM 99-47, 12.50% of RMG 492 x TM 99-47, 11.11% of Pusa vishal x PDM 54, 5% of RMG 344 x Pusa vishal and 4.76% of RMG 344 x ML 818 showed moderate resistance against YMV and can be utilized for breeding purposes effectively. The results also showed that crosses Pusa vishal x SML 589 and MUM 2 x Pusa vishal are most effective in transferring YMV resistance to the progenies.

Key words: Mungbean, diseases resistance, generation and YMV.

In India mungbean (*Vigna radiata* L. Wilczek) is one of the important pulse crop cultivated in *kharif, rabi, spring and, summer*. Primarily it is a *kharif* crop but with the development of short duration, photo-insensitive and synchronous maturing varieties, it has become popular in *spring* and *summer* seasons too. It is grown in almost all parts of the country in an area of 3.11 m ha with the production of 0.95 m tonnes and productivity 3.04 q/ha (2005-06). In India it is cultivated in the states of Maharashtra, Gujarat, Andhra Pradesh, Rajasthan, MP, Bihar, Karnataka, U.P., Tamil Nadu, Orissa, West Bengal, Punjab and

Haryana, whereas, Himachal Pradesh, J & K, Kerala and Tripura also contribute to some extent. It is an excellent source of easily digestible protein. Vitamin 'C' is synthesized in sprouted seeds of mungbean. Looking to the food habit of majority of Indian population which is vegetarian, it becomes more important because, it fulfills the protein requirement of the peoples. In the desert of Rajasthan it is the only media to fulfill the nutritional requirement of the poor farmers and labours. It improves soil health and also provides laxative and nutritive fodder to the cattles. Due to short duration behavior it fits well in so many crop rotations and crop mixtures easily. Though we have developed a number of improved varieties, production and protection technology but till date we the average productivity is very less as compared to genetic potential (1000-1500 kg/ha). Diseases are found to be the most important biotic constraint in pulse cultivation. The diseases like yellow mosaic virus, *cercospora* leaf spot, web-blight and leaf crinkle are the common diseases of mung bean prevalent in Rajasthan. Out of all the diseases of mungbean, yellow mosaic virus is a serious disease and some times occur in epidemic form, resulting in total failure of the crop. To avoid the yield losses due to YMV, development of resistant variety is an important task. An ideal way of YMV disease management lies in locating a resistant source, which could be used in future breeding programme [1].

MATERIALS AND METHODS

A study was conducted at hot spot of YMV, ARSS Hanumangarh, Rajasthan to screen most effective crosses for transferring diseases resistance in mungbean. Keeping this view in mind a field experiment was conducted to screen 270 F_3 lines of 44 crosses under the natural field condition during *kharif* 2006. Every genotype was planted in 2 rows of 4 meter length and 30 cm spacing. All the agronomic practices were adopted to grow better crop. There was sufficient disease pressure in the field. Screening was done at 2 stages (30 days after sowing and 50 days) under natural field condition. A genotype which was free from disease at 30 days stage but, showed disease symptom at 50 days stage was recorded as susceptible. The observations were recorded as resistant, moderately resistance, moderately susceptible, susceptible and highly susceptible for YMV.

RESULTS AND DISCUSSION

Out of 270 progenies none of the progeny showed resistance against YMV. 66.67 per cent progenies of Pusa vishal x SML 589, 50% of MUM 2 x Pusa vishal, 28.57% of PDM 11 x SML 589,25% of RMG 344 x PDM 54 and SML 589 x RMG 344 and 16.67% of RMG 62 x KM 2170 showed moderate resistance against YMV. Similar results were observed by Khedar *et al.* [2,3,4,5,6]. The crosses Pusa vishal x SML 589, and MUM 2 x Pusa vishal are most effective in transferring YMV resistance to the progenies and can be utilized for breeding purposes effectively.

Table 1: Cross progenies and their reactions to diseases

S. No.	Cross progenies	Total lines	Resistant	Moderately resistant (%)	Moderately susceptible	Susceptible	Highly susceptible
1.	MUM 2 x ML 5	4				2	2
2.	MUM 2 x RMG 492	3				3	
3.	MUM 2 x PDM 11	5				1	4
4.	MUM 2 x PDM 54	2				2	
5.	MUM 2 x SML 589	3				2	1
6.	MUM 2 x Pusa 9531	4				2	2
7.	Pusa vishal x RMG 62	8			1		7
8.	Pusa vishal x ML 5	6			2	1	3
9.	PDM 11 x PDM 54	4				1	3
10.	PDM 11 x SML 589	7		2 (28.57)		2	3
11.	PDM 11 x Pusa 9531	2					2
12	PDM 54 x Pusa 9531	1				1	
13.	PDM 54 x SML 589	5					5
14.	RMG 62 x KM 2170	24		4 (16.67)	5	10	5
15.	RMG 344 x ML 818	21		1 (4.76)	3	10	7
16.	RMG 344 x Pusa 0172	13			1	2	10
17.	RMG 344 x TM 99-47	15		2 (13.33)	5	3	5
18.	RMG 492 x Pusa 0172	24			6	14	4
19.	RMG 492 x TM 99-47	16		2 (12.5)	5	9	
20.	RMG 344 x Pusa vishal	20		1 (5.00)	2	14	3
21.	RMG 344 x RMG 62	7			2	5	
22.	RMG 344 x ML 5	5				1	4
23.	RMG 344 x RMG 492	1					1
24.	RMG 344 x PDM 11	7			1	2	4
25.	RMG 344 x PDM 54	4		1 (25.00)	1	2	
26.	SML 589 x RMG 334	4		1 (25.00)		1	2

27.	MUM 2 x Pusa vishal	4	2 (50.00)			2
28.	MUM 2 x RMG 62	1			1	
29.	Pusa vishal x SML 589	3	2 (66.67)			1
30.	Pusa vishal x Pusa 9531	2				2
31.	RMG 62 x ML 5	1			1	
32.	RMG 62 x RMG 492	1				1
33.	RMG 62 x SML 589	3		2	1	
34.	RMG 62 x Pusa 9531	1			1	
35.	ML 5 x PDM 54	8		2	6	
36.	ML 5 x SML 589	6		2	4	
37.	ML 5 x Pusa 9531	3		2		1
38.	RMG 492 x PDM 54	1				1
39.	RMG 492 x SML 589	2				2
40.	RMG 492 x PDM 11	2		1		1
41.	Pusa vishal x PDM 11	5		2	1	2
42.	Pusa vishal x PDM 54	9	1 (11.11)	1		7
43.	RMG 62 x PDM 54	2				2
44.	SML 589 x Pusa 9531	1			1	
	Total	270	19 (7.04)	46	106	99

REFERENCES

1. Jhamaria S.L., O.P. Khedar, R.V. Singh and Mahesh Shrimali 2003. Locating Resistant Source against Yellow Mosaic Virus Disease of Mungbean. 25th Annual Conference and Symposium of Mycology and Plant Pathology. October 8-10, 2003 at Agricultural Research Station, Rajasthan Agricultural University, Durgapura, Jaipur. J. Mycol. Pl. Pathol., Vol. No. 3: 500.
2. Khedar O.P., Mahesh Shrimali, R.V. Singh and V.L. Majumdar 2003. Breeding for Resistance against Common mungbean diseases in Rajasthan. 25th Annual Conference and Symposium of Mycology and Plant Pathology. October 8-10, 2003 at Agricultural Research Station, Rajasthan Agricultural University, Durgapura, Jaipur. J. Mycol. Pl. Pathol., Vol. No. 3: 489-490.
3. Khedar O.P., Singh R.V., Shrimali Mahesh, and. Majumdar V.L. 2006. Management of yellow mosaic virus by locating resistant source in mungbean. J. Arid Legumes 3 (2) 75.
4. Khedar O.P., Singh R.V. and Sharma Kuldeep. 2007. Management of yellow mosaic virus in mungbean. National Symposium on "Legume Research: Recent Trends and

Future Prospects in Post Genomic Era"on Jan. 13-14 2007 at Sardar Vallabh Bhai Patel University of Agriculture and Technology, Meerut. Abst.: p162

5. Khedar O.P., Singh R.V. Shrimali Mahesh and Sharma Kuldeep 2005. Breeding for resistance against yellow mosaic virus and leaf crinkle in black gram [*Vigna mungo* (L.) Hepper]. 4th International Food Legumes Research Conference. October 18-22, 2005 at IARI, New Delhi. Abst.: p A-562.
6. Singh R.V., O.P. Khedar, Mahesh Shrimali and V.L. Majumdar 2003. Breeding for Yellow Mosaic Virus Disease Management in Mungbean. 25th Annual Conference and Symposium of Mycology and Plant Pathology. October 8-10, 2003 at Agricultural Research Station, Rajasthan Agricultural University, Durgapura, Jaipur. J. Mycol. Pl. Pathol., Vol. No. 3: 471.
7. Khedar O.P., Shrimali Mahesh, Singh R.V., and. Majumdar V.L. 2006. Breeding for disease resistance against YMV in mungbean. J. Arid Legumes 3 (2) 71-721.

Effect of screening in breeding generations on yellow mosaic virus incidence in mungbean

O.P. Khedar, Mahesh Shrimali, N.P. Singh and Kuldeep Sharma

Rajasthan Agriculture University, Agricultural Research Station, Durgapura, Jaipur

ABSTRACT

Yellow mosaic virus is the major biotic stress causing heavy yield losses in mungbean. Growing disease resistant varieties is the only economic method for controlling the diseases. An ideal way of YMV disease management lies in locating a resistant source, which could be used in future breeding programme. Keeping this into consideration, a field study was conducted at hot spot of YMV, ARSS Hanumangarh, Rajasthan during *kharif* 2007 to see the effect of screening in breeding generations (F_3 & F_4). Seven cross progenies viz. RMG 344 x Pusa vishal, RMG 344 x RMG 62, RMG 344 x ML 5, RMG 344 x PDM 11, MUM 2 x ML 5, Pusa vishal x PDM 54 and PDM 11 x SML 589 were used for the study. The results indicated that in F_3 100% progenies of RMG 344 x Pusa vishal, RMG 344 x RMG 62, RMG 344 x ML 5, RMG 344 x PDM 11 and Pusa vishal x PDM 54 were highly susceptible against YMV whereas, in F_4 40% progenies of RMG 344 x PDM 11, 36.84% of RMG 344 x Pusa vishal, 33.33% of RMG 344 x RMG 62 & Pusa vishal x PDM 54 and 25% progenies of RMG 344 x ML 5 showed susceptible reactions against YMV. In MUM 2 x ML 5 all the progenies were susceptible to highly susceptible in F_3, whereas, in F_4 27.27% progenies were moderately susceptible against YMV. Similarly in PDM 11 x SML 589 none of the progeny showed any resistance in F_3 against YMV, whereas, in F_4 37.50% progenies showed moderate resistance against YMV. This indicates that regular screening in different breeding generations helps us in getting resistant material in later generations.

Key words: Mungbean, diseases resistance, generation and YMV.

Mungbean commonly known as 'green gram' is one of the third most important pulse in South Asia after chickpea and pigeonpea cultivated in all parts of the country over a wide range of agro-climatic condition by small and poor farmers on marginal lands in rain-fed condition. It is an important *kharif* crop cultivated in all parts of the country. It is also cultivated in *rabi,* and summer. India is the largest producer of mungbean in the world, accounts for almost 65% area and 54% production. In India it is grown in an area of 3.11 m ha with the production of 0.95 m tonnes and productivity 3.04 q/ha (2005-06). Maharashtra, Andhra Pradesh, Rajasthan, Bihar, Orissa, Karnataka, Chhattisgarh, Gujarat, MP, UP and Tamil Nadu are major mungbean growing states in the country. It is an ideal source of easily digestible proteins and amino acids for millions of peoples. Vitamin 'C' is synthesized in sprouted seeds of mungbean. It is able to fulfill the protein requirement of the vegetarian peoples up to some extent. It is consumed as *dal, halwa, namkeen,* snacks and many other preparations. It is easily digestible and good for patients. It also provides nutritive and laxative green and dry fodder to the cattle. It fixes atmospheric nitrogen to the soil through symbiosis and improves the fertility and health of the soil. Due to short duration behavior it fits well in so many crop rotations and crop mixtures easily. The average productivity is very less as compared to genetic potential (1000-1500 kg/ha). In general the productivity of mungbean is very low and unstable, primarily due to Poor economic condition of Rajasthani farmers, lack of irrigation facilities or mostly rainfed cultivation, rain fall uneven and uncertain, low seed replacement rate, negligible use of fertilizer, lack of scientific cultivation and heavy post harvest losses. Even than Biotic stresses play an important role in low production and productivity of mungbean. Diseases are found to be the most important biotic stress in mungbean cultivation. Yellow mosaic virus is the major biotic stress causing heavy yield losses. Growing disease resistant varieties is the only economic method for controlling the diseases. An ideal way of YMV disease management lies in locating a resistant source, which could be used in future breeding programme [1,2].

MATERIALS AND METHODS

Two breeding generations (F_3 & F_4) of seven cross progenies viz. RMG 344 x Pusa vishal, RMG 344 x RMG 62, RMG 344 x ML 5, RMG 344 x PDM 11, MUM 2 x ML 5, Pusa vishal x PDM 54 and PDM 11 x SML 589 were used for the study. The experiment was conducted at ARSS Hanumangarh, Rajasthan during *kharif* 2007. Sowing was done with 2 rows of 4 meter length and 30 cm spacing. There was more than sufficient disease pressure in the breeding generations. All the agronomic practices were adopted to grow better crop. The screening was done at 2 stages (30 days after sowing and 50 days) under natural field condition. The observations were recorded as resistant, moderately resistant, moderately susceptible, susceptible and highly susceptible in per cent (%)

for YMV. A genotype which was free from disease at 30 days stage but, showed disease symptom at 50 days stage was recorded as susceptible.

RESULTS AND DISCUSSION

In F_3 all the progenies of RMG 344 x Pusa vishal, RMG 344 x RMG 62, RMG 344 x ML 5, RMG 344 x PDM 11 and Pusa vishal x PDM 54 were highly susceptible against YMV, whereas, in F_4 40% progenies of RMG 344 x PDM 11, 36.84% of RMG 344 x Pusa vishal, 33.33% of RMG 344 x RMG 62 & Pusa vishal x PDM 54 and 25% progenies of RMG 344 x ML 5 showed susceptible reactions against YMV. In MUM 2 x ML 5 all the progenies were susceptible to highly susceptible in F_3, whereas, in F_4 27.27% progenies were moderately susceptible against YMV. Similarly in PDM 11 x SML 589 none of the progeny showed any resistance in F_3 against YMV, whereas, in F_4 37.50% progenies showed moderate resistance against YMV. This indicates that regular screening in different breeding generations helps us in getting resistant material in later generations.

Table: 1. Different generations and there reaction to Yellow Mosaic Virus.

S. No.	Generations	Total lines	Resistant (%)	Moderately resistant (%)	Moderately susceptible (%)	Suscep-tible (%)	Highly susceptible (%)
	F_3						
1.	RMG 344 x Pusa vishal	10					10 (100.00)
2.	RMG 344 x RMG 62	6					6 (100.00)
3.	RMG 344 x ML 5	4					4 (100.00)
4.	RMG 344 x PDM 11	10					10 (100.00)
5.	MUM 2 x ML 5	8				4 (50.00)	4 (50.00)
6.	Pusa vishal x PDM 54	4					4 (100.00)
7.	PDM 11 x SML 589	10			6 (60.00)	2 (20.00)	2 (20.00)
	Total F_3	52			6 (11.54)	6 (11.54)	40 (76.92)
	F_4						
1.	RMG 344 x Pusa vishal	19				7 (36.84)	12 (63.16)
2.	RMG 344 x RMG 62	6				2 (33.33)	4 (66.67)
3.	RMG 344 x ML 5	8				2 (25.00)	6 (75.00)
4.	RMG 344 x PDM 11	10				4 (40.00)	6 (60.00)
5.	MUM 2 x ML 5	11			3 (27.27)	6 (54.54)	2 (18.18)
6.	Pusa vishal x PDM 54	24				8 (33.33)	16 (66.67)

7.	PDM 11 x SML 589	16		6 (37.50)	6 (37.50)	4 (25.00)	
	Total F_4	94	0	6 (6.38)	9 (9.57)	33 (35.11)	46 (48.94)

REFERENCES

1. Faruqui O., Ajay Kumar Yadav and Naresh Chandra 2003. Mungbean evaluation for resistance to MYMV. National Symposium on "Pulses for crop Diversification and National Resource Management". held at IIPR, Kanpur, Dec.20-22, 2003. Abs. 245.
2. Jhamaria S.L., O.P. Khedar, R.V. Singh and Mahesh Shrimali 2003. Locating Resistant Source against Yellow Mosaic Virus Disease of Mungbean. 25th Annual Conference and Symposium of Mycology and Plant Pathology. October 8-10, 2003 at Agricultural Research Station, Rajasthan Agricultural University, Durgapura, Jaipur. J. Mycol. Pl. Pathol., Vol. No. 3: 500.

Reaction of Indian bean cultivars against pod borer insect pest in south Gujarat

S.B. Shewale, N.C. Desai, A.V. Gaike., and J.R. Patil

Department of Agricultural Botany, Navsari Agricultural University, Navsari-396450

ABSTRACT

The incidence of Indian bean pod borer (*Helicoverpa armigera Hubner*) started from October and reached to a peak during December. Out of 60 genotypes of Indian bean the genotypes NV 74, NV 206, NV 2005, NW 145, NW sweet white, NW 30 and NW 29/1 74, NV 206, NV 2005, NW 145, NW sweet white, NW 30 and NW 29/1 recorded the lowest population of pod borer and also recorded the lowest pod borer damaged.

Key words : Indian bean, Helicoverpa armigera Hubner

INTRODUCTION

Indian bean (*Dolichos lablab* L.) is one of the important commercial crops widely grown in dry, warm arid regions of South Gujarat. Among the insect pests such as aphids, bug, pod borer caused by *Helicoverpa armigera* Hubner is potentially the greatest insect pests hazard to Indian bean. The pod borer *Helicovera armigera* (Hubner), a major pest of Indian bean is one of the constraints in stepping up the production (1). The Indian bean has its own importance as tender immature green pods, cooked as vegetable alone. Dry seeds are used as pulse purpose. The foliage of Indian bean is used as hay, silage and green manure and it is a soil improver crop. In Gujarat, Indian bean is most important crop grown particularly in Navsari, Surat and Valsad districts.

For the control of insect pests by use of insecticides is one of the common and quick methods. Large number of newer chemicals are introduced in the market, to evaluate the bio-efficacy of these chemicals is need to study against major insect pests of Indian bean are to be studied under field conditions.

MATERIALS AND METHODS

Sixty cultivars of Indian bean were sown in RBD with three replications in *rabi* season at the Regional Pulse Research Station, Navsari Agricultural University, Navsari. Row to row spacing was 60 cm, while plant to plant distance was maintained at 20cm. Field experiment was conducted during *rabi* season in the year 2006. To find out the incidence of major insect pests on Indian bean genotypes, five plants were randomly selected and tagged from the entire area. Varietal screening of pod borers were recorded at weekly interval from tagged plants starting from one week after sowing till the harvest of the crop. The observations were recorded from each genotype of five selected and tagged plants. Number of larvae of pod borer were recorded.

Percentage of pod damage was worked out for pod borers at each picking. For this purpose pods of each treatment were harvested separately from all five plants and hundred pods selected at randomly were observed for pod damage (%) assessment. The pods with bored holes made by the larvae were considered as damage caused by pod borer complex.

RESULTS AND DISCUSSION

The data presented in table 1 indicated that none of the genotypes were found free from incidence of pod borers. However, the mean number of pod borer larvae/plant genotypes NV 74, NV 206, NV 2005, NW 145, NW sweet white, NW 30, NW 29/1 and NC 2 recorded lower number of pod borer larva (1.00 larva/plant) and it was at par with all other genotypes, respectively.

Table 1 : Mean population and per cent damage of pod borer on different genotypes of Indian bean

Sr. No.	Genotypes	Mean no. of pod pod borer larvae/plant	Per cent pod damage pod borer complex
	NV 38	1.03	18.00 (25.02)
	NV 74	1.00	12.17 (22.73)
	NV 104	1.09	25.07 (30.05)
	NV 2003/6	1.07	23.06 (28.92)
	NV 206	1.00	11.90 (20.14)
	NV 37	1.08	27.15 (31.40)
	NV 37/1	1.04	17.01 (24.31)
	NV 73	1.06	21.8 (27.83)
	NV 103	1.04	20.94 (27.23)

NV 91	1.02	17.61 (24.80)
NV 40	1.03	19.63 (24.27)
NV 2003	1.05	23.61 (29.10)
NV 72	1.03	18.21 (25.25)
NV 30	1.09	27.11 (31.36)
NV 81	1.09	31.03 (33.85)
NV 57	1.01	14.8 (22.62)
NV 2005	1.00	12.82 (20.97)
NV 101	1.05	20.02 (26.57)
NV 113	1.04	19.83 (26.43)
NV 6	1.07	24.24 (26.24)
NV 45/1	1.01	13.95 (21.93)
NV 2004	1.05	20.27 (12.05)
NV 55	1.05	20.99 (27.26)
NV 2002	1.06	23.34 (28.88)
NV 13/2	1.01	13.98 (21.95)
NV 33/1	1.01	13.77 (21.77)
NV 53	1.04	19.37 (26.10)
NW 69	1.04	20.69 (27.03)
NW 71	1.04	22.63 (28.39)
NW 145	1.00	12.10 (20.35)
NW sweet white	1.00	11.06 (19.42)
NW 30	1.00	12.39 (20.60)
NW 51	1.08	26.26 (30.83)
NW 68	1.08	27.42 (31.57)
NW 58	1.07	24.75 (30.83)
NW 2003/1	1.05	20.97 (27.25)
NW NC/1	1.06	22.25 (28.14)
NW 155	1.04	18.74 (25.65)
NW 67	1.05	19.53 (26.22)
NW 113/11	1.08	25.61 (30.42)
NW 156	1.01	14.24 (22.17)
NW 29/1	1.00	13.37 (21.44)
NW 49	1.05	20.42 (26.86)

NW 66	1.05	21.45 (27.60)
NW 173/1	1.06	22.45 (28.30)
NW 2004/m4	1.06	22.62 (28.39)
NW 48	1.07	26.08 (30.71)
NW 196	1.04	17.72 (24.88)
NC 2	1.00	12.30 (20.50)
NW 2004/m1	1.01	13.52 (21.57)
NW 20-4-5	1.01	13.31 (21.39)
NW 3	1.03	16.9 (24.27)
NW 80/1	1.07	28.43 (32.22)
NW 64	1.04	17.71 (24.88)
NW 82/2	1.05	19.03 (25.85)
NW 45	1.05	19.75 (26.37)
NW N/173	1.05	19.86 (26.46)
NW 109	1.05	19.44 (26.15)
NW N/108	1.06	20.53 (26.93)
NW 61	1.06	20.70 (27.10)
S.Em. ±	0.015	-
C.D. at 5%	-	-
C.V. %	2.48	-

Figures in parenthesis are retransformed values, these outside parentheses are $\sqrt{x + 0.5}$ transformed value and Arc sine transformed value

The per cent damaged of pod borer presented in table 1 indicated that none of the genotypes was free from attack of pod borer. However, lower pod damage was recorded on NW sweet white (11.06%) and it was at par with genotypes NV 74, NV 206, NV 57, NV 2005, NV 45/1, NV 31/2, NV 33/1, NW 145, NW 30 and NW 156, NW 29/1, NC 2, NW 2004m1 and NW 20-4-5 recording 12.17, 11.90, 14.8, 12.82, 13.95, 13.98, 13.77, 12.10, 12.39, 14.24, 13.37, 13.52 and 13.31 per cent pod damage, respectively. Maximum pod damage was recorded in NV 81 (31.03%) which was at par with NV 37, NV 30, NW 68 and NW 80/1, respectively. Similar findings was observed by (2) for pod borer complex.

Conclusively, the genotypes NV 74, NV 206, NV 2005, NW 145, NW sweet white, NW 30 and NW 29/1 recorded lower number of pod borer larvas (1.00 larva /plant, respectively).

REFERENCES

1. Vyas, N. C. and Kumar, A. (2005). Reaction of cluster bean cultivars against bacterial blight in Arid Rajasthan . *J. Arid legumes.,* **2** (2): 412-413.
2. Reddy, V. S. (2005). Population dynamics, varieties screening and bio-efficacy of newer insecticides against pest complex of Indian bean and biology of *Clavigralla gibbosa* spinala on Indian bean and pigeonpea. Thesis submitted to NAU, Navsari.

Population dynamics of *Poecilocerus pictus* (Fabr.) a pest on summer mung bean crop in arid region of Rajasthan

Shradha Katoch and F.S Poonia

Department of Zoology, Jai Narain Vyas University,
Jodhpur – 342 005, Rajasthan, India
E-mail: fspoonia_jnvu@yahoo.co.in

ABSTRACT

The Aak grasshopper, *Poecilocerus pictus* is a polyphagous pest and found to occur in all the crops and grasses grown among rainy and summer season in the Jodhpur district. Mungbean is attacked by more than 26 insect pests right from the date of sowing till harvest. Out of all major and minor insect pests the Aak grasshopper is the most important pest of the summer mungbean. This pest causing considerably damage both to the growth and yield of the crop. Therefore, for finding out an approximate time for conducting control measures of this pest in the Agro climatic condition prevailing around Jodhpur is assented. In the present study the investigations were carried out on the field population of pest at the farmer's field of Pal village near Jodhpur. Observations on summer mungbean crop were started when crop was 1 month old. The experiments were replicated 5 times with plot size and row-to-row distance being 5m X 5m and 15 cm respectively. In each experimental crop 10 plants were randomly selected and tagged in order to record all the observation. The record of grasshopper population was started for 2nd week of April to 1st week of May. Only adults stages were taken into consideration for recording pest population grasshopper population /50 plants were recorded at weekly intervals along with weekly average of maximum and minimum temperature and mean relative humidity. The data showed that the population of grasshopper remains low, ranging from 0-5 grasshopper/50 plants up to the middle of April. With slight increase in atmosphere increase and the maximum activities of this pest was recorded during last week of April. After 1st week of May the population started declining also might be due to slight rise in atmospheric temperature also due the maturity of crop. The above mention study indicated that grasshopper occurring during summer season of mungbean crop preferred moderate temperature and high humidity.

Key words: Population dynamics, poecilocerus pictus, mung bean, abiotic factor.

INTRODUCTION

The Aak grasshopper *Poecilocerus pictus* is a polyphagous pest and found to occur in all the crops and grasses grown among rainy and summer season in the Jodhpur District (1). Mungbean is attacked by more than 26 insect pest right form the date of sowing till harvest. Out of all major and minor insect pests the Aak grashopper is the most important pest of the summer mungbean. This pest causing considerable damage both to the growth and yield of the crop. The Aak Grasshopper has been recorded as a pest of agricultural and horticultural crops in the plains of India (2, 3, 4, 5 and 6).

Aak grasshopper represents the major group of insect pest in India. Grasshopper consume considerable amount of foliage during the nymphal development stage and as adults when they are abundant. They can damage economically important crops. Aak grasshopper requires warm, sunny conditions for optimal growth and reproduction. Food and optimal weather condition also facilitate growth and development of Aak grasshopper. Aak grasshopper is a very old denizen of grassland ecosystem. Grasshopper is an ecological role of providing food for wild up stimulating plant growth, creating plant liter for the soil and cycling elements and nutrients was developed as a functional part of whole ecosystem. Land managers should view grasshopper as a pest only when the insect increased the density that are clearly damaging to the rangeland plant cover and ecosystem.

The life cycle of 'AK' grasshopper comprises of 3 stages Egg, hopper and adult. In recent years due to advent of Indira Gandhi Canal and exploration of ground water; irrigation facilities has been improved in Jodhpur district, this provides constant vegetation cover to surface grasshopper for multiply and thus increase in its population uninterruptedly.

A long hot summer ensures a plentiful food supply and encourages early maturity of grasshopper and a long egg lying periods. It sheds its exoskeleton several times s it moves from one nymphal stage called an instar to another. The best time to control the insect pest is during mymphal development when it is vulnerable to disease, parasite, and predators and in appropriate weather. Grasshoppers are as such most active from July to October by sometimes according to the climatic conditions they are seen in early season (April-June). The Lab studies show that choice of food plant by *P. pictus* depends on the plant *sps.* available, the grasshopper intrinsic likes and dislikes and whether it has been previously starved.

The *P. pictus* breeds abundantly into areas where the host plant was available. When the host plant is defoliated, the population then migrated to cultivated crops as summer mungbean where it cause severe damage to growth and the yield. Mungbean (*Vigna radiata* (L). Wilczek) is one of the leading pulse crops in India and occupying a puvatol position in meeting the protein need of people. Summer, mung is the important pulse crop which is severely damaged by *P. pictus*. The population build up of any insect is very intimate associated with the weather parameters prevailing during preceedingly and corresponding periods. The pest status doesn't remain static through out the year but changes accordingly on the bases of abiotic factors out the year but changes accordingly on the bases of abiotic factors like Temperature, Relative humidity, Rainfall, Photoperiod etc.

The abiotic factors play substantial role in the biology of any pest of which temperature is the most crucial abiotic factor influencing the live economy of any organism. It is rather difficult to find the direct relationship between any single abiotic factor and pest activity because of the impact of weather parameters on pest is usually confounded. However, temperature, sunshine, rainfall, relative humidity and wind speed are the cheap weather parameters that largely direct the activity of given species of pest. The information is in time helpful in preparing productive models for forecasting the pest species. No information is available regarding the population dynamics of *P.pictus* a pest on summer mung bean crop in arid region of Rajasthan. Except for a brief report by (7, 8) against jassid and pod borer of mungbean and cowpea crops.

Keeping in mind the importance of population dynamics of Aak grasshopper on summer mung bean crop was carried out at farmer's fields at Pal village, Jodhpur.

MATERIAL AND METHODS

Topography of Jodhpur is undulating and rocky with isolated rhyolite and sandstone hills. Geologically the area is composed of Vindhyan sandstone and limestone resting on malani rhyolite and Aravalli state. On the basis of annual temperature, relative humidity and rainfall. Jodhpur has four distinct season *viz.* winter, summer, monsoon and autumn (post monsoon).

Mean, maximum and minimum value of abiotic factors during the experimental period was taken from the Agrometeorological observatory of CAZRI, Jodhpur. As the laboratory condition was inadequate for carrying out the activities so the field experiment was conducted at the farmer's field at village Pal near Jodhpur on the three varieties of mungbean mainly K-851, RMG-62 and RMG-868 during summer 2007.

To study the population dynamics of *P.Pictus* on three varieties by regular intervals the observations on the summer mung bean crop which was sown in the last week of March in the field. The experiments were replicated 3 times with plot size and row to distance being 10m x 10m x 30 cm respectively. In each plot, 10 plants of mung bean were randomly selected and tagged in order to record all the observation on them only. The record of Aak Grasshopper population was started from the first week of April to the last week of May '07. Only the adult stages were taken into consideration for recording the pest population. Aak grasshopper populations per 10 plants were recorded at weekly intervals along with weekly average of maximum and minimum temperature, relative humidity and average rainfall.

In order to study the influence of abiotic factors on population building simple correlation study was carried out between incidence of these insect pest and weather parameters.

RESULT AND DISCUSSION

In the present study the investigations were carried out on three varieties of summer mung bean in relation to various abiotic factors namely temperature, relative humidity and rainfall. The data on seasonal abundance & population fluctuation of 'AK' grasshopper from the first week of April '07 to the last week of May '07 is shown in above Table. 1. It can be inferred that the pest remained active April to May (Table 1). The result also demonstrate an apparent impact of climate conditions specially the minimum and maximum temperature ranged form 18.6°C to 42.4°C and relative humidity 7 to 70% respectively during the course of these investigations.

Table 1. Population dynamics of 'AK' Grasshopper in relation to ecological parameters of summer mung bean verities (K-851, RMG-62, RMG-868)

Day Interval	Population of summer mung verities			Average Temperature (°C)		Average Relative Humidity (%)		Rain fall (mm)
April '07	K-851	RMG-62	RMG 868	Max.	Min	Max	Min	
01-07	0.00	00.00	00.00	32.8	19.3	62.4	38.2	3.2
08-14	5.01	2.10	2.00	38.3	23.9	46.8	15	1.7
15-21	24.80	16.40	18.00	38.5	22.8	36.5	12.4	0.0
22-28	36.00	26.00	25.18	40.3	24.5	28.7	7	0.0
May '07								
29-5	22.10	18.00	19.81	42.4	26.5	46.2	18.2	0.0

06-12	11.08	10.00	11.11	40.0	23.8	70	35.4	8.9
13-19	8.00	5.60	4.80	38.32	26.8	69.4	35.14	1.0
20-26	5.11	2.8	1.8	39.8	22.8	65	36.14	0.0
27-31	0.01	0.0	0.0	36.5	18.6	62.6	32.2	4.3

The data summarized in figure 3 shows that the Aak grasshopper population in all the three verities was maximum on 22-28 week of April when average maximum and minimum temperature form 24.5 to 40.3°C and mean relative humidity form 7 to 29%. With the slight decrease in atmospheric temperature and increase in relative humidity due to the onset of monsoon rains. After the last week of May the Aak grasshopper population showing a slight decrease and the minimum activity of the Aak grasshopper was recorded form the 20th May to 31st May 2007. The Aak grasshopper population declined to 0 on all the three varieties of mung bean namely K-851, RMG-62, RMG-868 from 27 to 31st May 2007 respectively.

The Aak grasshopper population as observed on the mungbean on 3 (three) different varieties (K-851, RMG-62, RMG-868) reveal that the general trend of the population is similar in all the three varieties showing first the period of slow increase during ascending phase followed by a more rapid increase leading to a peak and then a sudden and steep decline under descending phase.

The above mention results indicated that 'AK' grasshopper occurring during summer season on mung bean crop preferred moderate temperature and high humidity to carry out the activities of its survival and quick multiplication as shown in Fig. 1 and 2.

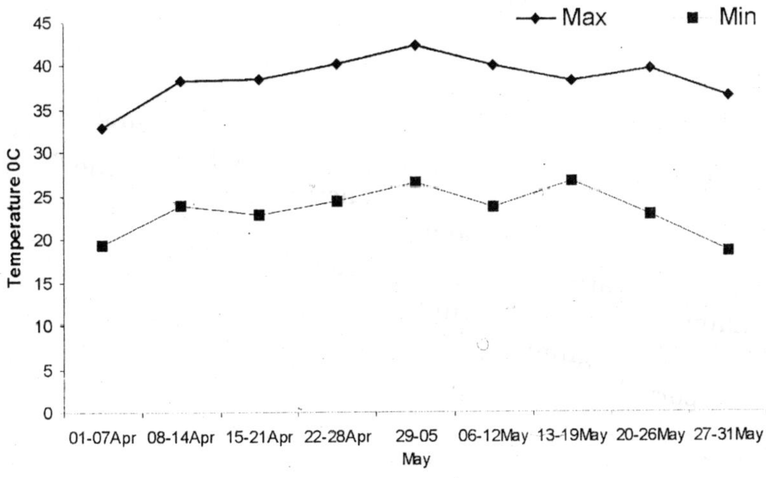

Fig. 1.

None of the abiotic factors singly was responsible for the multiplication and growth of the *P. pictus* and consequently its incidence on Mung bean crop. In nature, no environmental factor operates in isolation. All the factors operate in combination with others. The low temperature, high humidity and low rainfall during ascending phase of grasshopper population provided a very congenial condition for rapid build up and multiplication of grasshopper. The dry weather conditions accompanied by existing high temperature in 1st week of May proved fatal for the Aak grasshopper and set in a decline in its population. With further lower temperature but high relative humidity, the population started declining gradually.

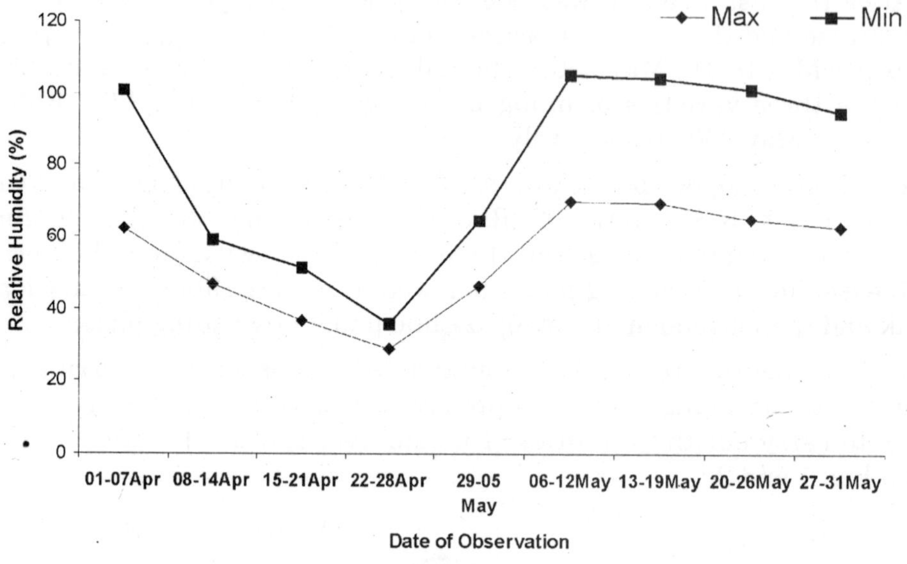

Fig. 2.

Based on the present results the three varieties of mung bean are arranged in their decreasing order of their resistance of Aak grasshopper K-851>RMG-62>RMG-868. The present results are in accordance with the findings of (7, 8) in case of jassids pest on cowpea and mung bean crop.

Similarly observations were reported earlier by (9) who recorded almost similar building of population in case of common cotton jassid during the end of summer and the beginning or rainy season at Ludhiana. Saini and Chhabra, 1968 (10) reported the jassid population on eight castor bean varieties and found similar relation of jassid population with meteorological factors. This indicates that most of the castor jassid occupying during summer and at the beginning of rainy season prefers moderate temperature, high humidity and occasional rainfall as similarly shown in results. Our findings concludes that Aak grasshopper population occurring during summer mung bean crop on three different varieties prefer moderate temperature, high humidity and uniformly distributed rainfall from April-May '07 for the quick multiplication for damaging the crop.

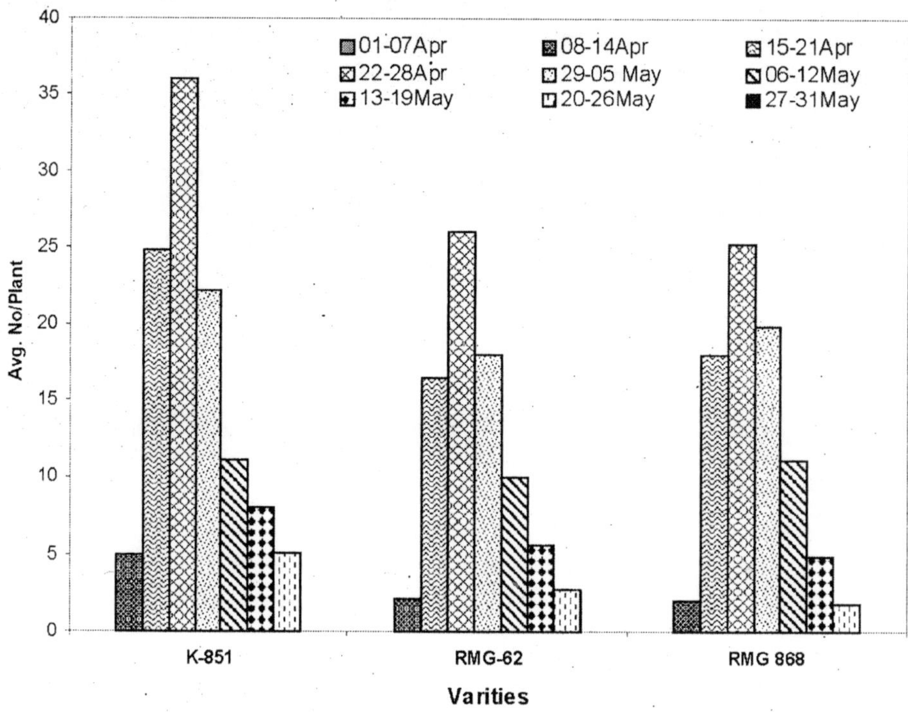

Fig 3. Population of *Poecilocerus pictus* on three mung bean varities

References

1. Parihar, D.R. (1974). Some observation on the life history of 'Ak' grasshopper, *P. Pictus* (Acridoidea: Pyrgomorphidae) at Jodhpur, Raj. *J. Zool. Soc*, India **26** (1&2): 99-129.
2. Bindra, O.S. (1958). Food preferences of *P.Pictus* (Faber)-A pest of some horticultural plants at Gwalior. *Indian J. Horti. Sabour* **15**, 80-86.
3. Khan, R.M. and Sharma, S.K. (1971): Unusual population of 'Ak' grasshopper. *Ento. News letter* **5**: 34
4. Butani, D.K. (1975): Insect pests of fruit crops and their control. *Pesticides* **9** (2), 23-24, **12**, 39-43.
5. Rai, P.S. and Nagesha Chandra, B.K. (1973): Note on grapevine as a new host of yakka grasshopper. *Poecilocerus Pictus* (Orth.: Acrididae). *Indian J. Agri. Sci.* **43**: 896-897.
6. Khurana, A.D. (1975): Alternative host plant of aak grasshopper, *Poecilocerus Pictus* (Fabr.). *Ento. News letter* **5**: 34

7. Poonia, F.S. (2005): Population dynamics of *Empoasca Kerri* pruthi, A jassid pest on cowpea in arid regions of Rajasthan. *J. Arid Legumes* **2***(1)*: 179-180.

8. Poonia, F.S. (2005): Influence of abiotic factors against jassid and pod borer of mung bean (*Vigna radiata L.*) in arid regions of Rajasthan. *J. Arid Legumes* **2***(1)*: 181-182.

9. Atwal, A.S., J.P. Chaudhary and B.S Sohi (1969): Effect of temperature and humidity on the development and population of cotton jassid, *Empoasca devastans. J. Res* (P.A.U, Ludhiana), **6,** 255-261.

10. Saini, M.L and K.C. Chhabra (1968): A technique for evaluating varietal resistance of castor-to-castor jassid, *Empoasca flavascens Fabricius. J. Res.* (P.A.U, Ludhiana). **1**, 72-74.

Progression of termite infestation in clusterbean field under arid conditions

M.P. Singh and A. Henry

Central Arid Zone Research Institute, Jodhpur 342 003

ABSTRACT

The initiation and spread of termite infestation in a large clusterbean field were monitored during kharif, 2007. Termite infestation began in the second week of August and progressed with passage of time in the untreated plots. The initial mean infestation was level increased three fold in a span of five weeks. The lateral movement of termites did not follow any definite pattern. The infestation was independent of total number of plants, but in percentage terms, lesser infestation was observed in densely populated fields. Positive correlation existed between termite infestation and temperature, sunshine hours and evaporation, while relative humidity and wind speed were found to bear a negative correlation.

Key words: Termites, progression, clusterbean field.

Termites are major pests of the arid region affecting the crop, grass and woody components of the production systems. The severity of infestation is influenced by the availability of moisture in the subsurface. The rainfed crops encounter greater termite infestation, especially under water stress conditions induced due to lack of rains or gap during the showers. Clusterbean being a relatively longer kharif crop faces the risk of termite injury. Across the field application of pesticide dust in soil as a preventive measure adds to cost of cultivation in arid regions, where productivity levels are low. Early detection of infestation and the knowledge about the pattern of its spread may help in curbing the injury through spot treatment. With this view, the initiation and progression of termite infestation in clusterbean field were monitored at the Central Research Farm of the Central Arid Zone Research Institute, Jodhpur during the kharif season of 2007.

MATERIAL AND METHODS

The crop of clusterbean variety RGM 112 was sown in the first week of July. No plant protection measures were applied in the area demarcated for the studies. Observations were recorded about the total number of plants and the termite-infested plants in clusters of six adjacent untreated plots of 6x5 m size. Each cluster of 180-m^2 area was widely separated from the other to delineate overlap due to lateral movement of the termites. The experiment was replicated four times, each replication represented by a cluster of six plots. The counts were made on three dates extending over a period of five weeks from the first detection of infestation.

The pattern of progression of infestation was studied by recording the direction of spread of infestation with respect to the first infested plant in one plot earmarked for the purpose. The position of infested plants within and between the rows and within this plot was mapped on completion of observations.

Data on the extent of infestation on each date of observation were correlated with the weather data for the corresponding dates obtained from the meteorological observatory of the Institute. Correlation was also worked out between plant density and extent of infestation on per plot basis.

RESULTS AND DISCUSSION

Termite infestation began in the second week of August and progressed with passage of time. A mean 7.6% of the plant population among the observed plots exhibited symptoms of termite attack in mid August. A fortnight later, an average 15.4% of the clusterbean plants in the experimental blocks were affected by termites. Three weeks from this point of time, the cumulative termite infestation rose to 21.8% of the total initial plant population of all the twenty-four plots (Table 1).

Table 1. The pattern of termite infestation in clusterbean field

Date	% Termite infestation* (Range)				Mean** % infestation
	Cluster 1	Cluster 2	Cluster 3	Cluster 4	
13.8.07	5.1 (4.1-6.4)	6.3 (2.1-10.1)	10.2 (7.6-10.9)	9.0 (4.0-16.6)	7.6
29.8.07	15.1 (8.3-18.7)	13.1 (8.1-15.5)	17.7 (12.6-21.0)	15.7 (9.3-25.7)	15.4
20.9.07	19.5 (14.2-23.6)	18.6 (13.9-23.7)	26.0 (19.9-33.3)	23.2 (15.2-36.5)	21.8
% Increase	14.4	12.3	15.8	14.2	14.2
% Variation	19.5	21.6	23.1	32.5	

* Mean of 6 plots, **mean of 24 plots

Whereas the level of infestation increased over time in all the clusters, the surge in terms of percentage increase in infestation was steeper in cluster 3, where there was high concentration of termites initially (10.2%). However, despite low initial termite activity in cluster 1 (5.1%), there was over 3.8 fold increase in level of infestation (19.5%) during the five-week interval, being the highest among all. Variation in extent of infestation among different plots within the cluster for the entire period of observations was maximum (32.5%) in cluster 4, even though the mean percentage of termite infestation in this cluster was lower than in cluster 3, indicating high termite activity in some plots under this cluster (Table 1). Assuming the four clusters together to represent fairly the termite activity across the field, an average initial infestation level of 7.6% had escalated to 21.8% in five weeks.

Observations on clusterbean plot demarcated for studying the progression of infestation revealed that the lateral movement of termites did not follow any definite pattern. The plants adjacent to infested ones did not exhibit infestation symptoms simultaneously, while distant plants were infested at the same time horizon. The current study finds support from those of Green *et al.* (1) who showed that termites do not always feed at a single area and will not necessarily concentrate at a single harborage. It has been shown that termite colonies distribute individuals to many different feeding locations within a given area (2). Abu Shama and Al Houty (3) and Haagsma and Rust (4) studied the infestation pattern of termites over time, but did not take into account the spread of termite infestation from a reference point in the field.

The extent of infestation was independent of the plant density in different plots, as denoted by small and insignificant correlation (r = 0.076) between total number of plants and the affected plants in each plot (Table 2). However in terms of affected plants as percentage of the plant population in different plots, a negative correlation (r = –0.559) was found existing, suggesting lesser infestation in densely populated fields. This may be due to greater availability of food resource in denser plots, which satiated the requirement of the infesting pest. Increasing plant population in field may thus serve to mitigate the pest injury to the remaining crop.

Table 2. Extent of termite infestation in different plots in clusterbean field

Total plants	Affected plants (%)	Total plants	Affected plants (%)	Total plants	Affected plants (%)	Total plants	Affected plants (%)
671	95 (14.2)	815	113 (13.9)	564	116 (20.6)	515	88 (17.1)
507	104 (20.5)	618	122 (19.7)	398	79 (19.9)	834	127 (15.2)
594	112 (18.9)	708	131 (18.5)	441	127 (28.8)	633	120 (18.9)

512	121 (23.6)	614	123 (20.0)	516	138 (26.7)	487	151 (31.0)
609	135 (22.2)	583	90 (15.4)	613	204 (33.3)	506	102 (20.2)
634	111 (17.5	493	117 (23.7)	419	113 (26.9)	490	179 (36.5)
r	0.076 (-0.559)						

The data presented in Table 3 reveal that a positive correlation existed between termite infestation and temperature. Sunshine hours exhibited similar relationship. Evaporation had positive but small correlation value. Relative humidity and wind speed were found to bear a negative correlation with mean termite infestation in the present studies. Rainfall events were few during the study period to exhibit any trend. Several studies have been conducted on the effect of weather conditions on the activity of termites, which are inconsistent with the present findings owing to the diverse conditions of experimentation and the parameters selected. Fei and Henderison (5) considered soil temperature instead of the ambient temperature as in present studies, but recorded similar effect on the foraging activities. Su and Puche (6) assigned high importance to moisture gradients in tunneling activity of subterranean termites in sandy soils.

Table 3. Correlation of termite infestation with different weather parameters

Date	Mean infestation	Max temp (ºC)	Min temp (ºC)	Relative humidity (%)	Rainfall (mm)	Wind speed (Km hr^{-1})	Sunshine hours	Evaporation (mm)
13.8.07	8.6	33.3	25.2	65	0	11.6	5.3	7.4
29.8.07	14.46	36	26.5	68	39.4	3.6	7.1	6.4
20.9.07	21.81	38.5	26.1	50	0	3.2	9.3	7.5
r	-	0.996	0.626	-0.816	-0.064	-0.854	0.999	0.146

REFERENCES

1. Green, J.M., Scharf, M.E. and Bennett, G.W. 2005. Impacts of soil moisture level on consumption and movement of three sympatric subterranean termites (Isoptera: Rhinotermitidae) in a laboratory assay. *Journal of Economic Entomology* 98: 933-937.
2. Thorne, B.L., Russek-Cohen, E., Forschler, B.T., Breisch, N.L. and Traniello, J.F.A. 1996. Evaluation of mark-release- recapture methods for estimating forager population size of subterranean termite colonies. *Environmental Entomology* 25: 938-951.
3. Abu Shama, F.T. and Al-Houty, W.A. 1988. The foraging activity of subterranean termites in the Kuwait desert. *Journal of Arid Environment* 14: 75-82.

4. Haagsma, K.A., and Rust, M.K. 1995. Colony size estimates, foraging trends, and physiological characteristics of the western subterranean termite (Isoptera: Rhinotermitidae). *Environmental Entomology* 24: 1520-1528.

5. Fei, H. and Henderison, G. 2004. Effects of temperature, directional aspects, light conditions, and termite species on subterranean termite activity (Isoptera: Rhinotermitidae). *Environmental Entomology* 33: 242-248.

6. Su, N.Y. and Puche, H. 2003. Tunneling activity of subterranean termites (Isoptera: Rhinotermitidae) in sand with moisture gradients. *Journal of Economic Entomology* 96: 88-93.

7. Vargo, E.L. 2004. Colony genetic organization and colony fusion in the termite *Reticulitermes flavipes* as revealed by foraging patterns over time and space. *Molecular Ecology* 13: 431-441.

Crop-weather disease interaction in *Cyamopsis tetragonoloba* L. (Taub.) in semi-arid environment

U.K. Chanderia and Reeti Singh

Department of Plant Pathology, JNKVV, College of Agriculture, Gwalior (MP)

ABSTRACT

Field experiments were conducted on the sandy loam soils of the research farm of the College of Agriculture, Gwalior (M.P.) during two Kharif seasons (2004-05 and 2005-06) to study the crop-weather disease dynamics in two *Cyamopsis tetragonoloba* cultivars HG-365 and local. The two cultivars were sown on 5 dates at weekly intervals starting from 13th July in the two years and the crops were grown following the standard recommended agronomic practices. This one month gap of sowing enable us to study the effect of same weather on different phenological events and also to study the effect of variable weather conditions on the same phenological events vis a vis the incidence and spread of Alternaria blight disease.

Key words: Cluster bean, crop-weather-disease interaction, semi-arid regions.

INTRODUCTION

Environmental factors play important role in the development of the disease, since they affect the pathogen, host or host pathogen interaction during pathogenesis. For disease prediction quantitative relationship between disease development and weather variables has been established in several host pathogen system. However magre information is available on the role of climatic factors on the development of Alternaria blight of clusterbean. (1, 2, 3). To overcome some of these problems, the present investigations were under taken to study the effect of sowing dates and weather factors on severity of the disease and to understand their practical utility in integrated disease management strategy of Alternaria blight.

MATERIALS AND METHODS

A field trial was conducted at JNKVV, College of Agriculture Farm, Gwalior with two clusterbean cultivars local and HG-365 during Kharif 2004-2005 and 2005-06. These were sown at weekly intervals on 5 dates starting from 22th June during 2004-05 and 2005-06 Kharif season following the recommended agronomic practices. Visual observations on the disease severity (percentage) were made at three days intervals following the scale given by Saharan (3) on 10 randomly selected and tagged plants. Data on daily maximum, minimum and temperatures, relative humidity and rainfall were taken from the Agrometeorological observatory of College of Agriculture, Gwalior.

RESULTS AND DISCUSSION

The maximum severity of Alternaria blight recorded in plants sown on these 5 dates in both the seasons were considered for the present study. It can be seen from the Fig. 1 and Fig. 2 that barring the first two sowings, the alternaria blight was more in the second season as compared to the first in HG-365. The same trend was also observed in case of local except that in local, only in the first two sown crop it was more in the first year. In both the varieties in the both seasons, the severity was found to be below 15 percent except on the 6th August sown crop where the severity was about 23 percent.

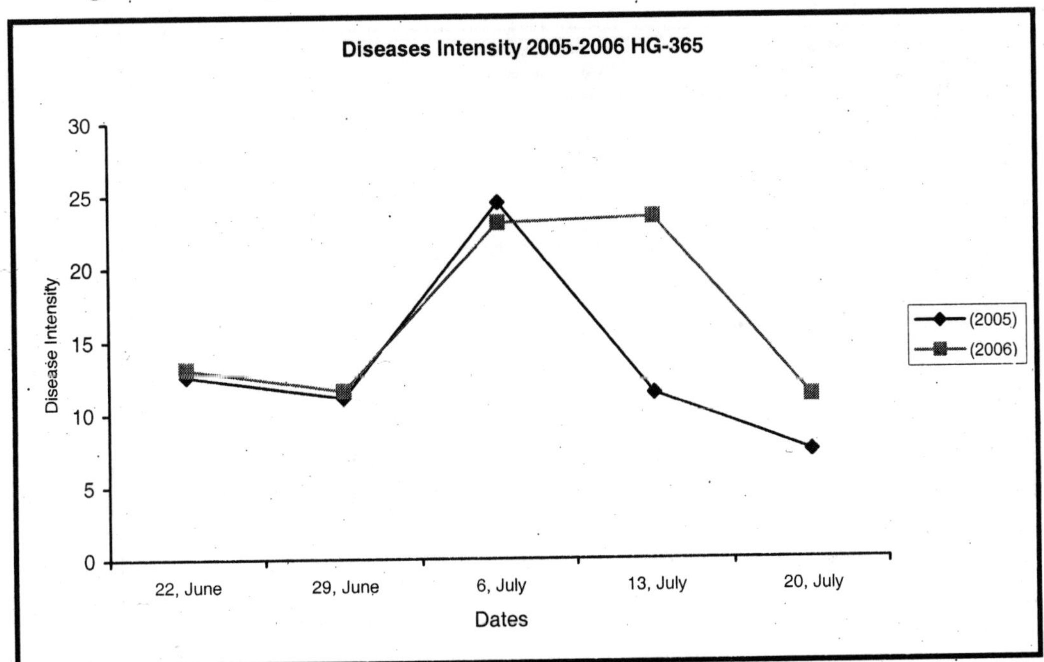

Fig. 1 : Diseases intensity in HG-365 (2005 and 2006)

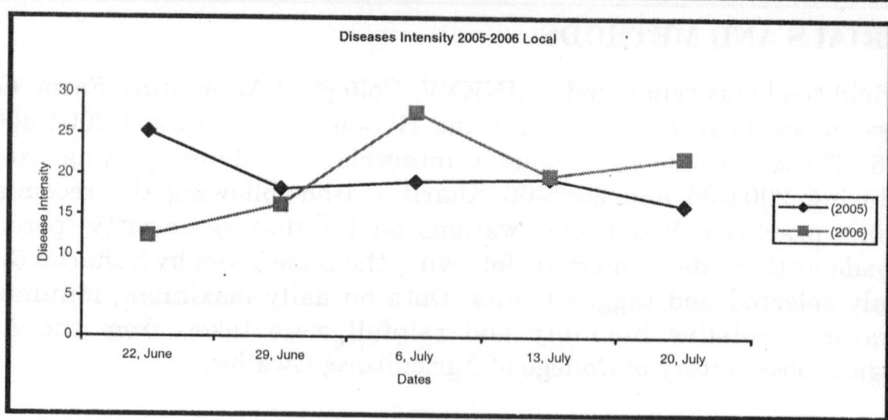

Fig. 2 : Diseases intensity in cultivar Local (2005 and 2006)

Weather

It is interesting to see that the maximum temperature during in the 3rd week of July in 2006 was found to be relatively lower than in the first season as can be seen in Fig. 3. The rainfall was found to be much more in this season as compared to the first season. The rainfall recorded in the second season was 32.3 m.m., while in the first year; it was much less nearly one third (10.8 mm) spread over six days.

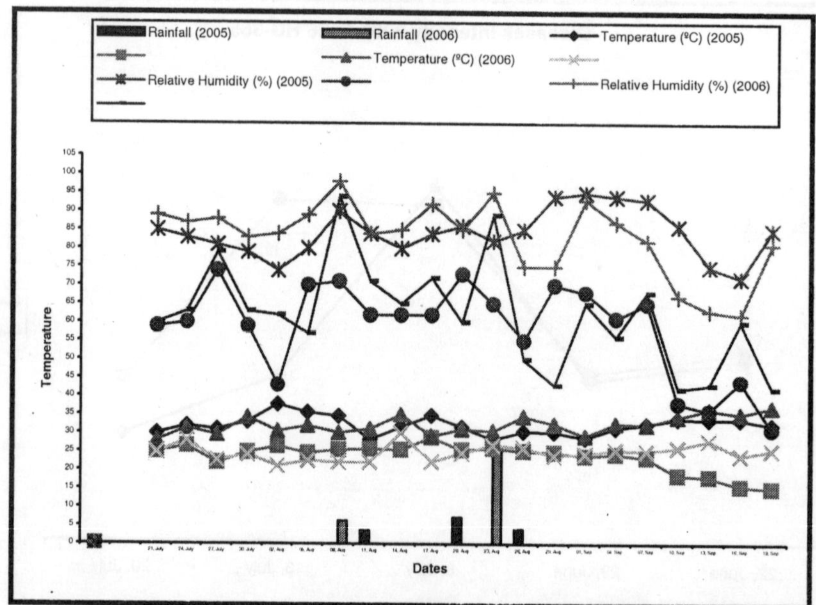

Fig. 3 : Maximum, minimum temperature, maximum, minimum RH and rainfall (2005 and 2006).

Crop Weather Disease Dynamics

It is readily seen that the disease was relatively higher in a season when the weather was humid with low maximum and minimum temperature along with higher rain fall. It is obavious that in the layer near the ground, the radiation was low coupled with humid, which might have facilitated the incidence and spread of the disease to a great extent in the second season as compared to the first crop season.

Thus the compound effect of all these weather parameters could be seen on the disease dynamics. Gupta *et al.* (1) found positive and significant correla-tion of disease index of Alternaria blight of clusterbean with temperature, relative humidity and rainfall, whereas with sushine hours it was negative and non-significant. While Saharan *et al.* (2) reported that minimum tempera-ture was significantly negatively correlated with the lesion size of Alternaria blight infected bean leaves. Chanderia (4) indicated that the cluster bean crop could be managed to be free from Alternaria blight if sown before 6th August in the Gwalior region.

REFERENCES

1. Gupta, P.P., Gupta, V., Singh, J.V. and Singh, V.P. (1998). Alternaria blight disease management strategy in clusterbean, *Proc. 7th Int. Cong. Pl. Pathol.,* Edinburgh, Scotland.
2. Saharan, M.S., Saharan, G.S. and Ram Niwas (2001). Progress of Alternaria blight on different varieties of cluster bean in relation to environmental conditions. *J. Mycol. Pl. Pathol.,* 31 : 220 – 226
3. Saharan, G.S. (1991). Assessment of loss, epidemiology and management of black spot disease of rapeseed mustard. GCIRC 8th Inst. Rapeseed Congr. July 9-11. Saskatoon, Canada 2(84) : 465 – 470.
4. Chanderia, U.K. (2008). Studies on Alternaria blight of Guar, Ph.D. Thesis. *Jiwaji University, Gwalior (MP).*

Field evaluation of certain newer molecules of insecticides against spotted pod borer, *Maruca vitrata* Fab. infesting cowpea

B.G. Prajapati, D.A. Dodia, S.B.S. Tikka and S.Acharya

Centre for Research on Seed Spices, S. D. Agricultural University, Jagudan (Dist. Mehsana)

ABSTRACT

A field experiment was taken up at Main Pulses Research Station, Sardarkrushinagar during *kharif* 2005 to 2007 using variety Gujarat Cowpea-4. Among them, the damage due to spotted pod borer *Maruca vitrata* Fab.was registered the least (2.45%) in profenophos 0.05 per cent and it was followed by indoxacarb 50g a.i. per ha (6.70%) and thiodicarb 0.04 per cent (8.94%). Highest grain yield of cowpea was harvested from the plots treated with profenophos 0.05 per cent (1026 kg/ha) followed by indoxacarb 50 g a.i.per ha (965 kg/ha), spinosad 50 g a.i. per ha (909 kg/ha) and thiodicarb 0.04 per cent (881 kg/ha). The highest cost benefit ratio was also registered in the plots sprayed with profenophos 0.05 per cent (ICBR = 1: 7.76). It was followed by monocrotophos 0.04 per cent (ICBR = 1: 4.47), lambda cyhalothrin 0.005 per cent (ICBR = 1:4.40) and thiodicarb 0.04 per cent (ICBR = 1: 2.95).

INTRODUCTION

Cowpea [*Vigna unguiculata* (L.) Walp.] is an important pulse crop suited to both arid and semi- arid regions. It is grown as sole crop, mix as well as inter crop in different cropping systems. As many as 21 insect pests have been found to be infested in cowpea crop right from seedling to maturity as well as storage conditions. Among them, sap feeding insects, foliage feeders, flower feeding insects, pod borers as well as root feeding insects are major limiting factor in the productivity of cowpea. Among different pod borers, spotted pod borer, *Maruca vitrata* Fab. is the most important pest of cowpea (1,2). It belongs to the family Pyralidae of order Lepidoptera. It is becoming a predominant pest of cowpea at flowering - podding stage and causes severe losses upto 60 per cent in the tropics and sub-tropics (3). Larvae feed on fowers, buds and pods by webbing them and this typical concealed feeding protects the larvae from natural enemies, human interventions or other adverse factors including insecticides(4). It also bores into the pods and feeds on the seeds. Owing to the larval feeding, there is an appreciable loss in seed yield of cowpea. With a view to manage the spotted pod borer, certain newer molecules as well as conventional insecticides were evaluated for their efficacy against *M. vitrata* in cowpea under field conditions and results are presented herein.

MATERIALS AND METHODS

A field experiment was undertaken during *kharif* 2005 to 2007 at Main Pulses Research Station, Sardarkrushinagar,Gujarat using variety Gujarat cowpea-4 in a randomized block design with ten treatments including untreated control replicated thrice. The size of each plot measured 4.00×1.35 m^2. Between rows and within rows distances were 45 cm and 10 cm, respectively. All the recommended agronomic practices were followed to raise the good crop of cowpea. The treatments consisted of profenophos 0.05 per cent, methomyl 0.05 per cent, indoxacarb 50 g a.i. per ha, spinosad 50 g a.i. per ha, monocrotophos 0.04 per cent, chlorpyriphos 50EC + cypermethrin 5EC @ 0.055 per cent, profenophos 40EC + cypermethrin 4 EC @ 0.044 per cent, lambda cyhalothrin 0.005 per cent and thiodicarb 0.04 per cent. Two foliar sprays by means of knapsack sprayer were given. First spray at the time of 50% flowering stage of the crop and the second spray was made at an interval of seven days the first spray. The number of pods infested by spotted pod borer was recorded and per cent pod borer damage was also worked out. These data were analyzed for its statistical interpretation with due transformation so as to compare the efficacy of different treatments. The grain yield from each net plot as well as economics of different treatments were also worked out.

RESULTS AND DISCUSSION

Per cent pod borer damage

Looking to the spotted pod borer damage, all the molecules of insecticides showed significantly superior over untreated control in all the three years as well as in pooled results (Table 1). It can be also evident from the pooled results that profenophos 0.05 per cent had exhibited the least pod borer damage (2.45%) and it was followed by indoxacarb 50g a.i. per ha (6.70%), thiodicarb 0.04 per cent (8.94%) and spinosad 50 g a.i.per ha (9.83%).

Table 1. Per cent spotted pod borer damage and grain yield of cowpea in different treatments

S. N.	Treatments	% spotted pod borer damage*				Grain yield (kg/ha)			
		2005	2006	2007	Pooled	2005	2006	2007	Pooled
1.	Profenophos 0.05%	7.62 (1.76)	11.69 (4.10)	7.71 (1.80)	9.01 (2.45)	439	1315	1328	1026
2.	Methomyl 0.05%	29.77 (24.6)	27.71 (21.62)	23.21 (15.53)	26.90 (20.47)	239	994	981	739
3.	Indoxacarb 50g a.i./ha	12.02 (4.34)	18.61 (10.18)	14.37 (6.16)	15.00 (6.70)	406	1228	1263	965
4.	Spinosad 50g a.i./ha	14.66 (6.40)	21.99 (14.02)	18.17 (9.72)	18.27 (9.83)	344	1154	1228	909
5.	Monocrotophos 0.04%	23.18 (15.49)	23.49 (15.89)	20.96 (12.79)	22.55 (14.71)	304	1115	1106	841
6.	Chlorpyriphos 50EC + cypermethrin 5EC@ 0.055%	23.26 (15.59)	18.51 (10.08)	14.57 (6.33)	19.44 (11.08)	300	1107	1096	835
7.	Profenophos 40EC + cypermethrin 4EC@ 0.044%	27.07 (20.71)	22.01 (14.04)	19.15 (10.76)	22.74 (14.94)	291	1083	1083	819
8.	Lambda cyhalothrin 0.005%	19.53 (11.17)	20.64 (12.42)	17.02 (8.57)	19.06 (10.66)	321	1167	1130	872
9.	Thiodicarb 0.04%	13.93 (5.79)	20.73 (12.53)	17.54 (9.08)	17.40 (8.94)	380	1124	1139	881
10.	Untreated control	33.39 (30.29)	33.72 (30.82)	30.20 (25.39)	32.46 (28.80)	198	920	815	644
	S.Em. ± T	1.64	1.83	2.54	1.65	19	37	37	24
	YXT	-	-	-	2.04	-	-	-	30
	C.D. at 5% T	4.86	5.43	7.54	4.90	56	93	93	74
	YXT	-	-	-	5.79	-	-	-	74
	C.V.%	13.74	14.45	24.01	17.40	10.41	4.91	5.19	5.86

* Aresin transformation Figures in the parenthesis are retransformed values

More or less similar observations on superiority of profenophos and thiodicarb against spotted pod borer, *M. vitrata* in cowpea were reported by Chanrayudu *et.al.*(5).Thus, the present findings corroborate the earlier reports.

Grain yield

The results pertaining to grain yield of cowpea showed that all the treatments were remained significantly superior over untreated control in all the three years as well as in pooled results (Table 1). Perusal of the pooled results indicated that profenophos 0.05 per cent gave the highest grain yield of cowpea (1026 kg/ha) and it was followed by indoxacarb 50 g a.i.per ha (965 kg/ha), spinosad 50 g a.i. per ha (909 kg/ha) and thiodicarb 0.04 per cent (881 kg/ha).

Economics

The highest cost benefit ratio (Table 2) was observed in the treatment of profenophos 0.05 per cent (ICBR = 1: 7.76). It was followed by monocrotophos 0.04 per cent (ICBR = 1: 4.47), lambda cyhalothrin 0.005 per cent (ICBR = 1:4.40) and thiodicarb 0.04 per cent (ICBR = 1: 2.95).

Table 2. Economics of different treatments

Sr. No.	Treatments	Qty. of insect-icide (L/Kg/ha)	Price of insec-ticide (Rs/ ha)	Labour cost (Rs/ ha)	Total cost of treat-ment (Rs/ha)	Yield (Rs/ha)	Gross realiz-ation (Rs/ha)	Net realiz-ation over control (Rs/ha)	Net gain (Rs./ ha)	ICBR
1.	Profenophos 0.05%	1.000	552	320	872	1026	20520	7640	6768	1:7.76
2.	Methomyl 0.05%	1.250	1550	320	1870	739	14780	1900	30	1: 0.02
3.	Indoxacarb 50g a.i./ha	0.690	2380	320	2700	965	19300	6420	3720	1:1.38
4.	Spinosad 50g a.i./ha	0.222	2442	320	2762	909	18180	5300	2538	1:0.92
5.	Monocrotophos 0.04%	1.100	400	320	720	841	16820	3940	3220	1:4.47
6.	Chlorpyriphos 50EC + cypermethrin 5EC@ 0.055%	1.000	650	320	970	835	16700	3820	2850	1:2.94
7.	Profenophos 40EC + cypermethrin 4EC@ 0.044%	1.000	675	320	995	819	16380	3500	2505	1:2.52
8.	Lambda cyhalothrin 0.005%	1.000	525	320	845	872	17440	4560	3715	1:4.40
9.	Thiodicarb 0.04%	0.550	880	320	1200	881	17620	4740	3540	1:2.95
10.	Untreated control	-	-	-	-	644	12880	-	-	-

Price of cowpea: Rs. 20/kg Labour charges: Rs. 80/day

Based on efficacy and economic viability, two foliar sprays of either profenophos 0.05 per cent (ICBR = 1 : 7.76) or thiodicarb 0.04 per cent (ICBR = 1 : 2.95) can be recommended in cowpea crop, first foliar spray made at 50% flowering stage and the second spray at 7days of the first spray for effective and economical management of spotted pod borer, *M vitrata*.

REFERENCES

1. Nair, M.R.G.K. 1975. Insect and mites of crops in India, ICAR, New Delhi, India. 404pp.
2. Srivastava, B.K. 1964. Pests of pulse crops. In Pant N.C. (Ed.) Entomology in India. Journal of Entomology, Silvar Jublilee Publication of Entomological Society of India. New Delhi. pp. 83 - 91.

3. Singh, S.R. and Van Emden, H. F. 1979. Insect pest of grain legumes. *Ann. Rev. Entomol.*, **14** : 255-78.
4. Sharma, H.C. (1998). Bionomics, host plant resistance and management of legume pod borer, *M vitrata* – A review. *Crop Protection* **17** : 373 - 386.
5. Chanrayudu, E., Srinivasan, S. and Venugopal Rao, N. 2006. Evaluation of certain new insecticides against spotted pod borer, *Maruca vitrata* in cowpea. *Indian· J. Plant Prot.* **34** (1): 118 -119

Legumes resistant starch as a functional food

A.B. Rodge

Department of Food Chemistry and Nutrition, Marathwada Agricultural University, Parbhani – 431 402

ABSTRACT

Food grain legumes are rich and less expensive source of dietary proteins. Legumes often fall below the popular grains and fruits and vegetables on the list of foods that people consume to meet the recommended intake of daily dietary fiber. Researchers are hoping to motivate greater consumption of legumes. Many researchers reported that legumes contain substantially higher levels of resistant starch than do cereal grains, flours and grain-based food products.

Resistant starches (RS) have drawn broad interest worldwide for both their functional properties and potential health benefits. Initial clinical studies demonstrated that resistant starches have properties similar to dietary fiber and promising physiological benefits in humans, which may result in prevention of various diseases. There are four types of resistant starches as based on the mechanism that contributes to their indigestibility are RS_1, RS_2, RS_3, and RS_4. Legumes come under RS_1 starches. RS_1 starch resists digestion because it is protected from digestive enzymes by other components normally present in a matrix of typical starch sources. Among the best-characterized forms of RS are those derived from legumes. Resistant starches provide many health benefits such as improving bowel health, blood lipid profile, increased feeling of satiety and micronutrient absorption (magnesium and calcium) in the colon. These factors may affect the risk of developing diseases such as colorectal cancer, cardiovascular diseases, osteoporosis and inflammatory bowel diseases. Resistant starch increases in faecal bulk are important in preventing diverticulosis constipation and hemorrhoids and in diluting potentially toxic compounds that might promote the formation of cancer cells. RS appears to affect lipid metabolism based on studies in rats where reduction in a number of measures of lipid metabolism are observed such as total lipids, total cholesterol, various lipoprotein fractions in blood. The major benefit of resistant starches in colon is modulation of flora and production of short chain fatty acids specifically propionate and butyrate's which lower the colonic PH and thus helps for maintenance of colonic integrity and prevention of colon cancer. RS may be of benefit to healthy individuals who are trying to achieve and maintain a healthy body weight. RS containing foods have low glycaemic index thus release glucose slow resulting in lowered insulin response and greater access to and use of stored fat. This helps in management of diabetes and impaired glucose tolerance. Hence it is used in the treatment of obesity.

Key words: Legumes, resistant starch, functional food.

INTRODUCTION

Resistant starch is generally considered to be represented by the total amount of starch and the products of starch degradation that are not digested in the small intestine and pass into the large intestine (or the colon). Therefore, resistant starch can be regarded as a component of dietary fibre. Legumes often fall below the popular grains and moisture-laden fruits and vegetables on the list of foods that people consume to try to meet the recommended intake of dietary fiber daily, about 25 g to 30 g. Now, researchers at the University of Illinois, are hoping to motivate greater consumption of legumes. They report in a recent study that legumes contain substantially higher levels of resistant starch than do cereal grains, flours and grain-based food products. The researchers say many legumes beans, lentils and peas should be on more plates.

Resistant starch, with a functionality of dietary fibre provides strengthful properties of low water holding capacity, small particle size, bland flavour with no energy and also exhibit several health benefits. Resistant starch is present mainly in cereals, tubers and its formation is influenced by processing the major ingredients i.e. starch, proteins, lipids and sugars. In the recent years Resistant starch (RS) has been the most sounded component in the functional properties of foods. Resistant starches are being examined for both their protential health benefits as well as functional properties to produce high quality foods. In simple terms, RS is defined as the starch fraction, which escapes enzymatic hydrolysis in the small intestine and gets fermented in the colon. Resistant starches (RS) have drawn broad interest world wide for both their potential health benefits and functional properties. Initial clinical studies demonstrated that resistant starches have properties similar to fiber and show promising physiological benefits in humans, which may results in prevention of various diseases.

Resistant starch (RS) can provide added values such as health benefits and fiber content to many processed foods without compromising taste and product quality. RS is defined as the total amount of starch and the products of starch degradation that resists digestion in the small intestine of healthy people, and the class is comprised of many chemically and physically distinct entities. Some forms of RS are present naturally in many foods and average global consumption is estimated at 3-10 g/day. Among the best characterized forms of RS are those derived from high-amylose maize (HAM). Animal and human studies confirm the health benefits and safety of HAM derived RS as a food ingredient. Legally, RS can be sold in most countries if it falls under current food classifications. However, approved fiber measurements vary through out the world and different methods are required to measure RS derived from specific sources. A particular RS can only be claimed as fiber on the nutrition label (and as a source for a high-fiber claim) if the method appropriate to that product is approved in the country of interest. Similarly each individual country's regulations must be consulted prior to developing health or other nutrition label claims, such as calorie content for RS-containing products, as these regulations vary widely around the world. Classification of RS is given in table 1.

Table 1. Classification of types of resistant starch (RS), food sources, and factors affecting their resistance to digestion in the colon (1).

Type of RS	Description	Food sources	Resistance minimized by
RS_1	Physically protected	Whole- or partly milled grains and seeds, legumes	Milling, chewing
RS_2	Ungelatinized resistant granules with type B crystallinity, slowly hydrolyzed by α- amylase	Raw potatoes, green bananas, some legumes, high amylose corn	Food processing and cooking
RS_3	Retrograded starch	Cooked and cooled potatoes, bread, cornflakes, food products with repeated moist heat treatment	Processing conditions
RS_4	Chemically modified starches due to cross-linking with chemical reagents	Foods in which modified starches have been used (for example, breads, cakes)	Less susceptible to digestibility in vitro

Goni *et al.* (2) classified the food materials according to the range of resistant starch *i.e.* Negligible (1%), Low (1-2.5%), Intermediate (2.5-5.0%), High (5.0-15%), Very high (15%). Raw legumes come under very high resistant starch content category.

Table 2. Resistant starch (RS) content in raw whole and decorticated legumes

Legume	RS content (g/100g)	
	Whole	Decorticated
Green gram	17.9	21.5
Kabuli Bengal gram	18.4	19.6

Desi Bengal gram	9.0	--
Red gram	10.5	14.2
Lentil	19.6	23.3
Cowpea	15.6	--
Field pea	12.6	--

Katyal *et al.* (3)

Table 3. Varietal differences in resistant starch (RS) content of legumes

Whole legume	RS Content (g/100g)	
	Wet basis	Dry basis
Green gram		
Pusa 105	17.2	19.2
Pusa Vishal	18.5	20.7
Pusa 9072	18.0	20.1
Lentil		
Lens 830	25.4	27.9
Lens 4076	17.7	19.6
Lens 4147	15.6	17.2
Bengal gram		
Pusa 267	19.8	21.8
Pusa 1053	16.9	18.7
Bengal gram (Desi)		
Pusa 362	8.7	10.0
Pusa 372	9.3	10.2
Red gram		
Pusa 33	19.6	21.8
Pusa 855	6.9	7.6
Pusa 945	7.9	8.8
Pusa 604	7.7	8.5
Cowpea		
V-16	12.4	13.8
V -585	19.9	22.1
V -240	14.7	16.4
Field pea		
DDR-23	14.0	15.0
DMR-7	11.1	12.3

Katyal *et al.* (3)

Beneficial physiological effects of RS

A number of physiological effects have been ascribed to RS (1), which have been proved to be beneficial for health.

a. RS as a component of DF

RS appears to be highly resistant to mammalian enzyme and may be classified as a component of fiber on the basis of the recent definitions of dietary fiber given by AACC (4). While part of the RS may consist of low-molecular-weight dextrins, the bulk consists of polymers, of which retrograded amylose often forms the major fraction (5). There is ample justification that RS behaves physiologically like fiber. RS assays as insoluble fiber, but has the physiological benefits of soluble fiber. Additionally, RS exhibits a level of slow digestibility and can be used as a vehicle for the slow release of glucose. Also, like soluble fiber, it has a positive impact on colonic health by increasing the crypt cell production rate, or decreasing the colonic epithelial atrophy in comparison with no-fiber diets. There is indication that RS like guar, a soluble fiber, influences tumorigenesis, and reduces serum cholesterol and triglycerides. Overall, since RS behaves physiologically as a fiber, it should be retained in the TDF assay (6).

b. Prevention of colonic cancer

Starch unabsorbed in the small intestine is fermented by the microflora of the large intestine. Generally, starch is not present in the faeces of humans or experimental animals, indicating more or less complete fermentation. In vitro experiments with human faecal inocula have shown that the butyrate yield from starch is high. As butyrate is a main energy substrate for large intestinal epithelial cells and inhibits the malignant transformation of such cells in vitro; this makes easily fermentable RS fractions especially interesting in preventing colonic cancer (7). Significant changes in faecal pH and bulking as well as greater production of SCFA in the cecum of rats fed RS preparations have been reported (8), which have been suggested to resemble the effects of soluble dietary fiber. However, when RS was combined with an insoluble dietary fiber like wheat bran, much higher SCFA levels, in particular butyrate was observed in the feces (9). In rats, when RS was combined with psyllium, the site of RS fermentation was pushed more distally. As the distal colon is the site where most tumors arise it may be of additional benefit for cancer protection if fermentation is further enhanced within the distal colon (10).

c. Hypoglycemic effects

Foods containing RS moderate the rate of digestion. The slow digestion of RS has implications for its use in controlled glucose release applications. The metabolism of RS occurs 5 to 7 h after consumption, in contrast to normally cooked starch, which is digested almost immediately. Digestion over a 5 to 7 h period reduces postprandial glycemia and insulinemia and has the potential for increasing the period of satiety (10 & 11). A study using 10 healthy normal-weight males fed test meals containing either 50 g starch free of RS (0% RS), or 50 g starch containing a high level of RS (54% RS) proved the ability of high RS meals to significantly lower the postprandial concentration of blood glucose, insulin, and epinephrine (10). Similarly, from a human study (11), using a commercial RS_3 ingredient (CrystaLean), the maximum blood glucose level was found to be significantly lower than that of other carbohydrates (simple sugars, oligosaccharides, and common starch). The RS_3-containing bar decreased postprandial blood glucose and may play a role in providing improved metabolic control in type II diabetes (non-insulin dependent).

d. As a prebiotic

RS has been suggested for use in probiotic compositions to promote the growth of such beneficial microorganisms as Bifidobacterium (12). Since RS almost entirely passes the small intestine, it can behave as a substrate for growth of the probiotic microorganisms.

e. Reduction of gall stone formation

The digestibility of starch in rice and wheat is increased by milling to flour (13). Digestible starch contributes gall stone formation via a greater secretion of insulin and insulin in turn leads to the stimulation of cholesterol synthesis. RS is found to reduce the incidence of gallstones (14). Gallstones are less in South India where, whole grains are consumed rather than as flour in Northern India. Notably, dietary intake of RS is 2 to 4 fold lower in the United States, Europe, and Australia, compared with populations consuming high-starch diets, such as India and China, which may reflect in the difference in the number of gallstone cases in these countries (15).

f. Hypocholesterolemic effects

Hypocholesterolemic effects of RS have been amply proved. In rats, RS diets (25% raw potato) markedly raised the cecal size and the cecal pool of short-chain fatty acids (SCFA), as well as SCFA absorption and lowered plasma cholesterol and triglycerides. Also, there was a lower concentration of cholesterol in all lipoprotein fractions, especially the HDL1 and a decreased concentration of triglycerides in the triglyceride-rich lipoprotein fraction. Results of feeding trials on rats using RS from Adzuki starch (AS) and tebou starch (TS) suggested that AS and TS had a serum cholesterol-lowering function due to the enhanced levels of hepatic SR-B1 (scavenger receptor class B1) and cholesterol 7 α –hydroxylase m RNA (16). The bean starches lowered the levels of serum total cholesterol and VLDL + IDL + LDL cholesterol, increased cecal concentration of short-chain fatty acids (in particular butyric acid concentration), and increased faecal neutral sterol excretion.

From studies on hamsters fed diets containing cassava starch extrudated with 9.9 per cent oat fiber or cassava starch extruded with 9.7 per cent RS, hypocholesterolemic properties of both were demonstrated suggesting their use in foods to improve cardiovascular health (17).

g. Inhibition of fat accumulation

Replacement of 5.4 per cent of total dietary carbohydrates with RS in a meal could significantly increase postprandial lipid oxidation suggesting reduction in fat accumulation in the long term (18).

h. Absorption of minerals

A study to compare the intestinal apparent absorption of calcium, phosphorus, iron, and zinc in the presence of either resistant or digestible starch brought out that a meal containing 16.4 per cent RS resulted in a greater apparent absorption of calcium and iron compared with a completely digestible starch (19). Thus RS could have a positive effect on intestinal calcium and iron absorption.

Food Applications of RS

Sajilata *et al.* (20) mentioned many food applications which are as follows

1. RS in bread –making for DF fortification

The physical properties of RS, particularly its low water-holding capacity, allow it to be a functional ingredient that provides good handling in processing and crispiness, expansion, and improved texture in the final product. Bread is commonly fortified with dietary fiber. However, dark color, reduced loaf volume, poor mouthfeel, and masking of flavor are all negative attributes that are often associated with high-fiber breads. A study was conducted at the American Inst. of Baking (AIB) to evaluate the effect of RS on bread characteristics and to compare their performance to traditional fibers. The study included cellulose, oat fiber, wheat fiber, and 2 commercially available RS with 23 per cent (Hylon VII starch) and 40 per cent TDF (Novelose 240 starch) and a blend of oat fiber with Novelose 240 starch in a 50/50 ratio based on TDF contribution. Compared with oat fiber, cellulose, and wheat fiber, both RS had a lower water-holding capacity, which is similar to that of flour. Although the water-holding capacity of the RS was lower than that of the other fiber sources, the quantity required to obtain the same level of TDF was greater. This raised the total water requirement of the RS dough. However, absorption of the RS doughs was less compared with the fiber doughs, despite a larger quantity of water used. The lower dough absorption consequently had less impact on dough rheology and was closest to the white pan bread dough. Bread containing 40 per cent TDF RS had greater loaf volume and better cell structure compared with traditional fibers tested (21).

2. RS as a texture modifier in baked goods

One way to ensure that the general population receives adequate amounts of fiber in the diet is to fortify good-tasting foods that normally do not come to mind with fiber fortification but are often eaten as breakfast items or snacks. RS were incorporated in a variety of baked goods, many of which include batter systems, such as in cakes, cake-like muffins, or brownies. In general, application tests showed that RS acts as texture modifier, imparting a favorable tenderness to the crumb. A low-fat, loaf cake was formulated with RS and various fibers to obtain approximately 3 per cent TDF or 2.5 g of fiber per 80 g serving. These included a 40 per cent TDF RS (Novelose 240 starch), oat fiber, a blend of oat fiber with Novelose 240 starch in a 50/50 ratio based on TDF contribution, and a 23 per cent TDF RS (Hylon VII starch). The baked cakes made with RS were similar to that containing oat fiber and the control in the amount of moisture loss after baking, height, specific volume, and density. A panel rated the 40 per cent TDF RS loaf cakes as the best for flavor, grittiness, moisture perception, and tenderness 24 h after baking.

3. RS as a crisping agent

Among other functional properties, RS can be used as an ingredient that improves crispiness in foods where high heat is applied to a product's surface during processing. French toast and waffles, especially frozen reheated types, represent foods in which surface crispiness is desired. Tests were conducted to compare the functionality of RS and various fibers in a buttermilk waffle formulation. Based on the evaluation of the toasted waffles for initial crispness, crispness after 3 min, moistness, and overall texture by a trained sensory panel RS waffle indicated greater crispness than control or traditional fiber.

4. RS as a functional ingredient in other foods

Along with textural enhancement, RS can improve expansion in extruded cereals and snacks. Various cereals were formulated to contain 40 per cent TDF RS (Novelose 240 starch) alone and in combination with oat fiber in ratios of

50/50 and 25/75 based on weight. The cereal with RS and no oat fiber had greater volumetric expansion than the control. In blends with oat fiber, the cereal containing 75 per cent of RS had better expansion than the one containing only 50 per cent. Dried pasta products containing up to 15 per cent RS can be made with little or no effect on dough rheology during extrusion. Although the resultant pasta was lighter in color, a firm "al dente" texture was obtained in the same cooking time as a control that had no added fiber. RS may also be used in thickened, opaque health beverages in which insoluble fiber is desired. Insoluble fibers generally require suspension and add opacity to beverages. Compared with insoluble fibers, RS imparts a less gritty mouthfeel and masks flavors less.

CONCLUSION

Legumes contain substantially higher levels of resistant starch than do cereal grains, flours and grain-based food products. Resistant starches have properties similar to dietary fiber and promising physiological benefits in humans, which may result in prevention of various diseases. Resistant starches provide many health benefits such as improving bowel health, blood lipid profile, increased feeling of satiety and micronutrient absorption (magnesium and calcium) in the colon. Resistant starch containing foods have low glycaemic index. This helps in management of diabetes and impaired glucose tolerance. Hence it is used in the treatment of obesity and weight management.

REFERENCES

1. Nugent, A.P. 2005. Health properties of resistant starch. Br Nutr Foundation Nutr Bull 30:27–54.
2. Goni, J., Garcia-Biz, Manas, E. and Calixto, S.F. 1996. Analysis of Resistant starch : A method for foods and food products. Food Chem. 56(4): 445-449
3. Katyal Deepti, Padmini S. Ghugre, Udipi, S.A. 2005. Resistant starch in selected raw and processed legumes. J. Food Sci. Technol. 42(6) : 506-510.
4. American Association of Cereal Chemists 2000. Approved methods of the AACC. 10th Ed. Method 44-15A. St. Paul, Minn.; AACC.
5. Ranhotra, G.S., Gelroth, J.A., Astroth, K, Eisenbraun, G.J. (1991a). Effect of resistant starch on intestinal responses in rats. Cereal Chem 68(2):130–132.
6. Haralampu, S. G. 2000.Resistant starch a review of the physical properties and biological impact of RS_3. Carbohydrate Polym 41:285–292.
7. ASP and Bjorck 1992. Resistant Starch, Trends Food Sci. Technol. 3(5) : 111-114.
8. Ferguson, L.R., Tasman-Jones, C., Englyst, H., Harris, P.J. (2000). Comparative effects of three resistant starch preparations on transit time and short-chain fatty acid production in rats. Nutr Cancer 36:230-237.

9. Leu, R.K., Hu, Y., Young, G.P. (2002). Effects of resistant starch and nonstarch polysaccharides on colonic luminal environment and genotoxin-induced apoptosis in the rat. Carcinogenesis 23(5):713–719.

10. Raben, A., Tagliabue, A., Christensen, N.J., Madsn, J., Holst, J.J., Astrup, A. 1994. Resistant Starch: the effect on postprandial glycemia, hormonal response and satiety. Am J Clin Nutr. 60:544–551.

11. Reader, D., Johnson, M.L., Hollander, P., Franz, M. 1997. Response of resistant starch in a food bar vs. two commercially available bars in persons with type II diabetes mellitus. Diabetes 46(1):254A.

12. Brown, I.L., McNaught, K.J., Ganly, R.N., Conway, P.L., Evans, A.J., Topping, D.L. and Wang, X. 1996. Probiotic compositions. Intl. Patent WO 96/ 08261/ A1. Issued Mar 21,1996; Univ New South Wales; Burns Philip and Co. Limited; Burns Philip Res and Dev Pty; Mauri Lab Pty Ltd; Commw Sci Ind Res Org; Arnotts Biscuits Ltd; Good man Fielder Ingredients Li; Goodman Fielder Ltd; Brown IL, McNaught KJ, Ganly RN, Conway PL, Evans AJ, Topping DL, Wang X.

13. Heaton, K.W. 1988. Gall stone prevention. In Bile acids and diseases. Lancester: MTP Press. p 57–169.

14. Malhotra, S.L. 1968. Epidemiological study of cholelithiasis among rail road workers in India. Gut 9:290–295.

15. Birkett, A.M., Mathers, J.C., Jones, G.P., Walker, K.Z., Roth, M.J., Muir, J.G. 2000. Changes to the quality and processing of starchy foods in a Western diet can increase polysaccharides escaping digestion and improve in vitro fermentation variables. Br J Nutr 84:63–72.

16. Han, K.H., Fukushima, M., Kato, T., Kojima, M., Ohba, K., Shimada, K., Sekikawa, M., Nakano, M. 2003. Enzyme-resistant fractions of beans lowered serum cholesterol and increased sterol excretions and hepatic mRNA levels in rats. Lipids 38(9):919–924.

17. Martinez-Flores, H.E., Yoon-Kil-Chang, Martinez-Bustos F., Sgarbieri, V. 2004. Effect of high fiber products on blood lipids and lipoproteins in hamsters. Nutr. Res. 24(1):85–93.

18. Higgins, J.A., Dana, H.R., Donahoo, W.T., Brown, I.L., Bell, M.L., Bessesen, D.H. 2004. Resistant starch consumption promotes lipid oxidation. Nutr. Met. 1:1–8.

19. Morais, M.B., Feste, A., Miller, R.G., Lifichitz, C.H. 1996. Effect of resistant starch and digestible starch on intestinal absorption of calcium, iron and zinc in infant pigs. Paediatr Res. 39(5):872–876.

20. Sajilata, M.G., Singhal, R.S. and Kulkarni, PR. 2006. Specialty starches for snack foods. Carbohydrate Polym. 59:131–151.

21. Baghurst, P.A., Baghurst, K.I., Record, S.J. 1996. Dietary fiber, non-starch polysaccharides and resistant starch a review. Food Aust. 48(3): S3–S35.

An assessment of guar gum as stabilizer in ginger RTS beverage

A.B. Rodge and S.S. Deshmukh

Department of Food Chemistry and Nutrition,
Marathwada Agricultural University, Parbhani – 431 402

ABSTRACT

In the food industry, gums have been used in various applications, primarily for their ability of imparting thickening, stabilizing, gelling and textural properties to individual food products to improve the overall quality of the products. Attempts have made to overcome the problem of cloudiness in Ginger Ready to serve (RTS) beverages. Galactomannan was exploited as an additive to stabilize the ginger ready to serve beverage for longer period. The gum was added in very small quantities (0.1, 0.2, 0.3, 0.4 and 0.5 %) and has a negligible influence on food value and flavour with overall acceptability as that of controlled sample. The gum stabilizes the system by avoiding the separation of two phases by combining with excess water. The guar gum used at the rate 0.1 per cent, 0.2 per cent, 0.3 per cent, 0.4 per cent and 0.5 per cent in the Ginger RTS beverage respectively. The sensory evaluation revealed that the RTS with 0.3 per cent guar gum has acceptable organoleptic characteristics as compared with controlled sample without cloudiness.

Key words : Stabilizer, phases, RTS.

INTRODUCTION

Guar *(Cyamopsis tetragonoloba)* is an important crop grown in various parts of the country for its fodder as well as it gum, extracted from its seed which has got an industrial importance (1). The gum is utilized in textiles, confectioneries, explosives, oil explorations and basically the gum industries. Gum and mucilage are generally categorized as hydrocolloids or viscogenic micro molecules. They are used enormously in different industries (2) namely paper, food textile, pharmaceuticals, cosmetics, fields of enhanced oil recovery, explosives etc. Guar gum is obtained from the ground endosperm of the guar plant, (*cyamopsis tetrogonoloba*) that is native to the arid climates of India and Pakistan. Guar is widely cultivated in India in Punjab, Harayana, Rajasthan, Gujarat and western utter Pradesh as grain crop for commercial exploration of gum. To a limited extent the plant has been cultivated in the U.S. in Texas and Arakansas. Guar gum first became of commercial interest in the 1950's following a shortage of locust bean gum. Guar gum is commonly used as a stabilizer in many food products like ice-cream, fruit beverage, chocolate, milk drinks, cake topping etc. The other stabilizers used are starches, dextrin, protein derivatives and other additives that stabilize and thicken the food by combining with water to increase viscosity and to form gel (3).

Ginger (*Zingier officinale*) which belongs to family zingiberaceae has a distinctive, spicy, penetrating flavour and is largely used in foreign countries in the manufacturing of ginger oil, ginger essence, ginger oleoresin or gingerin. Beside these, it is also used in spent ginger, gingers soft drinks or non-alcoholic beverages, also ginger powder is used in ginger candy, alcoholic beverage and lime ginger pickle. Ginger is flavourant in some food products and also known to be popularly used. Ginger oil is primary used as food flavour and acts as a flavoring agent in soft drinks like ginger ale, carbonated drinks, cordials, liquors etc. (4 & 5). Ready to serve beverage (RTS) is a type of beverage containing 10 per cent juice with 10°bx TSS and 0.3 per cent acidity, which is not diluted before serving (6). In present investigation work done on the utilization of Galactomannan in ginger RTS beverage as stabilizer from guar.

MATERIALS AND METHOD

The RTS from ginger was prepared by using standard FPO specifications. The RTS was prepared of 10 °bx TSS, containing 10 per cent juice and acidity 0.3 percent. For stabilization of RTS beverage the guar gum at different concentration (0.1%, 0.2%, 0.3%, 0.4% and 0.5%) were used. The effect of different guar gum concentration on organoleptic qualities of RTS beverage was studied. The organoleptic evaluation of RTS beverage was carried out by using departmental semitrained panel members on 9 point Hedonic scale (7). The process used for preparation of RTS beverage was as follows.

RESULTS AND DISCUSSION

In the present investigation, efforts have been made to study the physical characteristics of ginger and effect of different guar gum concentration on organoleptic characteristics of ginger RTS. Physical characteristics of ginger viz. weight of ginger, weight of peeled ginger, weight of waste and waste index are studies and average values are reported in the table 1.

Table 1. Physical Characteristics of ginger

Physical characteristics	Observations
Weight of ginger (gm)	55
Weight of peeled ginger (gm)	41
Weight of peel (gm)	14
Waste index (%)	25.45

The sensory evaluation of ginger RTS prepared by incorporation of guar gum having different concentration and mean score values of sensory attributes of ginger RTS is depicted in table 2. The score for colour, flavour, aroma, taste, appearance and over all acceptability for RTS beverage prepared by using 0.1 and 0.2 per cent guar gum rated highest compared to other sample. From table it was revealed that the incorporation of guar gum upto 0.3 per cent was acceptable. However, the ginger RTS with 0.5 per cent level of guar gum affect the sensory attributes of ginger RTS and scored lower values as compared to control

Table 2. Sensory evaluation of ginger RTS

Sample	Colour	Flavour	Aroma	Taste	Appearance	Over all Acceptability
Control	7.5	7.6	7.9	7.0	8.0	7.8
0.1 % gum	7.5	7.8	8.1	6.8	7.9	7.6
0.2 % gum	7.3	7.5	7.6	6.6	7.5	7.4
0.3 % gum	7.2	7.3	7.1	7.2	7.3	7.2
0.4 % gum	5.2	7.2	7.0	6.9	7.1	6.7
0.5 % gum	6.2	6.5	6.5	6.3	6.4	6.4

Cost estimation of ginger RTS beverage

Cost of production of ginger RTS beverage was estimated and presented in Table 3. The production cost of 1 lit beverage is about 15.00 Rs. It is evident from table that to prepare 3 liters of ginger beverage of the estimated cost is Rs. 45.0. Thus the prepared product is very cost effective and best suited for commercialization.

Table 3. Cost estimation of ginger RTS

Materials	Quantity	Amount (Rs)
Ginger	100 g	6
Sugar	500 g	10
KMS	0.3 g	2

Bottles	9	27
Production of 3 lit beverage	Total amount	45

Each value is an average of three determinations

Conclusion

The organoleptic characteristics of the RTS with 0.3 per cent level of guar gum were acceptable. The galactomannan exploited stabilize the ginger RTS for longer period.

References

1. Anon 1992. "Stabilizers properties and uses" The functional properties of stabilizer (guar gum, locust bean gum) commonly used in food industry are described dairy industries international, 57 (1) 25-26.
2. Whistler R.L. and Bemiller N.J. 1959. Industrial gums academic press, New York.
3. Dubois, M., Pass, G. and Phillips, G.O. 1977. The viscosity of dilute solution of guar gum with and without added sugars. J. Food Sci. 28 : 895-899.
4. Natarajan, C.P., Kuppaswamy, S. and Sankaracharya 1970. Product development of ginger. Indian spices 7(4) : 8-12 and 24-28.
5. Leverington, R. F. 1975. Ginger technology. Food techno. In Australia 27 (8) 309-311.
6. Srivastva, R.P. and Sajeev Kumar 2006. Fruits and vegetable preservation, principles and practices. IIIrd edition. International Book Distributing Co. Lucknow, India.
7. Amerine, M.A., Pangborn, R.M. and Roessler, E.B. 1965. Principles of sensory evaluation of food. Academic Press, New York.

Studies on fortification of guar germ meal on quality of bread

A.B. Rodge and M.H. Mandge

Department of Food Chemistry and Nutrition, Marathwada Agricultural University, Parbhani – 431 402

ABSTRACT

In present investigation attempts have been made to isolate the germ meal from guar *(Cyamopsis tetragonolbus)* by wet processing method. Guar germ meal was fortified in bread at various levels. Lysine and valine are two limiting amino acids of wheat proteins. Guar meal was good source of protein and also contain residual amount of guar gum. The germ meal is rich source of protein. Guar germ meal also exhibits good water holding capacity (390%) which was slightly decreased due to autoclaving treatment (280%). The fat absorption capacity of guar germ meal before and after autoclaving were (245%) and (203 %) respectively. It is found that water absorption capacity of mixture containing guar germ meal and wheat flour was increased with increasing proportion of guar germ meal. Breads containing upto 4 per cent guar germ meal are acceptable. Due to some residual gum content the guar germ meal had tenderizing and stabilizing effect on crumb structure of bread and acts as a antistaling agent.

Key words : Antistaling agent, germ meal

INTRODUCTION

Guar *(Cyamopsis tetragonoloba)* is an important industrial crop of the dry lands of Rajasthan, Haryana, Gujarat, Punjab and Uttar Pradesh as well as in Pakistan. The interest has created in recent years particularly in guar as it is a valuable industrial crop for the production of guar gum. Guar germ meal is a by product of guar gum industry and is a valuable source in unconventional protein as it contains 30 to 46 per cent protein on whole seed basis. It has been reported that the protein isolated from the guar germ meal by conventional alkali extraction and isoelectric precipitation was found to be more toxic than the native germ meal (1). The germ meal utilized in preparation of bread was extracted in food grade solvents.

Bread is one of the important processed bakery product in developed and under developed countries. It has been observed that consumption of bread is increased due to increase in population. Being a cheaper processed food, it will have good prospectus in under developed countries like India. The nutritional deficiencies of wheat proteins have been demonstrated earlier. So guar germ meal which is good source of protein and also may contain residual amount of guar gum is valuable food additive in bread. In the present investigation study was conducted on preparation of germ meal from guar, removal of antinutritional factors and effect fortification of guar germ meal on characteristics of bread.

MATERIALS AND METHODS

Guar seeds (*Cyamopsis tetragomoloba*) of varieties PLG-85 are obtained from experimental plot, College of Food Technology, MAU, Parbhani. Moisture, lipid, protein and carbohydrate contents were determined by the method of AOAC (2).

The preparation of germ meal was achieved by the simple wet processing method (Fig. 1), developed in the Department of Biochemistry and Applied Nutrition, College of Food Technology, MAU, Parbhani (3). The antinutritional factors in guar germ meal was removed by autoclaving treatment in typical experiment germ meal of guar (250 gm each) was moistured (20%) and sprayed in a tray separately. Then autoclaved at 1 kg/cm^2 (120°C for 20 min) and aerated. Then stored in desiccators at room temperature until further use. The functional properties of germ meal such as the water absorption capacity was determined by method of Savant (4). The fat absorption capacity was determined by the method of Deach et al. (5).

In this study the breads were prepared by standard recipe (table 1) and procedure given in Fig. 2. The organoleptic evaluation of bread was carried out by using departmental semitrained panel members on 9 point Hedonic scale (6). The variable levels of germ meal and wheat flour used for bread preparation was given in table 2.

Table 1. Variable levels of germ meal and wheat flour in bread

Germ meal (%)	Wheat flour (%)	Sugar (%)	Fat (%)	Salt (%)	Yeast (%)
0	100	2.0	1.5	2.0	1.0
2	98	2.0	1.5	2.0	1.0

4	96	2.0	1.5	2.0	1.0
6	94	2.0	1.5	2.0	1.0
8	92	2.0	1.5	2.0	1.0
10	90	2.0	1.5	2.0	1.0
12	88	2.0	1.5	2.0	1.0

Guar seeds

↓

Subjected to wet processing solvents : Alkaline solution (w/v)

↓

Vigorous boiling at 100°C for different time intervals (7-8 min).

Sieved through coarse sieve and sprinkled tap water till neutral pH

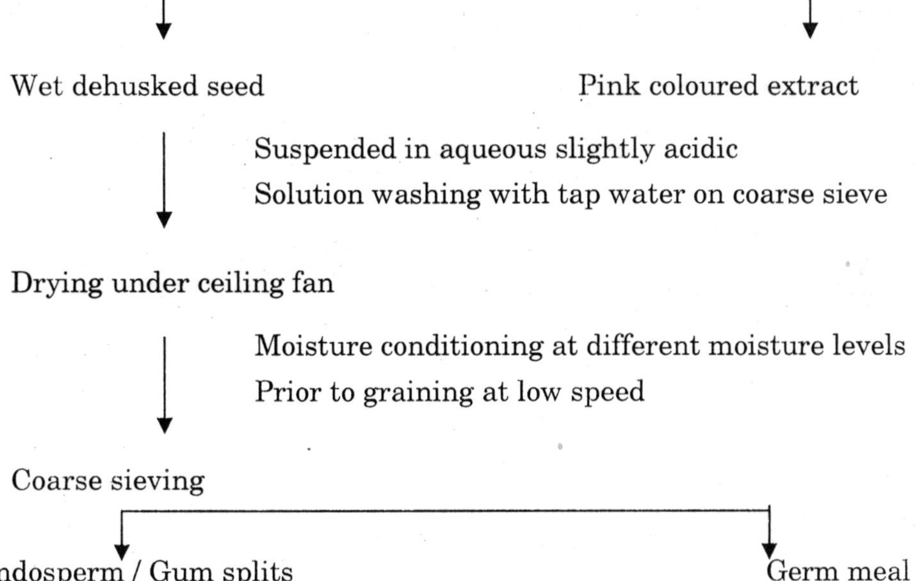

Fig. 1. Flow diagram of preparation of guar meal

RESULTS AND DISCUSSION

The chemical composition of guar seeds and guar meal has been investigated and results were presented in table 3. The results obtained showed that guar seeds are rich source of protein. They contain less moisture so can be stored for long period. It was evident from table that guar meal was rich in protein (56.8%) when compared to guar seeds (42.63%).

Table 2. Chemical composition of guar seeds and guar meal

Parameters	Guar seed (%)	Guar meal (%)
Moisture	9.00	6.20
Crude lipid	6.00	7.5
Crude protein	42.63	56.8
Total Carbohydrate	40.00	24.5
Ash	3.00	4.00

Each value is an average of three determinations

It was evident from Table 4 that the water absorption capacity of guar germ meal decrease from 390 per cent to 280 per cent after autoclaving treatment. This may be due to the defragmentation and disturbances to three dimensional structure of protein of guar germ meal. The data in Table 4 revealed that fat absorption capacity of guar germ meal was decreased from 245 to 203 per cent after autoclaving.

Table 3. Physical properties of germ meal

Sample	Fat absorption capacity (%)	Water absorption capacity (%)
Raw germ meal	245	390
Autoclaved germ meal	203	280

Each value is an average of three determinations

The data on effect of guar germ meal fortification on organoleptic evaluation of bread was presented in table 5. It was evident from table that with 2 and 4 per cent fortification of guar germ meal, grain, texture and crumb colour was slightly affected, but with higher percentage of fortification it becomes highly affected. Taste, aroma and mouthfeel characteristics were slightly decreases with increasing proportion of fortification of guar germ meal. Therefore guar germ meal incorporation upto 2 and 4 per cent in the bread was acceptable.

Table 4. Effect of fortification of guar germ meal on organoleptic characteristics of bread

Guar meal (%)	Grain	Texture	Crumb colour	Aroma	Taste	Mouth feel	Overall acceptability
0	9	9	9	9	9	9	9

2	8.5	9	8.5	8.0	8.	8.5	8.5
4	8.0	8.5	8.0	7.5	7.5	7.5	8.0
6	7.5	7.5	7.0	7.0	6.5	7.0	7.5
8	7.0	7.0	6.0	6.0	6.0	7.0	7.0
10	7.0	7.0	5.0	5.0	5.0	6.0	6.0
12	5.0	5.0	4	4	4	4	5.0

Each value is an average of three determinations

CONCLUSION

Water absorption capacity of mixture containing guar germ meal and wheat flour increased with increasing proportion of guar germ meal. Guar germ meal has tenderizing and stabilizing effect on crumb structure of bread due to some residual gum content. Bread prepared by incorporation upto 4 per cent guar germ meal is acceptable.

REFERENCES

1. Couch J. R. 1966. Trypsin inhibitor in guar meal .Pro. Soc. Exp. Bial. Med; 123- 163.
2. A.O.A.C. 1975. Official methods of analysis. 15th ed. Association of Official Analytical Chemist, Washington, D.C.
3. Rodge A.B., P.U. Ghatge, D.B. Wankhede and R.K. Kokate 2006. Isolation, purification and rheological study of guar genotypes RCG-1031 and RCG-1038. J. of Arid Legumes 3(2) : 41-43.
4. Savant R. R. 2007. Preparation of value added extruded product by using germinated horse gram flour (2007). M. Tech. Thesis, MAU, Parbhani.
5. Deach, J. E., Rivas, N.R. and Caygill, J.C. 1981. Selected functional properties of sesame (*Sesamum indicum* L.) flour and protein isolates. J. Sci. Food Agric. 23 : 557.
6. Amerine, M.A., Pangborn, R.M. and Roessler, E.B. 1965. Principles of sensory evaluation of food. Academic Press, New York.

Maltodextrin from cowpea starch as a fat replacer in cake

A.B. Rodge and V.D. Korde

Department of Food Chemistry and Nutrition,
Marathwada Agricultural University, Parbhani – 431 402

ABSTRACT

In present investigation, efforts have been made to extract starch from cowpea genotype and its conversion to maltodextrin. The cowpea starch was isolated from seeds using alkali solution (NaOH, 1.5%) as extractant with 57.20 per cent recovery. The production of maltodextrin were studied by using both acid and enzyme hydrolysis. It was found that the maximum amount of 94 per cent maltodextrin was liberated at 4.5 per cent concentration of hydrochloric acid from cowpea starch. Maltodextrin was used as a fat replacer in cakes. The results indicated that acceptable cakes were prepared by using maltodextrin with 35 per cent replacement of shortening.

Key word : Fat replacer, acid hydrolysis

INTRODUCTION

Food is a subject of vital interest to everyone in the world. Everyone wants high quality, convenience and more healthy food. Cereals supply the bulk of the food consumed by the human race. In response to increasing consumer's attention above health and nutrition, manufactures should go for production of low calorie food stuffs in incorporating various blends. Substances which taste and function like oil, but relatively lesser in calorie content. In recent years, the food industry has developed a wide variety of low calorie starch based materials to replace fat in processed foods (1). These fat replacer are used to deliver finished food product with the flavor, mouth feel, color, taste and appearance of their full at counter parts. Fat replacer are mostly carbohydrate based for protein based, used to replace or substitute fat in foods as they exhibit versatile textural and organoleptic properties. Carbohydrates based fat replacers are macromolecules, such as polydextrose, maltodextrin, starch derivatives, gums and mucilage's and dietary fiber (2).

Cowpea (*vigna unguiculata* L.) is also called as southern pea and black eyed pea is well adapted to the tropics (arid/ semiarid). The cowpea seed has maximum amount of starch about 45 per cent extracted from carbohydrates. This starch has been converted into creamy white hygroscopic powder which acts as a fat replacer in food products mainly called maltodextrin (3).

In this research project studies were carried out on isolation of starch, preparation of maltodextrin from cowpea and its utilization in cake as fat replacer.

MATERIALS AND METHODS

The starch was isolated from cow pea grains and maltodextrin prepared from it by enzyme hydrolysis (Fig. 1). The moisture, protein, carbohydrate, lipid and ash content of starch and maltodextrin were determined by the standard AOAC method (4). The high ratio cake formulation was adopted from the work of (5). A high ratio cup cake was selected as the test model, and 20 per cent, 30 per cent, 35 per cent and 40 per cent (w/w) of the shortening was replaced by maltodextrin. All cakes were prepared in bakery plant at College of Food Technology, M.A.U., Parbhani. Gel of maltodextrin was compared with the powder forms of the maltodextrin in the baking tests. Maltodextrin was mixed with dry ingredients (*viz.* maida, sugar, baking powder, fat and essence) during replacements and baked at 180°C. Effect of replacement of fat with maltodextrin on organoleptic attributes of cake was studied. Cake was organoleptically evaluated for colour, taste, flavour, consistency and over all acceptability by a panel of semi trained judges using 9 point hedonic scale scoring 9 for extremely like and 1 for extremely dislike (6).

RESULTS AND DISCUSSION

The cow pea was analyzed quantitatively for its chemical composition. The data in table 1 revealed that starch content was found to be 45.62 per cent in the sample. The results obtained regarding proximate composition of starch and maltodextrin during present investigation was presented in table 2. It has been observed that the starch has maximum carbohydrate content about 95.8 per cent which has been used for extraction of maltodextrin. The result revealed that maltodextrin has a creamy white hygroscopic powder, moderately sweet in taste, free flowing and consist of a mixer of saccharides without any starch.

Table 1. Chemical composition of Cowpea

Parameters	Percentage
Moisture	10.8
Ash	1.45
Crude Protein	27.2
Crude Fat	3.4
Carbohydrates	47.6
Starch	43.5
Crude fiber	9.8

Each value is an average of three determinations

Table 2. Proximate composition of starch and maltodextrin

Parameters	Starch (%)	Maltodextrin (%)
Moisture	9.1	8.0
Protein	1.2	0.7
Fat	0.6	0.3
Carbohydrate	95.8	96.5
Ash	0.33	0.45
Crude fiber	0.75	0.70
Colour	Bright white	White
Shape of granules	Polygonal to round	Irregular
D.E.		8-10

Each value is an average of three determinations

The data on organoleptic evaluation of cakes presented in Table 3. Cake prepared with different replacement levels with maltodextrin were subjected to organoleptic evaluation. The results were presented in table 3 given for taste sample C ranked first followed by sample G whereas sample A and B were at par and statistically superior over all the remaining samples for flavor. The organoleptic scale for texture indicated sample C was significantly superior over the rest of the samples studied. However, sample A and B were found to be at par with each other. The overall acceptability suggested sample C with significant superiority over remaining samples including control. In general it can be concluded that sample C was the best combination for all the organoleptic parameters evaluated. This particular sample combination with 35 per cent replacement by maltodextrin was acceptable. Good quality cakes were made by using maltodextrin replacing shortening in the range of 35 per cent. These cakes were more acceptable than the normal cake, with low in calories.

Table 3. Organoleptic evaluation of cake

Samples	Colour	Taste	Flavor	Texture	Overall Acceptability
G	9	8	9	9	9.00
A	9	8	9	8	8.75
B	8	9	9	7	8.50
C	8	9	8	9	8.50
D	7	8	8	8	7.75

Where,

G Experimental, control

A 20 per cent replacement with maltodextrin

B 30 per cent replacement with maltodextrin

C 35 per cent replacement with maltodextrin

D 40 per cent replacement with maltodextrin

Conclusion

Maltodextrin was prepared by enzyme hydrolysis method and used in cake preparation. Good quality cakes were prepared using maltodextrin with 35 per cent (w/w) replacement of shortening. Maltodextrin powder can be utilized in preparation of low fat cakes.

References

1. Smith, P.S. and Bell, H. 1986. New starches for food applications. Cereal Food World, 31 (10) : 724-726.
2. Anon 1990. Fat substitute updated. Food Tech. 44 (3) : 92-97.
3. Conforti, F.D. and Archillia, L. 2001. Evaluation of a maltodextrin gel as a partial replacement for fat in a high ratio white layer cake. International J. of Consumer Studies. 25(3):238-245.
4. A.O.A.C. 1975. Official methods of analysis. 15th ed. Association of Official Analytical Chemist. Washington, D.C.
5. Barh D.E., Shelke, K.S. and Hoseney, R.C. 1992. Fat replacers in high ratio layer cakes. Cereal food world, 37: 495-500.
6. Amerine, M.A., Pangborn, R.M. and Roessler, E.B. 1965. Principles of sensory evaluation of food. Academic Press, New York.

Technology dissemination in clusterbean – impact in arid and semi-arid districts of Rajasthan

R.S. Dohare, M.K. Mandape, R.N. Prasad and P.K. Satapathy

Zonal Coordinating Unit-VI, CAZRI Campus, Jodhpur – 342005

ABSTRACT

Keeping in view of the importance of guar, promotional efforts through technological dissemination process which includes conducting Frontline Demonstration (FLDs) on improved technological interventions, trainings and extension activities were carried out in different districts of Rajasthan namely, Jodhpur, Barmer, Bikaner, Jaisalmer, Sikar, Nagaur, Jhunjhunu, Jalore, Ajmer, Pali, Dausa, Tonk, Alwar, Bharatpur, Karauli, Dholpur, Udaipur and Sawai Madhopur during kharif, 2001-02 to 2006-07. The technological components such as improved varieties and Production Technology were included to know the extent of production and feedback for further stepping up production. In varietal component, the performance of 730 demonstrations covering 275.75 ha area reveals that weighted average yield of demonstration was found to be increased ranging from 33.84 to 71.13 per cent higher over local checks. The feed back for researchers, development departments and policy makers also invited for immediate attention to make concerted efforts to evolve technological interventions for value addition, more promising varieties having more guar gum content, drought tolerant and disease resistant, suitable package of practices for both rainfed and irrigated farming situations. Further, it is emphasized that input required should be in access to the farmers and to ensure easy marketing of produce for remunerative prices.

Key words: Clusterbean, technology, dissemination, arid, semi-arid, districts.

Introduction

Clusterbean or guar (*Cyamopsis tetragonoloba*) is a drought tolerant, highly adopted in arid and semi-arid summer annual legume. It is mainly cultivated in north-western states of India and covers approximately 20-25 lakh hectare area which is widely concentrated in states, namely, Rajasthan, Haryana, Gujarat and Punjab. The cultivation of clusterbean has its roots since ancient time in India, which was introduced into the United States in 1903, as it was being grown there for export, as a vegetable for human consumption, as cattle feed and as a green manure crop.

Recently, clusterbean has been recognized as a commercial crop due to rise of its industrial worth having gum in endosperm of its seed. The guar gum contains galactomannans, which are polysacchrides stored in the seeds, used in wide range of foods products. The galoctomannan gum which forms a viscous gel with other polysaccharide in cold water that prevents formation of ice crystal and quick meltdown, or otherwise best known use of guar gum is as stiffer in soft ice cream, instant pudding and whipped cream substitute. In such products, most highly refined food grade guar gum which accounts for a small portion of total production is used. The larger volume of guar gum is used in textile, paper sizing, oil well drilling, printing, cosmetics and pharmaceuticals. The remaining volume after extraction of gum contains about 35 per cent proteins. Of this, about 95 per cent is digestible, making it an excellent portion supplement for ruminants which is an equal or superior to cotton seed meal in amino acids as enough gum still remains in the meal to make it an excellent feed pelleting materials. Further, toasting improves its palatability to livestock.

The guar gum product as a diet tablet, is also being marketed which swells in stomach to alleviate sensation of hunger. It has dietary value in soluble fibres. For diabetics, guar gum have also benefits which are widely accepted.

The clusterbean is a popular unconventional vegetable in India since ancient time in arid and semi-arid region. The green and tender edible pods of clusterbean are used as favourite vegetable in rural areas and also in many other parts. Pods are quite nutritious and contain moisture (82.5%), protein (3.7%), fat (0.2%), fibre (2%), carbohydrates (9.9%), mineral matter (1.4%), calcium (0.13%) and phosphorous (0.25%). They also have iron 5.8 mg/100 gm. and vitamin 491 μg/100 g.

In world production scenario of clusterbean cultivation, it is done in semi-arid regions as it is extremely drought resistant and thrive well in such conditions where most plants get damaged/destroyed. It grows in sandy soils and area of west and north-west India and parts of Pakistan. With the breakthrough of the Technology of guar gum extraction which was commercialized in 1953 in the USA and after a approximately a decade in India attracted attention of industries. Presently, Jodhpur city in north western state of India stood as the most ideal and important processing centre of guar gum and contributes approximately 40% of the world guar gum supply.

Further, India is the major producer of guar seed followed by Pakistan and US. In India, guar seed production fluctuates between years and has been around 2-6 lakh tons in the recent years. Indian guar production in 2003 was around 6 lakh tonnes. The world market of guar is estimated to be around 150000 tones/year. India accounts for 80% of the total guar produced in the world. 70% of India's production comes from Rajasthan. Seeing the facts, clusterbean has emerged out as one of the important commercial crop which is well adopted to the production parameters of low soil fertility, economically viable crop in arid and semi-arid agro climatic regions of rainfed farming situation. More recently, the clusterbean has been an important agri-commodity that invited attention of researchers, development departments and policy makers. The State Govt. of Rajasthan has made a big stride in its promotional efforts to declare exempt of entry tax in import of guar in its budget of 2007-08. Keeping in view the importance of clusterbean crop in different agro-climatic region of western Rajasthan, it was felt to make all concerted efforts through technology dissemination process so as to have its impact on production and productivity. Further, it was visualized to obtain feedback and constraints that hindering the rapid dissemination and adoption of clusterbean technologies.

Methodology

The Krishi Vigyan Kendras (KVKs) are farm science based an innovative institution of ICAR, functioning on the principles of collaborative participation of scientists, subject matter specialists, extension workers and farmers. The technology dissemination process has taken place in clusterbean producing area of Rajasthan state. Presently, the KVKs have been established in all 32 districts of Rajasthan and functioning for technology dissemination in agricultural and other allied aspects. The details of Agro-Climatic Zone in Rajasthan state including KVKs, rainfall, climate, soil types, important and crops grown are given in Table 1. The technology dissemination process was carried out through conducting Front Line Demonstration (FLDs) at farmers' field on improved clusterbean technological intervention. The technological interventions were identified based on the technology gap between existing and new/improved varieties, way of cultivation and improved recommended technology, gap in knowledge and skills and its use in application of various activities and also in use of different critical inputs such as seed of improved variety, fertilizers, seed inoculants, seed treatment with insecticide and fungicide and plant protection measures both, monetary and non-monetary. The KVKs provided technical knowledge and skills to selected farmers through trainings and practical exercises, laid out frontline demonstration involving farmers at all steps of its conductance. The farmers were closely supervised and advised regularly throughout the crop season. The demonstrations were also evaluated through organizing field days at farmers' field involving scientists, subject matter specialists, extension workers and fellow farmers. The feedback and suggestion, reaction of the farmers on each critical inputs and constraints were obtained based upon the performance of frontline demonstration.

Table 1 : Agro-climatic zone wise location of KVKs in Rajasthan

Zone & Agro-climatic zone	KVK District	Rainfall (mm)	Climate	Soil	Important crops grown
I a : Arid western plains zone	Jodhpur Barmer	100-370	Arid	Aridisol, Entisol	Pearl-millet, greengram, moth bean, cluster bean, ber, lasoda and ker
I b. Irrigated north western plain zone	Hanumangarh Sriganganager	100-350	Arid	Aridisol, Entisol	Wheat, gram, mustard, cotton, kinnow, citrus
I c. Hyper arid and partially irrigated western plain	Bikaner Jaisalmer Churu	100-350	Arid	Aridisol, Entisol	Pearl millet, greengram, moth bean, cluster bean, gram, ber, datepalm and aonla
II a. Transitional plain of inland drainage	Sikar Nagaur Jhunjhunu	300-500	Semi arid	Aridisol, Entisol	Pearl millet, green gram, moth bean, cluster bean, groundnut, wheat, barley, ker, lasoda
II b. Transitional plain of luni basin	Jalore Sirohi Pali	300-500	Semi arid	Aridisol, Entisol, Hills and inceptisols	Cumin, green gram, moth bean, cluster bean, wheat, barley, cotton, gram, pomegranate, aonla
III a. Semi arid eastern plain	Jaipur Ajmer Tonk Dausa	500-700	Semi arid	Entisols and alluvium soils	Groundnut, jowar, mustard, gram, cluster bean green gram, moth bean, wheat and barley, ber, citrus
III b. Flood prone eastern plain zone	Alwar Bharatpur Karauli Dholpur	500-700	Semi arid	Inceptisols, Entisols and Vertisols	Mustard, pigeon pea, gram, wheat, cluster bean, barley, mango, citrus
IV a. Sub humid southern plain & Aravalli hill zone	Bhilwara Chittorgarh Rajsamand Udaipur	500-900	Sub-humid	Inceptisols, Entisols , Vertisols and Hills	Groundnut, jowar, maize, gram, green gram cotton, cluster bean, mustard, black gram, and soybean, papaya, guava
IV b. Humid southern plain	Banswara Dungarpur	525-1000	Humid	Vertisols, Alfisols and hills	Cotton, soybean linseed, maize paddy, gram mustard, guava

V. Humid south eastern plain	Kota ,Bundi, Jhala-war Baran S.Madhopur	650-1000 Humid	Vertisols	Sorghum, maize, cotton, paddy, sugarcane, soybean, cluster bean wheat, gram and mustard. Guava, citrus

In Rajasthan state, 32 districts are spread over in ten agro-climatic zones. As per identified thrust areas, clusterbean is grown in Jodhpur, Barmer, Bikaner, Sikar, Nagour, Jhunjhunu, Jalore, Pali, Ajmer, Tonk, Dausa, Alwar, Bharatpur, Karoli and Sawai Madhopur districts. The KVKs of different agro-climatic zones demonstrated production potential of different new/improved varieties through FLDs are detailed in Table 2. The performances of FLDs conducted by KVKs in different districts during 2001-02 to 2006-07 were processed to ascertain the impact of technology disseminated in clusterbean of arid and semi-arid agro-climatic farming situation. The KVKs also initiated truthful seed (TFL) Production at their farm to meet out requirements of different varieties of clusterbean.

Table 2 : Improved varieties of clusterbean demonstrated under FLDs in different agro-climatic zones of Rajasthan

Zone & Agro-climatic zone	District	Rainfall (mm)	Climate	Soil	Varieties
I a : Arid western plains zone	Jodhpur Barmer	100-370	Arid	Aridisol, Entisol	RGC-936, RGC-1003, RGM-112, Maru Guar
I c. Hyper arid and partially irrigated western plain	Bikaner Jaisalmer	100-350	Arid	Aridisol, Entisol	RGC-936, RGC-1003
II a. Transitional plain of inland drainage	Sikar Nagaur Jhunjhunu	300-500	Semi arid	Aridisol, Entisol	RGC-197, RGC-936,
II b. Transitional plain of luni basin	Jalore Pali	300-500	Semi arid	Aridisol, Entisol, Hills and inceptisols	RGC-936, RGM-112,
III a. Semi arid eastern plain	Ajmer Tonk Dausa	500-700	Semi arid	Entisols and alluvium soils	RGC-1003, RGC-1002, RGC-936, RGC-1029, RGC-1031, RGC-1066, RGC-1055,
III b. Flood prone eastern plain zone	Alwar Bharatpur Karauli Dholpur	500-700	Semi arid	Inceptisols, Entisols and Vertisols	RGC-936, RGC-1003, RGC-1033, RGC-1038, RGC-1025, RGC-1055
IV a. Sub humid southern plain & Aravalli hill zone	Udaipur	500-900	Sub-humid	Inceptisols, Entisols , Vertisols and Hills	RGC-936

V . Humid south eastern plain	S.Madhopur	650-1000 Humid	Vertisols	RGC-936

Impact of Technology Dissemination

The Krishi Vigyan Kendras (KVKs) have made their concerted efforts to promote clusterbean cultivation using improved varieties and production technologies in their districts. In the process of technology dissemination, the KVKs took up enormous responsibility of conducting front line demonstrations, trainings and extension activities on clusterbean for fulfilling the following basic objectives:

1. Creation of awareness among the farmers about the new varieties of high yield potential and production technologies.
2. Motivation amongst the farmers through demonstrations worth of technologies for adoption in their field.
3. Assisted the farmers, field level extension functionaries in latest technical know how of clusterbean production.
4. Development of suitable communication support by field day, convention, extension literature, discussion on radio, TV, etc.

As an effort of technology dissemination in clusterbean, the impact has been observed in the following areas :

1. Increase in area and number of demonstrations

The FLDs conducted and area covered in demonstrations presented in Table 3 revealed that number of front line demonstrations and area covered were increased over a period of time. The numbers of FLDs in first two years (2001-02 and 2003-04) were less in number (54 demonstrations covering 16.25 ha). As a result of creating awareness amongst the farmers, the number of FLDs during 2004-05 and 2006-07 were increased around nine times i.e., 422 demonstrations covering 167.25 ha and the unaccounted fellow farmers also horizontally increased spreading over the state.

Table 3 : Distribution of Front Line Demonstration conducted during 2001 -02 to 2006-07

Year	Area (in ha)	No. of FLDs
2001-02	10.50	31
2002-03	5.75	23
2003-04	37.50	119
2004-05	55.00	135
2005-06	96.25	217
2006-07	70.75	205
Total	275.75	730

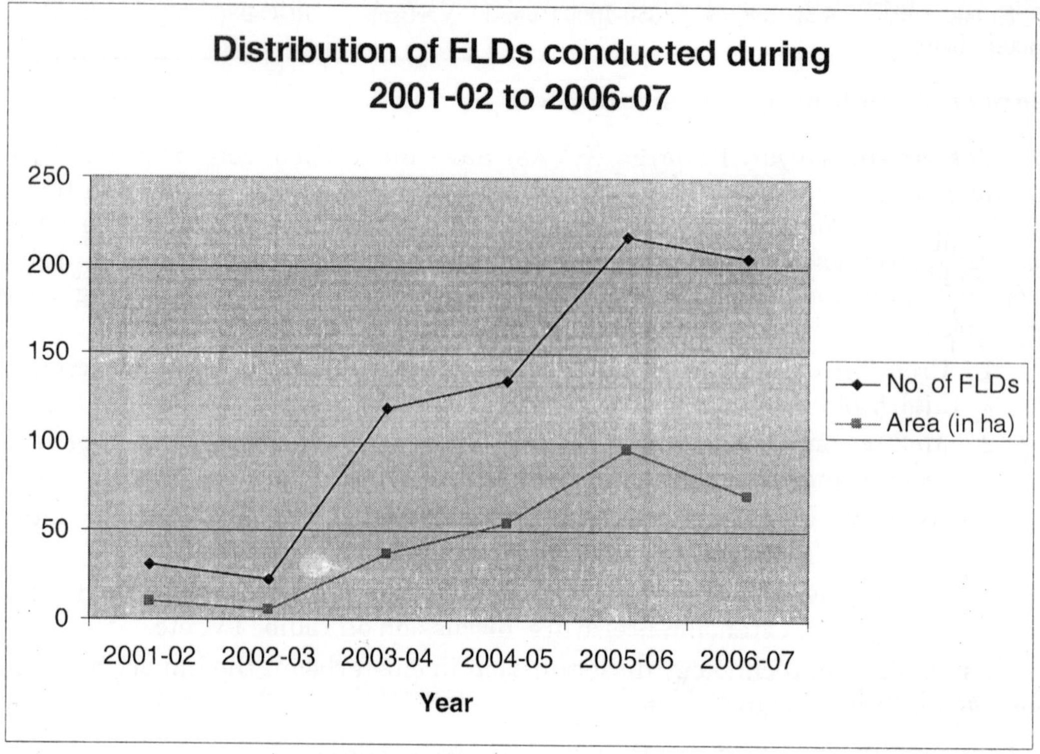

2. Increase in latest technical know-how amongst the farmers and extension personnel

The farmers and extension functionaries were imparted need based practical trainings and as they acquired technical know-how about scientific cultivation of clusterbean. In addition to that they were also provided an opportunity to assess and evaluate latest technical know-how demonstrated at farmers' field through organizing field days. KVKs also published their accomplishment through daily news paper for wide coverage amongst the farmers. The details of trainings and extension activities conducted are presented Table 4.

Table 4: Training and extension activities conducted during 2001-02 to 2006-07

Year	Trainings		Field days		News coverage
	No. of Courses	Participants	No.	Participants	
2001-02	1	31	2	121	3
2002-03	1	23	1	78	4
2003-04	4	119	4	251	10
2004-05	5	135	5	258	12

2005-06	9	219	7	275	11
2006-07	7	167	5	287	13
Total	27	694	24	1270	53

3. Increase in yield over local check

The worth of improved varieties and production technologies has registered remarkable impact. The performance of FLDs presented in Table 5 revealed that yield of demonstration was increased throughout state over a period of six year i.e., 2001-02 to 2006-07. The increase in yield over local checks was in range of 33.84 to 71.13 per cent.

Table 5: Performance of Front Line Demonstrations conducted during 2001-02 to 2006-07.

Year	Wt. Av. Yield q/ha		% increased
	Demo.	Local check	
2001-02	8.06	5.91	53.26
2002-03	10.25	6.16	71.13
2003-04	8.82	6.63	34.49
2004-05	6.81	4.93	39.78
2005-06	9.67	8.41	33.84
2006-07	8.48	6.05	39.49

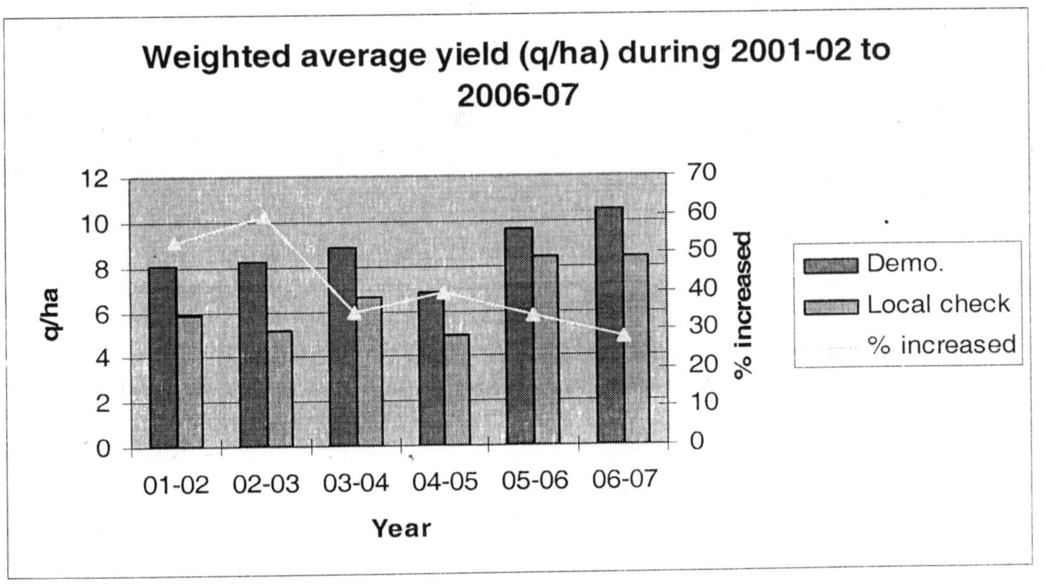

4. Impact of FLDs in different agro-climatic zone

The yield performance of demonstrations over local check as presented in Table 6 indicates that there was more than 23 per cent increase in yield over local check recorded in all agro-climatic zone of the Rajasthan state. However, the better performance of FLDs was observed in agro-climatic zones, i.e., IIa, IIb, IIIa, IIIb and V. The increase in yield might be due to semi-arid nature of agro-climatic zone, soil type i.e., entisol, inceptisol and alluvium soil, occurrence of rainfall ranging between 300 to 700 mm, more opportunities to use critical inputs and to perform cultural practices properly.

Table 6 : Performance Front Line Demonstrations in different Agro-Climatic zones of Rajasthan

Agro-climatic zone	No. of demo	Area (ha)	Wt. Av. Yield q/ha		Average cost of critical input (Rs.)		% increase
			Demo	Local	Demo	Local	
I a : Arid western plains zone	95	50.00	2.74	1.98	651.20	210.30	46.29
I c : Hyper arid and partially irrigated western plain	28	8.00	3.66	2.98	540.00	225.00	23.49
II a :Transitional plain of inland drainage	131	60.50	11.97	9.82	604.88	215.59	23.14
II b :Transitional plain of luni basin	93	25.00	9.23	5.62	548.32	322.12	65.22
III a :Semi arid eastern plain	142	60.00	10.64	8.41	1163.17	450.00	28.36
III b :Flood prone eastern plain zone	212	59.25	11.74	8.78	786.46	291.24	36.73
IV a :Sub humid southern plain & Aravalli hill zone	23	10.00	5.50	3.50	1612.00	1111.00	57.14
V :Humid south eastern plain	6	3.00	10.00	7.15	1006.00	846.00	39.86

5. Increase in interest for using latest and various improved cultivars

After realizing the importance of improved varieties by virtue of their inherent characteristic related to yield potential, suitability in existing farming condition, the farmers intended to use cultivars having specific quality. The use of improved varieties in chronological order is presented in Table 7 indicates that more number of cultivars were demonstrated during subsequent years of conductance. The varieties used during 2001-02 to 2004-05 were two to three in the crop season which increased in subsequent years as 7 to 8 in numbers. This indicated that farmers' interest was placed on priority by scientists and extension functionaries engaged with technology dissemination. They introduced timely newer technological options to solve the problems of the farmers.

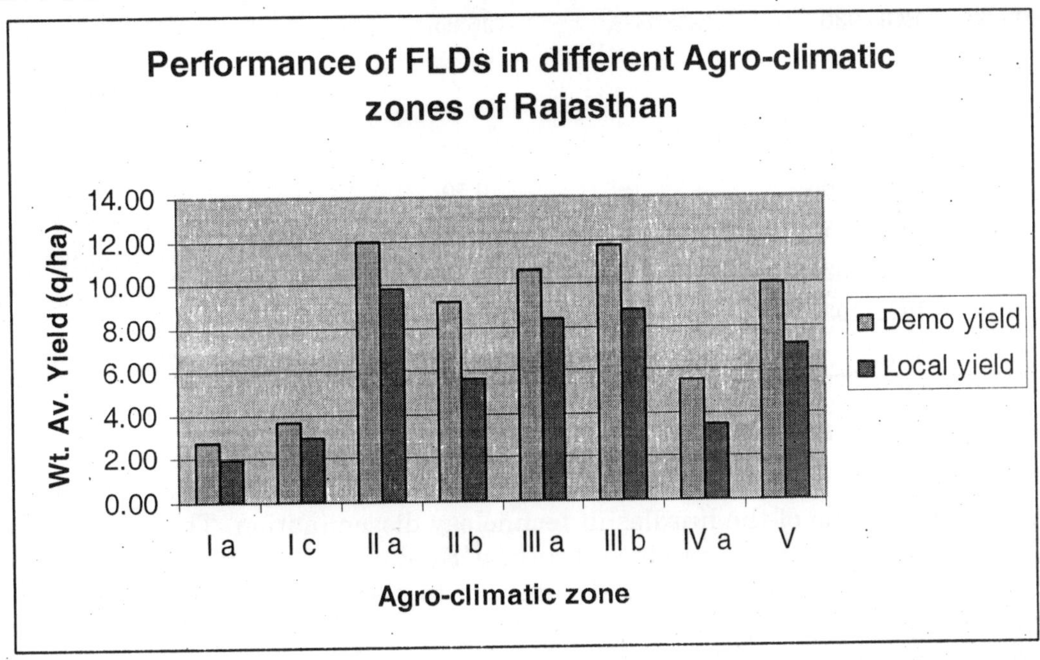

Table 7. Demonstration of varieties, area covered and number of demonstrations in FLDs during 2001-02 to 2006-07

Year	Varieties	Area (in ha)	No. of Demonstrations
2001-02	RGC-936	8.00	26
	RGC-197	2.50	5
2002-03	RGC-1003	5.75	23
2003-04	RGC-936	20.00	59
	RGC-1003	15.50	56
	Maru Guar	2.00	4
2004-05	RGC-936	40.00	95
	RGC-1003	15.00	40
2005-06	RGC-936	40.50	83
	RGM-112	3.50	2
	RGC-1002	30.00	62
	RGC-1003	21.50	67
	RGC-1025	0.25	1
	RGC-1033	0.25	1
	RGC-1038	0.25	1

2006-07	RGC-936	26.25	75
	RGC-1003	27.75	79
	RGM-112	10.00	30
	RGC-1029	1.00	4
	RGC-1066	0.50	2
	RGC-1055	1.00	4
	RGC-1033	1.25	5
	RGC-1031	3.00	6
	Total	275.75	730

6. Quality seed production

The necessity of required quantity and quality of seed was very much experienced as one of the hurdles in technology dissemination. The scientists of KVKs made an effort to produce Truthful Seed (TFL) at their instructional farms of KVKs to provide desired cultivars to the farmers. The KVKs have produced more than 1000 q seed of worth Rs.2318948.00 of different varieties and distributed/sold to the farmers and different seed distributing agencies. The details of seed production made during 2001-02 to 2006-07 are given in Table 8.

Table 8 : Seed production of different varieties by the KVKs during 2001-02 to 2006-07.

Year	Variety	Quantity produced (in q.)	Amount (Rs.)
2001-02	RGC-936	76.10	93937.00
	RGC-197	1.52	1102.00
	RGC-1003	27.08	35210.00
	Maruguar	2.00	1980.00
Total		106.70	132229.00
2002-03	RGC-936	12.04	18530.00
	M-83	0.8	250.00
	PNB	0.84	1428.00
Total		13.68	20208.00
2003-04	M-13	0.78	9750.00
	Maruguar	6.07	6137.00
	PNB	0.2	254.00
	RGC-1002	5	6000.00

	RGC-1003	76.56	117150.00
	RGC-1017	5	6000.00
	RGC-936	119.43	584429.00
Total		213.04	729720.00
2004-05	RGC-1002	18.80	28200.00
	RGC-1003	54.19	81285.00
	RGC-1017	10.00	15000.00
	RGC-936	104.65	156975.00
Total		187.64	281460.00
2005-06	M-83	0.47	500.00
	RGC-1002	32.85	61966.00
	RGC-1003	42.45	92610.00
	RGC-936	213.78	448371.00
	RGM-112	0.9	1800.00
2006-07	RGC-1002	19.06	46455.00
	RGC-197	74.38	211462.00
	RGC-936	103.63	289567.00
	RGM-112	0.3	600.00
Total		197.37	548084.00
Grant Total		1008.88	2316948.00

7. Feedback and suggestions

As a process of transfer of technology dissemination, the scientists encountered with many field issues related to research, development and policy. As a result of close interaction with farmers, production system, marketing and disposal system the feedback was obtained and provided to concerned systems.

For Researchers

- There is a need to evolve suitable varieties with determinate growth habits.
- Package of practices especially for rainfed farming situation need to be standardized, and popularized.
- The varieties having more content of gum in its endosperm of seed need to be evolved.

- The varieties of high yield, drought tolerant and disease resistance be evolved.
- Value addition of guar produce be explored to make people friendly crop on both accounts i.e., domestic as well as for export.
- Transgenic guar be developed for therapeutic use for diabetes following bio-technological approaches.
- Some varieties having vegetable utility should be evolved.
- Suitable farm implement for seed cum fertilizers drilling, weed control and plant protection measures should be developed and popularized.

For Development Departments

- The seed of improved variety along with all other inputs should be made available in the outlets located in rural areas.
- Extensive efforts should be made to cover larger area involving more number of farmers.
- Commercial production of clusterbean like other crops such as cotton, sugarcane ,etc, should be promoted.
- Rapid replacement of seed of local cultivars with improved varieties alongwith recommended package of practices should be targeted.
- Transport facility and other infrastructure should be created for easy marketing by the common producer/farmers.
- Monopoly of itinerant merchant/stockists could be abolished by opening a larger number of authorized sale outlets.
- The crop is mainly grown in rainfed situation of Rajasthan where size of holding is large. Hence, mechanized farming for clusterbean should be promoted.

For Policy Makers

- The crop should be promoted in all respect by providing insurance cover.
- Support price for clusterbean be fixed by the government .
- Promotion of processing units should be done to protect the farmers.

8. Constraints

All the efforts were made to make a big success. However, some of the constraints create hindrance in technology dissemination of clusterbean. Some of the important constraints of different nature were identified and presented.

Technical

1. The value addition in guar has limited opportunities for the common farmers being a product of industries which require high investment infrastructure and equipments with machinery supported by highly qualified skilled personnel.
2. Clusterbean cultivation mainly depends upon rainfall in arid and semi-arid region where uncertainity of rainfall is always faced. In such situation, use of critical inputs as advised by the experts may not be undertaken or performed.
3. Being a larger holding size in rainfed farming situation of Rajasthan, the proper placement of seed with basal application fertilizers becomes difficult as seed-cum-fertilizer drilling equipments are not available with farmers.
4. The clusterbean crop is generally grown during Kharif season. The weed infestation becomes very prevalent during rainy season. The weed control is time and money consuming as experienced by the farmers.
5. The effective control measures with low cost for yellow vein mosaic, wilt, blight disease are still to be provided.
6. Seed of recommended newly released varieties are not available. The physical market of the commodity involves speculators and stockists. The commodity is subjected to a long storage period based upon the demand and market prices.
7. The demand is almost constant over the years, supply varies largely between years.
8. The production is directly related to monsoon. The volatility in production and consequently on prices is experienced due fluctuation of rainfall in Rajasthan.

Administrative

1. The policy of clusterbean production should be well stated as in other crops.
2. Farm machinery like, seed-cum-fertilizer drill are not available, hence, proper placement of seed and fertilizer is not followed as recommended.
3. There are no government rules and regulations governing the production, distribution, marketing, export or imports of the commodity. In such conditions, the market forces determine the prices.

Conclusion

The clusterbean has been recognized as one of the important commercial crop by virtue of its industrial worth because of gum content in the endosperm of its seed which is mainly used in food industries, as a feed ingredient having rich protein content and uses in textiles, printing, oil well drilling, pharmaceutical and also as vegetables. It is grown mainly in arid and semi-arid agro-climatic zones of north western states under rainfed farming situation with low cost of inputs.

Recently, a big stride has been made to increase production and productivity of the crop through the process of technology dissemination. The awareness has been created amongst the farmers to use improved technological options. However, it invites the attention of researchers, development departments and policy makers towards issues like value addition, more promising varieties (having high yield, drought tolerant and pest and disease resistant), suitable package of practices for rainfed farming situation, availability of critical inputs in time, creation of marketing facility, crop insurance and minimum support price. Further, promotion for more processing units of guar gum to protect the farmers needs urgent attention.

References

1. Seasonal Reports on FLDs of concerned KVKs on clusterbean (2001-02 to 2006-07)
2. Annual Progress Report of concerned KVKs (2001-02 to 2006-07)
3. Action Plans of concerned KVKs (2001-02 to 2006-07).

AUTHOR INDEX